Mechanical Behavior of Diamond and Other Forms of Carbon

MATERIALS RESEARCH SOCIETY
SYMPOSIUM PROCEEDINGS VOLUME 383

Mechanical Behavior of Diamond and Other Forms of Carbon

Symposium held April 17-21, 1995, San Francisco, California, U.S.A.

EDITORS:

Michael D. Drory
Crystallume
Santa Clara, California, U.S.A.

David B. Bogy
University of California, Berkeley
Berkeley, California, U.S.A.

Michael S. Donley
Air Force Wright Laboratory
Wright Patterson Air Force Base, Ohio, U.S.A.

John E. Field
University of Cambridge
Cambridge, United Kingdom

PITTSBURGH, PENNSYLVANIA

Single article reprints from this publication are available through
University Microfilms Inc., 300 North Zeeb Road, Ann Arbor, Michigan 48106

CODEN: MRSPDH

Published by:
Materials Research Society
9800 McKnight Road
Pittsburgh, Pennsylvania 15237
Telephone (412) 367-3003
Fax (412) 367-4373

Library of Congress Cataloging in Publication Data

Mechanical behavior of diamond and other forms of carbon: symposium held
held April 17–21, 1995, San Francisco, California, U.S.A. / editors,
M.D. Drory, D.B. Bogy, M.S. Donley, J.E. Field.
p. cm. -- (Materials Research Society symposium proceedings; v. 383)
Includes bibliographical references and index.
ISBN: 155899-286-3
1. Diamonds, Artificial--Mechanical Properties--Congresses. I. Drory, M.D.
II. Bogy, D.B. III. Donley, M.S. IV. Field, J.E. V. Series: Materials Research Society Symposium Proceedings; v. 383.

TA455.C3M43 1995 95-25205
620.1'93--dc20 CIP

Manufactured in the United States of America

CONTENTS

PART I: PROCESSING AND STRUCTURE

PART II: ELASTIC PROPERTIES AND DEFORMATION

*Invited Paper

PART III: RESIDUAL STRESSES

PART IV: FRACTURE AND ADHESION

*Invited Paper

PART V: FRICTION AND WEAR

*Invited Paper

PART VI: APPLICATIONS

*Invited Paper

PREFACE

These are the proceedings of Symposium I of the 1995 MRS Spring Meeting, held in San Francisco during April 17–21. The focus of the symposium was to assess the current understanding of the mechanical behavior of diamond and other forms of carbon. An interdisciplinary approach was taken to bring together scientists from various disciplines, including diamond synthesis, mechanics, and materials science, in order to examine this topic from several points of view.

Carbon has numerous structures produced by nature or synthetic methods. This leads to considerable interest in relating the structure of carbon to mechanical properties, which is the focus of this symposium. The mechanical properties of carbon span a very broad range of values, as illustrated in the following figures compiled by Ashby.[1] In many cases, diamond possesses the extreme value. These charts are also included to serve as reference to the various properties and structures described herein.

We hope that these proceedings serve as a useful reference in the mechanical behavior of carbon and identify further avenues of research on the topic of diamond synthesis-structure-properties relationships.

M.D. Drory
D.B. Bogy
M.S. Donley
J.E. Field

September 1995

1. M.F. Ashby, Materials Selection in Mechanical Design, Appendix C. Pergamon Press, 1992.

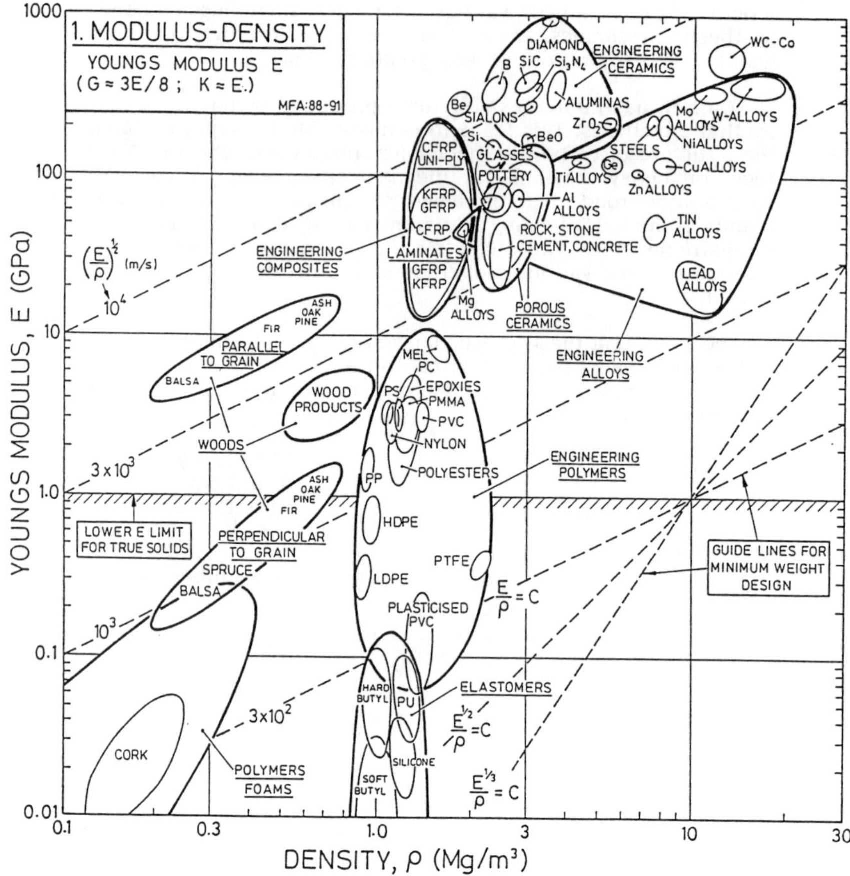

Figure 1: Young's modulus vs. density[1]. (Courtesy of M. F. Ashby, Engineering Department, Cambridge University, U.K.)

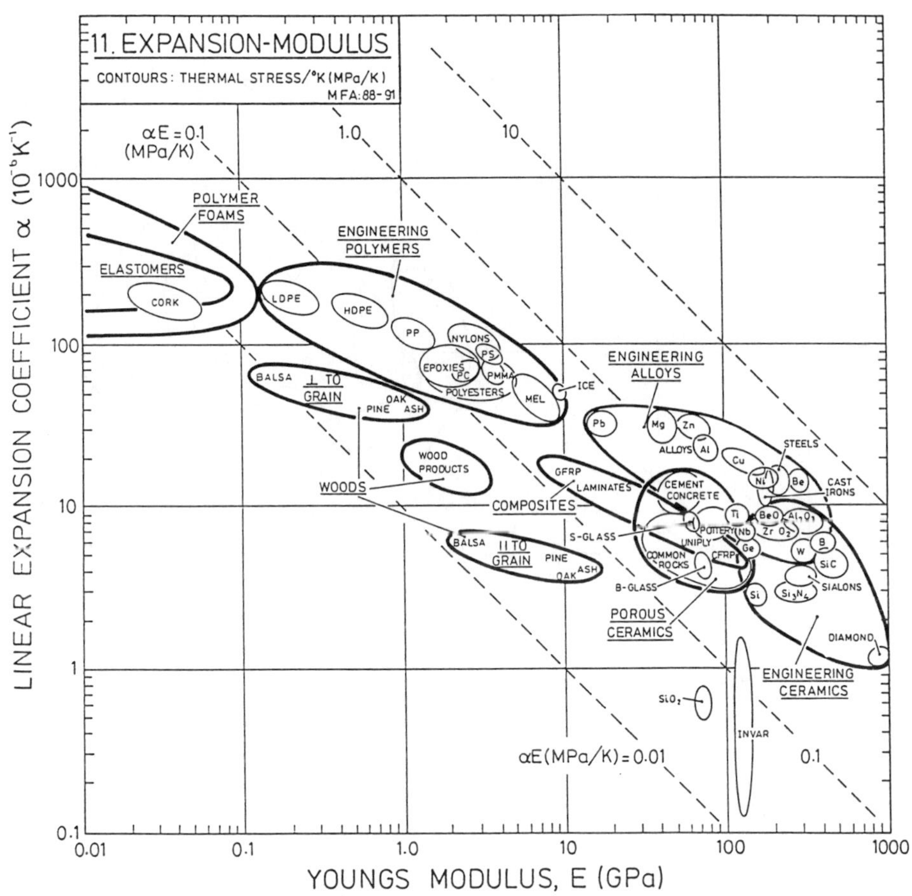

Figure 2: Thermal expansion coefficient vs. Young's modulus[1]. (Courtesy of M. F. Ashby, Engineering Department, Cambridge University, U.K.)

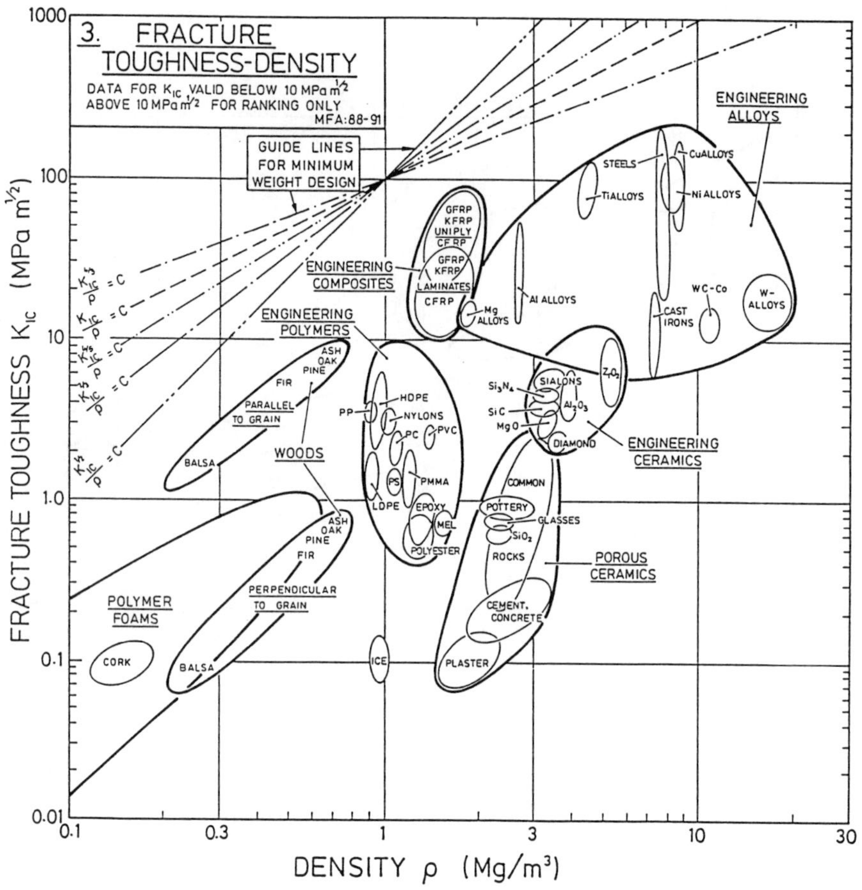

Figure 3: Fracture toughness vs. density[1]. (Courtesy of M. F. Ashby, Engineering Department, Cambridge University, U.K.)

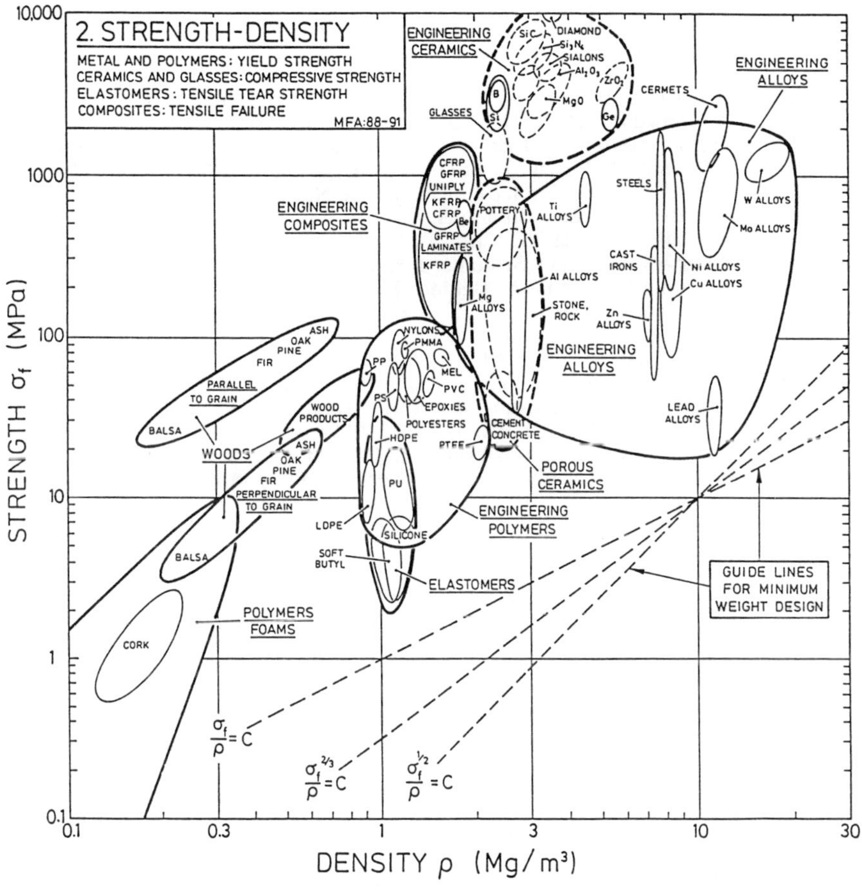

Figure 4: Strength vs. density[1]. (Courtesy of M. F. Ashby, Engineering Department, Cambridge University, U.K.)

MATERIALS RESEARCH SOCIETY SYMPOSIUM PROCEEDINGS

Volume 352—Materials Issues in Art and Archaeology IV, P.B. Vandiver, J.R. Druzik, J.L. Galvan Madrid, I.C. Freestone, G.S. Wheeler, 1995, ISBN: 1-55899-252-9

Volume 353—Scientific Basis for Nuclear Waste Management XVIII, T. Murakami, R.C. Ewing, 1995, ISBN: 1-55899-253-7

Volume 354—Beam-Solid Interactions for Materials Synthesis and Characterization, D.E. Luzzi, T.F. Heinz, M. Iwaki, D.C. Jacobson, 1995, ISBN: 1-55899-255-3

Volume 355—Evolution of Thin-Film and Surface Structure and Morphology, B.G. Demczyk, E.D. Williams, E. Garfunkel, B.M. Clemens, J.E. Cuomo, 1995, ISBN: 1-55899-256-1

Volume 356—Thin Films: Stresses and Mechanical Properties V, S.P. Baker, P. Børgesen, P.H. Townsend, C.A. Ross, C.A. Volkert, 1995, ISBN: 1-55899-257-X

Volume 357—Structure and Properties of Interfaces in Ceramics, D.A. Bonnell, U. Chowdhry, M. Rühle, 1995, ISBN: 1-55899-258-8

Volume 358—Microcrystalline and Nanocrystalline Semiconductors, R.W. Collins, C.C. Tsai, M. Hirose, F. Koch, L. Brus, 1995, ISBN: 1-55899-259-6

Volume 359—Science and Technology of Fullerene Materials, P. Bernier, D.S. Bethune, L.Y. Chiang, T.W. Ebbesen, R.M. Metzger, J.W. Mintmire, 1995, ISBN: 1-55899-260-X

Volume 360—Materials for Smart Systems, E.P. George, S. Takahashi, S. Trolier-McKinstry, K. Uchino, M. Wun-Fogle, 1995, ISBN: 1-55899-261-8

Volume 361—Ferroelectric Thin Films IV, S.B. Desu, B.A. Tuttle, R. Ramesh, T. Shiosaki, 1995, ISBN: 1-55899-262-6

Volume 362—Grain-Size and Mechanical Properties—Fundamentals and Applications, N.J. Grant, R.W. Armstrong, M.A. Otooni, T.N. Baker, K. Ishizaki, 1995, ISBN: 1-55899-263-4

Volume 363—Chemical Vapor Deposition of Refractory Metals and Ceramics III, W.Y. Lee, B.M. Gallois, M.A. Pickering, 1995, ISBN: 1-55899-264-2

Volume 364—High-Temperature Ordered Intermetallic Alloys VI, J. Horton, I. Baker, S. Hanada, R.D. Noebe, D. Schwartz, 1995, ISBN: 1-55899-265-0

Volume 365—Ceramic Matrix Composites—Advanced High-Temperature Structural Materials, R.A. Lowden, J.R. Hellmann, M.K. Ferber, S.G. DiPietro, K.K. Chawla, 1995, ISBN: 1-55899-266-9

Volume 366—Dynamics in Small Confining Systems II, J.M. Drake, S.M. Troian, J. Klafter, R. Kopelman, 1995, ISBN: 1-55899-267-7

Volume 367—Fractal Aspects of Materials, F. Family, B. Sapoval, P. Meakin, R. Wool, 1995, ISBN: 1-55899-268-5

Volume 368—Synthesis and Properties of Advanced Catalytic Materials, E. Iglesia, P. Lednor, D. Nagaki, L. Thompson, 1995, ISBN: 1-55899-270-7

Volume 369—Solid State Ionics IV, G-A. Nazri, J-M. Tarascon, M. Schreiber, 1995, ISBN: 1-55899-271-5

Volume 370—Microstructure of Cement Based Systems/Bonding and Interfaces in Cementitious Materials, S. Diamond, S. Mindess, F.P. Glasser, L.W. Roberts, J.P. Skalny, L.D. Wakeley, 1995, ISBN: 1-55899-272-3

Volume 371—Advances in Porous Materials, S. Komarneni, D.M. Smith, J.S. Beck, 1995, ISBN: 1-55899-273-1

Volume 372—Hollow and Solid Spheres and Microspheres—Science and Technology Associated with their Fabrication and Application, M. Berg, T. Bernat, D.L. Wilcox, Sr., J.K. Cochran, Jr., D. Kellerman, 1995, ISBN: 1-55899-274-X

Volume 373—Microstructure of Irradiated Materials, I.M. Robertson, L.E. Rehn, S.J. Zinkle, W.J. Phythian, 1995, ISBN: 1-55899-275-8

Materials Research Society Symposium Proceedings

Volume 374—Materials for Optical Limiting, R. Crane, K. Lewis, E.V. Stryland, M. Khoshnevisan, 1995, ISBN: 1-55899-276-6

Volume 375—Applications of Synchrotron Radiation Techniques to Materials Science II, L.J. Terminello, N.D. Shinn, G.E. Ice, K.L. D'Amico, D.L. Perry, 1995, ISBN: 1-55899-277-4

Volume 376—Neutron Scattering in Materials Science II, D.A. Neumann, T.P. Russell, B.J. Wuensch, 1995, ISBN: 1-55899-278-2

Volume 377—Amorphous Silicon Technology—1995, M. Hack, E.A. Schiff, M. Powell, A. Matsuda, A. Madan, 1995, ISBN: 1-55899-280-4

Volume 378—Defect- and Impurity-Engineered Semiconductors and Devices, S. Ashok, J. Chevallier, I. Akasaki, N.M. Johnson, B.L. Sopori, 1995, ISBN: 1-55899-281-2

Volume 379—Strained Layer Epitaxy—Materials, Processing, and Device Applications, J. Bean, E. Fitzgerald, J. Hoyt, K-Y. Cheng, 1995, ISBN: 1-55899-282-0

Volume 380—Materials—Fabrication and Patterning at the Nanoscale, C.R.K. Marrian, K. Kash, F. Cerrina, M. Lagally, 1995, ISBN: 1-55899-283-9

Volume 381—Low-Dielectric Constant Materials—Synthesis and Applications in Microelectronics, T-M. Lu, S.P. Murarka, T.S. Kuan, C.H. Ting, 1995, ISBN: 1-55899-284-7

Volume 382—Structure and Properties of Multilayered Thin Films, T.D. Nguyen, B.M. Lairson, B.M. Clemens, K. Sato, S-C. Shin, 1995, ISBN: 1-55899-285-5

Volume 383—Mechanical Behavior of Diamond and Other Forms of Carbon, M.D. Drory, M.S. Donley, D. Bogy, J.E. Field, 1995, ISBN: 1-55899-286-3

Volume 384—Magnetic Ultrathin Films, Multilayers and Surfaces, A. Fert, H. Fujimori, G. Guntherodt, B. Heinrich, W.F. Egelhoff, Jr., E.E. Marinero, R.L. White, 1995, ISBN: 1-55899-287-1

Volume 385—Polymer/Inorganic Interfaces II, L. Drzal, N.A. Peppas, R.L. Opila, C. Schutte, 1995, ISBN: 1-55899-288-X

Volume 386—Ultraclean Semiconductor Processing Technology and Surface Chemical Cleaning and Passivation, M. Liehr, M. Hirose, M. Heyns, H. Parks, 1995, ISBN: 1-55899-289-8

Volume 387—Rapid Thermal and Integrated Processing IV, J.C. Sturm, J.C. Gelpey, S.R.J. Brueck, A. Kermani, J.L. Regolini, 1995, ISBN: 1-55899-290-1

Volume 388—Film Synthesis and Growth Using Energetic Beams, H.A. Atwater, J.T. Dickinson, D.H. Lowndes, A. Polman, 1995, ISBN: 1-55899-291-X

Volume 389—Modeling and Simulation of Thin-Film Processing, C.A. Volkert, R.J. Kee, D.J. Srolovitz, M.J. Fluss, 1995, ISBN: 1-55899-292-8

Volume 390—Electronic Packaging Materials Science VIII, R.C. Sundahl, K.A. Jackson, K-N. Tu, P. Børgesen, 1995, ISBN: 1-55899-293-6

Volume 391—Materials Reliability in Microelectronics V, A.S. Oates, K. Gadepally, R. Rosenberg, W.F. Filter, L. Greer, 1995, ISBN: 1-55899-294-4

Volume 392—Thin Films for Integrated Optics Applications, B.W. Wessels, D.M. Walba, 1995, ISBN: 1-55899-295-2

Volume 393—Materials for Electrochemical Energy Storage and Conversion—Batteries, Capacitors and Fuel Cells, D.H. Doughty, B. Vyas, J.R. Huff, T. Takamura, 1995, ISBN: 1-55899-296-0

Volume 394—Polymers in Medicine and Pharmacy, A.G. Mikos, K.W. Leong, M.L. Radomsky, J.A. Tamada, M.J. Yaszemski, 1995, ISBN: 1-55899-297-9

Prior Materials Research Society Symposium Proceedings available by contacting Materials Research Society

MATERIALS RESEARCH SOCIETY SYMPOSIUM PROCEEDINGS

Part I

Processing and Structure

THE PRESSURE–TEMPERATURE PHASE AND REACTION DIAGRAM FOR CARBON

Francis P. Bundy, (retired) General Electric R & D Center; (home) 4607 Swallow Court, Lebanon, OH 45036-9541.

ABSTRACT

Carbon atoms form very strong bonds to each other, yielding materials like: (i) crystalline graphite, diamond and their many "amorphous" hybrids; (ii) crystalline forms of giant closed–surface molecules such as the fullerenes; and (iii) liquid and gas phases which have molecular contents which are complicated and not yet defined or understood. Because of the high bonding energy the melting and vaporization temperatures of the solid forms are very high, and the activation energies required to transform one solid form to another are large. One consequence is that at lower temperatures the different solid phases may continue to exist metastably far into a P, T region in which another solid phase is the thermodynamically stable one.

In the thermodynamic sense the vapor pressure line of graphite, the graphite/liquid/vapor triple point, the graphite melting line, the graphite/diamond equilibrium line, and the graphite/diamond/liquid triple point are quite well established. Data for the melting temperature of diamond vs. pressure are sparse and rough, but they indicate that the melting temperature increases with pressure,–in agreement with some theories. Although carbon should transform to a solid metallic state at very high pressures, experimental evidence shows diamond to be stable to over 400GPa, and theoretical calculations indicate that it could be the stable form up to pressures of 1200 to 2300GPa. Attention is given to the solid state transformations which can take place when graphite is compressed and heated along different P, T paths under different conditions.

INTRODUCTION

In respect to the abundance of the chemical elements in our universe hydrogen ranks first, followed by helium, carbon, nitrogen, oxygen, etc. There is a lot of carbon in our universe; a lot in our solar system; a lot in our Earth.

Carbon is the lightest of the Group IV elements. Each atom has two

Mat. Res. Soc. Symp. Proc. Vol. 383 © 1995 Materials Research Society

electrons in the inner shell and four valence electrons in the outer shell. This valence electron combination makes it possible to link together carbon atoms with very strong bonding to form very stable and strong crystals such as graphite and diamond, "amorphous" solid forms such as "carbon blacks" and "glassy carbon," and complicated "cage molecules" such as the recently discovered "fullerenes." This very large bonding energy between carbon atoms causes the melting temperatures of graphite and diamond to be very high (~5000K), their vapor pressure at temperatures of incandescence to be remarkably low (Edison's carbon filament lamps), and their mechanical strength and rigidity to be very high (diamond tools, carbon fibers). Another consequence of these very strong binding modes of carbon, viz. graphitic sp^2–type, and diamond sp^3–type, is that when the atoms get locked into one mode of bonding together a large amount of "activation energy" is required to disrupt a given assemblage to transform it into an assemblage of the other type. Examples of this, which will be shown later in this paper, are that at temperatures below 2000–3000K graphite specimens can persist metastably far into the P, T region of thermodynamic stability of diamond, and diamond persists far into the graphite stability zone (as at room pressure and temperature).

Before presenting and discussing the P, T Phase and Reaction Diagram of Carbon the reader is reminded that for the sp^2 graphitic bonding each atom is bonded to three others, 120° apart, in the same plane forming a flat grid of hexagons, and the perfect graphite crystal consists of layer upon layer of these grids of atoms bonded together quite weakly. For the sp^3 diamond–type bonding each atom is strongly bonded to four equidistant neighbors in symmetrical tetrahedral array. The diamond crystal can be pictured as layers of puckered hexagons with the layers bonded to each other by the very strong sp^3 bonds. There are two possible layering sequences in graphite and in diamond. For the sequencing $ABAB\cdots$ in which the third layer exactly superimposes the first, the crystal cell geometry is hexagonal and the substances are referred to as "hexagonal graphite" (the common kind), and "hexagonal diamond" (the uncommon kind). For the layer sequencing $ABCABC\cdots$, in which the fourth layer superposes the first, the crystal cell is rhombohedral for graphite and cubic for diamond;–the corresponding substances being called "rhombohedral graphite" (rare) and "cubic–type diamond" (the common kind).

THE P, T PHASE AND REACTION DIAGRAM [1,2,3]

In 1901, on the basis of thermodynamics he had helped develop, Roozeboom, the Netherlands, constructed one of the earliest P, T phase diagrams of carbon,–Fig. 1. He recognized that at a given T the vapor pressure of diamond would be slightly greater than that of graphite, as shown by the lines AD (for diamond) and AB (for graphite). For P's and T's greater than B carbon would exist in the liquid phase. The line BC would be the vapor pressure of liquid carbon as a function of T. The line BE would be the melting line of graphite as a function of P.

In 1909, Tammann, Germany, modified the diagram to that shown in Fig. 2. He postulated a region MBL in which diamond and graphite would be in a state of pseudo-equilibrium. The line MB represented the P, T's at which graphite would transform to diamond, and LB the threshold of the reverse reaction. When carbon is in solution,–as in molten iron, silver or silicates,–he assumed that diamond and graphite would be in equilibrium and hence quick cooling could produce some diamond, while slow cooling would yield only graphite. This concept provided an explanation for the then–current and accepted Moissan method of making diamond by rapid quenching of carbon–saturated molten iron.

In 1938, Rossini and Jessup of the US Bureau of Standards reported more accurate measurements of the heats of formation of graphite and of diamond and using these values along with other known thermodynamic properties they calculated that at zero K the lowest P at which diamond could be stable relative to graphite would be about 13,000 atmospheres,–and at 500K about 20,000 atmospheres.

In 1939, Leipunskii, in Russia, made a thorough review of the problem of diamond synthesis. Using the thermodynamic data of Rossini and Jessup he proposed the phase diagram shown in Fig. 3. Using rough estimates of the kinetics of reaction as a function of T he suggested that a direct recrystallization from graphite to diamond would require conditions as severe as 55,000 atmospheres and 1750K. He proposed that upon cooling, a solution of carbon in a material like molten iron, for example, diamond might crystallize out (in preference to graphite) at conditions of at least 40,000 atmospheres and 1250K. Later, in 1955, when at the General Electric Research Laboratory diamond synthesis had been accomplished and understood, and the P, T position of the diamond/graphite equilibruim line had been determined experimentally, it was found that Leipunski's P, T and reaction conditions were not adequate or correct. The successful process is isothermal, not one of cooling;–and the iron–carbon system at high P, T conditions is more compli-

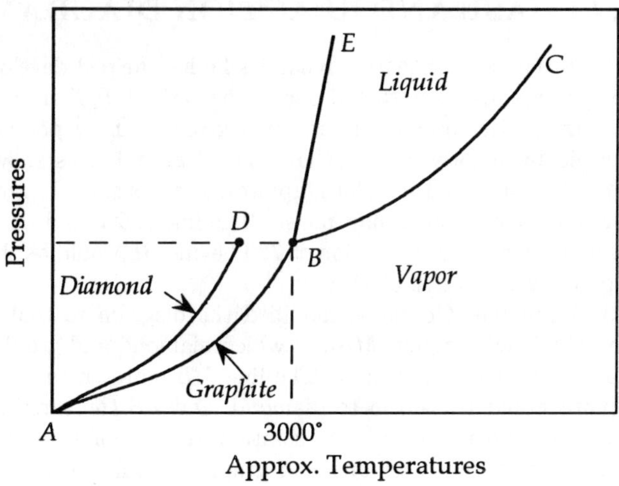

Figure 1: Phase diagram for carbon as conceived by Rooseboom in 1901.

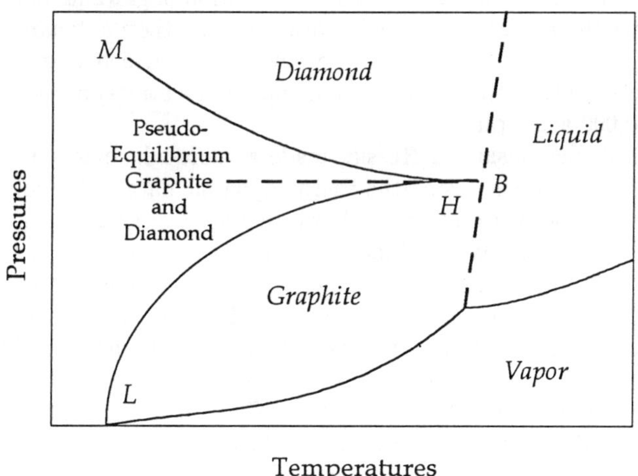

Figure 2: Phase diagram for carbon according to the ideas of Tammann 1909; 1921.

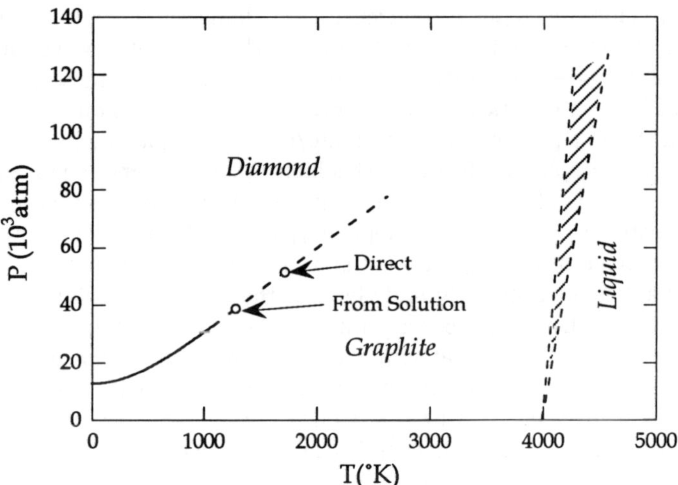

Figure 3: Phase and reaction diagram for carbon proposed by Leipunskii in 1939.

cated than Leipunski visualized it. His assessment of the position and slope of the graphite melting line was much better.

The melting line of graphite at high P's was established experimentally in 1963 by Bundy. Fig. 4 shows sections of specimens from the melting experiments. The radial crystal texture, which shows very clearly in those specimens which were most completely melted (90–100V), is due to the thermal conductivity of graphite in the a,b crystal plane being much better than that in the c–axis direction. Solidification has to proceed from the "cool" wall radially inward and the crystals which grow most rapidly are those whose a,b planes lie more nearly in the radial direction, thus removing the latent heat of fusion (~25kcal/mole) most rapidly.

The first "direct transformation" of graphite to diamond, using shock compression, was reported by DeCarli and Jamieson in 1961. Later, in 1963, Bundy reported the first "direct transition" of graphite to diamond at very high static pressure, heating the specimen by a strong pulse of dc current from a bank of electric capacitors,–Fig.5. This experimental work also established the P,T position of the graphite/diamond/liquid triple point at about 120,000 atmospheres/4000K. In this high P,T work some rough experiments on the melting of diamond were made, but they were not extensive or accurate enough to determine the slope, dT_m/dP, of the diamond melting line. It was assumed, by analogy with the known melting behavior of the "sister substances" diamond–cubic Si and Ge, that the dT_m/dP for diamond would be negative. In the P,T phase diagram published in 1969,–Fig. 6–, it was shown with a negative slope. Later, in the mid–1980's, as theory and experiment developed a better understanding of how carbon should, and does, behave at extremely high P,T's (Yin & Cohen, 1983[4]; Biswas, Martin, Needs & Nielson, 1984[5]; Shaner, Brown, Swenson & McQueen, 1984[6]; Weathers & Bassett, 1987[7]; etc.) it became evident that diamond is the stable form of carbon up to at least 13 million atmospheres, possibly as high as 23 million atmospheres, and that the dT_m/dP of diamond is positive. This would mean that the melt is less dense than solid diamond.

The high–pressure high–temperature diagram, based upon our best experimental and theoretical knowledge today, is as shown in Fig. 7. There are three regions of thermodynamic stability: graphite, diamond, and liquid. These regions are separated by three phase boundary lines: (i) the graphite/diamond equilibrium line, (ii) the graphite melting line, and (iii) the diamond melting line. There are two "triple points":– the graphite/liquid/vapor one at 0.011GPa/5000K and the graphite/diamond/liquid one at about 12GPa/5000K (1GPa ≅ 10,000 atmos.). It is interesting that the melting temperature of graphite has a maximum at about 5 to 6GPa. The melting

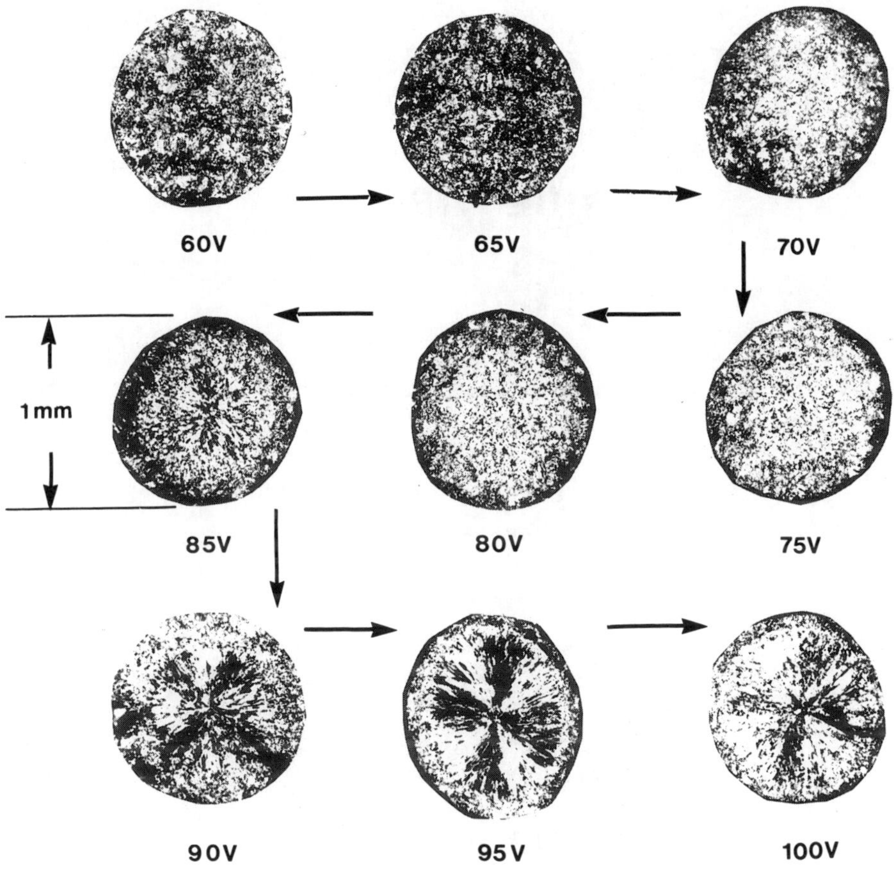

Figure 4: Photos of polished cross sections of graphite–melting specimens flash–heated at successively higher capacity voltages.

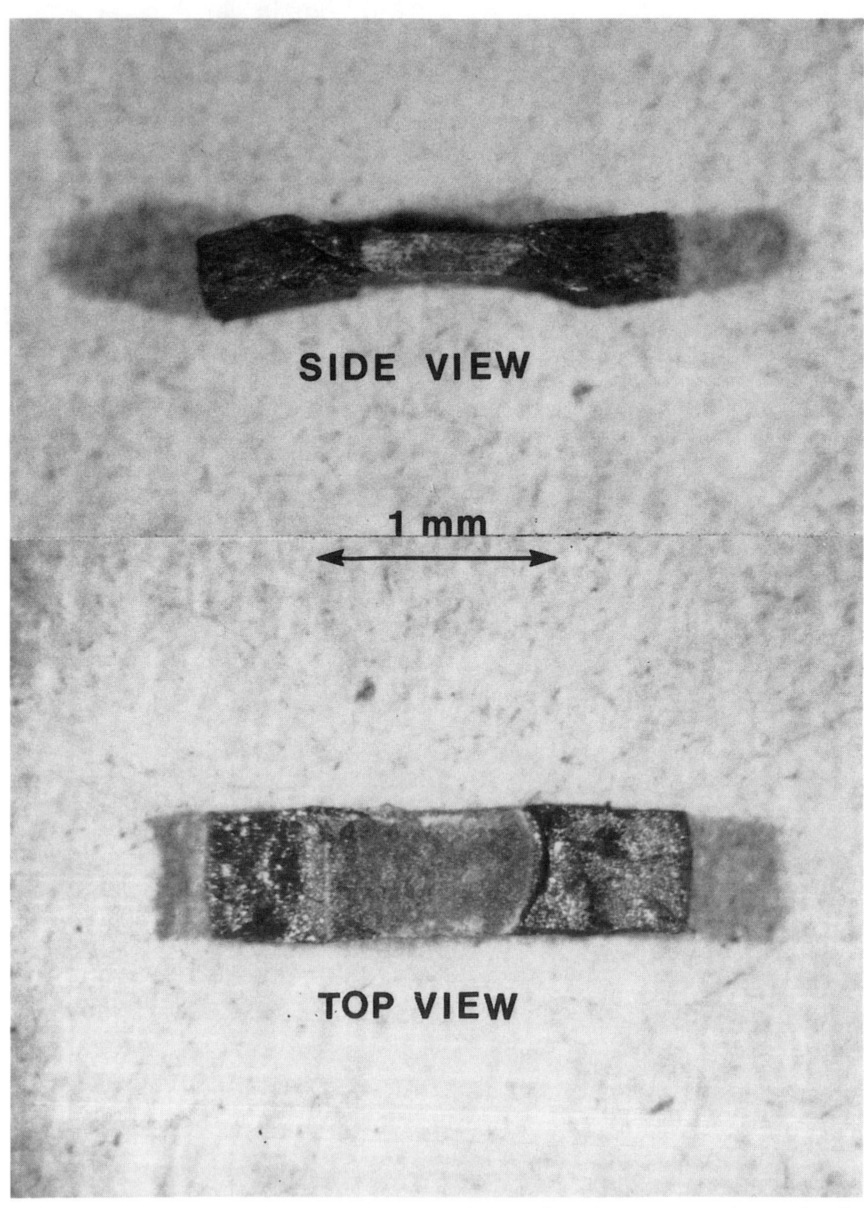

Figure 5: Photo showing center section of a graphite bar converted completely to polycrystalline diamond.

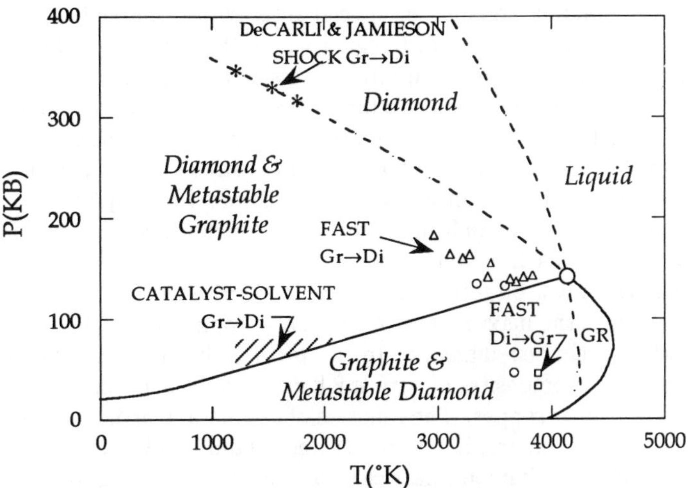

Figure 6: Carbon phase diagram showing graphite and diamond melting lines and threshold for direct graphite to diamond transitions. (Year 1969)

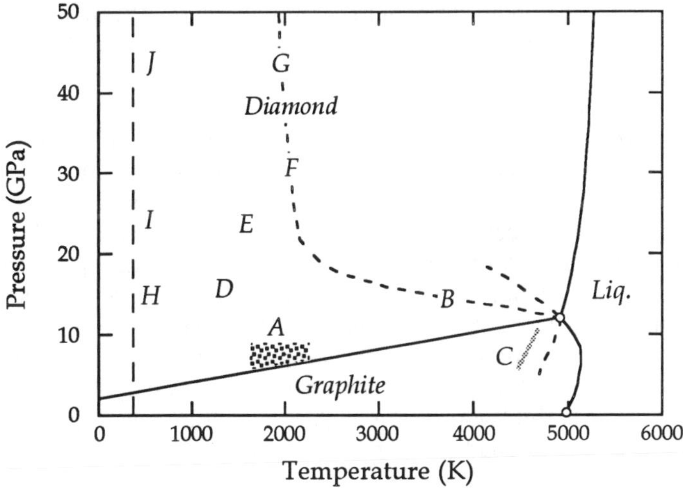

Figure 7: High P, T phase and reaction diagram of carbon according to best of knowledge through Year 1994.

temperature of diamond increases with pressure.

On the diagram, regions of special interest are marked *A, B, C,* etc. The region *A* indicates the *P, T* region where the commercial high pressure synthesis of diamond from graphite process is accomplished with the use of molten catalyst–solvent metals to flux the reaction. The dashed line *B* marks the *P, T* threshold of the "fast (i.e. *m*sec) direct transformation" of graphite to diamond (cubic–type). The dotted line *C* marks the *P, T* threshold of the "fast (i.e. *m*sec) direct transformation" of diamond to graphite.

The general area, *D*, indicates the *P, T* region in which hex–graphite, compressed in its *c*–axis direction and heated to ~1200-1500K, transforms to hex–diamond which can be retrieved after cooling and decompression. The *P, T* region *E* marks the maximum *P, T*'s of shock–compression, thermal–quench, cycles which yield hex–diamond. In this process the bodies which are shock–compressed are composites of very small grains of hex–graphite embedded in a matrix of iron or copper. The function of the metal matrix is to provide the pressure field on the graphite particles and, mainly, to thermally quench the newly–formed hot diamond particles so that they do not graphitize during decompression. *F* and the region to the right (higher *T*'s) marks the upper ends of shock–compression cycles, using matrix metal for thermal quenching, in which hex-graphite particles convert to cubic–type diamond. The dashed line *B,F,G* marks the reaction threshold for fast (*m*sec, μsec) transformation of compressed graphite or hex–diamond to cubic–type diamond by rapid heating (electrical, laser, shock compression). The dashed line *H,I,J* corresponds to slow, room *T*, compression of graphite. When single–crystal graphite is compressed in the *c*–axis direction its electrical resistance, measured in *a,b* plane, decreases until at *H* (~12–14GPa) it abruptly increases rapidly. Coincident with the rise of resistance the optical reflectivity of the specimen decreases and its transmissivity increases, indicating transformation to a more insulating structure,–such as diamond. Xray diffraction patterns at this stage show diffuse lines corresponding to compressed, strained, graphite and hex–diamond. When the applied pressure reaches *I* (~25GPa) the specimen is almost completely transparent, its Raman spectrum looks like "amorphous carbon," and the xray diffraction pattern remains as diffuse "lines" indicating more strain and very small domains of coherent diffraction. This condition continues on up through *J*, and even to 100GPa. Upon decompression from *J, I,* or *H* the specimen reverts to graphite.

Fig. 8 presents the low-pressure, high–temperature phase diagram. Here we see the areas for solid graphite (including perhaps carbynes?), liquid (including perhaps a low pressure insulating phase?), and vapor. The liquid/vapor boundary line extends to a "critical point" roughly estimated to

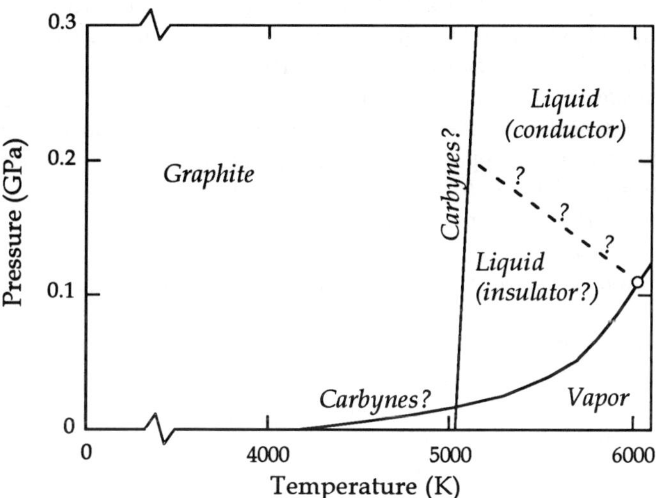

Figure 8: Low-pressure, high temperature phase diagram of carbon according to best of knowledge through Year 1994.

be at about 0.2GPa/6800K. At temperatures near the melting line there is some vague experimental evidence for a transformation from solid graphite to another solid phase which has been proposed to consist of linear–bonded carbon molecules, carbynes,–but this remains controversial. There is some old experimental evidence that can be interpreted to indicate that low-pressure liquid carbon is electrically insulating. However, none of the experiments in recent years, employing much better instrumentation and heating circuit control, yield such results. It is highly probable that there is no low–pressure insulating phase, and hence no phase boundary line between a supposed insulating phase and the conducting one.

A reaction path that cannot be shown well on the P, T phase diagram goes from the gas phase to the solid phase at pressures less than atmospheric. The free energy of carbon in the vapor phase is very high compared to that in the graphite or diamond states. The free energies of graphite and diamond at sub–atmospheric pressure and ~1200K (conditions favorable to good crystalline condensation) differ by a small amount (~1 kcal/mole). When this small difference is compared to the difference of the free energy of carbon vapor at say 5000K and that of graphite or diamond (i.e. ~200 kcal/mole) it is obvious that condensation of carbon vapor to solid graphite or diamond would have nearly equal probability. Serious attempts to grow diamond from carbon vapor go back at least as far as 1911, but lack of control of graphite nucleation frustrated the efforts. Early in the 1980's it was found that having atomic hydrogen, as well as carbon vapor and various hydrocarbon species in the gaseous plasma over the hot deposition surface, eliminated graphite nuclei preferentially to diamond ones so that continuous deposition as crystalline diamond could take place. Because this process will be discussed in detail in other papers in this Symposium it will not be discussed further here,–except to state that it is now possible by this reaction path to deposit films and sheets of polycrystalline diamond that have commercial utility.

A second vapor–to–solid reaction path at low pressure, is the tailored thermal quenching of carbon vapor to form fullerene molecules. No attempt will be made in this paper to go into any details of this interesting reaction path and its product materials.

SOME CHARACTERISTICS AND PROPERTIES OF DIFFERENT PHASES OF CARBON MATERIALS

Graphite: Graphite is very anisotropic, mechanically, electrically, and thermally. The strength, rigidity, electrical and thermal conductivity are much greater in the *a,b* crystal plane direction than in the *c*–axis direction. In the *a,b* plane direction graphite is the strongest and most rigid of all known materials. Rods and tubes of graphite in which the *c*–axis of the composite crystallites have radial orientations (relative to the axis of the rod or tube) are extremely strong and rigid in the axial direction of the rod or tube. There are processes for growing "nano–tube" carbon whiskers which may be pictured as concentric "wrap-around" layers of graphite hexagonal sheets forming giant, multi–layered "molecular whiskers." They grow endwise and can be relatively long. Their wall thickness can be increased by pyrolytic deposition of carbon on their outside walls continuing the c–axis radial orientation. Incorporation of such carbon fibers in composite structures enhances the average strength and stiffness of the composite in the direction of the fibers.

Referring back to Fig. 1 regarding the cool (room T) compression of graphitic carbon along the path *H,I,J*, the changes in the specimen depend upon a number of factors such as the isotropy of the compression, (unidirectional or hydrostatic), the crystal texture of the specimen, the direction of compression relative to the crystal axes, etc. For single–crystal specimens compressed in the *c*–axis direction the transition to the high–electrical–resistance, optically–transparent nanosized–domain material (sometimes described as "amorphous") begins sharply at about 12GPa. For the specimen consisting of randomly oriented crystallites the transition does not start until 20 to 25GPa. For poorly graphitized, sooty, or glassy specimens pressures of 40–45GPa may be required. As stated earlier, when this "high pressure," densified, carbon is decompressed it reverts back to graphite, or effectively to its original state. In the compressed, dense, state the free conduction electrons associated with the sp^2 graphitic bonding of the initial uncompressed material must have become tied up in covalent, sp^3, diamond–like bonding. The fact that the material decompresses back to graphite means that the atom displacements are relatively small and reversible. This high–pressure chaotic carbon is extremely hard and strong. It dents and cracks the pressure faces of the single–crystal diamond anvils which press against it in the experiments.

Diamond: In Region A of Fig. 1 there are two ways of growing diamond using molten metal "solvent/catalysts" for transporting the carbon atoms to the new–growing diamonds[8,9]. First is the "constant $T,\Delta G$" process in which at the diamond–stable P,T conditions diamond nuclei form at the interface of graphite and the molten metal, and the diamond grows at the expense of graphite as carbon atoms diffuse through the molten metal film between the graphite and diamond. The driving force is the greater solublity of graphite compared to diamond at the ambient T (due to the ΔG between graphite and diamond at the P, T conditions). Second is the "ΔT,ΔC" process, carried out at P, T's higher in the A region, in which a diamond carbon source is located at a higher T part of a pressurized bath of molten solvent metal and a diamond seed located at a lower T part. The higher solubility at the higher T provides a concentration gradient of carbon in the bath, making the net flow of carbon go from the hotter source to the cooler seed. In the ΔG process, the concentration gradient is dependent upon the pressure at the growing site. Because the loss of graphite and gain of equal weight of diamond involves a net decrease in local volume the P and hence ΔG, changes. To keep the transformation going the apparatus has to maintain the pressure. In practical apparatus for diamond synthesis it is difficult to monitor and control the P at the growth sites. Thus good, accurate, control of the process is not practicable. By contrast, in the ΔT process there is no net change of volume (diamond $\rightarrow$ diamond) and the temperature field in the cell can be tailored and controlled quite accurately. For this reason the ΔT process is used for growing gem–quality large diamond crystals which require slow steady growth conditions. The ΔG,const. T process is used for faster, less–critical, growth of the various grades of industrial abrasive diamond material.

In the ΔG,const. T (graphite–to–diamond), process the size, shape, and quality of the diamond material produced is controlled mainly by selecting the P, T part of Region A in which the run is made. In most cases the process cell is brought up to pressure at temperatures well below the melting temperature of the catalyst/solvent metal. Then the temperature is brought up to the desired operating value and controlled during the run. If the P, T operating point is very close to the diamond/graphite equilibrium line ΔG is very small, diamond nucleation sparse, and growth rate relatively slow. If the P, T operation point is well above the equilibrium line nucleation and growth rate are both large. If very small, intergrown, crystals are desired the process would be run with considerable "over pressure." To produce larger, independent, well–formed crystals the P, T operating point must be near the equilibrium line and for a longer time. The crystal shape obtained roughly depends upon the temperature;–lower temperatures favor cubic morphology, mid temperatures,–cuboctahedral, and higher temperatures–octahedral. The graphite and the catalyst/solvent metal used also affect the nucleation and the character of the diamonds produced.

Another "diamond process" carried out under the P, T conditions of Region A is the fabrication of sintered polycrystalline diamond compacts which are used as the cutting tips of tools for machining/drilling of hard strong materials and for wire–drawing dies[10]. In the process very clean diamond powder is placed in a capsule on top of a disc of cemented tungsten carbide and subjected to conditions corresponding to the upper parts of Region A. Some of the cobalt metal from the cemented carbide wicks through the diamond powder and promotes sintering of the diamond grains to each other, diamond–to–diamond. The strength and toughness of these compacts is due to the diamond–to–diamond bonding of the grains, not to the small amount of cobalt which remains in the interstices of the compact. In fact, in applications in which the compact gets hotter than about 1000K, the cobalt has a deleterious effect. When such sintered compacts are to be used in applications where the temperature gets very high,–as in dressing or shaping grinding wheels,–the cobalt must be removed by chemical leaching before the dressing tool is made. The random orientation of the individual grains of the polycrystalline compacts causes them to be much more wear resistant than single–crystal diamond because in the compact there are always some grains so oriented as to present their "hard" faces to the material being worked. This feature provides outstanding service life in wire–drawing dies, for example.

Electronically, cubic–type diamond crystals are like cubic–type crystals of Si or Ge which are semiconductors that can be doped to be n–type or p–type, and are used widely as transistors in solid–state electronic devices. In principle, diamonds, also should be capable of serving in this capacity if they could be effectively doped with electron donor or acceptor elements. In such applications diamond solid–state electronic devices could offer the advantage of operating satisfactorily at much higher temperatures due to the high Debye temperature of diamond. Single–crystal diamond growth experiments have shown that boron and nitrogen atoms will go in substitutionally for carbon atoms in the lattice of a growing diamond crystal. Boron doping introduces electron acceptors and makes the diamond p–type semiconducting with an activation energy of $\sim$0.18eV, and causes the crystal to be optically blue. Substitutional nitrogen, with its five valence electrons would be expected to make the diamond n–type semiconducting. Diamond growth experiments show that nitrogen atoms do indeed grow in substitutionally for carbon, making the crystal yellow in color, but it remains a good electrical insulator. Theory and experiments have shown that the extra electrons are there but at an energy level $\sim$2eV below the conduction band,–which is far too much for practical solid–state device applications. However, non-conducting diamond crystal does serve a useful purpose as thermal heat sinks for keeping solid–state elements cool. This use is based upon the very high thermal conductivity of high–quality diamond crystal (four to five times that of copper or silver at room temperature).

The most recognized and applied properties of diamond are its strength, hardness, and rigidity. In these properties it exceeds all other known materials. Because of its hardness and strength it is used widely in industry for grinding, cutting, sawing, drilling, and crushing of other hard strong materials. Currently over 85 percent of the diamond used in industry is man–made. In science, diamond anvil apparatus is used to subject tiny specimens to static pressures up to over 400GPa (4 million atmospheres).

Going to Region B of Fig. 1 let us consider the properties of the diamond made by fast direct–transition from graphite,–an extremely high P, T process[11]. Using a 1000–ton "high compression belt" apparatus specimens ~1x2x4mm have been made. They are black and glassy. Xray diffraction shows them to be cubic–type diamond with polycrystalline texture in which the crystallites or "coherent domains" are in the order of 500Å (0.05μm) in size. In torsion shear tests these specimens showed the highest shear strength of any specimens tested, including single–crystal diamond and natural carbonado. It fails by conchoidal fracture,–like glass or chinaware. It could be characterized as a "diamond ceramic." Unfortunately, because of the extreme pressure required for its synthesis it is prohibitively expensive and difficult to make.

For about a century the idea of making diamond by freezing liquid carbon has been entertained. Actually this has been done on a small scale in the laboratory[11]. There are basic problems due to the great strength and rigidity of diamond, to the relative densities of diamond and liquid at melting/freezing conditions, and to the extremely high P and T required for the process. The fact that T_m of diamond increases with P means that liquid carbon is less dense than diamond at the melt/freeze condition. In the actual physical situation a body of carbon melt can freeze only from the outside inward. Thus a strong cage of nearly constant volume forms around the enclosed liquid. Continued freezing of the contained liquid eventually reduces the ambient P to below the diamond–stable value and the remainder freezes as graphite. In the laboratory experiments, done at about 14GPa, the core of the diamond–shelled body always contained a mixture of glistening diamond and graphite crystals.

CONCLUDING REMARKS

The known states of carbon, and their properties, show carbon to be one of the most versatile and useful of the chemical elements. No doubt there is still much more to be learned about the material states in which it can exist, and the physical properties it can possess. New reaction paths may be discovered and old ones improved.

Several scientific questions remain to be resolved: One is whether there is a "carbyne phase" at high T's between graphite and the melting line, as mentioned in the text. Another is the electrical conductivity of liquid carbon as a function of P. Definitive experiments are needed to determine beyond question whether or not there is a non-conducting liquid phase at the lower P's. Still another scientific need is the accurate determination of the slope, dT_m/dP, of the diamond melting line, and/or the densities of diamond and liquid at melting line conditions. Of theoretical and astrophysical interest would be the properties of *metallic carbon* and its corresponding melt at extremely high pressures (densities).

There are a number of pertinent materials and process problems. The ideal randomly–oriented, polycrystalline, fine–grained, completely bonded diamond–to–diamond compact has not been fully achieved as a commercial product. *NATURE* has come close to such a material in *ballas* and the better *carbonadoes*. Such diamond bodies would be the hardest, strongest, toughest pieces of matter imaginable. Perhaps it was achieved in the "direct transition" diamond bodies mentioned in the text, but, as mentioned there, a less expensive and technically less demanding process is needed.

Applications involving the semi–conduction and thermal conductivity properties of high–quality single–crystal diamond remain to be developed and exploited satisfactorily. Perhaps a combination of high P, T synthesis plus CVD techniques could be directed to this opportunity? Or, if a process for growing high–quality single–crystal cubic BN could be discovered/developed, then possibly usable p– and n– type electronic devices could be achieved.

Certainly an interesting future lies ahead for carbon and for iso–electronic BN–pair materials.

REFERENCES

General:

1. F. P. Bundy, Physica **A156**, 169 (1989).

2. P. Gustafson, Carbon **24** 169 (1986).

3. D. A. Young, Phase Diagram of the Elements, (University of California Press, 1991), p. 97.

Specific:

4. M. T. Yin and M. L. Cohen, Phys. Rev. Lett. **50**, 2006 (1983).

5. R. Biswas, R. M. Martin, R. J. Needs and O. H. Nielson, Phys. Rev. **B30**, 3210 (1984).

6. J. W. Shaner, J. M. Brown, C. A. Swenson and R. G. McQueen, J. Phys. **45**, suppl. C8, 235 (1984).

7. M. S. Weathers and W. A. Bassett, Phys. Chem. Minerals **15**, 105 (1987).

8. F. P. Bundy, H. P. Bovenkerk, H. M. Strong and R. H. Wentorf, Jr., J. Chem. Phys. **35**, 383 (1961).

9. F. P. Bundy, H. M. Strong and R. H. Wentorf, Jr., Chem. and Phys. of Carbon **10**, 213-261 (1973).

10. R. H. Wentorf, Jr., R. C. DeVries and F. P. Bundy, Science **208**, 873 (1980).

11. F. P. Bundy, J. Chem. Phys. **38**, 631 (1963).

SHOCK WAVE SYNTHESIS OF DIAMOND AND OTHER PHASES

PAUL S. DECARLI
Poulter Laboratory, SRI International, Menlo Park, CA 94025

ABSTRACT

Shock wave synthesis of diamond was an unexpected result of experiments designed to explore the effects of shock waves on a variety of materials. The initial announcement in 1959 was controversial; shock synthesis of diamond had been shown to be unlikely, on the basis of kinetic arguments. Jamieson confirmed the identification and suggested a diffusionless mechanism, c-axis compression of rhombohedral graphite. Subsequent work has provided strong evidence that shock wave synthesis of cubic diamond is a conventional thermally activated nucleation and growth process. Thermal inhomogeneities provide the requisite high temperatures; quenching via thermal equilibration is implicit in the process. Shock synthesis of adamantine BN phases appears to be quasi-martensitic; a martensitic mechanism may partially account for the Lonsdaleite (hexagonal diamond) observed in some meteorites and in some artificial shock products. Diamond is also formed as a detonation product in oxygen-deficient explosives. The poly-crystalline product of shock synthesis is similar to natural carbonado. The association of carbonado with an ancient giant impact crater is noted.

INTRODUCTION AND HISTORICAL BACKGROUND

This paper is largely devoted to work that was performed between 1955 and 1970. Since relevant patents have long since expired, along with my moral and legal obligations to the companies that supported this research, it is now possible to more adequately document this early work. The pre-1960 archival literature was thoroughly searched; I found only three references to shock wave synthesis of diamond. I also found one relevant patent, to be discussed in the section on commercialization. Between 1970 and the present, numerous papers on shock synthesis of diamond and other materials have appeared. Although I do refer to a few recent papers, I do not mean to imply that I have thoroughly reviewed the post-1970 literature.

The earliest reference that I have found to shock wave synthesis of diamond describes experiments performed by Noble and Abel of the British Admiralty in 1908. They exploded cordite in sealed steel containers. The carbonaceous residue was examined by Crookes, who reported the presence of recognizeable diamonds. [1] Crookes also confirmed that Moissan's method made diamond; neither identification was ever confirmed by X-Ray diffraction. Furthermore, the peak pressure claimed by Noble and Abel was only 8 kilobars, well within the stability field of graphite.

Mat. Res. Soc. Symp. Proc. Vol. 383 © 1995 Materials Research Society

In his 1918 Bakerian Lecture, Parsons summarized the results of thirty years of research on diamond synthesis. [2] Among numerous techniques, he describes projectile impact experiments which may represent the first scientific attempt at shock wave synthesis of diamond from graphite. Although he reported failure to detect diamond in the shocked graphite, my reanalysis of his well-documented experiments indicates that he was operating in an appropriate pressure range, above 150 kbar, for shock diamond synthesis. However, his techniques for identification of diamond were inadequate. I think it unlikely that he could have detected the typical product of shock synthesis. In a 1956 publication, Riabinin also reported failure to shock synthesize diamond. [3] Riabinin was also operating in an appropriate pressure range, and it is possible that some of his experiments succeeded and he simply failed to detect the diamond.

Of course, there were many unsubstantiated and poorly documented claims for successful shock wave synthesis of diamond prior to 1960. Among a certain class of amateur inventors, diamond synthesis competed for popularity with perpetual motion machines. The history of diamond synthesis has been marked by numerous mistaken identifications and possibly by cases of deliberate fraud. Until quite recently, respectable scientific journals were extremely wary of publishing any account of diamond synthesis without substantial corroboration.

EARLY SRI WORK

During the mid-1950s, my colleagues and I in Poulter Laboratory were involved in studies of shock effects, including microstructural evidence of phase transformations, in metals. John Jamieson, a summer visitor, suggested performing experiments with minerals. Our first experiment yielded an interesting effect, shock amorphization of quartz. [4] Feldspar behaved similarly. After noting similarities between radiation damage and shock wave damage, I requested permission (in 1959) to perform a few shock loading experiments on materials, including a commercial nuclear-grade graphite, known to be susceptible to radiation damage. I did not intend to make diamond; I had not even considered the possibility. I was therefore very surprised to find that this first sample of shocked graphite contained hard particles that could scratch sapphire. No hard particles were found in unshocked control samples (National Carbon Company grade CS-312). X-ray diffraction examination indicated the presence of very fine-grained polycrystalline diamond in a matrix of graphitic carbon.

My colleagues and supervisors were justifiably skeptical when I presented the evidence for diamond in this sample. They knew that H.T. Hall had recently presented the results of a theoretical analysis of the possibility of the direct transition of graphite to diamond. His conclusion that no condition of temperature would yield an observable reaction rate appeared to explain the failure of numerous attempts at direct static synthesis. [5] My supervislors also knew that Weeks and Goranson had presented arguments that shock wave synthesis of diamond was not likely. [6] Their reasoning was that direct diamond synthesis should take place via a conventional nucleation and growth mechanism. If the shock temperature were high enough to permit significant carbon diffusion over the sub-microsecond time when the shock pressure was within the stability field of diamond, the newly grown diamond would revert to graphite with equal rapidity as the shock pressure dropped below the diamond stability line. Thus, the arguments that diamond could not possibly be present overwhelmed the physical evidence for the presence of diamond. Since the X-ray diffraction lines of shock synthesized diamond are very broad, optimal X-Ray techniques must be used if one wishes to detect the presence of 5 % diamond in shocked graphite. Even then, the evidence does not appear very convincing to anyone who is not experienced in X-Ray diffraction techniques.

Pieces of the shock loaded material were therefore sent out to various researchers, including John Jamieson at the University of Chicago, for independent examination and confirmation of the presence of diamond. Jamieson was the only person who could confirm my X-ray diffraction evidence for the presence of diamond in the sample. Of course, I could not have detected diamond myself without Jamieson's earlier instruction in good X-ray diffraction specimen preparation techniques. Additional carefully controlled and monitored shock synthesis experiments were performed during Jamieson's subsequent summer visit. Jamieson could thus vouch that the synthesis was reproducible and that we were not observing artifacts of contamination. Jamieson also suggested a diffusionless mechanism for the shock synthesis, c-axis compression of rhombohedral graphite, and my supervisors finally granted permission to publish. [7] A patent application was also prepared, in the hope of eventual commercialization of the process.

Contemporaneously, Alder and Christian (Lawrence Livermore Laboratory) were performing shock wave equation of state measurements on carbon [8]. We were aware of this work; it implied that graphitic carbon could transform to diamond (or to another dense phase of carbon) on a sub-microsecond time scale. Lipschutz and Anders (U. of Chicago) were preparing the first of a series of papers on the shock wave origin of meteoritic diamond. [9] It is interesting to note that their original mechanism, shock induced decomposition of Cohenite (meteoritic iron carbide), was proposed because they considered the direct graphite-diamond transition to be kinetically unlikely.

During this period between 1959 and 1961, we were plagued by the fact that some of the reputable scientists who examined our samples were unable to detect shock synthesized diamond even when it was present at the level of about 5%. Furthermore, other shock wave workers privately reported that they were unable to duplicate our results. In every case the problem was with identification of diamond. This problem was eventually solved after Eric Lundblad of ASEA taught us how to purify diamond by selective oxidization of the graphitic carbon component in refluxing sulfuric-nitric acids. Since there is no problem in identifying purified shock-synthesized diamond, we simply disseminated the purification technique. Lundblad also provided welcome support when he reported that he had independently confirmed the shock synthesis of diamond. [10]

SCIENTIFIC RESULTS

We begin this section by noting that shock wave propagation is an orderly process; modern shock wave theory is derived from Newton's laws of motion and from the laws of thermodynamics.[11] Shock wave theory was essentially complete before 1960, and techniques for shock wave measurement were well established. However, it is only in recent years that we have had the computational power to be able to make detailed calculations of shock wave propagation in various complicated materials such as composites and porous materials. If this study were to be performed today, it would probably be a predominantly computational effort.

The goals of our experimental program were to understand the mechanism of shock synthesis of diamond and to determine the optimal conditions for shock wave synthesis of diamond. If we could understand the mechanism, we could predict the optimal conditions, wheras a trial-and error approach might yield only a local optimum. Our tools were a good understanding of shock wave theory and technology; we knew how to perform controlled shock-loading experiments on samples of carbon. The yield of diamond was determined by Lundblad's analytic technique, selective oxidation of non-diamond carbon. Every sample was examined by X-ray diffraction, to insure that the purified sample was indeed diamond; additional chemical treatments were performed as needed. Quantitative measurement of diamond yield was possible for yields above about 0.5 %. However, we could qualitatively detect the presence of smaller yields, down to a mere trace, by making an X-ray powder camera specimen from a few tens of micrograms of material scraped from the surface of a filter.

Between 1959 and 1964, hundreds of shock loading experiments were performed and many kilograms of diamond powder were synthesized, but we made little progress in understanding the mechanism of shock diamond synthesis. We did manage to soon convince ourselves that the diffusionless mechanism, c-axis compression of rhombohedral graphite, could account for only a small fraction of the diamond yield observed in many experiments. Analyses of the experimental data qualitatively supported hypotheses of a thermallly activated mechanism, but we were unable to obtain firm quantitative interpretations.

We worked very hard to design and perform sets of shock loading experiments in which only a single parameter was sensibly varied. The shock wave parameters that could be varied included peak pressure, peak pressure duration, pressure pulse shape (thermodynamic loading path), and peak internal energy increase. The sample parameters that could be varied included crystallinity (ranging from "amorphous" to well-crystallized graphite) and sample porosity. Note that peak pressure, peak internal energy and initial sample porosity are coupled; one can vary one of these parameters independently of the other two... but only over limited regions of parameter space. We extended the range of accessible conditions by performing a few experiments at initial sample temperatures of 77 K and 1300 K.

Variations in peak pressure duration were largely limited by cost considerations to the range between about 0.5 μs and 20 μs at peak pressures below about 600 kbar. At the maximum shock pressure of our experiments, about 1.5 megabars, cost considerations limited us to a pressure pulse duration of about 0.5 μs.

In one of our earliest parameter variation experiments, we studied the effect of initial carbon crystallinity on diamond yield and diamond properties. To discuss carbon crystallinity without getting bogged down in details important to specialists, it may be useful to consider crystallinity as a continuum bounded by end-members. At one end there are ideal single crystals of graphite in which perfect layers are stacked in ideal (hexagonal or rhombohedral) sequences. The other end-member is so-called amorphous carbon, which might be more accurately desribed as stacking-disordered nano-crystalline graphite. Commercial carbon and graphite products (e.g., electrodes, bricks, bar stock, etc.) are made from petroleum or anthracite coke particles, bonded by pitch and molded to shape, and subsequently heat-treated. If the heat-treatment is brief, the product retains the "amorphous" properties of the source coke and is designated carbon. If the heat-treatment is extensive, the crystallinity can improve and the product is designated graphite. However, in the crystallinity continuum, these commercial graphites occupy the middle region. In the early 60s, natural graphite samples most closly approached the single crystal ideal; available pyrolytic graphites might be described as quasi-single-crystals having very poor long-range stacking order.

Samples of anthracite and petroleum coke carbon, various grades of commercial "graphite", pyrolytic graphite, and natural graphite were shock loaded under identical conditions. Since the internal energy increase is a function of sample compression, it was important that all of the carbon samples have about the same initial density. The commercial graphites and carbons had bulk densities in the range of 1.6 to 1.7 g/cc. Samples of these materials were machined from as-received electrode or bar stock. The pyrolytic and natural graphite materials were initially too dense, ranging fron 2.2 to 2.26 g/cc. Samples of these latter materials were prepared by grinding the bulk material and pressing the resultant powder to a density of 1.7 g/cc. Since grinding does affect crystalline perfection and stacking order, natural graphite samples were made up from from both coarsely ground and extensively ground samples.

Analysis of the results of the experiments indicated that the the crystallinity of the starting material had no sensible effect on diamond yield; any effects on diamond properties was too subtle for us to detect. All of the diamond comprised polycrystalline particles with a crystallite size of about 20 nm and with particle sizes in the range below about 10 μm. These results appeared incompatible with the originally proposed diffusionless mechanism.

In experiments with commercial graphite products of similar crystallinity, the initial density of the carbon, the loading path, and the peak shock pressure were varied independently. The yield of diamond generally increased with increasing pressure (for constant initial density and qualitatively similar loading paths). The yield also increased with decreasing initial sample density (for constant peak pressure and nearly identical loading paths). Finally, the yield was extremely sensitive to loading path. About a 25% yield of diamond could be obtained by shocking graphite in a single step to 250 kbar. If the same material were loaded via multiple shock reflections to the same peak pressure, the yield of diamond was only about 0.2%.

These results implied that the most significant parameter was the increase in internal energy during shock compression. Higher internal energies imply higher shock temperatures. These results thus supported the hypothesis of a thermally activated mechanism. The internal energy increase in shock compression can be determined with adequate accuracy; the first step in calculation of shock temperature is to partition the energy between mechanical and thermal. The problem was that even if one went to absurd lengths (e.g. assuming the mechanical energy to be zero) to maximize the temperature estimates, the results seemed to imply that we were making diamond in less than a μs at peak temperatures of less than 2000 K. Furthermore, it was difficult to reconcile a thermally activated mechanism with the observation that the yield of diamond was insensitive to pressure pulse duration.

In 1963, Bundy reported direct (static high pressure) synthesis of diamond from graphite.[12] The graphite was compressed to pressures above 130 kbar and then heated by a millisecond duration capacitor discharge. Bundy's diamond was polycrystalline, much like shock synthesized diamond, but with an order-of-magnitude larger crystallite size of about 100 nm. He found that diamond formation required the insertion of sufficient energy to heat the graphite to a temperature of about 3000 K. In subsequent work, Wentorf inferred that the activation energy for direct growth of diamond from graphite is about 200 kcal/mol.[13]

If one makes the simple assumption that shock synthesis of diamond occurs by the same mechanism as the static direct direct transition, a shock temperature of about 3700 K is needed to account for growth of the shock product within a μs. We need only a two order-of-magnitude increase in growth rate because of the smaller crystallite size of the shock product. The problem is that the diamond would graphitize at a comparable rate as soon as the shock pressure decayed into the graphite stability field.

The phenomenon of hot spot formation in shock compressed solids is well known.[14] Mechanisms for hot spot formation include adiabatic shear and jetting in initially porous solids. Either mechanism could yield spots of sufficient hotness to account for diamond growth in our experiments. Survival of hot diamond would be possible if it were rapidly quenched by surrounding cooler graphite before the pressure decayed into the diamond stability field. The evidence that a hot spot mechanism might be important included the observation that relatively large yields of diamond were obtained from initially porous graphites, while initially dense pyrolytic graphite samples yielded negligible diamond when shocked to higher pressures and similar or even higher internal energy increases.

One crucial question is whether the energy content of the hot spots can account for observed yields of diamond. Given appropriate Hugoniot data, one can accurately calculate the energy increase of shock compression. Hugoniot data for ATJ graphite at 230 kbar indicates that the net energy increase is about 1900 J/g.[15] The average peak temperature of the graphite (if the available energy were homogeneously distributed) would be only about 1600 K.

However, our thesis is that the energy is distributed heterogeneously. We need a formula to partition the available energy between hotspot energy and homogeneously distributed energy.

We note that both Hugoniot measurements and recovery experiments showed that no (or negligible) diamond is formed from non-porous pyrolytic graphite over the shock pressure range up to 400 kbar. [15,16] We therefore assume the energy increase of shock compression to be homogeneously distributed in the pyrolytic graphite. (This latter assumption is non-controversial). We then hypothesize that the homogeneously distributed component will be the same in an initially porous graphite as in pyrolytic graphite shocked to the same peak pressure. Noting further that the internal energy increase for initially porous carbons is much higher than for pyrolytic graphite shocked to the same pressure, we argue that the difference should be a measure of the energy available for hot spots. Note that this energy balance approach gave us a way around the problem of calculating the details of hot spot formation on a nanosecond time scale. The hot spot can be initially very hot, perhaps above 10000 K, in the first nanoseconds of its formation. On the time scale of 10s of nanoseconds, the hot spot will become larger and cooler. Eventually, perhaps within as little as 0.1 μs after its initial formation, the hot spot becomes too cool to support further diamond growth.

At 231 kbar, the internal energy difference between pyrolytic graphite and initially porous ATJ graphite is about 1600 J/g.[15] This is sufficient energy to heat about 30 % of the graphite to a temperature of about 3500 K. Our analysis of the dynamic data of ref. 15 indicates that the graphite was shocked to a density corresponding to about 30 % diamond at pressure and that most of the diamond was retained through pressure release. We conducted diamond production experiments at somewhat higher pressures of about 250 kbar; up to 25 % diamond was recovered. Other sets of comparisons between McQueen's precision dynamic data and our relativly crude production experiments also tended to support the hotspot mechanism.

Furthermore, crude heat flow calculations indicate that a pressure pulse duration of a microsecond is sufficient to permit quenching of a 10 micron diameter diamond from about 3500 K down to about 2200 K (sufficient to avoid graphitization). These heat flow calculations used handbook data on thermal conductivity of graphite at high temperature; they did not invoke the high conductivity of perfect diamond at low temperature. If the thermally activated mechanism were correct, the heat flow calculations accounted for the fact that we did recover polycrystalline diamonds as large as 10 μm diameter.

As a crucial test of the thermally activated mechanism, we designed an experiment with a mixture of 50 volume % copper powder (5μm) and 50 volume % -200 mesh ground natural graphite.

Experimental conditions (sample density, peak pressure, loading path) were designed to provide a final temperature, after thermal equilibration of carbon and copper, of 1900 K. The internal energy increase in the carbon was calculated to be about 4500 J/g. This energy was sufficient to heat all of the graphite to about 2800 K or to heat about 70% of the graphite to about 3500 K. Our analysis indicated that the recovered carbon (determined after acid dissolution of the copper) contained 60% diamond.

At this point, the thermally activated mechanism appeared to be satisfactorily confirmed. All of our past experimental results now made perfect sense, when reanalyzed in terms of a thermal activation merchanism, and when hot spots were properly taken into account. In many of our parameter variation experiments, we were working in the wrong range. If the diamond were formed and quenched in our shortest-duration experiment, we were obviously going to see no effect of longer duration experiments. Although our scientific understanding was essentially complete by 1969, commercial considerations delayed publication of the results until 1979. [17]

It is now a simple matter to design an experiment to produce mm-size diamonds. One would simply scale up the dimensions of porous carbon and solid quenchant particles and use enough explosive to provide a long pressure pulse. The principal design constraints on the explosive system are that the peak pressure be sufficient to heat all of the carbon to a shock temperature af about 4000 K and that the pressure pulse duration be long enough to permit to permit quenching of the hot diamond to a temperature below about 2000 K before pressure decayed below the diamond stability line.

The only practical problem is cube-root scaling; one must use a thousand times as much explosive to increase the pressure pulse duration by a factor of ten. Preliminary and very crude design calculations indicate that one might require as much as 4 kilotons of a military explosive (a RDX or PETN based plastic explosive such as C-4 or Semtex would work). One might make 20 tons of diamond (100 million carats) in such a shot, but the cost of the shot would be about $ 100 million (assuming one could buy surplus military explosive cheaply enough). The design parameters could be optimized via detailed computer code calculations, but I think it unlikely that the cost per shot could be brought down to less than about $ 20 million.

We note a recent paper describing dynamic observations of an apparently martensitic phase change to a diamond-like density at a shock pressure of about 200 kbar in very highly ordered pyrolytic graphite. [18] We point out that diamond was not recovered from these experiments. However, the results do seem to support the view that there is nothing magical about shock pressures; as the paper points out, similar results are obtained under static high pressures.

COMMERCIALIZATION OF SHOCK WAVE SYNTHESIS

Between 1959 and mid-1962, the diamond shock synthesis program was supported by internal SRI funds. The first patent application was filed in January 1961, a continuation-in-part was filed in October 1963, and U.S. Patent 3,238,019 was granted on 1 March 1966. Early in the patent process, the examiner called our attention to prior art, British patent 822,363, granted in 1959 to a Dutch inventor, Jan Van Tilburg. This patent cited two methods; an exploding wire method and the use of a shaped charge to shock load graphite. It was claimed that large diamonds could be made, in the shape of a tube. It was also claimed that the shock synthesized diamond was sometimes harder (and sometimes softer) than natural diamond.

It was obvious to us that the exploding wire method could not possibly have worked in the manner described, and the patent examiner accepted our evidence that exploding wire technology (as of 1961) was unsuitable. It was equally clear to us that the shaped charge technique could indeed yield small quantities of diamond (but not a tube), provided that the jet was directed into porous carbon. However, the method was clearly inefficient and it appeared to have a major drawback; any diamond made by the action of the very high pressure leading edge of the shaped charge jet would probable be eroded away by the low pressure tail. We tested the method exactly as described in the patent, and we were able to report that we recovered no diamond. We were not required to report that we could have designed improvements to the shaped charge geometry which would have made it possible to recover small amounts of diamond.

In mid-1962, SRI's patent rights were assigned to Allied Chemical Corporation, which supported further work at SRI and began pilot plant production of shock synthesized diamond. In the production facility, the shock loaded graphite fragments were captured in a water tank. The material recovered from the tank was a mixture of coarser mm-size fragments and very much finer material. The chemist in charge of diamond purification soon learned that most of the diamond was in the fine fraction and could be purified in a few hours of treatment via the selective oxidation method. The mm-size fragments contained less than 5 % diamond in a matrix of dense graphite that was resistant to oxidation. When the chemist learned that it could take days of treatment to purify the diamond in the larger fragments, he made a unilateral decision to screen out the coarse material and discard it. The fine fraction alone contained diamond in a quantity equal to about 25% of the initial weight of the graphite feedstock. However, all of the diamond contained in this fine fraction was in the form of nanoparticles, about 5 nm in diameter. Allied made many kilograms of these nanodiamonds, but could not find a market for them. The discarded material, on the other hand, contained marketable abrasive-grade particles as large as 10 μm in diameter. In view of recent results showing that nanodiamond is a detonation product of oxygen-deficient explosives, one may ask whether the source of Allied nanodiamond was actually the explosive.[19] My best estimate is that less than 20 % of Allied nanodiamond originated in detonation products. I base the estimate on the fact that Allied's diamond yields increased with increasing shock pressure, but the amount and type of explosive remained relatively constant over the range of different explosive geometries that were used to obtain different pressures.

DuPont was contemporaneously working on shock diamond synthesis. Their U.S. Patent, 3,401,919, was granted in 1968. They explicitly cited the importance of quenching. However, we could show that quenching was implicit in our method of shocking porous graphites. I could also show witnessed notebook entries to prove that I had performed experiments with metal heat sinks in early 1962. Dupont, on the other hand, had bought rights to the British patent mentioned above. The stage was set for a bloody court fight, but Allied and Dupont decided to settle the matter by signing mutual cross-licencing agreements. After some market research, Allied decided that the abrasive market was too small to justify remaining in the diamond business.

SRI also supported early work on shock synthesis of BN high pressure phases. This work was successful. [20] The yield of BN high pressure phases was found to be highly sensitive to the crystalline perfection of the feedstock. We obtained high-pressure phase yields of up to 70 % by shocking well-ordered hexagonal BN to pressures as low as 200 kbar, wheras only trace yields were obtained in experiments with poorly crystalline hexagonal BN. Despite the evidence that the BN mechanism was not thermal, our patent aplication was rejected on the grounds that our diamond patent was sufficiently broad to cover synthesis of dense BN phases.

GIANT IMPACTS

It is now generally accepted that the diamonds found in some meteorites are impact-synthesized. The presence of Lonsdaleite, also known as hexagonal diamond, is generally considered diagnostic evidence of impact synthesis. However, it is well established that Lonsdaleite can only be made from highly crystalline graphites. DuPont made Lonsdaleite by shocking graphitic cast iron. [21] In my experience, however, the usual product of shock synthesis is ordinary cubic diamond. [22] In recent years, it has been recognized that giant impacts may result in the conversion of terrestrial carbon to diamond. [23]

In 1984, Smith and Dawson suggested that carbonado, a type of polycrystalline diamond found both in Central Africa and in Brazil, might have formed in giant impacts. [24] Magnetic and photographic data obtained from spacecraft have revealed a possible 500 mile diameter fossil impact crater in Central Africa. [25] If the "structure" is actually a crater, geologic evidence indicates that it would have to date back to the early Pre-Cambrian, an era in which Brazil was contiguous with Central Africa. We are currently investigating the question of whether the detailed characteristics of carbonados are compatible with a shock synthetic origin. This work is being done in collaboration with many others, including H.J. Milledge of University College London, R.J. Girdler and D.W. Collinson of Newcastle-upon-Tyne, C.T. Pillinger of the Open University, and Monica Grady of the British Museum.

ACKNOWLEDGEMENTS

I particularly thank George Duvall, former Scientific Director of Poulter Laboratory for excellent scientific guidance and for his insistence that I make an air-tight case for shock synthesis of diamond. Francis Bundy was very generous with information and with samples of flash-synthesized diamond. T.C. Poulter, Laboratory Director, arranged for in-house support of my early efforts. Finally, I thank my wife, Anne, for moral support and for patient acceptance of my workaholic period.

REFERENCES

1. W. Crookes, Diamonds, London, 1909
2. C.A. Parsons, Phil Trans Roy Soc (London) **A220**, 67 (1920)
3. Iu.N. Riabinin, Sov Phys Tech Phys **1**, 2575 (1956)
4. P.S. DeCarli and J.C. Jamieson, J Chem Phys **31**, 1675 (1959)
5. H.T. Hall, Proceedings of Symposium on High Temperature, Stanford Research Institute, 1956, p 161
6. I.F.Weeks and R.N. Goranson, The Feasibility of Producing Diamonds by Nuclear Explosions, U Cal Res Report Number UCRL-5253, 1958
7. P.S. DeCarli and J.C. Jamieson, Science **133**, 182 (1961)
8. B.J. Alder and R.H. Christian, Phys Rev Lett **7**, 367 (1961)
9. M.E. Lipschutz and E. Anders, Geochim et Cosmochim Acta **24**, 83, (1961)
10. European Scientific Notes, ONR London Branch, No. 15-8, 171 (1961)
11. G.E. Duvall and G.R. Fowles in High Pressure Physics and Chemistry, Vol 2, R.S. Bradley (ed.), Academic Press, 1963
12. F.P. Bundy, J Chem Phys **38**, 618 (1963)
13. R.H.Wentorf, Jr., J Phys Chem **69**, 3063, (1965)
14. J.H. Blackburn and L.B. Seeley, Nature **194**, 370 (1962); **202**, 276, (1964)
15. R.G. McQueen and S.P. Marsh, in Behavior of Dense Media Under High Dynamic Pressure, Gordon and Breach, New York, 1968
16. D.G. Doran, J Appl Phys **34**,844 (1963)
17. P.S. DeCarli in High Pressure Science and Technology, Vol 1, Timmerhaus and Barber (ed), Plenum, New York, (1979)
18. D.J. Erskine and W.J. Nellis, Mat Res Soc Proc **270**, 470 (1982)
19. A.I. Lyamkin. E.A. Petrov, A.P. Ershov, G.V. Sakovich, A.M. Staver, and V.M. Titov, Sov Phys Doklady **33**, 705 (1988)
20. P.S. DeCarli, Bull Am Phys Soc II, **12**, 127 (1967)
21. L.F. Trueb, J Appl Phys **39**, 4707 (1968)
22. P.S. DeCarli in Science and Technology of Industrial Diamonds, Vol 1, p49, Burls (ed), Industrial Diamond Information Bureau, London, 1967
23. F.V. Kaminsky, in Proceedings of the Fifth International Kimberlite Conference, Vol 2, Brazil 1991, p 136
24. J.V. Smith and J.B. Dawson, Geology **13**, 342 (1985)
25. R.W. Girdler, P.T. Taylor, and J.J. Frawley, Tectonophysics **212**, 45 (1992)

THEORY OF SOME CARBON SOLIDS

MARVIN L. COHEN
Department of Physics, University of California, and Materials Sciences Division, Lawrence Berkeley Laboratory, Berkeley, CA 94720

ABSTRACT

Starting from properties of the carbon atom, it is possible to explore the structural stability and properties of various forms of carbon. A discussion of the theoretical approach and the relevant atomic characterisitcs will be given followed by example of applications to existing and hypothetical materials formed from carbon and related elements.

INTRODUCTION

Robust physical models and modern computers have enabled accurate predictions and explanations of solid-state properties. Although the focus of much of the research has been in the realm of basic theory and development of models to improve precision, applications have been made and some of them have been useful.

A major goal of theory in this area has been the determination of solid-state properties using *ab initio* theory with only some input knowledge (such as atomic number) about the properties of the atoms making up the solid. Other empirical and semi-empirical approaches use theoretical models and a minimum of measured properties (such as the lattice constant) of the solid as input. The latter approaches are somctimes useful to explain trends in the properties of materials. They also can help in establishing numerical limits on these properties and in making specific predictions.

Applications in this area of research are often focused on prototype materials such as Si and GaAs representing homopolar and heteropolar semiconductors, Na and Al which are typical simple free-electron-like metals, and diamond as a representative insulator. Although the band gap and resistivity of pure diamond imply that this material is an insulator, it has much more in common with Si, Ge and grey Sn than with the more classic insulators such as NaCl. Hence research on the group IV solids as a whole has had the largest influence on establishing the microscopic basis for understanding the properties of diamond. The same is not true for other forms of carbon. For example, in the case of graphite, Si, Ge, and Sn are not found in the graphitic structure so that comparisons, analogies and theoretical connections based on atomic structure calculations for graphite are better made with other layer materials such as the $B_xC_yN_z$ compounds.

It can be argued that the ability of carbon to form sp^2 bonds in graphite is related to the electronic structure of its core. Unlike Si and the heavier group IV atoms, there are no p-electrons in the C core. From the point of view of pseudopotential theory [1], a p-like valence

Mat. Res. Soc. Symp. Proc. Vol. 383

electron will not experience a repulsive core potential arising from the Pauli exclusion principle. As a result, these valence electrons are not pushed out into the bond as effectively as in the case of Si, and they concentrate in roughly the same region with respect to the core as they would in the free atom. The result is a double-hump bond charge density [2]. Since B and N have the same $1s^2$ core as C they have similar characteristics and BN can exist in a graphite-like structure. Layer structures are also found for BC_2N [3] and BC_3 [4]. All of these materials are candidates for fullerene-like structures such as buckyballs, buckytubes, bucky onions, and so on. Experimental and theoretical studies in this area are very active at present and much of the research is motivated by the possibility of producing novel and useful materials [5-11].

Other paths to novel sp^2 materials using C and BN exist, but at this point the research reported has been theoretical. In particular, suggestions have been made [12-14] for candidate three dimensional structures which are based on sp^2 bonding. These open lattice-like structures such as H-6 carbon [12] are appealing as sp^2 prototypes, but up to this point none have been made and theoretical calculations [12-14] suggest they are unstable with respect to a transition to diamond.

The focus of the remainder of this paper will be explaining the models and theoretical methods in this area to provide a description of what modern theory can do. Applications of the theory will be divided between basically sp^3 carbon and related compounds such as BN and carbon nitride [15, 16, 17]. Here emphasis will be placed on structural properties such as hardness and compressibility. The other area of applications will be to sp^2 carbon and related compounds BN, BC_2N and BC_3 with emphasis on novel materials having novel electronic and structural properties.

THEORETICAL APPROACHES AND PHYSICAL MODELS

Although it is clear that the Bohr-Rutherford model of the atom including its modern extensions is consistent with experiment and serves as a basis for *ab initio* calculations on solid phases, there is some merit in examining earlier models. In particular, although Rutherford's experiments disproved J.J. Thomson's plum pudding atomic model, the latter picture has many similarities to the modern jellium model of a metallic solid. Thompson's model utilized a spherical positive pudding embedded with negative electron "plums". The jellium model assumes that a solid metal is composed of a positive jelly background made by smearing the ions throughout the solid providing a medium similar to Thompson's pudding. The electrons, however, are not considered to be plums, but are treated quantum mechanically with the appropriate wavefunctions to describe their properties. This jellium model is an excellent first approximation to metallic solids and is applicable to about a fifth to a quarter of the elements in the periodic table and to metallic clusters [18].

Newton's atomic model was probably motivated by efforts to explain the existence of solids. The Newton atoms had hooks for binding atoms together. Modern electron density plots [1] for covalent solids like diamond, Si, and Ge illustrate the pile up of electronic charge in the bond region and little imagination is needed to visualize Newton's hooks holding the ionic cores together. Hence when one moves across the columns in the periodic table for a given row such as the Na, Mg, Al, Si, series, the elements Na, Mg and Al form free-electron-like metals which can be described using jellium models or Thomson-like atoms. However, when one reaches Si, Newton's hooks show up. The sp^3 electron configuration gives directional bonds and covalency.

A requirement of a robust *ab initio* theory of the electronic structure of solids is to reproduce these trends and give electronic energy levels and eigenfunctions with good precision for Thomson-like or Newton-like systems. One approach which has had great success is based on a model introduced by Fermi [19] in 1934 and re-invented and expanded often since then. Fermi used this approach, now called the pseudopotential model, to explain the properties of highly excited atoms in a gas. He argued that the inner oscillations of the valence electron wavefunctions were not relevant for calculating the properties of the outer shell and produced a potential to yield a nodeless wavefunction. Because solid-state effects are dominated by the outer portions of valence electron wavefunctions, this approach is also useful for calculations on solids.

Modern pseudopotential calculations begin with computer generated potentials constrained to produce nodeless wavefunctions which are identical to precise atomic wavefunctions everywhere except in the regions of the atom near the core. The pseudowavefunctions are properly normalized and yield electronic densities which are consistent with observed X-ray scattering experiments.

The path from Fermi's first calculations to the modern approaches was not a direct one. A useful diversion was the development of a semi-empirical theory which utilized experimental input about the solids under study for developing the pseudopotentials. For example, in the case of the group IV diamond structure solids, it was possible to fit the lowest three Fourier components $V(\vec{G})$ of the pseudopotential to optical reflectivity data and in turn then compute energy bands and wavefunctions capable of describing a host of other solid-state properties. Thus the Empirical Pseudopotential Method (EPM) represented the pseudopotential $V(\vec{r})$ as

$$V(\vec{r}) = \sum_{\vec{G}} S(\vec{G})V(\vec{G})e^{i\vec{G}.\vec{r}} \quad (1)$$

where $S(\vec{G})$ is the structure factor. The successes of the EPM in explaining electronic properties, photoemission spectra, and optical properties of dozens of solids [1] and the transferability of the potentials between compounds and for different structures gave credibility to the approach and provided information about the underlying physics which was very useful for developing first-principle methods.

Because the EPM pseudopotential was treated as a single potential containing electron-core and electron-electron interactions, it was not easily adapted to calculations of surface properties where charge rearrangements occur or for calculations of structural properties where different candidate structures having different electronic charge distributions are compared. Studies of this kind required the separation of electron-core and electron-electron interactions in the potential. This separation led to schemes [20] which were similar to Fermi's original approach where, as described earlier, atomic wavefunctions are used to generate the core pseudopotential. The electron-electron interactions are evaluated using a density functional formalism [21,22].

Although the EPM gave accurate electronic charge densities which represented the bonding properties of crystals, the *ab initio* approach together with a total energy formalism [23,24] allowed detailed structural studies. The approach involved the comparison of the total energy E_{tot} of solids in different structural configurations

$$E_{tot} = E_{cc} + K_e + E_{ec} + E_{ee} \tag{2}$$

where E_{cc} is the core-core contribution evaluated using Madelung sums. The electronic energies include the kinetic energy K_e which is obtained from the electron wavefunction, the electron-core contribution E_{ec} which is calculated via the pseudopotential and the electron-electron interaction containing a Hartree term and the exchange-correlation potential which is evaluated using a local density approximation [21,22].

The early successes for Si using this approach were followed by applications to dozens of crystalline solids. A particularly dramatic series of applications [24] to high pressure Si predicted the existence of new phases, superconductivity, electronic structure, lattice parameters, phonon spectra, and structural properties. This study demonstrated conclusively that the existence of new pressure induced solid-solid phase transitions, transition volumes, and transition pressures can be predicted using this total energy method.

The above approach together with the empirical and *ab initio* models and applications provided a background for the conceptual model used in many current electronic and structural calculations. This "Standard Model of Solids" envisions a collection of cores in a periodic array together with a sea of valence electrons which arrange themselves dictated by a minimum energy principle with E_{tot} given by Eq(2). Although the pseudopotential approach together with the LDA approximation is the most popular version of the standard model, one can also consider "all-electron" calculations in which core electron rearrangements are included. It is also possible to go beyond the LDA approximation which is known to be inappropriate for calculating excited states in solids. One such approach in the GW method [25] which has been very successful for computing quasiparticle energies and properties.

The standard model is robust and readily implemented. Applications are numerous. Here we describe applications of the model to carbon and related systems together with empirical and semi-empirical approaches.

APPLICATIONS TO DIAMOND AND RELATED SOLIDS

Once E_{tot} from Eq(1) is evaluated for a given structure as a function of volume the resulting $E_{tot}(V)$ curve is fit to an equation of state [26]. This procedure yields lattice constants and the bulk modulus for a given structure. Comparison of different structures yields the most stable structure as a function of volume (or pressure) and allows the determination of transition volumes and transition pressures associated with the pressure induced solid-solid phase transitions. For example, under hydrostatic pressure diamond will convert to other structures. For the limited number of test structures which have been examined [27-29], it is expected that a simple cubic structure can be stable at 23 Mbar. A lower pressure structure for C has been predicted to be BC8 [28] around 11 Mbar and even lower transition pressures occur for non-hydrostatic pressure [30]. It should be emphasized that these studies are far from being exhaustive since there may be (and probably are) many other possible low energy structures. A striking difference between C and Si or Ge or Sn is the fact that the ß-Sn structure which is one of the lowest pressure structures in the latter materials is not a stable structure [30] for C. In addition, there is the obvious absence of a graphite phase for Si, Ge, and Sn.

The first-principles total energy approach accurately reproduces the large bulk modulus of diamond. This property is obtained through an equation of state fit to the $E_{tot}(V)$ calculation or by a direct evaluation of the curvature of the E_{tot} versus V curve near the minimum. An interesting question can be posed at this point. Does diamond represent the ultimate low compressibility crystal or can the bulk modulus of diamond be exceeded?

To study this question it is useful to introduce a semi-empirical theory of bulk moduli [15]. In this approach the bulk modulus in GPa is expressed as a function of the bond length d in Å and the coordination number N_c,

$$B = \frac{N_c}{4} \; \frac{1972 - 220I}{d^{3.5}} \quad , \qquad (3)$$

where I=0 for group IV solids, I=1 for III-V compounds, and I=2 for II-VI's. This expression has been applied to semiconductors and insulators formed from atoms near the center of the periodic table as shown in Table 1

TABLE 1

TABLE OF BULK MODULI

MATERIAL	d	B(exp)	B(calc)	%Diff.
Group IV				
diamond	1.54Å	443GPa	435	1.8
Si	2.35	98	99	1.0
Ge	2.45	78	85	8.1
III-IV				
AlP	2.36	86	87	1.1
AlAs	2.43	77	78	1.2
AlSb	2.66	58	57	1.7
GaP	2.36	89	87	2.2
GaAs	2.45	75	76	1.3
GaSb	2.65	57	58	1.7
InP	2.54	71	67	5.6
InAs	2.61	60	61	1.6
InSb	2.81	47	47	0
II-VI				
ZnS	2.34	77	78	1.3
ZnSe	2.46	62	66	5
ZnTe	2.64	51	51	0
CdS	2.52	62	60	3
CdSe	2.62	53	53	0
CdTe	2.81	42	41	2

The accuracy of Eq(3) is impressive since it is not straight forward to calculate bulk moduli even with an input from experiment. First-principles calculations using the standard model approach yield fairly high precision estimates as do potentials for rare gas solids. However, simple models such as those based on the free electron gas models are not very successful. For example, the free electron gas model gives the following expression for B (in GPa)

$$B = \left(\frac{6.13}{r_s}\right)^5 \tag{4}$$

In Eq(4) r_s is the electron gas parameter

$$r_s = \left(\frac{3}{4\pi n}\right)^{1/3} \tag{5}$$

where n is the electron concentration and a_o is the Bohr radius. Eq(4) works well for K, Rn, and Cs, but not for most metals. For example in the case of Al which is a free electron-like metal with r_s=2.07, the calculated B is 228 GPa whereas the measured value is 76 GPa. It is interesting to note that Eq(4) gives B=2160 GPa for r_s=1.32 which is appropriate for diamond. This seems counter-intuitive since it is often argued that the concentration of electrons into strong directional covalent bonds is likely to be the origin of the large incompressibility of diamond. Here we have an alternative picture. For a covalent model of diamond, Eq(3) yields a bulk modulus of 435 GPa which is close to the experimental value of 443 GPa. However if the electrons in diamond were free and the free electron model is assumed, then the bulk modulus would increase by a factor of five. It is unclear at this point what is the most desirable distribution of charge for maximizing the bulk modulus. One is not generally free to choose the distribution since it is determined by structural constraints. However, the general question of the best compromise for structure and large moduli is an interesting one.

Returning to Eq(3), it can be seen in Table 1 that the semi-empirical approach yields estimates for B which are comparable in precision with modern *ab initio* calculations. In fact, this approach has predictive power. For BN, there were several indirect experimental estimates of B, and some of these suggested values which exceeded that of diamond. The use of Eq(3) however gave a value of 367 GPa and a subsequent measurement yielded 369 GPa [31].

Another use of Eq(3) is the determination of trends in properties and the development of an approach for obtaining insight for increasing B which for many cases implies an increase in hardness. The functional dependences on the material parameters suggest that it is important to minimize d and I and maximize N_c. Since the minimization of d has the largest influence on B, one is led to the first row of the periodic table for small atomic radii. In particular the radii for B,C, and N are approximately 0.88Å, 0.77Å, and 0.70Å respectively. This suggests [15] that solids based on C-N bonds should have large B - perhaps larger than diamond. However if one assumes a 4-fold coordinated zincblende structure for CN, it can be argued that this IV-V compound will have an "extra" electron per primitive cell which will occupy antibonding orbitals and hence lead to an instability. However Si_3N_4 is known to exist in at least two structures, α and β. Hence, a hypothetical β-structure was chosen for an *ab initio* calculation [16,17] for carbon nitride.

Because CN in the β-Si_3N_4 structure is not 4-fold coordinated, there is a decrease in B coming from the factor N_c=3.43. In addition it is reasonable to assume d~1.47Å from atomic radii and a value for I~0.5. The uncertainties are not large and yield an estimate for B from Eq(3) of 410-440 GPa. *Ab initio* calculations [16,17] yield similar results. Experimental data have suggested even shorter bond lengths which could raise B(CN) above that of diamond. Some of the current experimental studies are given in references [31-35]; however this is an active area with many more studies in progress. Recently [36] superlattice structures containing

CN and TiN have been explored and have yielded materials with hardness comparable with diamond. These materials may prove to be very useful as coatings [36].

At this point, it is not clear whether covalent materials such as ß-C_3N_4 or other forms of carbon nitride will be produced which are less compressible or harder than diamond. However, the theoretical studies indicate that: Eq(3) is accurate for trends and can be used for predicting low compressibility solids; there is a possibility that the bulk modulus of diamond can be exceeded; carbon nitride in some structure is an excellent candidate for a wide band-gap [37] hard material with a high thermal conductivity; and bulk diamond or diamond surfaces can be strengthened with nitrogen implantation.

APPLICATIONS TO GRAPHITE AND RELATED SOLIDS

The graphitic forms of C rely on the characteristic sp^2 bonding nature of the C atom. Pressure induced transitions between rhombohedral graphite and diamond [38] and between hexagonal graphite and diamond [39] illustrate the change from layer sp^2 bonding to sp^3 three dimensional bonding as the graphitic layers are pushed close together. Although the LDA approach does not accurately describe the van der Waals bonding between graphitic layers, it does reproduce the stronger repulsive interactions and bonding between layers, and good estimates for structural properties of graphite and compressed graphite are obtained.

Because of the dependence of the bulk modulus on the bond length given by Eq(3), one expects that the shorter bond length in graphite may lead to high B solids. In, in fact, it were possible to compress an unbuckled graphite sheet it is expected that this material would be less compressible than diamond. One attempt to take advantage of the short strong bonds of graphite and graphitic-like BN is to try to construct three dimensional solids with sp^2 bonding using these materials. As mentioned earlier much of this work is still theoretical [12-14] and at this point is not very encouraging since these hypothetical solids appear to be unstable.

Another class of materials based on sp^2 bonding is the fullerene family with C_{60} and other clusters [40] and C nanotubes [5]. The C_{60} solids doped with alkali metals such as K_3C_{60} and Rb_3C_{60} have received considerable attention since they are metals and superconducting. The standard model has been applied to these systems with considerable success [41,42]. It also can be argued [43] that the BCS theory of superconductivity with electron-phonon interactions and standard Bloch-Boltzmann transport theory are applicable for these interesting systems. At present most of the experimental data appear consistent with these standard theories.

The case for theory being quantitative and predictive for nanotubes is still being made. Since the discovery of C nanotubes [5], questions related to their electronic structure [6] have been explored using theoretical models having different levels of sophistication. Band folding has given a basis for discussing the electronic structure and *ab initio* theory has been applied.

Just as structural models have been based on rolling graphite sheets into tubes, electronic band structures for graphite sheets have been mapped onto Brillouin zones appropriate for tubes. Calculations for metal atoms in tubes such as K in C tubes [44] explore potentially interesting systems having novel properties. Questions related to restricted dimensional conductivity, Pierels distortions, and superconductivity are being posed. The types of interaction mechanisms between the chain of atoms and the graphitic walls of the tubes have been explored [44]. The dependence of these interactions on the particular atom intercalated and the nature of the confining tubes remains to be studied. Growth mechanisms for buckyballs, onions, and tubes are also open questions to be explored using modern theoretical approaches.

One example of success for theory in this area is the successful prediction of tubes formed from $B_xC_yN_z$ [9-11]. Although the BN analogue of C_{60} isn't possible because of structural frustrations, the same restrictions are not true for sheets of graphitic BN [7]. An interesting predicted property [45] of BN tubes is that these wide band gap materials are semiconducting or insulating for different helical geometries. This in in contrast to C tubes which can behave as semiconductors or semimetals. Predictions for BC_2N and BC_3 [8,9] yield semiconductor and metallic tubes with interesting properties. For BC_2N, chiral structures and nanocoils are predicted. For BC_3, the conductivity may depend on interactions between walls of neighboring tubes. Doping of $B_xC_yN_z$ tubes should allow a variety of electronic properties and device applications.

In general the study of nanotubes represents a fruitful and possibly technologically important area of materials science. The variety of structures, their unusual electronic and mechanical properties, and their use as hosts for intercalation and doping make these systems interesting for use as structural, electronic, and optical materials. From the point of view of theory, they provide novel materials to use for tests of the standard model and flexible systems which may exhibit new and interesting physical phenomena.

ACKNOWLEDGEMENTS

This work was supported by National Science Foundation Grant No. DMR-9120269 and by the Director, Office of Energy Research, Office of Basic Energy Sciences, Materials Sciences Division of the U.S. Department of Energy under Contract No. DE-AC03-76SF00098.

REFERENCES

1. M. L. Cohen and J. R. Chelikowsky, Electronic Structure and Optical Properties of Semiconductors (Springer-Verlag, Berlin, 1988).

2. M. L. Cohen, Science **234**, 549 (1986).

3. A. Y. Liu, R. M. Wentzcovitch, and M. L. Cohen, Phys. Rev. B **39**, 1760 (1989).

4. D. Tomanek, R. M. Wentzcovitch, S. G. Louie, and M. L. Cohen, Phys. Rev. B 37, 3134 (1988); and R. M. Wentzcovitch, M. L. Cohen, and S. G. Louie, Phys. Lett. **A131**, 457 (1988).

5. S. Iijima, Nature **354**, 56 (1991).

6. N. Hamada, S.-i. Sawada, and A. Oshiyama, Phys. Rev. Lett. **68**, 1579 (1992).

7. A. Rubio, J. L. Corkill, and M. L. Cohen, Phys. Rev. B **49**, 5081 (1994).

8. Y. Miyamoto, A. Rubio, M. L. Cohen, and S. G. Louie, Phys. Rev. B **50**, 4976 (1994).

9. Y. Miyamoto, A. Rubio, S.G. Louie, and M.L. Cohen, Phys. Rev. B **56**, 18360(1994).

10. Z. Weng-Sieh, K. Cherrey, N.G. Chopra, X. Blase, Y. Miyamoto, A. Rubio, M.L. Cohen, S.G. Louie, A. Zettl, and R. Gronsky, Phys. Rev. B **51**, 11229 (1995).

11. N.G. Chopra et al, Science (in press).

12. A. Y. Liu, M. L. Cohen, K. C. Hass, and M. A. Tamor, Phys. Rev. B **43**, 6742 (1991).

13. A. Y. Liu and M. L. Cohen, Phys. Rev. B **45**, 4579 (1992).

14. J. L. Corkill, A. Y. Liu, and M. L. Cohen, Phys. Rev. B **45**, 12746 (1992).

15. M. L. Cohen, Phys. Rev. B **32**, 7988 (1985).

16. A. Y. Liu and M. L. Cohen, Science **245**, 841 (1989).

17. A. Y. Liu and M. L. Cohen, Phys. Rev. B **41**, 10727 (1990).

18. M. L. Cohen and W. D. Knight, Physics Today **43**, 42 (1990).

19. E. Fermi, Nuovo Cimento **11**, 157 (1934).

20. J. R. Chelikowsky and M. L. Cohen, in Handbook on Semiconductors, completely revised edition, Vol. 1, ed. T. S. Moss (Elsevier Science Publishers B. V., Amsterdam, 1992), p. 59.

21. P. Hohenberg and W. Kohn, Phys. Rev. B **136**, 864 (1964).

22. W. Kohn and L.J. Sham, Phys. Rev. A **140**, 1133 (1965).

23. J. Ihm, A. Zunger, and M. L. Cohen, J. Phys. C **12**, 4409 (1979). Erratum: J. Phys. C **13**, 3095 (1980).

24. M. L. Cohen, Physica Scripta **T1**, 5 (1982).

25. M. Hybertsen and S.G. Louie, Phys. Rev. Lett. **55**, 1418 (1985).

26. F.D. Murnaghan, Proc. Nat. Acad. Sci. U.S.A. **30**, 244 (1944).

27. M. T. Yin and M. L. Cohen, Phys. Rev. Lett. **50**, 2006 (1983).

28. S. Fahy and S.G. Louie, Phys. Rev. B **36**, 3373 (1987).

29. M. T. Yin and M. L. Cohen, Phys. Rev. B **29**, 6996 (1984).

30. M. P. Surh, S. G. Louie, and M. L. Cohen, Phys. Rev. B **45**, 8239 (1992).

31. E. Knittle, R. M. Wentzcovitch, R. Jeanloz, and M. L. Cohen, Nature **337**, 349 (1989).

32. K. M. Yu, M. L. Cohen, E. E. Haller, W. L. Hansen, A. Y. Liu, and I. C. Wu, Phys. Rev. B **49**, 5034 (1994).

33. M.Y. Chen, D. Li, X. Lin, V.P. Dravid, Y.-W. Chung, M.-S. Wong, and W.D. Sproul, J. Vac. Sci. Techn. A **11**(13), 521 (1993).

34. F. Fujimoto and K. Ogata, Jpn. J. Appl. Phys. **32**, L420 (1993).

35. C. Niu, Y.Z. Lu, and C. Lieber, Science **261**, 334 (1993).

36. Y.-W. Chung (private communication).

37. J. L. Corkill and M. L. Cohen, Phys. Rev. B **48**, 17622 (1993).

38. S. Fahy, S. G. Louie, and M. L. Cohen, Phys. Rev. B **34**, 1191 (1986).

39. S. Fahy, S. G. Louie, and M. L. Cohen, Phys. Rev. B **35**, 7623 (1987).

40. H.W. Kroto, J.R. Heath, S.C. O'Brien, R.F. Curl, and R.E. Smalley, Nature **318**, 162 (1985).

41. M. L. Cohen, in Proceedings of the 4th NEC Symposium on Fundamental Approaches to New Material Phases, Physics and Chemistry of Nanometer Scale Materials, Mat. Sci. & Eng. B **19**, 111 (1993).

42. E.L. Shirley and S.G. Louie, Phys. Rev. Lett. **71**, 133 (1993).

43. M.L. Cohen, Phil. Mag. B**70**, 627 (1994).

44. Y. Miyamoto, A. Rubio, X. Blase, M.L. Cohen, and S.G. Louie, Phys. Rev. Lett. (in press).

45. X. Blase, A. Rubio, S.G. Louie, and M.L. Cohen, Europhys. Lett. **28**, 335 (1994).

RELATIONSHIP OF PROCESSING CONDITIONS TO GROWTH RATE AND QUALITY OF DIAMOND GROWN BY CHEMICAL VAPOR DEPOSITION

John C. Angus, William D. Cassidy, Long Wang, Yaxin Wang, Edward Evans, and Christopher S. Kovach
Chemical Engineering Department
Case Western Reserve University
Cleveland, OH 44016-7217

Michael A. Tamor
Research Laboratories
Ford Motor Company
Dearborn, MI 48121-4053

ABSTRACT

Diamond quality is strongly coupled to growth rate. Incorporation of non-diamond (sp^2) carbon and morphological instabilities both increase with increasing growth rates. The intersection of twins with the growth surface produces re-entrant corners that enhance growth in the plane of the twin. Morphology and the development of texture both depend on substrate temperature and methane concentration and hence on growth rate. Experimental evidence and modeling results that relate growth rates and quality to controllable process parameters are reviewed.

INTRODUCTION

The relationship of diamond quality to growth rate is a critical issue for the chemical vapor deposition of diamond. Many applications for diamond are not feasible because current processes grow diamond too slowly, of too poor quality, in reactors that are too costly. Addressing and finding solutions to this issue will be critical if diamond is to find wide application in the market place.

The crystalline quality of diamond grown by chemical vapor deposition, in common with most other crystals, decreases with increasing growth rate. Most attention has been paid to the incorporation of non-diamond (sp^2) carbon[1,2]. Other processes, e.g., twinning[3-6], texturing[6,7] and morphological instabilities[8-10] can play equally important roles in determining the final quality of the diamond. In this paper we review the factors influencing growth rate and diamond quality. We emphasize that quality is not a uniquely defined term and depends on the nature of the application. For example, some diamond films with low optical absorption are brittle and unsuitable for mechanical applications.

FACTORS INFLUENCING GROWTH RATE

All processes for the growth of diamond by chemical vapor deposition involve three steps; 1) generation of active species, 2) transport of active species to the growth surface, and 3) incorporation of the growth species into the growth surface. The overall

Mat. Res. Soc. Symp. Proc. Vol. 383 © 1995 Materials Research Society

growth rate depends on all three of these series steps. However, the strong correlation between growth rate and power density[11] implies that the generation and transport of species are rate limiting in most processes. Detailed modeling of the entire deposition process is required to fully understand all of the factors influencing growth rates. See, for example, the study of the RF torch reactor by Girshick[12]. However, much can be learned by order of magnitude estimates of the key process parameters.

The relative importance of forced convective mass transport to diffusive mass transport is given by the Peclet number,

$$Pe \equiv \frac{uL}{D} \tag{1}$$

where u is the convective velocity, L is a characteristic length, and D is the gas phase diffusion coefficient of the growth species.

For hot-filament and microwave reactors, the Peclet numbers are much less than one, indicating that mass transport is almost entirely by diffusion. In these reactors there is no boundary layer. The characteristic distance, L, in a hot-filament reactor can be estimated as the filament to substrate distance, typically from 0.5 to 1.0 cm. Because of the low pressures used in hot-filament and microwave reactors, typically 20-40 torr, the diffusion coefficienct of active species are large. Setting $D \approx 500$ cm^2/sec and $L \approx 1$ cm, the time constant for transport, L^2/D, is on the order of several milliseconds. Gas phase transport is therefore rapid; however, fluxes to the surface are limited by the low gas densities at the low operating pressures. Growth rates in hot-filament and microwave reactors are typically from 0.5 to 5 μm/hr.

The combustion, RF torch and DC arc-jet processes operate at near atmospheric pressure and at high gas velocities. The Peclet numbers, Pe, are much larger than one and transport from the activation zone to the vicinity of the substrate is primarily by forced convection. Final transport of active species to the growth surface is through a boundary layer. To first order, one can estimate the importance of these two processes by estimating the time constants for transport in the free stream by convection, L/u, and the time constant for diffusion across the boundary layer, δ^2/D. The boundary layer thickness, δ, varies approximately as $L(Re)^{-1/2}$, where Re is the Reynolds number, $Lu\rho/\mu$. Therefore, increasing the velocity, u, decreases δ as $u^{--1/2}$ and the time constant for transport across the boundary layer as u^{-1}.

Using representative values for the RF plasma reactor at 1 atm (L_{rf} = 1 cm, u = 10^3 cm/sec, $\mu = 3 \times 10^{-4}$ g/cm sec, $\rho = 2 \times 10^{-5}$ g/cm^3, and D = 10 cm^2/sec) we find the time constants for transport in the free stream and across the boundary layer are both on the order of a millisecond, on the same order as in hot-filament and microwave reactors. The higher growth rates in the RF plasma reactor, approximately 10 μm/hour, arise from the higher species concentrations at the higher operating pressure. Still further increases in plasma velocities, e.g., ~ 10^4 cm/sec, can be obtained in DC plasma jet reactors and lead to still higher growth rates, e.g., ~ 100 μm/hr.

The combustion, RF torch and DC arc-jet methods also provide higher equivalent gas temperatures so the distribution of growth species reaching the surface will be richer in the more reactive species, CH_2, CH and C; the primary growth specie in the hot-filament and microwave methods is believed to be CH_3. Not enough is known about the details of the attachment mechanism to estimate quantitatively how much the observed growth rates depend on the rates of surface reactions.

Some quantitative measurements of growth rate as a function of concentration of hydrocarbon source gas have been reported[13-21]. Representative results obtained by *in situ* microbalance measurements in a hot-filament reactor are shown in Fig. 1. Accurate first order kinetics are observed for methane concentrations up to approximately 1.0%; above 1% the rate approaches zero-order. This rate behavior is of the classic Langmuir-Hinshelwood form and is suggestive of a mechanism in which the surface becomes saturated with a surface-adsorbed intermediate. It is of interest that some of the earliest rate studies were interpreted using Langmuir-Hinshelwood kinetics.[13] A tendency to zero-order kinetics at high methane concentrations was also reported by Kobashi *et al.*[19] using microwave plasma assisted deposition and by Windischmann[22] for DC arc jets.

INCORPORATION OF NON-DIAMOND CARBON

Graphite is the stable solid carbon phase at the conditions where diamond is grown by chemical vapor deposition. Therefore, the possible incorporation of non-diamond, graphitic (sp^2) carbon within the diamond is not surprising. Incorporation of the non-diamond carbon is believed to depend on the relative rates of growth and etching of diamond and non-diamond carbon. Details of the growth and etching mechanisms are still speculative; however, very simple reduced reaction sets give predictions in general agreement with observations. For example, the following reduced mechanism is sufficient to illustrate the principles involved.

r1	$C_d + CH_X \rightarrow G_O$
r2	$G_O + H \rightarrow$ Diamond
r3	$G_O + CH_x \rightarrow G_1$
r4	$G_1 + H \rightarrow CH_4, C_2H_2$

Here C_d is a bare surface site and CH_x is the growth species. G_O is a surface-adsorbed intermediate that either reacts with atomic hydrogen to form diamond or with CH_x to form a non-diamond surface species, G_1, which is subsequently gasified by reaction with atomic hydrogen. When there is a large excess of atomic hydrogen, a steady state analysis of this reaction set shows that

$$[G_1] \,\alpha\, \frac{r}{[H]^2} \tag{2}$$

where r is the growth rate of diamond, [H] is the local concentration of atomic hydrogen and $[G_1]$ is the steady state concentration of non-diamond carbon. This and other similar models[1,2,10] are based on a simple competition between diamond growth and the removal of non-diamond carbon by gasification with atomic hydrogen. The growth rate, r, is believed to be proportional to $[CH_x]$. Therefore, formation of non-diamond carbon is

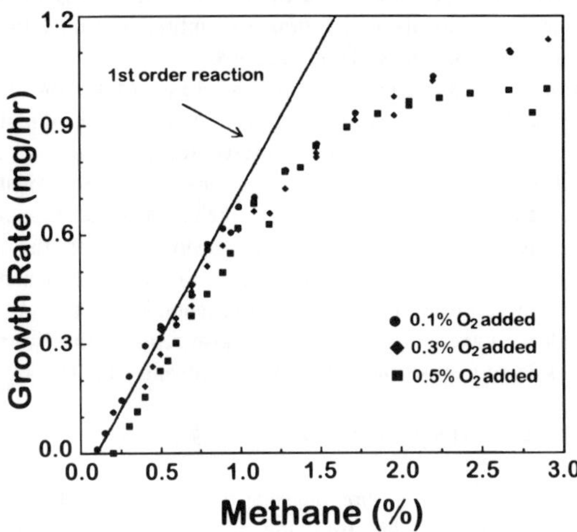

Figure 1. Diamond growth rate versus methane concentration. The reaction is first order in methane concentration below approximately 1% methane and shifts to approximately zero order at higher methane concentrations.

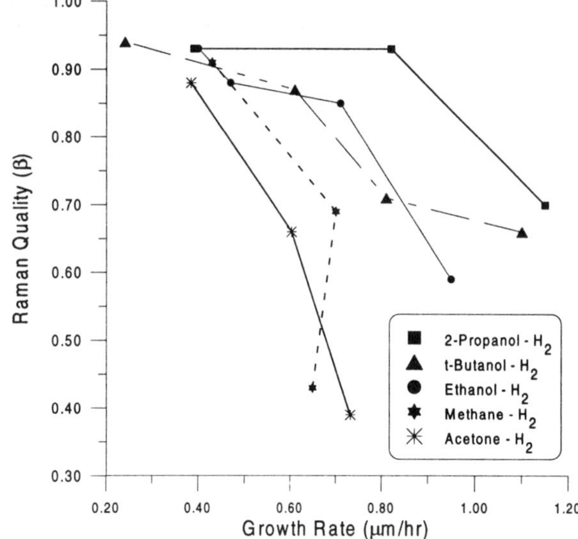

Figure 2. Raman quality, β, versus linear growth rate for a series of source gas molecules. β is given by D/(D + ND) where D and ND are the Raman peak intensities for diamond and non-diamond carbon respectively. β ranges from 0 (no detectable diamond peak) to 1 (no detectable non-diamond peak).

suppressed at high [H]/[CH_x] ratios and favored at low [H]/[CH_x] ratios. Because of the high bond energy of H_2, high equivalent gas temperatures are required to increase the [H]/[CH_x] ratio. For example, for a source gas containing 1 mole percent CH_4 in H_2 at 20 torr, the equilibrium [H]/[CH_3] ratio is 5200 at 2400K and 420 at 2000K[23].

Cassidy[23,24] has shown that the amount of non-diamond carbon increases with growth rate, in general agreement with the predictions of these elementary models. A summary of his results is shown in Fig. 2. Note that in all cases the Raman quality decreases with growth rate. Note also, however, that differences are observed between the different source gases. The highest quality is observed with 2-propanol; the lowest with methane and acetone. The reason for these differences is not entirely clear. The nature of the non-diamond structures that might be present during growth is also not known. However, recent theoretical work by Davidson and Pickett[25] and Jungnickel *et al.*[26] indicates that graphitic-like sp^2 layers may be present during diamond growth.

MORPHOLOGY AND TEXTURE

The morphology of an isolated crystal depends on the relative growth velocities normal to the principal crytsallographic planes. The fastest growing surfaces disappear and the crystal is bounded by the slower growing surfaces. For diamond grown by chemical vapor deposition, cubes, cubo-octahedra and octahedra bounded by {100} and {111} faces are observed; {110} surfaces are rarely, if ever, seen. These empirical observations indicate that the morphology of isolated diamond can be adequately described by the growth velocities, v_{100} and v_{111}, normal to the {100} and {111} surfaces respectively. It is customary to define the growth parameter, α, by

$$\alpha \equiv \sqrt{3}\ v_{100}/v_{111} \tag{3}$$

The relationship between α and the various shapes of untwinned crystals is determined solely by geometrical construction. For $\alpha = 1$, perfect cubes are formed; for $\alpha = 3/2$, perfect cubo-octahedra; and for $\alpha = 3$, perfect octahedra.

The situation for polycrystalline films is quite different. The well known van de Drift model[27] shows that the crystals with greatest velocity normal to the growth surface will eventually crowd out the other crystals. Therefore, from randomly oriented nuclei, a film evolves in which the surviving crystals are oriented with the direction of most rapid growth normal to the surface of the film. Therefore, films grown under conditions where $\alpha = 3$ will give a polycrystalline film with <100> texture; $\alpha = 1$ will lead to <111> texture. The special case of $\alpha = 3/2$ deserves mention. In this case the vector sum of v_{111} and v_{100} is exactly along a <110> direction and thus will give <110> texture.

Highly oriented polycrystalline films can be produced by a two-step process in which oriented nuclei are formed and then permitted to grow under conditions that favor growth along the direction of orientation. For example, diamond nuclei with preferential <100> orientation can be formed on {100} silicon wafers by applying a negative bias to the substrate[28,29]. Growth conditions are fixed so that α is slightly less than three, i.e., conditions that favor <100> texture. The result is a mosaic film covered with highly oriented crystals with flat {100} faces facing up. An example is shown in Fig. 3. Similar

Figure 3. Highly oriented mosaic diamond film produced by microwave assisted chemical vapor deposition. Square facets are {100} surfaces bounded by <110> edges; side facets are {111} surfaces.

results can be achieved along the other two principal orientations by using {110} or {111} silicon wafers and choosing growth conditions so that $\alpha \approx 3/2$ *or* $\alpha \approx 1$ respectively.

The simultaneous presence of strong <100> texture and {100} faceting, as shown in Fig. 3, may appear to be inconsistent since the conditions that lead to <100> texture, i.e., α slightly under 3.0, will lead to isolated crystals in the shape of octahedra with truncated tips. This apparent inconsistency is resolved by recognizing that the {100} facets seen in Fig. 3 are the truncated tips of (virtual) octahedra. The square {100} facets visible in Fig. 3 are bounded by <110> edges, not <100> edges, and the adjacent faces are {111} facets. The micrograph in Fig. 3b also shows that these {111} facets are of poor quality with many secondary nuclei and penetration twins (see the next section). The conditions that lead to strong <100> texture, i.e., high methane concentration and low substrate temperature, are unfavorable for growth on {111} facets.

TWINNING

Diamond crystals grown by chemical vapor deposition are often highly twinned. These are growth twins that arise when a nucleus of improper orientation forms on a {111} growth surface. The stacking error that gives the twinned nucleus corresponds to a 60° rotation from the correct orientation. At the atomic level, the six-membered rings in

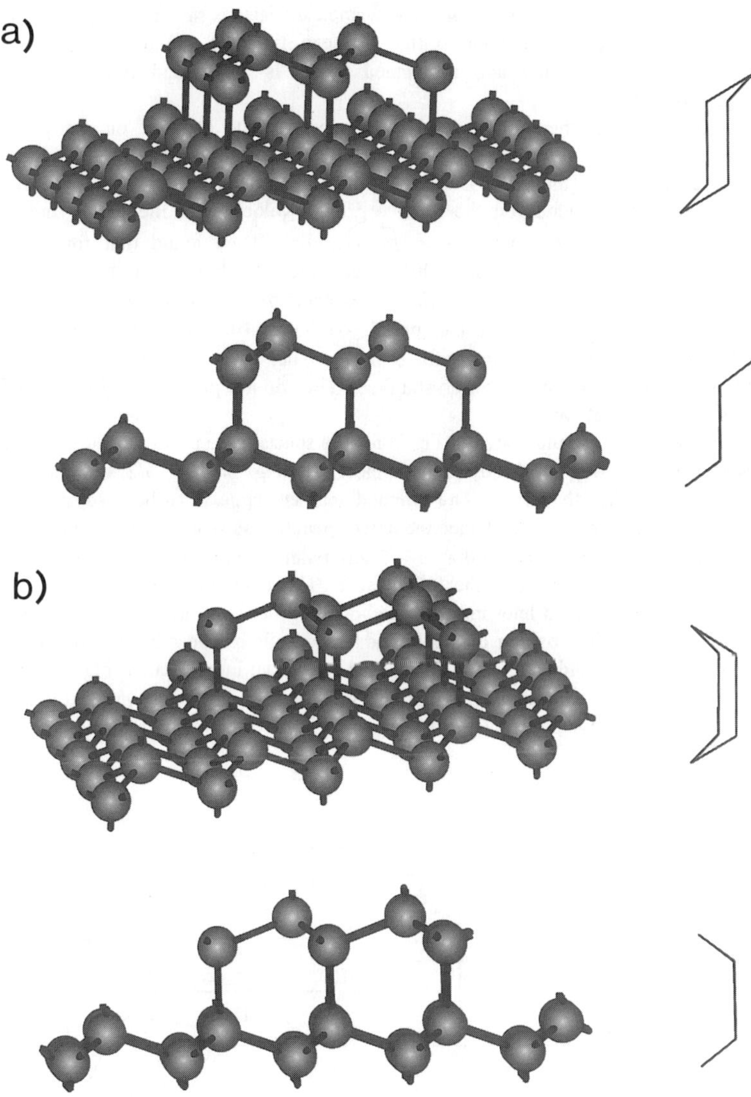

Figure 4. a) untwinned and b) twinned nuclei on a {111} surface. Oblique view and view along a <110> direction are shown. Figures on right show interface geometry, i.e., chair and boat conformations.

the plane of the error are in the "boat" rather than the "chair" conformation. The geometric orientations of twinned and un-twinned nuclei are shown in Figs 4a and 4b. Since only next-nearest neighbors differ between the two orientations, the energy difference between twinned and untwinned nuclei is small, and twin formation is consequently easy.

Once formed, the evolution of a twinned nucleus depends on many factors, especially the growth velocities v_{111} and v_{100} normal to the {111} and {100} growth surfaces respectively. Tamor and Everson[6] and Koidl *et al.*[7] showed how the evolution of a twinned segment depends on the relative growth velocities of the {100} and {111} faces, i.e., on the growth factor $\alpha \equiv \sqrt{3}\ v_{100}/v_{111}$. They found that for $\alpha < 3/2$, penetration twins persist only on {100} facets; they are buried on the more rapidly advancing {111} faces. For $\alpha > 2$ twinned segments persist only on {111} facets. For $3/2 < \alpha < 2$ twinned segments can persist on both {100} and {111} facets. It is important to note that these criteria for α only predict when twins persist, i.e., are not buried by the advance of the parent crystal face. They do not predict conditions that favor the initial formation of twins.

These points are illustrated in Fig. 5 for the special case of a twinned segment that forms on a {111} step on {100} growth facet. In Fig. 5a $\alpha < 3/2$ and the twinned segment grows faster than v_{100}. The twinned segment appears to be a separate crystal "penetrating" the parent crystal; hence the name "penetration twin". An example is shown in Fig. 6. When v_{100} increases so that $\alpha > 2$ any twinned segment that forms on a {100} facet is buried by the advance of the {100} facet. This case is shown in Fig. 5b. In both Figs. 5a and 5b the dashed lines indicate the boundary between the parent crystal and the twinned segment. The position of this boundary depends on the relative growth rates of the twinned segment and the parent crystal and can be an incoherent or high order grain boundary. The orientation of the two lattices is, however, of the $\Sigma 3$ type and is fixed by the formation of the twinned nucleus on the {111} facet where the twin formed.

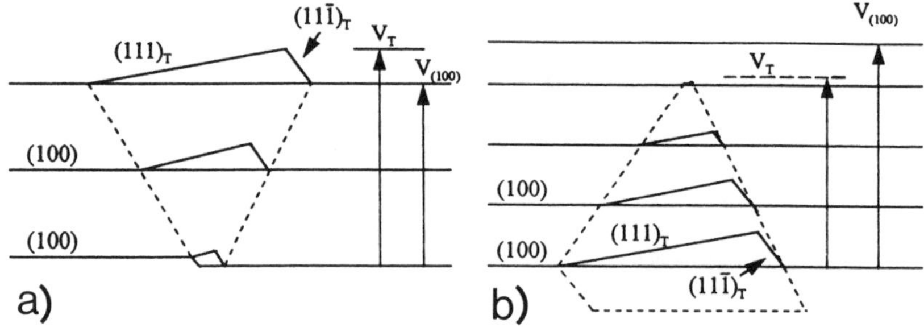

Figure 5. Cross sectional views of evolution of growth twins formed on a step on a {100} parent facet. a) Growth parameter $\alpha < 3/2$ so twinned segment grows faster than {100} parent face. b) Growth parameter $\alpha > 2$ so twinned segment is buried by {100} parent face.

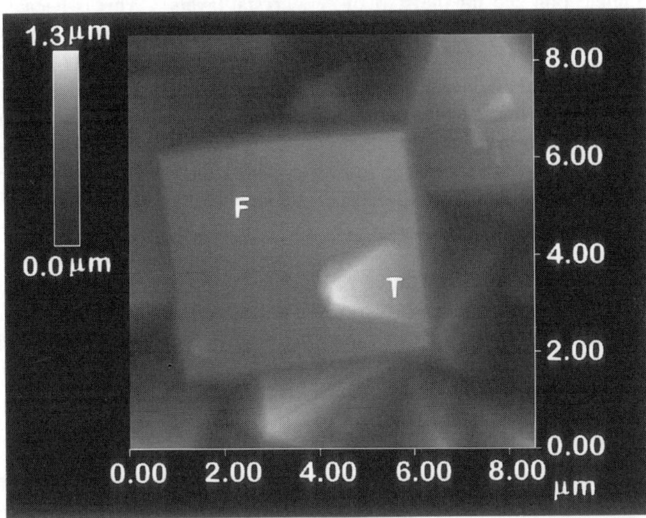

Figure 6. Atomic force microscope image of penetration twin on a {100} facet. See Fig. 5a.

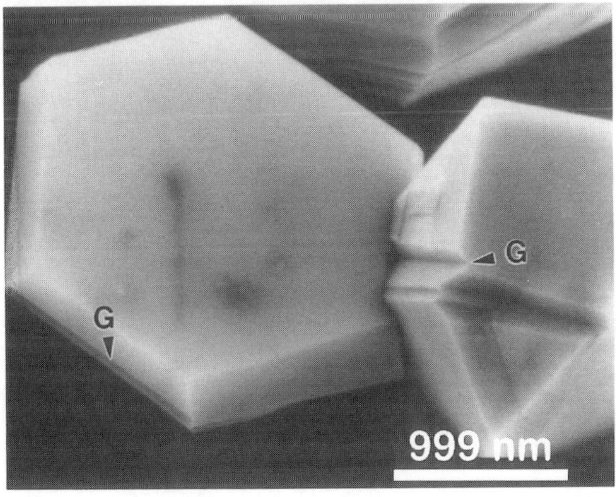

Figure 7. Scanning electron micrograph of twinned crystals. Hexagonal platelets are generated by re-entrant corners that form when two parallel stacking errors on {111} planes intersect the surface. Twinning on other {111} planes leads to the multiply-twinned three-dimensional crystals. Re-entrant corners are indicated by G.

A stacking error that intersects the surface of the crystal is believed to be a favorable nucleation site for the start of new crystal layers[3]. This is because the "boat" conformation requires only two atoms to form a surface nucleus rather than three that are required on a smooth {111} surface. Two parallel stacking errors bound a twinned element of the crystal. When this twinned element of the crystal intersects the surface, a re-entrant corner is formed that persists, no matter how many atoms are added. The formation of twinned segments therefore accelerates the growth of the crystal within the plane of the twin. There are many consequences of this effect. For example, multiply twinned crystals grow more rapidly than their untwinned counterparts and are therefore often the dominant form. Examples of twinned crystals are shown in Fig. 7. Hexagonal platelets have been observed by several workers[30,3] and are formed when growth is accelerated by a single re-entrant corner.

When a twinned layer meets a non-twinned segment of the crystal, a defect is formed. Examples of several types of defect structures found within diamond are shown in Fig. 8. Schectman[31] has presented other high resolution images of defects in diamond.

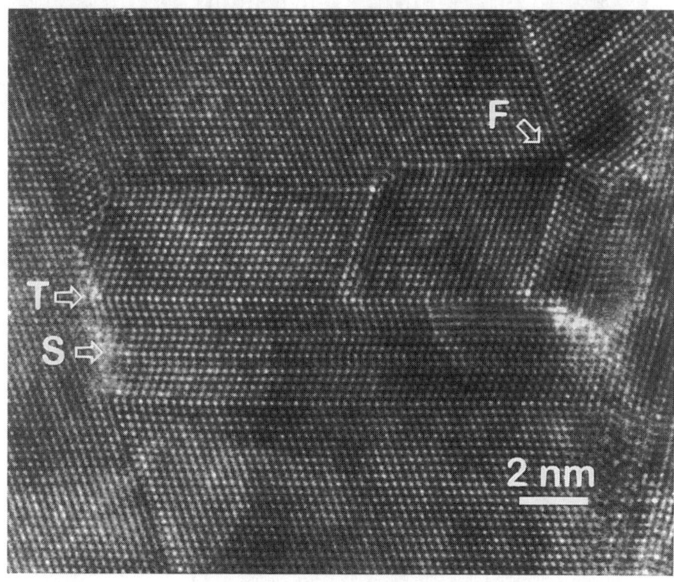

Figure 8. High resolution transmission electron micrograph of diamond. Stacking error (S), micro-twin (T), five-fold twin (F) and an incoherent boundary (above T) are shown.

MORPHOLOGICAL INSTABILITY AND GROWTH NON-UNIFORMITY

If the growth rate of a crystal is limited, even in part, by mass transfer resistance in the gas phase, morphological instabilities can arise. Elements of the growth surface, e.g., corners or protuberances, that have better access to gas will grow at a faster rate than other parts of the surface. This leads to still greater access to the gas and still greater growth rates. If no damping mechanism, e.g., surface diffusion or enhanced etching, is present, an instability occurs and the local growth rate becomes uncontrolled. This effect, well known in many systems, was noted in diamond by Ravi[8] and Kovach *et al.*[9]

A heuristic criterion for growth non-uniformities can be obtained from the Thiele modulus, m_T.

$$m_T \equiv \frac{kL}{D} \tag{4}$$

where k is a first order rate constant for attachment, L is a characteristic length along the crystal surface, and D is the gas phase diffusion coefficient of active species.

For $m_T << 1$, i.e., very efficient mass transfer, the tendency for non-uniform growth is minimized; for $m_T >> 1$, non-uniform growth is likely. Values of the rate constants, k, for carbon attachment are not known, but are likely to be similar for the various diamond deposition processes. D varies approximately inversely with pressure. Therefore, for substrates of similar dimensions, one can expect the atmospheric pressure processes, e.g., RF plasma and combustion, to have values of m_T significantly larger than the lower pressure processes, e.g., hot-filament and microwave. For example, assuming the pressures are 760 and 30 torr respectively we find $m_{T,760} \cong 25\, m_{T,30}$. The higher rate, atmospheric pressure processes are therefore much more prone to exhibit growth non-uniformities than hot-filament or microwave synthesis. This conclusion is in agreement with experimental observations.

ACKNOWLEDGMENT

The support of the National Science Foundation Materials Research Group grant and the Ford Motor Company is gratefully acknowledged.

REFERENCES

1. D.G. Goodwin, J. Appl. Phys. **74**, 6888 (1993).
2. J.C. Angus and E.A. Evans, Mat. Res. Soc. Symp. Proc. Vol. **349**, Materials Res. Soc., Pittsburgh, PA, pp. 385-390 (1994).
3. J.C. Angus, M. Sunkara, S. Sahaida and J.T. Glass, J. Mater. Res. **7**, 3001 (1992).
4. R.E. Clausing, L. Heatherly, L.L. Horton, E.D. Specht, G.M. Begun and Z.L. Wang, Diamond and Related Materials **1**, 411 (1992).
5. M.P. Everson and M.A. Tamor, J. Mater. Res. **7**, 1438 (1992).
6. M.A. Tamor and M.P. Everson, J. Mater. Res. **9**, 1839 (1994).

7. C. Wild, R. Kohl, N. Herres, W. Muller-Sebert and P. Koidl, Diamond and Related Materials **3**, 373 (1994).
8. K.V. Ravi, J. Mater. Res. **7**, 384 (1992).
9. C.S. Kovach, B. Roozbehani, T. Suzuki and J.C. Angus, Proc. 2nd Int. Conf. on the Applications of Diamond Films and Related Materials, M. Yoshikawa, M. Murakawa, Y. Tzeng and W.A. Yarbrough, Eds., MYU Tokyo, 1993.
10. Y. Wang, E.A. Evans, C.S. Kovach, U. Landau, and J.C. Angus, Mat. Res. Soc. Symp. Proc. Vol. 363, B.M. Gallois, W.Y. Lee and M.A. Pickering, Eds., Materials Res. Soc., Pittsburgh, PA, pp. 127-138 (1995).
11. J.C. Angus, F. A. Buck, M. Sunkara, T.F. Groth, C.C. Hayman and R. Gat, Materials Res. Soc. Bull. **XIV** (10), 38 (1989)
12. B.W. Yu and S.L. Girshick, J. Appl. Phys. **75**, 3914 (1994).
13. S.P. Chauhan, J.C. Angus and N.C. Gardner, J. Appl. Phys. **47**, 4746 (1976)
14. D.V. Fedoseev and B.V. Deryagin, Zh. Fiz. Khimii. **53**, 752 (1979)
15. S.J. Harris and A.M. Weiner, J. Appl. Phys. **70**, 1385 (1991)
16. C.J. Chu, M.P. D'Evelyn, R.H. Hauge and J.L. Margrave, J. Appl. Phys. **70**, 1695 (1991).
17. K.A. Snail and C.M. Marks, Appl. Phys. Lett. **60**, 3135 (1992).
18. B.R. Stoner, B.E. Williams, S.D. Wolter, K. Nishimura and J.T. Glass, J. Mat. Res. **7**, 257 (1992).
19. K. Kobashi, K. Nishimura, Y. Kowate and T. Horiuchi, Phys. Rev. **B38**, 4067 (1988)
20. Y. Wang and J.C. Angus, Proc. 3rd Symposium on Diamond Materials, Proc. Vol. 93-17. Electrochemical Society. Pennington. NJ (1993), pp. 249-255.
21. Y. Wang, E.A. Evans, L. Zeatoun and J.C. Angus, Proc. Third IUMRS Int. Conf. on Adv. Materials, M. Wakatsuki et al., Eds., Nikkam Kogyo Shimbum, Ltd., Tokyo (1993).
22. H. Windischmann, personal communication.
23. W.D. Cassidy, M.S. Thesis, Case Western Reserve University, Cleveland, OH, 1995.
24. W.D. Cassidy, P.W. Morrison, Jr., and J.C. Angus, Proc. 4th Int. Symp. on Diamond Materials, Reno, NV, May 21-26, 1995; Electrochemical Society, Pennington, NJ.
25. B.N. Davidson and W. Pickett, Phys. Rev. B **49**, 14770 (1994).
26. G. Jungnickel, D. Porezag, Th. Frauenheim, W.R.L. Lambrecht, B. Segall and J.C. Angus, MRS Symposium I, Novel Forms of Carbon, San Francisco, CA, April, 1995 (This volume).
27. A. van der Drift, Philips Res. Rep. **22**, 267 (1967).
28. X. Jiang, C.P. Klages, R. Zachai, M. Hartweg and J.J. Fusser, Appl. Phys. Lett. **62**, 3438 (1993).
29. S.D. Wolter, B.R. Stoner, J.T. Glass, P.J. Ellis, D.S. Buhaenko, C.E. Jenkins and P. Southworth, Appl. Phys. Lett. **62**, 1215 (1993).
30. M.P. Everson, and M.A. Tamor, J. Vac. Sci. Tech. **139**, 1570 May/June 1991.
31. D. Schectmann, J.L. Hutchison, L.H. Robins, E. Farabaugh and A. Feldman, J. Mater. Res. **8**, 473 (1993)

Part II

Elastic Properties and Deformation

THE EFFECT OF TEMPERATURE ON THE DEFORMATION OF DIAMOND SURFACES

CHRIS A. BROOKES, E.J. BROOKES AND G. XING
Department of Engineering Design and Manufacture
University of Hull, HU6 7RX, UK.

ABSTRACT

A brief review is given of recently published work on the fracture and plastic flow of natural and synthetic diamond crystals when deformed by softer impressors at elevated temperatures. Under these conditions, a brittle-ductile transition (BDT) temperature for the different types of diamond has been established.

In this paper we show that below the BDT temperature, brittle 'chatter' cracking on {111} cleavage planes accounts for anisotropic wear provided that the mean contact pressure exceeds about 10 GPa. Above the BDT temperature, titanium diboride sliders develop contact pressures sufficient to cause extensive plastic deformation of the diamond specimens but insufficient to produce cleavage fracture. However, repeated sliding with TiB_2 does lead to cumulative plastic flow preceding fatigue type fracture on {110} planes. Using cubic boron nitride and diamond sliders above the BDT temperature leads to a combination of both plastic deformation and {110} cracking. The coefficient of friction is a maximum in <100> and a minimum in <110> directions. and the measured scratch hardness, which confirms that the <100> directions are softer than <110>, is shown to be consistent with the predictions of a resolved shear stress model which was developed for explaining anisotropy in all crystals.

INTRODUCTION

In this paper, we shall concentrate on the effect of temperature on the surface deformation of (001) surfaces of natural type I_a and type II_a diamonds and synthetic type I_b diamond. Briefly: type I_a is a natural diamond with a typical nitrogen content of about 500 ppm which may form a variety of point defects in the lattice; type I_b is a synthetic diamond formed at high pressure:high temperature with a similar level of nitrogen impurities but, mostly, in substitutional sites; type IIa natural diamond has a low level of nitrogen impurities but has a high density of 'grown-in' dislocations. A complete account on the defect structure of various types of diamond can be found in the 'The Properties of Natural and Synthetic Diamond' [1].

In order to maximise the degree of plastic deformation and to minimise fracture we have exploited the advantages of deforming hard crystalline solids by cones made from softer solids. This technique has previously been particularly informative when applied to the study of the mechanical properties of hard ceramic materials. For example, it has been used to measure the effect of temperature on the critical resolved shear stress [2]; the shear strength of polycrystalline materials [3]; impression creep [4,5]; the integrity of surface coatings [6]; fatigue of bulk ceramics and coatings [7] and the threshold temperature for fusion between a given workpiece and a variety of cutting tool materials [8].

Mat. Res. Soc. Symp. Proc. Vol. 383 © 1995 Materials Research Society

SUMMARY OF PREVIOUS WORK

When a cone of softer material is loaded against a harder surface, it will tend to plastically deform to make a circular contact area which intimately conforms to the topography of the surface. In the absence of creep, the load is supported elastically and the resultant **effective** mean contact pressure (P_m') is directly related to the hardness of the softer material - in the case of metal cones P_m' is about one third of the indentation hardness [9]. The contact area of a given cone at a fixed load may be increased by creep of the cone material or by wear. In that event, we distinguish the **nominal** mean contact area (P_m) as the load divided by the area at a given moment in time. The selection of a range of different materials for the cone enables P_m to be controlled and varied. When the hard solid behaves in a brittle manner and P_m is above a certain level, classical Hertzian ring cracks are formed around the periphery of the contact area. The tensile stress in the region of crack initiation is around 20% of P_m and, again, varying P_m enables the tensile fracture stress to be determined [10]. When the behaviour is entirely plastic, permanent impressions are made at a sufficiently high level of P_m and the volume of those impressions will increase as P_m increases. At temperatures between that where elastic:brittle deformation ends and fully plastic behaviour prevails, there is a range of mean contact pressures where, at the lower end, dislocation movement is initiated and, at the top end, the tensile cracks are formed. The data from a number of measurements on diamond, illustrating these three temperature regimes, are summarised in the schematic diagram shown in Figure 1.

Whilst some limited dislocation movement may be associated with deformation under special conditions in regime I, such as is thought to have been observed beneath Knoop indentations in diamond at room temperature [11], the response to soft impressors is

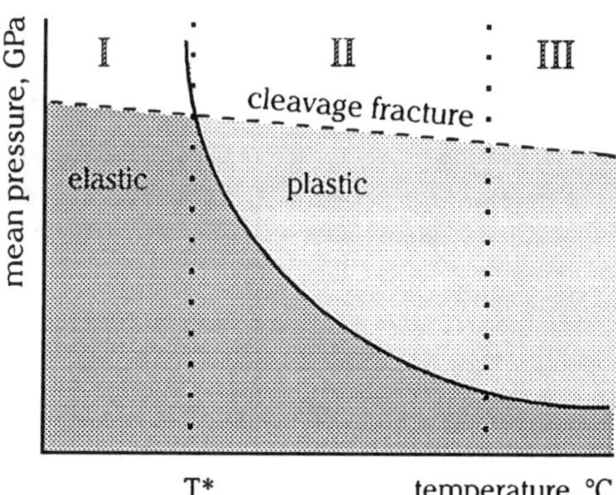

Figure 1. Schematic to show the effect of temperature on the deformation of diamond. I - elastic deformation precedes brittle fracture; II - above the brittle:ductile transition temperature multiple slip is initiated, the flow stress is temperature dependent, and creep becomes significant; III - where the flow stress is temperature independent.

essentially elastic until the tensile stress exceeds the cleavage stress and cracking occurs predominantly on the {111} cleavage planes (Figure 2). At a specific transition temperature (T*), significant dislocation movement is developed at values of P_m less than those required to form the ring cracks. We have recognised this to be the brittle:ductile transition (BDT) temperature and we have shown that it is dependent on the type of diamond - i.e. the BDT temperature for type I_b, I_a and II_a corresponds to 750°, 900°, and 1100° C respectively [2]. Also, we have observed that impression creep of the hard solid is measureable above this BDT temperature. Thus, if P_m is initially between that required to initiate cracks and dislocation movement, as the dwell time is increased at a given normal load, the impressor creeps increasing the contact area whilst, consequently. reducing the nominal mean contact pressure. Concurrently, impression creep of the harder solid continues until the nominal mean contact pressure beomes equal to that required to initiate dislocation movement and then the creep process of the harder solid stops - i.e. the P_m value is no longer sufficient to exceed the critical resolved shear stress [4].

(a)

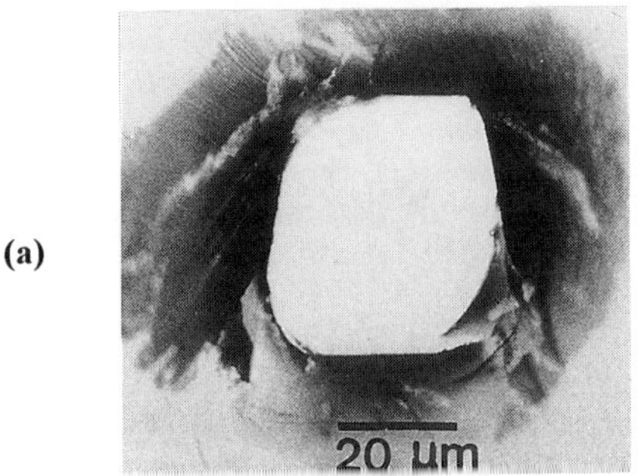

Figure 2. Typical impressions in type I_a diamond, (001) plane using a cBN cone, a normal load of 100 N, and a dwell time of 300 seconds. (a) I - at 20°C showing ring cracking due to tensile failure. (b) II - at 1100°C showing slip lines and {110} cracking. (c) III - at 1400°C illustrating rosette formation. Interferograms are shown alongside to emphasise the increasing pile-up. NB - fringe spacing is 276 nm

(b)

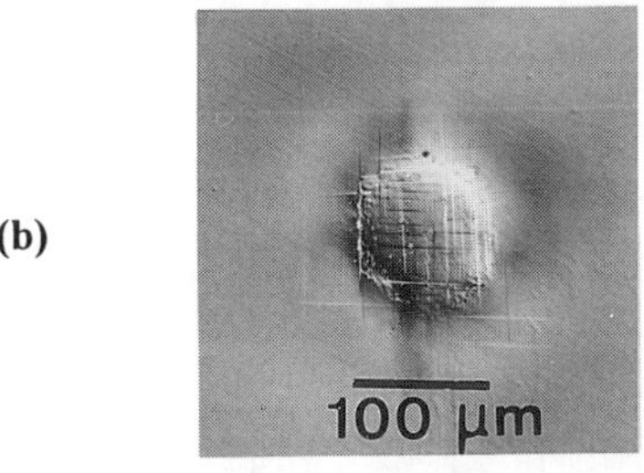

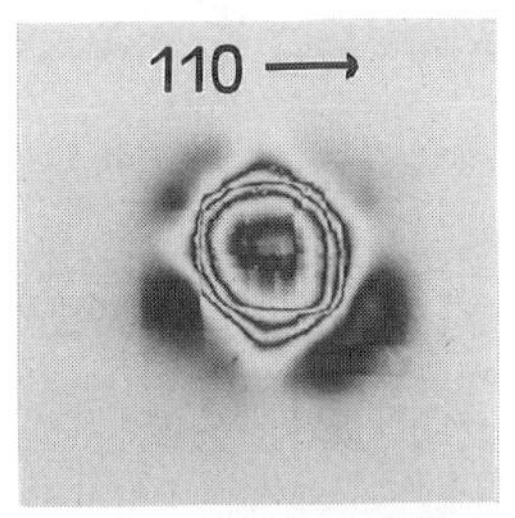

(c)

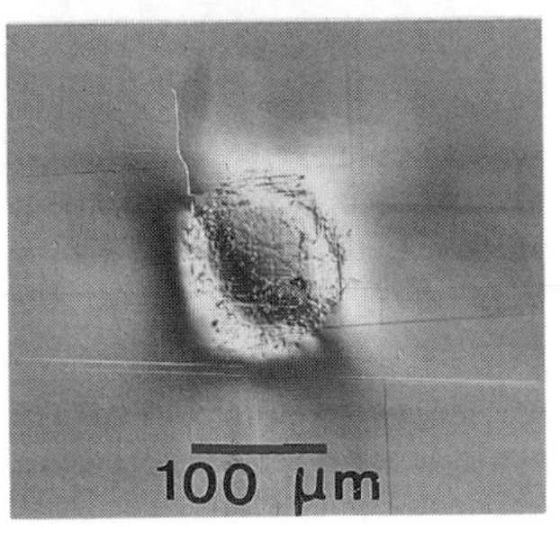

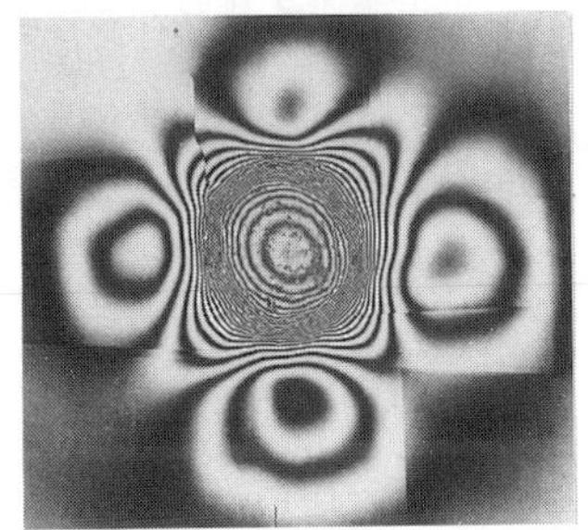

DEFORMATION OF THE SURFACES DUE TO SLIDING

Regime 1 - As a frictional force is applied to a stationary cone, the tensile stresses on the trailing side are proportionately increased [1] and Hertzian ring cracks are generated by a material which would not, under static loading conditions, develop a sufficiently high contact pressure to do so. Thus, we have found that titanium diboride ($P_m \sim 10$ GPa) will not cause ring cracking when used as a static impressor but readily forms the characteristic chatter cracks when sliding in a vacuum at room temperature. The control afforded by these experimental conditions enables us to demonstrate that subsequent rates of wear are strongly influenced by the direction of sliding and by the type of diamond being worn.

Figure 3 shows that for <100> sliding on the (001) surface of a type I_b diamond, the chatter cracks are formed on intersecting {111} cleavage planes and that, as the result of multiple traversals over the same track, these cracks intersect and interact to give the classic tetrahedral wear debris first suggested by Tolkowsky [12]. In contrast, whilst also formed on {111} cleavage planes, the chatter cracks when sliding in <110> directions are less likely to intersect and give rise to a lower rate of wear. The Talysurf profiles reproduced in Figure 4 give clear testimony to this anisotropy in the room temperature wear rate of a type I_b diamond.

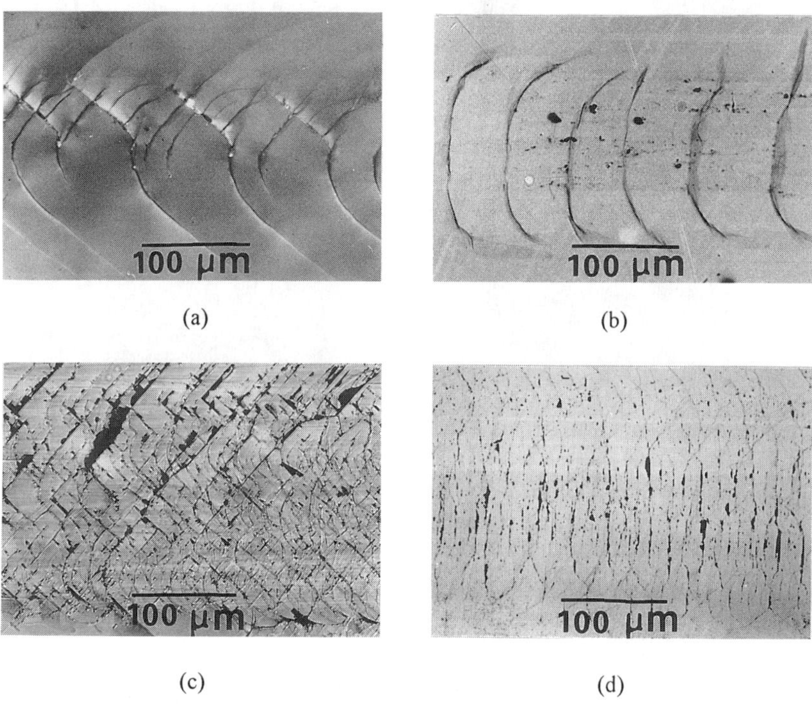

Figure 3. Showing the nature and anisotropy of wear, due to a TiB_2 slider at room temperature, on the (001) surface of a type I_b diamond: (a) one traversal in [100]; (b) one traversal in [110]; (c) 3500 traversals in [100] and (d) 3500 traversals in [110].

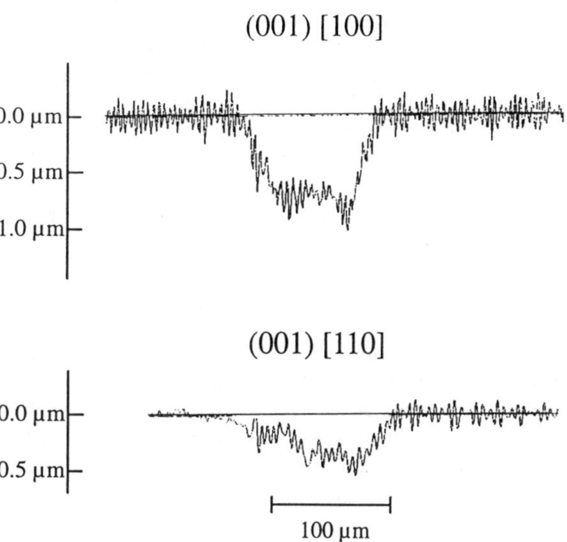

Figure 4. Typical Talysurf profiles of the wear tracks shown in Figure 3c (left) and Figure 3d (right).

We have repeated this type of experiment on the (001) surfaces of type I_a and II_a diamonds and have calculated a wear coefficient as the volume of wear per unit length of sliding (m) and normal load (N). These results are summarised in Table I and establish that whilst the nature of anisotropic wear is the same for all diamond types, the synthetic material is less resistant than the natural diamonds.

Using sliders harder than titanium diboride, such as cubic boron nitride (cBN) or diamond, increases the degree of cracking but wear remains anisotropic with cracking largely on the {111} cleavage planes.

Table I

type:direction	**cross-sectional area (mm^2)**	**wear rate (mm^3 / mN)**
I_b : [100]	1.51×10^{-4}	4.31×10^{-8}
I_b : [110]	0.72×10^{-4}	2.06×10^{-8}
I_a : [100]	0.86×10^{-4}	2.46×10^{-8}
I_a : [110]	0.42×10^{-4}	1.37×10^{-8}
IIa : [100]	0.78×10^{-4}	2.23×10^{-8}
II_a : [110]	0.40×10^{-4}	1.14×10^{-8}

Regime II - The effect of temperature on the surface deformation of a (001) type I_b diamond using a slider made from the octahedral tip of a type I_a natural diamond, within this regime, is illustrated by the micrographs in Figure 5. Since type I_a is significantly harder than type I_b in this temperature range, this combination was used to make the scratch hardness measurements presented later. Each track was made with a normal load of 80N and represents one traversal, left to right with respect to the figure, in the [110] direction. For comparison, similar measurements were made in a [100] direction. At 1000°C, there is some plastic grooving accompanied by a significant amount of 'chatter' cracking on the {111} cleavage planes. At 1200°C, there is much more plasticity with a reduction in the amount of cleavage cracking and, at 1400°C, there is again a significant amount of cracking but now it is primarily on the {110} planes. The interferogram of the track formed at 1400°C confirms that its depth is of the order of 1.5 micrometers (6 fringes).

Using a blunted cubic boron nitride cone, but with otherwise identical experimental conditions, the anisotropic plastic deformation of diamond can be demonstrated - as in Figure 6. At a temperature of 1400°C, there is a little cracking on the {110} planes and a great deal of pile-up of plastically deformed material. When sliding takes place in the [110] direction, material is displaced immediately in front of the cone whilst, in the [100] direction, it is displaced to the side of the groove. The interferograms indicate that the depth of the grooves are approximately 1.38 micrometers and 2.48 micrometers in the [110] and [100] directions respectively.

By ensuring that the mean contact pressure is kept below a certain threshold level, i.e. insufficient to produce either {111} or {110} cracking, relatively large volumes of material can be heavily plastically deformed as the result of multiple traversals over the same diamond surface as shown in the micrographs of Figure 7. It is apparent that after 1000 traversals with a titanium diboride cone (Pm ~ 3.0 GPa) at 1200°C, there is much greater degree of plastic deformation and none of the cracking associated with a rigid diamond slider after one traversal. In this observation lies the possibility of using plastic flow to produce conforming diamond surfaces and, perhaps, to heal surface defects [13].

Finally, a direct comparison of the resistance to plastic deformation for type I_a and type I_b diamond, under these experimental conditions, can be made by reference to Figure 8. Whilst the deformation of the (001) type I_a surface is apparent only by the appearance of a relatively small number of slip lines parallel to the [110] direction of sliding after 100 traversals of a titanium diboride cone at 1400°C, a groove with a depth of about 1.2 micrometers and extensive pile-up is produced on the type I_b surface.

Regime III - At temperatures of around 1400°C and higher, where the flow stress is essentially temperature independent, the dislocations are very mobile and extend well beyond the actual contact area of the impression or friction track. This is evident, in the former, by the formation of a classical 'rossette' of slip lines and, in the latter, by slip steps which extend to a distance beyond the groove at least equal to its width. Cleavage cracks on {111} planes are not formed in any of the various type of diamond but strain induced cracking on {110} planes occurs for type Ib diamonds under impression conditions. Also, the high level of plastic flow during sliding at these temperatures results in the formation of shear cracks within the grooves and just beyond - as shown in the 1400°C track of Figure 5. These cracks tend to be on {110} cracks in type I_b but are shorter and less crystallographic in the natural diamonds bearing a striking resemblance to those formed within the grooves made by harder sliders on ductile metals.

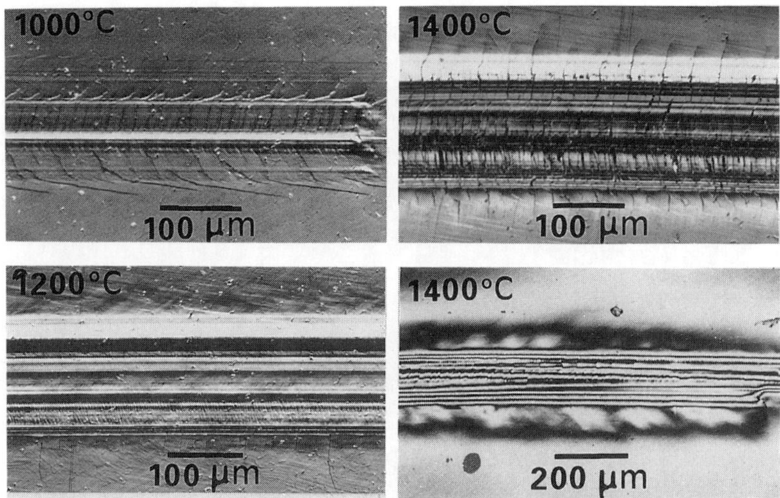

Figure 5. Illustrating the effect of temperature in regime II on the deformation of a (001) surface of a type I_b diamond after one traversal in the [110] direction with the tip of a type I_a natural octahedron.

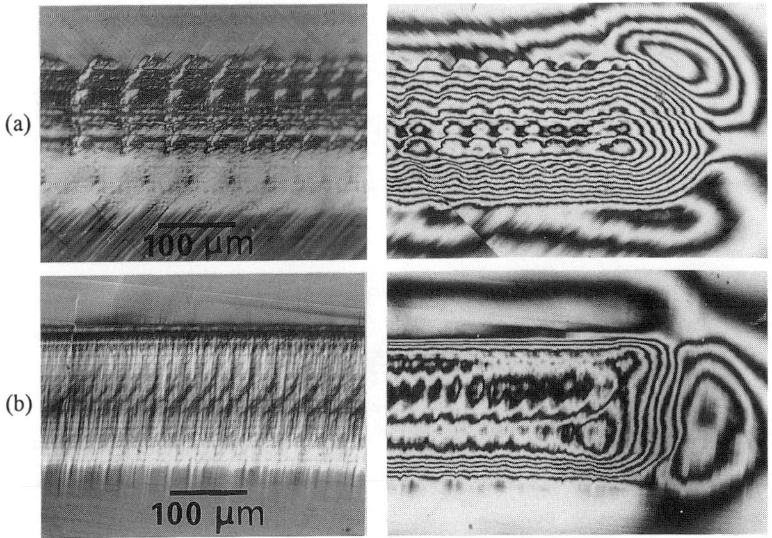

Figure 6. Showing the anisotropic nature of plastic flow and pile-up due to a cubic boron nitride blunted cone sliding in (a) [100] and (b) [110] directions on the cube plane of a type Ib diamond at 1400°C.

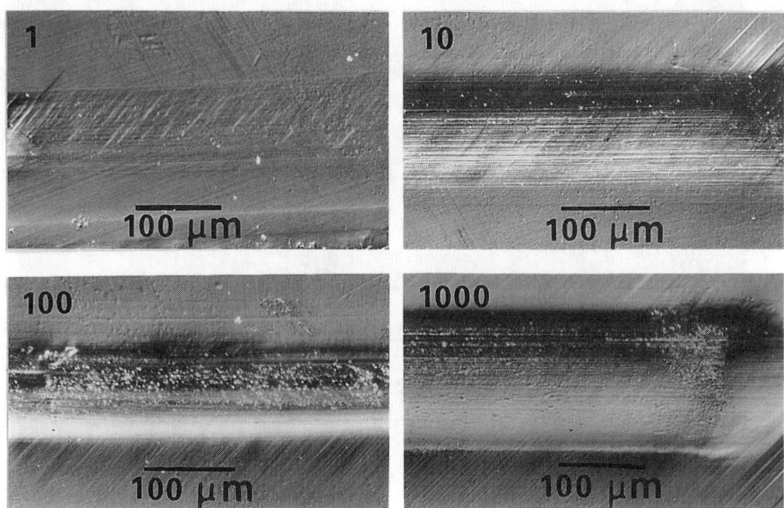

Figure 7. Illustrating how, by cumulative deformation due to multiple traversals, i.e. 1, 10, 100 and 1000, extensive plasticity is induced in a type I_b diamond surface at 1200°C.

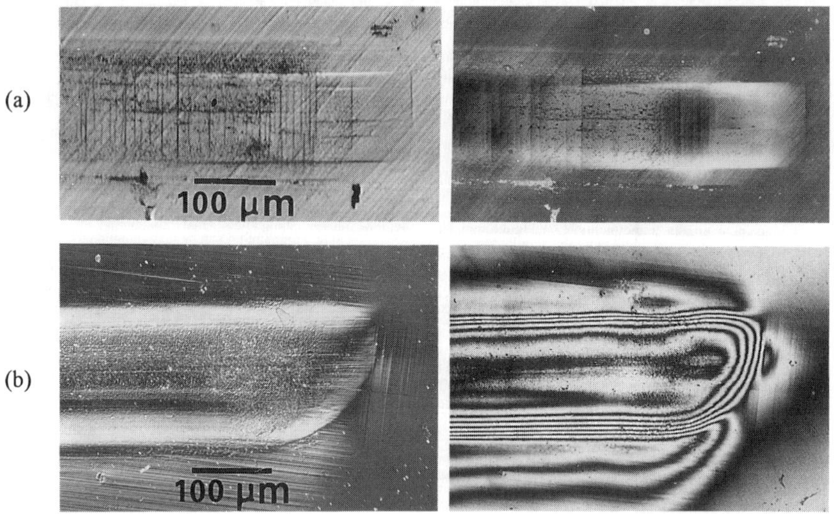

Figure 8. Contrasting the resistance to plastic deformation of (a) type I_a and (b) type I_b diamonds when subjected to 100 traversals of a titanium diboride cone in a [110] direction on the (001) surface at 1400°C.

EXPERIMENTAL RESULTS

The coefficient of friction - Typical values of the coefficient of friction between the sliders used in this work and the (001) surfaces of type I_b diamond, during a single traversal at a speed of about 10 mm/min and for a distance of about 10 mm, are summarised in Table II.

Table II

slider:direction	20°C	1000°C	1200°C	1400°C
cBN:[100]	0.41	0.28	0.25	0.28
cBN:[110]	0.22	0.28	0.22	0.22
Si_3N_4:[100]	0.36	0.26	0.28	0.26
Si_3N_4:[110]	0.45	0.16	0.17	0.13
TiB_2:[100]	0.51	0.17	0.15	0.21
TiB_2:[110]	0.26	0.13	0.13	0.20

When making repeated traversals at temperatures in regimes II and III, the measured friction tended to increase somewhat with time. For example, μ increased from 0.19 to 0.40 during the cumulative deformation of the specimens shown in Figure 7.

It is apparent that μ does not change significantly with temperature in Regimes II and III but tends to increase as the hardness of the slider increases. Also, the nature of anisotropy is consistent and the friction in the <100> is invariably higher than in <110> directions.

Scratch hardness - In earlier work [14], we assumed that only the front half of the slider is supported by the specimen material as the groove is formed. Similarly, we have measured the scratch hardness (H_s) by dividing the applied normal load (N) by the projected area of the front half of the slider and based on the full width (w) of the groove: i.e. $H_s = 4\,N / w^2$.

In using the full width of the groove, we have reasoned that the material piled-up in front of the slider is contributing to the support of the normal load. The extent of the pile-up shown in the above figures would seem to justify that approach. Nevertheless, we would recommend that the results shown in Table III should not be used in an absolute sense but, essentially, to verify the nature of anisotropy for comparison with the behaviour of other crystals.

Table III

Temperature - °C	H_s - GPa [100]	H_s - GPa [110]
1000	32.00	50.00
1200	16.33	22.22
1400	9.88	14.22

DISCUSSION

The nature of anisotropy in the abrasive wear rate of diamond, at room temperature, is well documented [15] and these results are consistent with the earlier work. With a normal load of 2N and under conditions of much higher sliding speeds, i.e. 88 mm/s, Crompton and Hirst [16] established a linear relationship between the rate of wear (mm^3/mN), in <100> directions on the (001) surface of natural diamond, and the indentation hardness of the abrasive material. Those materials included a number of steels, of differing hardness, and carborundum. Whilst the wear rates in this work fit that linear relationship, their experimental conditions would have resulted in the development of appreciable temperatures at the sliding interface and it appears unlikely that the same mechanisms of wear would prevail. The difference in abrasion resistance for the synthetic and natural diamonds, shown in Table 1, has not been reported previously and, since measurements of fracture do not distinguish between these types [1], raises the question of whether a mechanism based solely on brittle fracture can be justified.

Although we are not concerned primarily with the coefficient of friction and the measured anisotropy for <100> and <110> directions on the cube face, it is of interest to note that the nature of anisotropy persists, whilst the coefficient of friction remains approximately constant for a given crystallographic direction of sliding, despite the very significant changes in the modes of deformation with increasing temperature. Above the BDT temperature the flow stress decreases most markedly; dislocation mobility and cross slip is facilitated; cleavage (chatter) cracking stops; strain induced cracking on {110} planes may occur. An explanation of these effects based on the ploughing and deformation component of friction, rather than adhesion, would seem to be the most appropriate approach. It may be considered significant that the coefficient of friction increases with the mean contact pressure for the sliders used here (see Table 2) indicating that ploughing and the depth of penetration of the diamond surface is important.

All of the common methods for measuring scratch hardness are based on the deformation of the specimen by a harder slider. For example, the hardness measurement may be determined by the volume of material removed; the critical normal load required to produce the first evidence of permanent deformation; the minimum hardness of sliders of other materials which cause the formation of a scratch (Moh's scale). All of these methods reflect the basic anisotropy in the mechanical properties of crystalline solids but, perhaps surprisingly, the phenomenon is most marked in those crystals with a tendency to brittle behaviour. In such cases, some explanations of anisotropy have been based on the orientation of the cleavage planes with respect to the sliding direction [17]. However, solids that have common slip systems but different cleavage planes may have the same anisotropy in scratch hardness - e.g. calcium fluoride and lead sulphide both have {100}<110> slip systems but calcium fluoride cleaves on [111] planes whilst lead sulphide cleaves on [100} planes. This fact indicates that plastic deformation is more likely to determine the nature of anisotropy than fracture [14].

For those conditions where plastic deformation controls the formation of a groove, Brookes and Green [14] proposed a model based on the active slip systems for a given crystal and the crystallographic displacement of material by the slider. Their model has two components: one which resolves shear stresses on to the operative slips systems (the Schmid:Boas component); and the other was intended to reflect the degree of pile-up limiting the penetration of the slider (the constraint component). First, the frictional force F and the normal load W are resolved into a resultant L inclined at an angle, θ, to the vertical axis of the slider - where $\mu = \tan \theta$. The axis of L represents the principal compressive stress axis as used in the treatment of frictional anisotropy by Bowden and Brookes [18]. There will of course be a gradient of stress and strain along this axis and, for that reason, Brookes and Green preferred to use an axis *JJ'* which is normal to *L* and lies in the same plane as *F*, *W* and *L*. When slip takes place on a given system, rotation of the slip plane occurs about an axis normal to the slip direction and contained in that plane. Slip on a system where the axis of rotation is orthogonal to the sliding direction, and parallel to the specimen surface, will produce slip steps normal to the sliding direction and lead to pile-up of material having been displaced from within the bulk of the crystal. This continuous process of pile-up ahead of the slider resists its penetration and reduces the groove width - i.e. this is the hardest direction. The softest direction then corresponds to one where the axis of rotation is parallel to the sliding direction and the material is displaced on either side of the groove allowing deeper penetration and a wider groove. Their constraint term was then quantified through γ which was the angle between the axis of rotation of a given slip plane and an axis *MM* which was normal to the sliding direction and contained in the surface of the specimen. On the basis that a high effective resolved shear stress would lead to a greater penetration, and therefore a lower scratch hardness, Brookes and Green proposed the following equation:

$$\tau = (J/A) \cos \varphi \ \cos \lambda \ 0.5 \, (1 + \sin \gamma) \quad (1)$$

where: τ = the effective resolved shear stress

ϕ = the angle between the slip plane normal and *JJ'*

λ = the angle between the slip direction and *JJ'*

γ = the angle between the axis of rotation and the axis *MM*.

A = the area of material supporting the tensile force *J*.

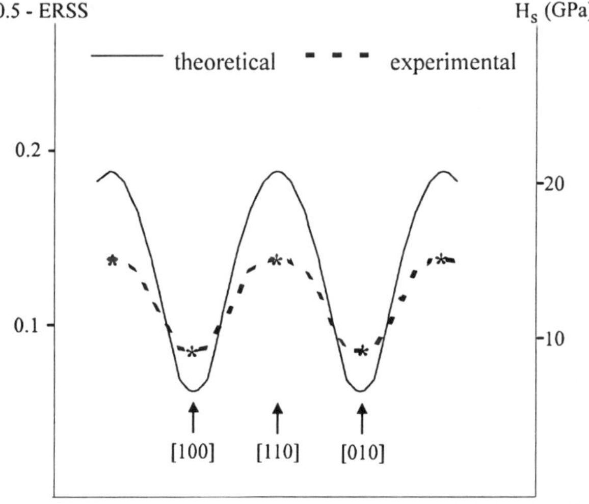

Figure 9. The theoretical model for scratch hardness and the measured values of scratch hardness for type I_b diamond using a type I_a diamond slider at 1400°C.

The change in τ with direction of sliding can be determined for a given crystallographic plane of sliding and knowing the operative slip systems for that crystal. In Figure 9 we have plotted $0.5 - \tau$, for a given J/A, effectively inverting the curve to give an indication of the hardness of a given direction and to illustrate the predicted behaviour for a (001) surface of a cubic crystals having {111} <110> slip systems. The original model was verified by using metal and ionic single crystals to ensure a high level of plasticity to demonstrate the effect of the constraint term and the resultant pile-up. It is interesting that our current results on diamond, although under conditions which still involve an element of fracture, are consistent with the predictions of that earlier model.

CONCLUSIONS

The results of this work have shown that, similar to earlier work using softer impressors, a marked change occurs in the deformation of diamond when subjected to sliding friction at temperatures above a certain temperature - identified as the brittle-ductile transition temperature. This temperature corresponds to that required to enable dislocations to move readily; creep mechanisms to occur and may lead to strain (dislocation) induced cracking on {110} planes. It also becomes possible to develop extensive plastic deformation and changes in surface topography, without inducing cracks, by repeatedly rubbing the diamond surface with a softer material. Other more specific conclusions can be summarised:

- the soft slider method used here is capable of quantifying anisotropy in the abrasion resistance of diamond, i.e. wear in <100> is about twice that in <110> directions on the cube face, and has established that natural diamonds (types I_a and II_a) have a greater wear resistance than synthetic (type I_b).

- the nature of the anisotropy in the coefficient of friction, i.e. a maximum in <100> and a minimum in <110> directions, remains constant over the whole temperature range despite the significant changes in the mechanisms of deformation and fracture.

- above the BDT temperature, anisotropy in the scratch hardness of synthetic type Ib diamond is shown to be consistent with a resolved shear stress model in that <110> directions are harder than <100> due to the pile-up of material in front of the slider in the former directions.

- above the BDT temperature, natural diamond crystals, both type I_a and II_a, have a greater resistance to plastic deformation than synthetic type I_b under conditions of sliding friction.

ACKNOWLEDGEMENTS

The authors are pleased to acknowledge the generous support of De Beers Industrial Diamond Division Ltd. in support of this work and a research studentship for one of us (GX).

REFERENCES

1. C.D. Clark, A.T. Collins and G.S. Woods, in The Properties of Natural and Synthetic Diamond, J.E. Field, (editor), Academic Press, London (1992), pp35-81.

2. E.J. Brookes, PhD Thesis, University of Hull, 1992.

3. A. Al-Watban, unpublished work.

4. C.A. Brookes, E.J. Brookes and G. Xing, Proceedings of International Colloquium 'Mechanics of Creep Brittle Materials 2' (eds: A.C.F. Cocks and A.R.S. Ponter), Elsevier Science Pubs., (1991), 345-355.

5. E.J. Brookes and C.A. Brookes, Proceedings of International Conference on Plastic Deformation of Ceramics, Snowbird, Utah, (1994).

6. C.A. Brookes, E.J. Brookes and L.Y. Zhang, Proceedings of 2nd International Conference on the Applications of Diamond Films and Related Materials, Tokyo, August 1993, pp737-744.

7. C.A. Brookes, E.J. Brookes and L.Y. Zhang, Proceedings of International Conference on Plastic Deformation of Ceramics, Snowbird, Utah, (1994).

8. C.A. Brookes, R.D. James, F. Nabhani and A.R. Parry, Proceedings IMechE Seminar Tribology and Metal Cutting and Grinding, MEP, (1992).

9. C.A. Brookes, M.P. Shaw and P.E. Tanner, *Proc Royal Soc.*, **A409**, (1987), 141-159.

10. C.A. Brookes in The Properties of Natural and Synthetic Diamond, J.E. Field, (editor), Academic Press, London (1992).

11. P. Humble, and R.H.J. Hannink, *Nature* (London) **273**, (1978), 387-39.

12. M. Tolkowsky, PhD Thesis, University of London, 1920.

13. C.A. Brookes, International Patent: WO 89/04239 (1989).

14. C.A. Brookes, and P. Green, *Proc. Royal. Soc.* Lond., A **368**, (1979), 37-57.

15. J. Wilks and E.M. Wilks, in The Properties of Natural and Synthetic Diamond, J.E. Field, (editor), Academic Press, London (1992), pp573-605.

16. D. Crompton, W. Hirst and M.G.S. Howes, *Proc Roy Soc.*, London **A333,** 1973, 435-454.

17. L.G. Tsinzerling, E.S. Berkovich, L.A. Sysoev and M.P. Shackol'skaya, *Soviet Physics Crystallogr*, **14**, (1970), 897-906.

18. F.P. Bowden and C.A. Brookes, *Proc. Royal. Soc.*, Lond. A **295**, (1966) 244-266.

REFERENCES

1. [illegible] Cohen [illegible] Wierda in *The Physics of* [illegible] (Academic Press, London 19[illegible]) p. [illegible]

2. [illegible]

3. [illegible]

4. [illegible]

5. [illegible]

6. [illegible]

7. [illegible]

8. [illegible]

9. [illegible]

10. [illegible]

11. [illegible]

HIGH-TEMPERATURE INDENTATION OF NATURAL DIAMOND AND THE QUEST FOR LONSDALEITE

P. PIROUZ, A. GARG*, X. J. NING, J. W. YANG**, AND S. Q. XIAO***
Department of Materials Science and Engineering, Case Western Reserve University, Cleveland, OH 44106
*Now at: NASA Lewis Research Center, Cleveland, OH 44135
**Now at: APA Optics, Inc., Blaine, MN 55434
***Now at: National Center for Electron Microscopy, Lawrence Berkeley Laboratory, Berkeley, CA 94720

ABSTRACT

Attempts were made to produce bands of hexagonal diamond (Lonsdaleite) by high-temperature indentation of cubic diamond. $(01\bar{1})$ wafers of type I diamond were indented over the temperature range 1000-1300°C and the microstructure of the indentation plastic zone investigated by transmission electron microscopy (TEM). No hexagonal diamond was produced; instead, the plastic zone consisted of arrays of $\frac{1}{2}$<110> dislocations lying on the {001} planes. A possible mechanism for the generation of such dislocations, and reasons for the absence of hexagonal diamond, are discussed.

INTRODUCTION

The strain produced by the indentation of elemental semiconductors such as silicon and germanium can be relieved in a number of different modes which, primarily, depends on the deformation temperature. At low temperatures, where the material is brittle (usually less than $\sim\frac{1}{3}T_m$, where T_m is the melting point), indentation fracture (occurrence of cracks) is predominant[1]. At temperatures above $\sim\frac{2}{3}T_m$, where the material is ductile, dislocation generation occurs readily and material is pushed away from the indentation site by dislocation glide to produce a plastic zone and surface uplift around the indent. At intermediate temperatures, $\sim\frac{1}{3}T_m > T > \sim\frac{2}{3}T_m$, both fracture as well as generation and glide of perfect (or weakly dissociated) dislocations take place and, in addition, twinning can become an important mode of strain accommodation. Deformation twinning may be considered as generation and glide of partial dislocations on adjacent {111} slip planes, and a twin band of thickness nd_{111} may be thought of as n neighboring stacking faults on adjacent {111} planes (d_{111} is the interplanar spacing of {111} planes).

An interesting phenomenon, first observed by Eremenko and Nikitenko [1, 2], was that in this intermediate temperature range, narrow bands of hexagonal Si and Ge can form. This phenomenon was later studied in Si by high resolution electron microscopy

[1]Because of the hydrostatic component in the stress field of an indent, the brittle-ductile transition temperature, T_c, for the incidence of cracking is much lower than the conventional T_c measured in a uniaxial tensile test. The difference can sometimes be hundreds of °C.

Mat. Res. Soc. Symp. Proc. Vol. 383 © 1995 Materials Research Society

(HREM) and analytical microscopy and two mechanisms for the formation of the hexagonal phase were proposed [6]. Both of these mechanisms involved the interaction of twins on different habit planes. In the first case, the hexagonal phase forms at the intersection of two twin bands with, say, $(111)_m$ and $(1\bar{1}1)_m$ habit planes as in Fig. 1(a) for Ge [8].[2] In the second case, a secondary twin with a $(1\bar{1}1)_t$ habit plane nucleates within a primary twin band that has a $(111)_m$ habit plane. When the secondary twin propagates outside the primary twin band into the matrix, a ribbon of hexagonal material forms. The ribbon of hexagonal Ge shown in Fig. 1(b) is thought to form by this mechanism [8].

Based on these observations in Si and Ge, it was hypothesized in [7] that hexagonal diamond, Lonsdaleite, forms by a similar mechanism. This phase was first detected in Canyon Diablo and Goalpara meteorites [9, 10], and soon after produced during laboratory experiments on c-axis compression of graphite followed by high pressure annealing above $1000°C$ [11],

In all the three cases (diamond, Si and Ge), it has been found that the hexagonal phase is tetrahedrally coordinated with an ideal c/a ratio, $(8/3)^{1/2}$. Thus. the structure has been termed *diamond-hexagonal*, dh, in contrast to the normal *diamond cubic*, dc, phase [7]. In the case of dh diamond, the orientation relationship (OR) between the hexagonal and cubic phases was not determined. Instead, using x-ray diffraction, Bundy and Kasper [11] determined the following OR between the graphite crystal, which partially converted upon compression, and the product dh diamond:

$$(10\bar{1}0)_{dh}//(0001)_{graphite}$$

$$[0001]_{dh}//[11\bar{2}0]_{graphite}$$

In the case of Si and Ge, the OR between the cubic and hexagonal phases was determined by electron diffraction, as well as by HREM, and found to be [1, 2, 5, 8]:

$$(0001)_{dh}//(110)_{dc}$$

$$[1\bar{2}10]_{dh}//[1\bar{1}0]_{dc}$$

In addition the habit plane between dh and dc Si and Ge was found to be $(115)_{dc}$ [see Fig. 1(b)]. The unexpected OR between the two phases and the high-index habit plane are all characteristics of a martensitic transformation whereby the dc material (partially) converts to the dh phase through a diffusionless transformation such as twinning [5-7].

The objective of the present study was an attempt to see if Lonsdaleite could also be formed by indentation at the appropriate temperature range. The deformation temperature should correspond to the range at which twinning becomes a mode of strain relaxation. As mentioned before, this is the intermediate range between high temperatures at which slip takes place by the glide of perfect dislocations (e.g. $\geq 1700°C$ [12]) and low temperatures (e.g. $\leq 1000°C$) where indentation cracking occurs. The actual temperature range chosen for this purpose was a compromise; the minimum temperature was the value below which no indentation marks could be observed on the diamond surface (implying that no observable plastic deformation had taken place) and the maximum temperature was determined by the high temperature capabilities of the indentation machine. Specifically, above $1300°C$, the diamond indenter blunted rather rapidly on contact because of the faster graphitization rate of the diamond tip. The indented specimens were subsequently investigated by electron diffraction as well as by conventional and high resolution TEM.

---

[2]The subscripts "m" and "t" refer to the matrix and twin lattices, respectively.

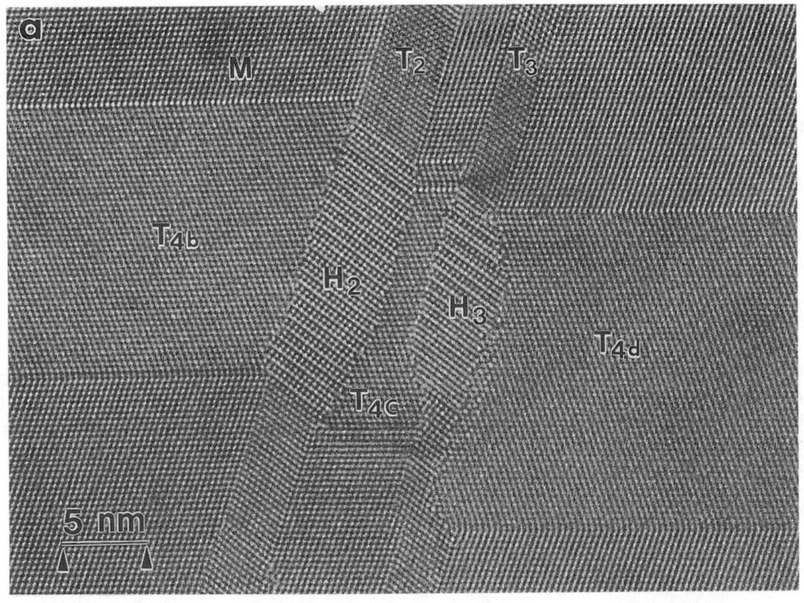

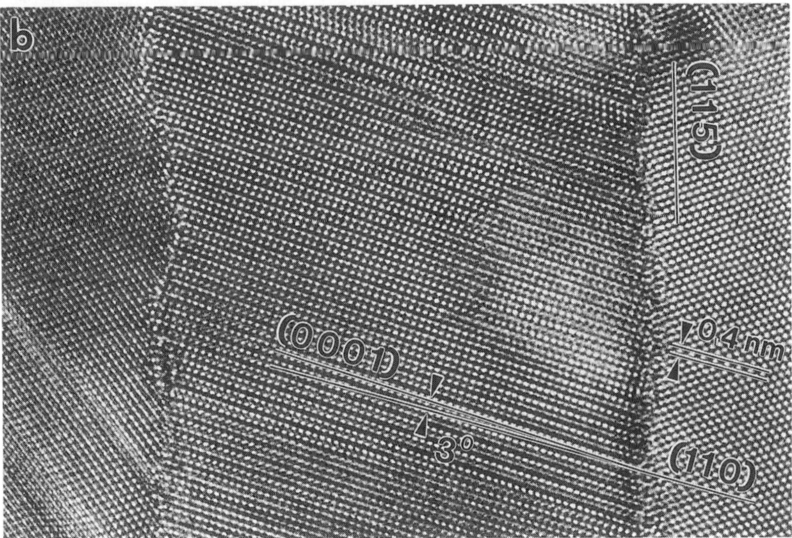

Fig. 1. Diamond-hexagonal, dh, Ge produced by 330°C indentation of a $(1\bar{1}0)$ wafer of diamond-cubic, dc, material, (a) dh Ge produced at the intersection of two twin bands in the cubic phase; (b) dh Ge ribbon in the dc matrix; note the parallelism of $(0001)_{dh}$ and $(110)_{dc}$ planes and the $(115)_{dc}$ interface between the two phases (from [8]).

The results of this study were unexpected in that the slip systems activated in electron-transparent areas of the indentation plastic zone were found to be <110> {001} instead of the usual <1$\bar{1}$0>{111}.

EXPERIMENTAL

100 *μm* thick square-shaped wafers of type I diamond[3] with an edge length of 1 *mm* and [011] orientation were indented in a vacuum of ~10^{-5} *Torr* using a Vickers diamond indenter in a high-temperature indentation machine[4] with a load in the range 200-300 *g* at 1000, 1240, and 1300°*C*. The (011) surface of diamond was found to be very hard and the diamond indenter was often damaged (blunted or fractured) after a few indentations. In addition, the impressions usually had a distorted lozenge shape, and often were irregularly curved at the indent edges. On each wafer, 5-10 indentations were made and the deformed specimens were mechanically polished from the back (opposite face to the indentations) to a thickness of ~50 *μm*. The specimens were then dimpled and ion-milled to perforation. In regions of the thin foil where an indentation was close to the edge of the hole, the electron-transparent areas were electron-optically investigated by a Philips CM 20 and a JEOL 200CX transmission electron microscopes both operating at 200 *kV*.

RESULTS

In contrast to the indentations on (011) faces of Si and Ge, where there was profuse generation of microtwin bands and occasional bands of the hexagonal phase, indentations on the (011) face of diamond mostly produced arrays of perfect dislocations. An example of such arrays, taken from a region a few microns away from the indentation center, is shown in Fig. 2(a). This is a multi-beam bright-field (BF) micrograph close to the [011] zone axis. Four dislocation arrays may be observed which traverse the micrograph nearly vertically. Despite the fact that the dislocation density was very high in each array, they can still be resolved individually in this region. A dark-field (DF) micrograph of a nearby region is shown in Fig. 2(b); it should be noted that the dislocations in all the arrays lie along <110> directions, which are the Peierls valleys in tetrahedrally-coordinated materials with a cubic structure.

In addition to these arrays of individually-resolvable dislocations, there are also nearly edge-on dark bands in Fig. 2 which cross the arrays at 90°; these bands are nearly horizontal in Fig. 2(a), and extend from the top left corner of the micrograph to the bottom right hand edge in Fig. 2(b). On tilting the TEM specimen in the microscope, these dark bands were also found to contain dislocation arrays similar to the nearly vertical arrays in Fig. 2(a).

In Fig. 3, a weak-beam DF image of another region of the indented zone is shown. On the left hand side of this micrograph, two dislocation loops can be observed, the left hand segments of which have been truncated possibly by the top or bottom surface of the TEM specimen. There is also the possibility that these two dislocations are actually half-loops (instead of full loops) and the micrograph in Fig. 2(b) shows surface nucleation of dislocation half-loops [13].

---

[3]Drukker International, Amsterdam, Holland.

[4]QM High-Temperature Microhardness Tester, Nikon, Inc., Tokyo, Japan.

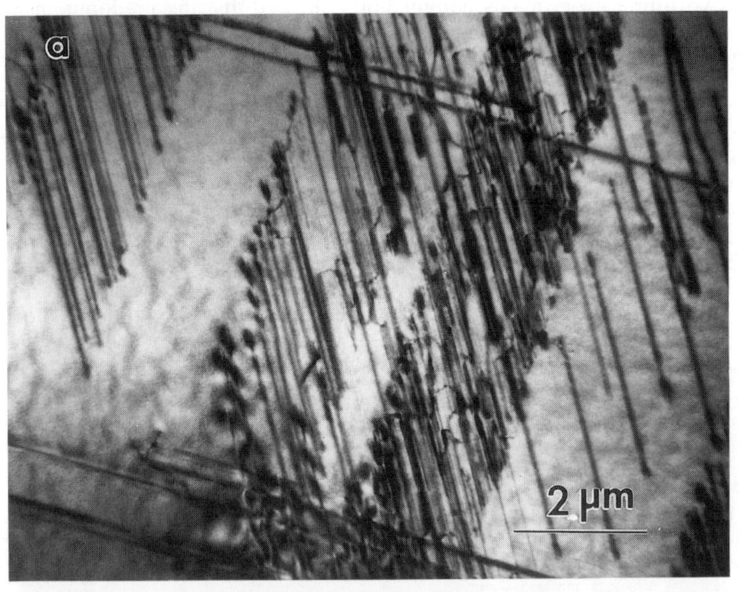

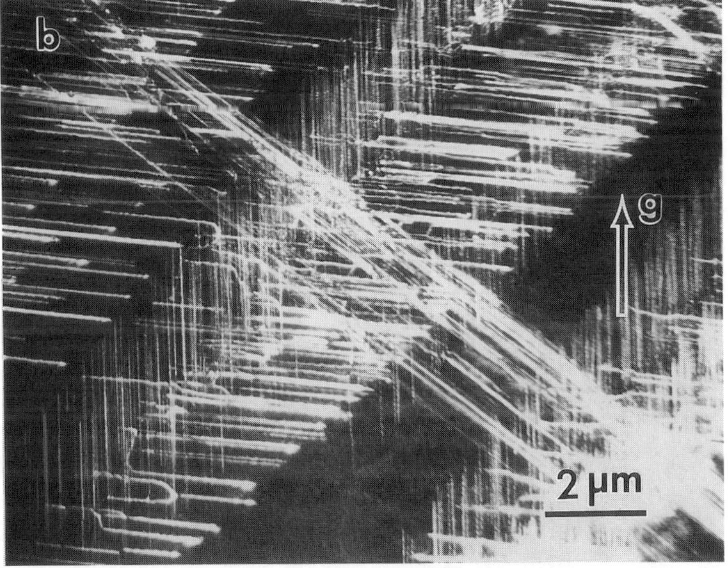

Fig. 2(a). Multi-beam bright-field (BF) micrograph of an area within the indentation plastic zone near the [011] zone axis, Note the dislocation arrays running approximately vertically and the orthogonal dark bands running approximately horizontally; (b) 2-beam dark-field (DF) micrograph of a nearby region close to the [21$\bar{1}$] zone axis, (**g**=022).

Extensive tilting experiments convincingly showed that the predominant majority of dislocation arrays in the plastic zone of indented (011) diamond wafers lie on the three orthogonal sets of {100} planes [14]. In TEM observations with the beam close to the [011] zone axis, the dislocations are in the form of elongated half- or complete loops when they lie on one of the two {001} planes that have a 45° inclination to the foil surface, e.g. on (010) or (001). On these planes, the elongation of the loops is along one of the <110> directions in the corresponding {001} plane. On the other hand, the dislocations that lie on the (100) planes normal to the foil surface, appear as sets of straight parallel lines. It is possible that the dislocations on these planes are also in the form of half- or full loops,

The Burgers vectors of the dislocations were determined by the **g.b** technique and found to be parallel to <110> directions. Attempts were made to apply the LACBED technique [15] in order to find the absolute sense and the magnitude of the Burgers vectors. These attempts proved unsuccessful mainly because of the high dislocation density in observable regions. However, since the shortest lattice vector in materials with an fcc lattice is $\frac{1}{2}$<110>, we shall henceforth assume this to be the Burger vector of the dislocations. In all cases in which the Burgers vector of a dislocation was determined, it was found to lie on the same {001} plane as the dislocation loop, i.e. the dislocations could presumably glide on these planes (see the next section).

The technique of weak-beam dark-field imaging was used to determine the dissociation of the dislocations. Within the resolution of this technique, no evidence for the dissociation of {001} dislocations was found.

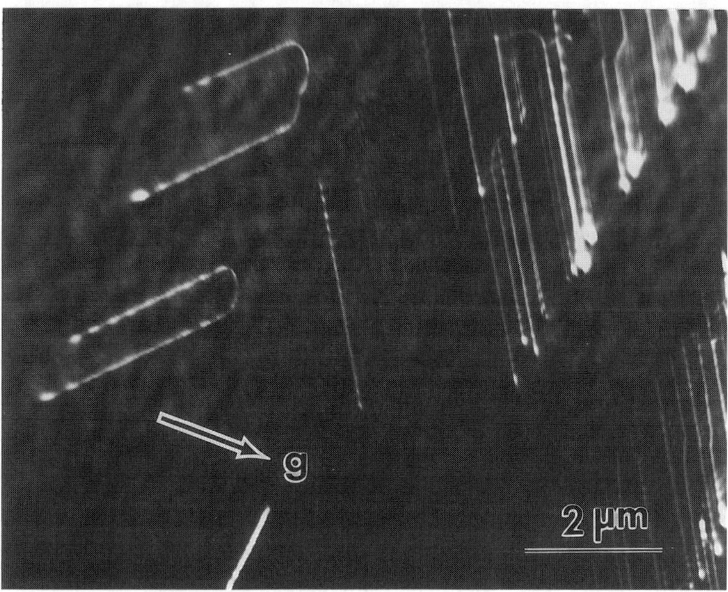

Fig. 3. **g/3g** weak-beam DF micrograph of dislocation arrays and two dislocation loops lying on different sets of {001} planes (**g**=$\bar{2}$20).

Occasional $\frac{1}{2}<1\bar{1}0>$ dislocations lying on {111} planes were also observed. Some of these are shown in Fig. 4 where they are mostly curved and, sometimes are connected to dislocations lying on {001} planes.

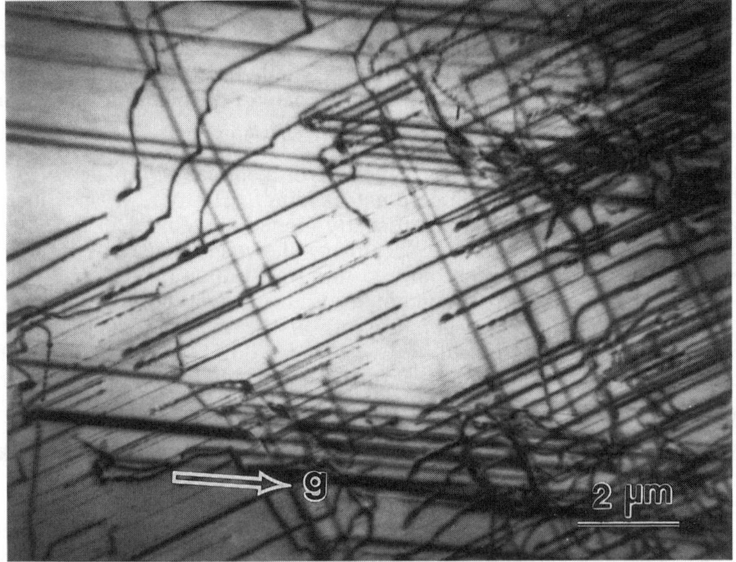

Fig. 4. 2-beam BF micrograph of another region of the plastic zone near the [111] zone axis (**g**=$\bar{2}$20). Note the curved dislocations which lie on (111) planes and, in places, are connected to the dislocations lying on {001} planes.

DISCUSSION

The diamond cubic structure is based on the fcc lattice and dislocations in this structure are known to lie on {111} planes and to have $\mathbf{b}=\frac{1}{2}<1\bar{1}0>$ Burgers vectors. In natural IIa diamond, this has been confirmed by TEM investigation of pre-existing dislocations [16]. In addition, dislocations in this type of diamond, as in other cubic semiconductors, were found to be dissociated into two $\frac{1}{6}<11\bar{2}>$ Shockley partials with a relatively high stacking fault energy, $\gamma \approx 280\ mJ/m^2$ [16]. It is thus very surprising to find $\frac{1}{2}<110>$ dislocations that lie on {001} planes since the latter are not the usual slip planes in this structure. The fact that no evidence for dissociation of these dislocations was found is not, however, surprising since there is no reasonable and energetically favorable reaction, $\mathbf{b}=\mathbf{b}_l+\mathbf{b}_t$, for the dissociation of a perfect $\mathbf{b}=\frac{1}{2}<110>$ dislocation on a {001} plane into partial dislocations with shorter Burgers vectors $\mathbf{b}_l$ (for the leading partial) and $\mathbf{b}_t$ (for the trailing partial).

Since the dislocations lie on {001} planes and, moreover, they cannot dissociate, it follows that twins cannot form. Also, since no deformation twins could form by the indentation of the sample and, as shown above, hexagonal bands form by the interaction of twin bands on different {111} planes (e.g., Fig. 1), it is not surprising that no evidence of hexagonal diamond was detected in the present experiments.

The question then arises as to why indentations at homologous temperatures in {011} Si and Ge, which have isomorphous structures to diamond, produce profuse twinning and occasional hexagonal bands. We shall come back to this question later, but it should be mentioned that deformation of natural diamond has shown mostly to generate the expected $\frac{1}{2}<1\bar{1}0>$ dislocations lying on {111} planes. Thus, Evans and Wild [12] carried out extensive deformation experiments on thin plates of type Ia and type II diamond at 1800°*C* by three-point bending. The type Ia specimens, which contain (nitrogen?) platelets, had significantly higher strengths than type II diamond. TEM investigation of the deformed specimens showed dislocations lying on {111} planes which interacted with each other to form dislocation tangles or dipoles; in the case of type Ia specimens, the diamond platelets acted as obstacles to the movement of dislocations. Although the Burgers vectors of the dislocations produced by the deformation were not characterized by Evans and Wild [12], the geometry of the slip lines on the surface of the deformed specimens, and the configuration of dislocations in the TEM micrographs, indicates that they had $\frac{1}{2}<1\bar{1}0>$ Burgers vectors.

Indentation experiments on diamond have been done by a number of workers (see, e.g., [17] in these proceedings). An interesting set of experiments, which is very appropriate to the present study, is the work of Humble and Hannink [18] who indented (101) surfaces of type IIb diamond at room temperature with a Knoop indenter using a load of ~150 *g*. Observations of the impression site by TEM indicated that when the specimen was indented with the Knoop indenter oriented with its long axis parallel to the [$\bar{1}$01] direction, deformation of the specimen occurred as a discontinuous series of individual *planar shear events* on the (001) set of planes. The highly deformed nature of the indented region made it difficult to resolve and unambiguously identify these shear events with individual dislocations. However, based on *in-situ* heating experiments, the authors suggested that the room-temperature *shear events* are probably shear cracks which rearrange into dislocation arrays on heating and possibly propagate further in the form of dislocation loops [18]. It is interesting that when Humble and Hannink [18] indented (101) diamond with the long axis of the Knoop indenter approximately parallel to the [010] direction (i.e. orthogonal to the previous case), resolvable dislocation loops could be observed by TEM which lay on {111} planes.

Generation and glide of dislocations on {001} planes have also been reported in indentation of another tetrahedrally-coordinated material. This is the work of Qin and Roberts [19] who indented GaAs at low (20°C) and high (400°C) temperatures. Subsequent chemical etching and TEM studies of the indented regions revealed the activation of the {001}<110> slip system over a wide range of experimental conditions [19].

In fcc metals, the possibility of shear on {001} planes was suspected since the early work of Schmid and Boas on high temperature deformation of Al [20] (also see [21]). Cottrell [22], commenting on the work of Lomer [23] on the reaction of dislocations on intersecting {111} planes in the fcc lattice, actually suggested that the product $\frac{1}{2}<110>$ dislocations are not truly sessile (in the sense that their line direction and Burgers vector are not in the same plane) and might be able to glide on {001} planes. More recently, there have been a number of reports on the observation of such dislocations in fcc metals and alloys which have been deformed into stage II of the work-hardening curve [24]. These observations suggest that $\frac{1}{2}\langle 110\rangle$ dislocations glide on the cubic planes not only in high stacking fault energy materials such as Al or Ni, but also in Cu and Ag which have moderate and low values of γ.

The usual interpretation for the formation and glide of $\frac{1}{2}\langle 110\rangle$ dislocations on {001} planes of the fcc metals has been that they are actually Lomer dislocations [23] formed by the interaction of $\frac{1}{2}\langle 1\bar{1}0\rangle$ dislocations gliding on two intersecting {111} planes [24]. If the dislocation splits on the intersecting {111} planes, it will be sessile and will form a Lomer-Cottrell (LC) lock [22]. On the other hand, if it is not split, then it would be able to glide on the {001} plane on which it (and its Burgers vector) lies. Karnthaler [24] has suggested that the $\frac{1}{2}\langle 110\rangle$ dislocations that are mobile on {001} planes are actually constricted nodes which occur in the Lomer-Cottrell reaction. According to this model, the process can start with a very short piece of a composite dislocation, say $\frac{1}{2}[110]$, which can glide on the (001) plane. However, for the continuation of glide, the assumption has to be made that the reacting dislocations (say, $\frac{1}{2}[10\bar{1}]$ and $\frac{1}{2}[011]$) on the intersecting {111} planes, (111) and $(11\bar{1})$, are able to cross-slip on the common cross-slip plane, $(1\bar{1}1)$, and maintain the mobile node on the same (001) plane. In this mechanism, stacking fault energy does not play a very important role nor are large stresses required to move the total $\frac{1}{2}[110]$ dislocation on the (001) plane.

Karnthaler's mechanism [24] is reasonable for metals where the Peierls energy is low and dislocations do not align parallel to the Peierls valleys. In diamond cubic materials, e.g. Si, Ge and diamond, the Peierls energy is high and, at low or moderate temperatures, the dislocations are straight and well aligned with the $\langle 110\rangle$ Peierls valleys. It is worth noting that, assuming that dislocations lie in the Peierls valleys, they can have only two characters on the {001} plane: pure screw or pure edge. In addition, indentation activates a number of slip systems from the beginning of the deformation [25]. Hence, interactions between dislocations on different slip systems are inevitable and Lomer or Lomer-Cottrell dislocations can easily form in the plastic zone of the indentation [25]. Simple consideration of dislocation reactions also shows that the product $\frac{1}{2}[110]$ dislocations on the (001) plane have an edge character. This is also consistent with the observations of Qin and Roberts [19] on indentation deformation of GaAs. Thus, we propose that initially, $\frac{1}{2}\langle 1\bar{1}0\rangle$ dislocation half-loops are nucleated on inclined {111} planes from the surface of the specimen, or from the indentation facets. In the relatively low temperature range under consideration, these half-loops are hexagonally-shaped with the edges of the hexagon parallel to the $\langle 110\rangle$ Peierls valleys. Following glide and interaction on intersecting {111} planes, the parallel segments of these half-loops produce LC locks. In the case of diamond, which has a very high energy stacking fault, the locks often do not split (i.e., they will be Lomer dislocations), and, as a result, they will be able to slip on {001} planes. Moreover, the $\frac{1}{2}[110]$ Lomer dislocations may multiply on the (001) plane and form dislocation loops. In fact, it is very unlikely that all the dislocations in an {001} array (such as the ones in Figs. 2) form by interaction of $\frac{1}{2}<1\bar{1}0>$ dislocations on intersecting {111} planes. It would be more reasonable to assume that once a Lomer dislocation forms on a {001} plane (which is automatically pinned at two nodes to the other segments of the half-loops on {111} planes), dislocation multiplication takes place under the applied stresses to give rise to an array of dislocations.

The formation of the first Lomer dislocation is basically a recombination process where a sessile LC barrier breaks down and the resulting total $\frac{1}{2}<110>$ dislocation becomes mobile on the less densely packed {001} planes. The high stacking fault energy of diamond makes the recombination of an LC lock to a Lomer dislocation simpler than in Si or Ge which have much lower stacking fault energies (~60 mJ/m^2 and ~75 mJ/m^2, respectively [26]).

Once the Lomer dislocations form, their motion and multiplication require relatively high stresses. These are the conditions that are satisfied in the present experiments: formation of Lomer dislocations is favorable because the resolved shear stress on two sets of intersecting {111} planes are identical, the stacking fault energy of diamond is very high, and the stresses produced in an indentation test are very large.

The fact that glide of $\frac{1}{2}<110>$ dislocations on {001} planes has not been observed in Si and Ge may be partly because their formation is more difficult because of the lower stacking fault energy in these materials and, more importantly, it could be because they were obscured by the extensive occurrence of "rosette" dislocations which glide on inclined {111} planes and have $\frac{1}{2}<1\bar{1}0>$ Burgers vectors parallel to the surface. It could be that it is these latter dislocations in Si and Ge that dissociate in the moderate temperature ranges under consideration and give rise to twin formation. Lower-temperature indentation experiments in Si and Ge are under way to see whether dislocation glide on the cubic planes can be observed. It also follows that indentation at higher temperatures and on other faces of diamond, where the resolved shear stresses on intersecting {111} planes are not equal, may provide a greater likelihood for the formation of twin bands and, possibly, hexagonal regions.

CONCLUSION

Attempts were made to produce hexagonal diamond by twin interactions in a similar fashion to that used in producing hexagonal Si or Ge. The experiments were carried out by indentation of $(01\bar{1})$ type Ia diamond in the temperature range 1000-1300°C. No twin bands and consequently no hexagonal regions were observed in a TEM investigation of the deformed indentation zone. Arrays of undissociated $\frac{1}{2}<110>$ dislocations were present lying on all three sets of {100} planes. The absence of twin and hexagonal bands is attributed to the fact that dislocations lying on {100} planes cannot dissociate and produce twins on {111} planes. It would be interesting to perform high-temperature indentations on other faces of diamond where the normally dissociated $\frac{1}{2}<1\bar{1}0>$ dislocations may be generated on the usual {111} slip planes.

ACKNOWLEDGMENTS

The authors would like to thank Dr. P. M. Hazzledine for useful discussions and comments, and Professor J. C. Angus for criticism of the manuscript. This work was partially supported by National Science Foundation Materials Research Group, Grant DMR-9121479.

REFERENCES

1. V. G. Eremenko and V. I. Nikitenko, phys. stat. sol. (a) **14**, 317-330 (1972).
2. V. G. Eremenko, Sov. Phys. Solid State **17**, 1647-1648 (1976).
3. P. Pirouz, R. Chaim and J. Samuels, Izvestia Nauka S.S.S.R., Ser. Fiz. **51**, 753-762 (1987) (In Russian). In English see Bulletin of the Academy of Sciences of the U.S.S.R., Physical Series, 51, #9 (1987).).
4. P. Pirouz, R. Chaim and U. Dahmen, in Defects in Electronic Materials, edited by M. Stavola, S. J. Pearton and G. Davies (Mater. Res. Soc. Proc. **104**, Pittsburgh, PA., 1988), pp. 133-138.
5. P. Pirouz, R. Chaim, U. Dahmen and K. H. Westmacott, Acta metall. mater. **38**, 313-322 (1990).
6. U. Dahmen, K. H. Westmacott, P. Pirouz and R. Chaim, Acta metall. mater. **38**, 323-328 (1990).
7. P. Pirouz, U. Dahmen, K. H. Westmacott and R. Chaim, Acta metall. mater. **38**, 329-336 (1990).
8. S.-Q. Xiao and P. Pirouz, J. Mater. Res. **7**, 1406-1412 (1992).
9. R. E. Hanneman, H. M. Strong and F. P. Bundy, Science (New York) **155**, 995-997 (1967).
10. C. Frondel and U. B. Marvin, Nature (London) **214**, 587-589 (1967).
11. F. P. Bundy and J. S. Kasper, J. Chem. Phys. **46**, 3437-3446 (1967).
12. T. Evans and R. K. Wild, Phil. Mag. **12**, 479-489 (1965).
13. X. J. Ning, T. Perez and P. Pirouz, Phil. Mag. In press. (1995).
14. A. Garg, X. J. Ning and P. Pirouz, In preparation. (1995).
15. D. Cherns and A. R. Preston, Proceedings of the XIth Int. Cong. on Electron Microscopy, Kyoto, Japan, (1986).
16. P. Pirouz, D. J. H. Cockayne, N. Sumida, P. B. Hirsch and A. R. Lang, Proc. Roy. Soc. Lond. A **386**, 241-249 (1983).
17. C. A. Brookes, E. J. Brookes and G. Xing, in Mechanical Behavior of Diamond and Other Forms of Carbon, edited by M. Drory, D. Bogy, M. Donley and J. Field (Mater. Res. Soc. Proc. **383**, Pittsburgh, PA, 1995), pp.
18. P. Humble and R. H. J. Hannink, Nature (London) **273**, 37-39 (1978).
19. C. D. Qin and S. G. Roberts, in Structure and Properties of Dislocations in Semiconductors, edited by S. G. Roberts, D. B. Holt and P. R. Wilshaw (Inst. Phys. Conf. Ser. No. **104**, Bristol, 1989), pp. 321-326.
20. E. Schmid and W. Boas, Kristallplastizitat, (Springer Verlag, Berlin, 1935).
21. P. Lacombe and L. Beaujard, J. Inst. Met. **74**, 1- (1947).
22. A. H. Cottrell, Phil. Mag. **43**, 645-647 (1952).
23. W. M. Lomer, Phil. Mag. **42**, 1327-1331 (1951).
24. H. P. Karnthaler, Phil. Mag. A **38**, 141-156 (1978).
25. P. B. Hirsch, P. Pirouz, S. G. Roberts and P. D. Warren, Phil. Mag. B **52**, 761-784 (1985).
26. H. Alexander, H. Eppenstein, H. Gottschalk and S. Wendler, Journal of Microscopy **118**, 13-21 (1980).

INDENTATION CHARACTERISATION OF CARBON MATERIALS

JOHN S. FIELD AND M.V. SWAIN
CSIRO Division of Applied Physics, Lindfield, NSW 2070, Australia, and
Department of Mechanical Engineering, University of Sydney, NSW 2006, Australia

ABSTRACT

High precision force-displacement measurements of various forms of glassy and graphitic carbon have been made with a Berkovich and spherical tipped indenter. The force-displacement data show almost complete recovery despite contact pressures exceeding a considerable fraction of the elastic modulus. With a spherical tipped indenter the transition from elastic to inelastic response could be readily identified. It was found that the extent of the hysteretic response with a spherical indenter increased with increasing load. Only minor differences in the force-displacement response of crystalline graphite indented normal to the basal plane and the glassy carbon was identified. A simple model for interpreting the force-displacement data is developed. These results are compared with indentations made on diamond like carbon films on various substrates.

INTRODUCTION

Carbon materials exist in a myriad of forms ranging from the strongest and hardness to amorphous and far more compliant structures. These materials have and continue to have a major role within the technological developments over the last few decades. Current interest is particularly focussed on materials such as high strength carbon fibres for composite applications, diamond like carbon films for protective coatings and glassy carbon for heart valves.

The two major bonding types of carbon materials are the very strong covalent SP^3 in diamond and the weaker Van der Waals forces (SP^2) between graphite sheets. The stiffness or modulus of carbon based materials is a reflection of these two bonding types and is most exemplified in the highly anisotropic graphite crystalline form. In the direction of the SP^2 bonding perpendicular to the lattice plane the stiffness is C_{33} = 36.5 GPa whereas parallel to the basal planes with the SP^3 bonding the value of the stiffness is C_{11} = 1060 GPa [1].

Despite such characterisation of the elastic constants for diamond and graphite, relatively limited information exists for the remaining materials, even less on the non-linear response of these materials. Most of the data that does exist beyond the elastic limit is for carbon/graphite materials for nuclear moderators, etc. In the course of material characterisation for that industry some interesting non-linear behaviour was observed. This aspect is featured in a review article by Kelly [2] who reports that when such materials were loaded in compression both permanent set and hysteresis occurred. Pyrolytic graphite exhibited stress-strain curves that when aligned to enable basal plane shear revealed a critical yield stress and a post yield response sensitive to strain history. There were also indications that the stiffness was not constant with strain. Jenkins [3] proposed an explanation based on basal plane shear dependence on slippage of internal elements opposed by frictional forces. This resulted in a non-linear constitutive relationship given by,

Mat. Res. Soc. Symp. Proc. Vol. 383 © 1995 Materials Research Society

$$\epsilon = A\sigma + B\sigma^2 \ (loading) \tag{1}$$

and

$$\epsilon_m - \epsilon = A(\sigma_m - \sigma) + B(\sigma_m - \sigma)^2 \ (unloading) \tag{2}$$

where ε_m is the maximum strain achieved at the maximum stress σ_m. On reloading it was proposed that

$$\epsilon - \epsilon_o = A\sigma + B\sigma^2 \tag{3}$$

where $\varepsilon_o = B\sigma_m^2$ is the permanent set. The physical basis for such relationships was not satisfactorily defined. Blackslee et al [1] measured the shear strength of a pyrolytic graphite as a function of normal applied stress and found that there was a linear increase with applied normal stress. Kotlensky and Martens [4] measured the deformation response of a pyrolytic graphite and proposed a model to explain the behaviour. Skinner and Gane [5] investigated the indentation response of graphite and found that significant interlaminar shearing as well as twinning of the planes took place.

In the case of glassy carbon materials the situation is even more complex as the structure consists of micro crystalline graphite like particles which have been modelled by different authors (e.g. Jenkins and Kawamura [6] and Shiraishi [7]). Previous hardness measurements of these materials reported by Jenkins and Kawamura [6] indicate a direct relationship between hardness and modulus (with $E/H \approx 18$) over a wide range of materials ($E = 0$ to 60 GPa). These authors also reported that the hardness was strongly dependent upon pyrolysing temperature and hydrogen content. They further indicated that following indentation glassy carbon materials showed complete recovery and virtual absence of a residual impression. A similar response was recently reported by Sakai et al [8,9] for Vickers indentations made with instrumented hardness impressions, although he did observe hysteretic response.

The aim of this paper is to provide a limited investigation of the behaviour of a range of carbon materials, particularly diamond-like films, glassy carbon, cokes and pyrolytic graphite. Indentations predominantly with small spherical tipped diamond indenters have been made with a precision instrumented micro-mechanical probe system. As discussed below these systems enable a more analytical interpretation of data than for pointed indenters.

BACKGROUND

As all the work presented here will focus on interpretation of force-displacement data, the relevant relationships for pointed and spherical indentation will now be considered.

Pointed indenters

For micro hardness determination based on depth measurement, a pointed indenter with an equilateral triangular base such as the Berkovich indenter is preferred to the four sided bases of the Vickers or Knoop indenters. With the Berkovich a sharper tip is possible with only three non-parallel planes intersecting at a single point.

Hardness

Interpretation of the continuous load-unload cycle data generated with Berkovich indenters is critically summarised by Mencik and Swain [10]. A schematic diagram of a typical loading and unloading force displacement curve as well as a cross-section of an impression at maximum load and upon unloading is shown in Figure 1.

The relation between depth of penetration and projected area is given by

$$A = kh_p^2 \tag{4}$$

where h_p is the effective plastic penetration depth, k = 24.5 for Berkovich and Vickers indenters. The simplest means of estimating the extent of permanent deformation is to project the tangent to the unloading curve to intercept the abscissa at h_p.

As the edge recovery is expected to be less than the central contact area, Oliver and Pharr [11] suggested the following procedure to account for this:

$$h_p' - h_p - 0.25(h_t - h_p) \tag{5}$$

From the value of h'_p, the maximum load and knowledge of the indenter geometry, the hardness is then calculated from the expression

$$H = \frac{P}{A} = \frac{P}{24.5(h_p')^2} \tag{6}$$

where P = the applied load. A more thorough consideration of this situation must also consider estimating the indenter penetration at the minimum load the testing instrument can detect as well as allowance for the non-perfect form of the indenter tip.

Elastic modulus

Loubet et al [12] proposed that the elastic modulus could be estimated using a model based upon Sneddon's [13] analysis of surface displacement for a flat punch on a semi-infinite half space. This approach has been the subject of further development [11] which enables determination of the elastic properties of the indented material by relating the composite elastic modulus of the indenter and the substrate to the slope of the unloading curve. The basic assumption of the data analysis is that the area of contact between the indenter and the specimen remains initially constant as the indenter is unloading. The

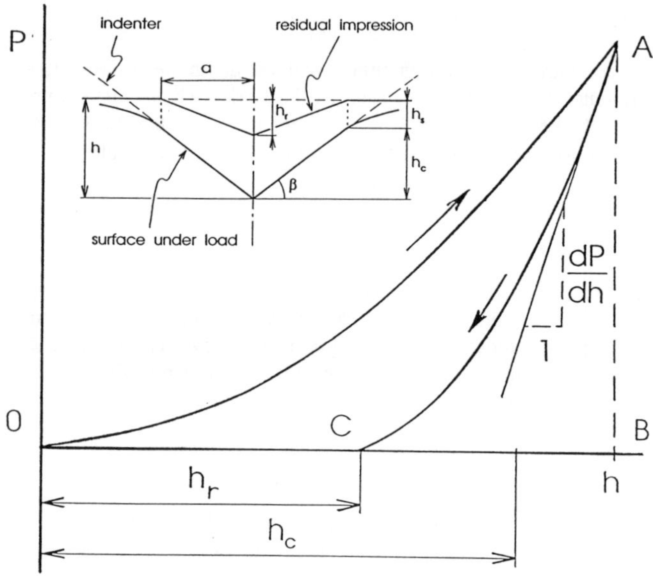

Figure 1. Schematic cross section of the deformation beneath a pointed (Berkovich) indenter and a typical force-displacement curve.

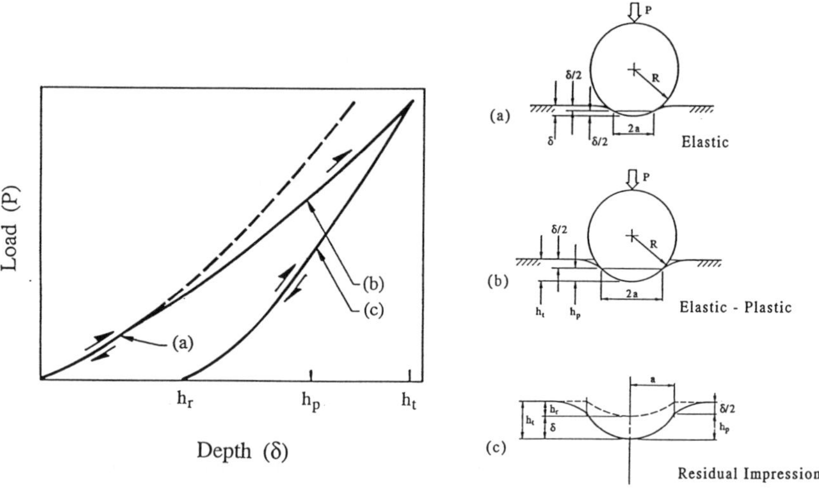

Figure 2. Schematic cross section of the deformation beneath a spherical tipped indenter and a typical force-displacement curve. a) Elastic contact, b) elastic-plastic, and c) residual impression.

slope of the unloading curve is given by:

$$\frac{dP}{dh} = \alpha E^* \sqrt{A} \tag{7}$$

where α is a constant for axi-symmetric indenters; for a Berkovich indenter α = 1.167 and E* is the effective modulus given by

$$\frac{1}{E^*} = \frac{(1-\nu_i^2)}{E_i} + \frac{(1-\nu_m^2)}{E_m} \tag{8}$$

where υ is Poisson's ratio and the subscripts i and m indicate the indenter material and the material of the infinite half space respectively. The modulus and Poisson's ratio of the diamond indenter are usually known whereas that of the indented material are not, hence only $E_m/1-\upsilon_m^2$ may be estimated for a material.

Spherical Indenters

For indentation with spherical indenters the load-partial unload cycle is used. The contact area and therefore the contact pressure (or Meyer's hardness) is calculated for every partial unloading step and provides a plot of hardness versus depth. The theory of spherical indentation is based on the early work of Hertz [14]. The use of a spherical indenter enables one to follow the transition from elastic to elastic-plastic behaviour in the material. The model for spherical indentation has been developed by Field and Swain [15]. A schematic illustration of the force displacement response and cross-sections of the resultant elastic and elastic-plastic behaviour during spherical indentation of a half sphere is shown in Figure 2a.

Hardness

With spherical indenters the hardness is defined as Meyer's hardness, P_m, that is the applied force over the circle of contact, as in

$$P_m = \frac{P}{\pi a^2} \tag{9}$$

where a = radius of the contact circle.

Hertz's solution is only valid for the initial stage of the indentation process, when the indentation is fully elastic. The radius of elastic contact is given by:

$$a^3 = \frac{(4kPR)}{(3E_m)} \tag{10}$$

where P = the applied normal force, R = the radius of the indenter, E_m = Young's modulus for the indented material and k = a dimensionless constant given by

$$k = \frac{9}{16}\left[(1 - \nu_m^2) + (1-\nu_i^2)\right]\frac{E_m}{E_i} \tag{11}$$

The penetration (see Figure 2b) of the indenter into the surface of the infinite half space is given by Johnson [14]

$$\delta = \left[\left(\frac{9}{16}R\right)(PE^*)^2\right]^{1/3} \tag{12}$$

where E* is given by equation (5). A critical feature of elastic spherical indenter deformation is that the diameter of contact area lies at a depth of penetration $\delta/2$, as shown in Figure 1b. The analysis of contact dimensions upon exceeding the yield stress is somewhat more complex. The simplest approach for estimating the radius of contact is based upon similar assumptions to those used for pointed indenters, namely that the force-displacement response during unloading is fully elastic and pile-up does not occur. As shown in Figure 2 the force-displacement curve has an inflection upon the onset of yielding, however around the contact area there is still elastic deflection (Figure 2b) of the surface. Upon unloading (Figure 2c) the response is elastic and results in a residual impression depth of h_r. The form of the unloading, or reloading, of the residual plastic impression is elastic and the expression relating force and displacement is given by equation (9) with the radius of the indenter replaced with the effective radius:

$$\frac{1}{R^*} = \frac{1}{R} + \frac{1}{R_{imp}} \tag{13}$$

where R_{imp} is the radius of the residual impression.

The elastic recovery of the residual impression is then half the unloading recovery, shown in Figure 2c. This then enables the contact radius to be estimated from Figure 2b as

$$h_p = h_t - \frac{1}{2}(h_t - h_r) \tag{14}$$

and the radius of contact is given by

$$a = \sqrt{2Rh_p - h_p^2} \quad (15)$$

The contact pressure is then estimated from equation (6).

Equations (11) and (12) show that a knowledge of h_t and h_r enables the contact radius and therefore contact pressure to be estimated. However, since the unloading response for a sphere is elastic and, as shown by Field and Swain [15], follows the relationship given by equation (9), that is $\delta \propto P^{2/3}$, the residual depth of the impression may be estimated from any two points of the unloading curve with the simple proportionality ratio:

$$(h_t - h_r)/(h_x - h_r) = (P_m/P_x)^{2/3} \quad (16)$$

where h_t is the displacement at maximum load P_m and h_x is the displacement at intermediate load P_x on the unloading curve. This enables h_r to be established for each step in the load partial unload curve. More recently Field and Swain [16] have extended this analysis to incorporate piling-up or sinking-in around the impression.

The total depth of the elastic displacement and the radius of contact a are, for small penetrations, related by

$$\delta = \frac{a^2}{R} \quad (17)$$

The mean pressure in equation (6) is then given by

$$P_m = \frac{P}{\pi \delta R} \quad (18)$$

Elastic modulus

The use of the multiple load-partial unload indentation cycle enables any deviation from elastic behaviour to be determined. The elastic modulus can be obtained with the aid of equations (9) and (14) and leads to the simple expression

$$E^* = \frac{3}{4} \cdot \frac{P}{ad} \quad (19)$$

where d is the elastic depth recovery (h_t - h_r).

Stress-strain behaviour

The load-partial unload cycle enables determination of the stress-strain behaviour of the material, where the stress is simply the contact pressure or the Meyer's hardness and where the representative strain, ε_r, following Tabor [17] is obtained from the expression

$$\epsilon_r = \frac{a}{R} \tag{20}$$

Tabor [17] showed experimentally that tensile or compressive strain was related to the representative strain by $\varepsilon = 0.2\varepsilon_r$.

MATERIALS AND METHODS

Materials

A range of carbon materials were investigated including: pyrolytic graphite, glassy carbon, coke and various diamond like carbon films with thickness from 50 nm to 8 μm on substrates of germanium, silicon and tungsten carbide. The latter were deposited by various techniques and were not well characterised.

Methods

Samples were of high quality with surface roughness typically less than 5 nm r.m.s. value. The materials were indented with spherical and Berkovich indenters attached to a UMIS-2000 micro-mechanical probe system. The maximum loads for investigation were varied to suit the material being indented, its thickness and type of indenter. Both continuous and load-partial unload indenting procedures were used to determine force-displacement data.

OBSERVATIONS

Typical observations of Berkovich indentations made on the various materials are shown in Figure 3. The results are shown in terms of increasing resistance to penetration. In all instances no specific evidence for cracking about the impressions was detected; in fact for many of the materials it was difficult to identify a residual impression after the indentation. The results were only slightly influenced by load and maximum depth of penetration, with most influence being seen at very shallow depths of penetration. For the very stiff materials the response in this regime was almost entirely elastic or reversible.

Observations with spherical indenters, the radius being chosen by the resistance to penetration, are shown in Figure 4. The order of results are the same as for Figure 3. All of the materials showed an initial elastic response followed by some non-linear behaviour. The extent of the hysteretic response for glassy carbon increased with increasing maximum load as shown in Figure 5. A clear feature of all these spherical indentations is that for most of the carbon materials there is complete or near complete recovery upon unloading, although the energy loss (area within loading and unloading curve) increases

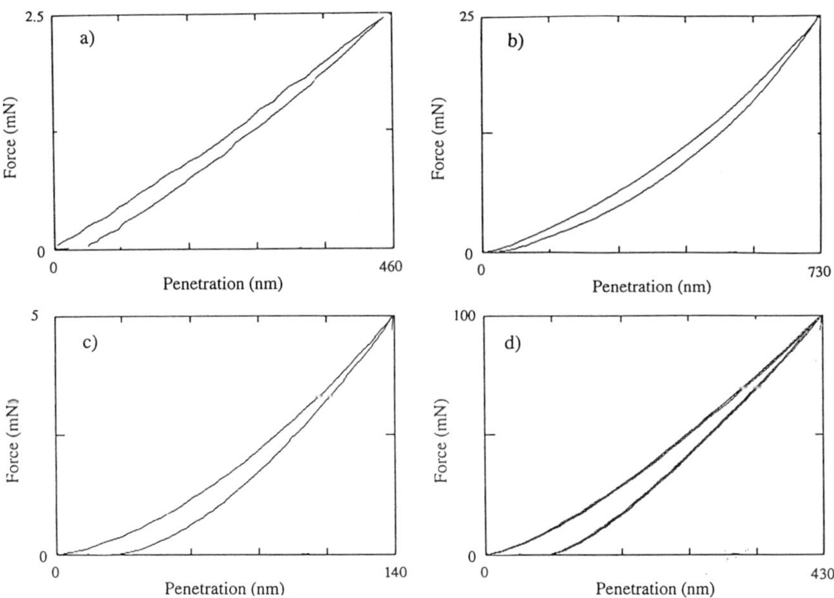

Figure 3. Force-displacement curves obtained with a Berkovich indenter on various forms of carbon, a) pyrolytic carbon, b) glassy carbon, c) diamond like carbon (DLC) films on germanium and d) WC-Co hardmetal.

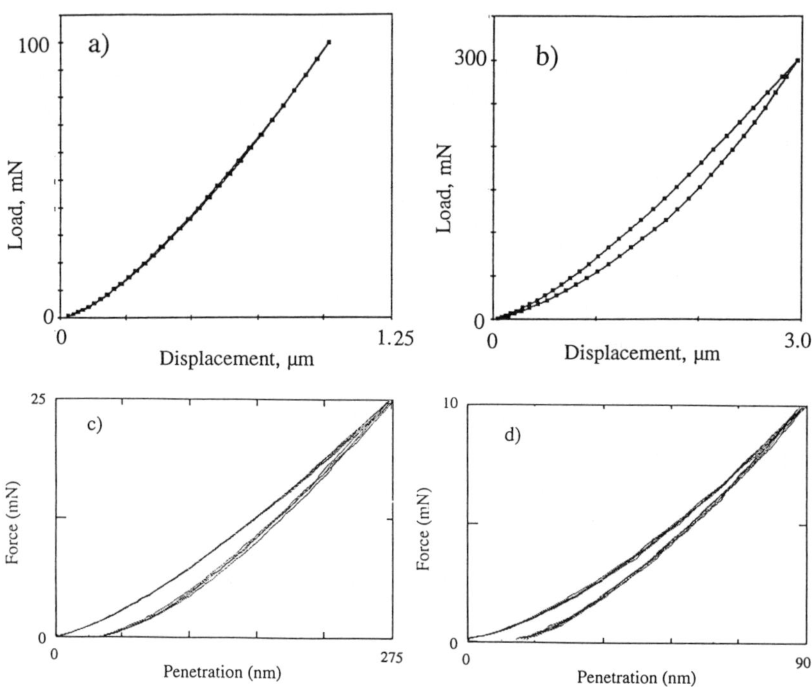

Figure 4. Force-displacement curves obtained with a small spherical tipped indenter on various forms of carbon, a) pyrolytic carbon, b) glassy carbon, c) diamond like carbon (DLC) films on germanium, and d) WC-Co hardmetal..

with maximum load. For the DLC materials there is a slight but permanent set detectable, the extent of which does not change appreciably with increasing load upon exceeding the elastic limit.

DISCUSSION

The present observations with both pointed and spherical indenters show common trends for all the carbon materials. In all instances there is very little evidence of plastic deformation. The near complete recovery upon unloading shows very limited energy loss throughout the load/unload cycle. The only difference between the materials is that of the stiffness. The virtual absence of any permanent set indicates that the H/E ratio is very high and that the usual relationships and approximation used to analyse elastic/plastic materials must be treated with caution. In both instances, considering the contact as entirely elastic would be a better approximation than that associated with elastic/plastic response. In the case of pointed indenters whose depth of penetration significantly exceeds the tip radius influence (> 100 nm for a Berkovich indenter) the relevant expression for a conical indentation approximation (for a Berkovich indenter) is given by a simple elastic analysis, namely the depth of the contact diameter is given by

$$h_p = h_r + 2/\pi[(h_t - h_r)]$$

where h_t is the total depth of penetration and h_r is the residual penetration depth. based upon this expression suggests that the contact pressure at maximum load range from 0.8 GPa, 3.0 GPa (glassy carbon), 16.4 to 38 GPa (DLC films on Ge and WC/Co). These values compare with 1.3 GPa, 3.2 GPa 17.3 and 58 GPa, respectively for the same materials using the traditional analysis for the Berkovich impression force-displacement data given by equation 6. A very simple procedure to accommodate the rounded nature of the "rounded" tip is described by Mencik & Swain [10]. Analysis of the force-displacement data may also be pursued following the energetic approach outlined by Sakai [8]. However, this approach does not lead to a more basic appreciation of the observed hysteretic response, which was also noted by Sakai for glassy carbon materials.

The spherical tipped indentation force-displacement responses for the various carbon materials do provide a basis for a more critical interpretation of the results. These observations show a number of distinct trends. At modest loads the response is virtually completely elastic and readily described by the classic Hertzian relationship (equation 10).

With increasing maximum load the extent of the hysteretic response becomes more significant, but despite extremely high pressures being achieved at maximum load there was little to no indication of permanent set. This feature is clearly identified in Figure 5, which shows a family of curves made on glassy carbon at increasing loads. If we consider just one of these curves in more detail (Figure 6) we can see that the loading curve initially follows the elastic response before deviating at ~ 100 mN. One can draw another "elastic" loading line (indicative of a lower E modulus) through this upper point that although not shown here bisects the load and unloading curves. It is tempting to ascribe the response of the glassy carbon material as that of a continuously decreasing modulus during the loading and unloading. The modulus may be estimated directly from equation 12, and assuming constant R leads to a continuously decreasing value of E beyond the "yield" point. This change in modulus with strain is in accord with that proposed by Jenkins [3]. It was also considered that this hysteretic response may have been due to a time dependence or creep response - however, creep tests showed no discernible increased penetration. Furthermore, reloading of the same location generated precisely the same

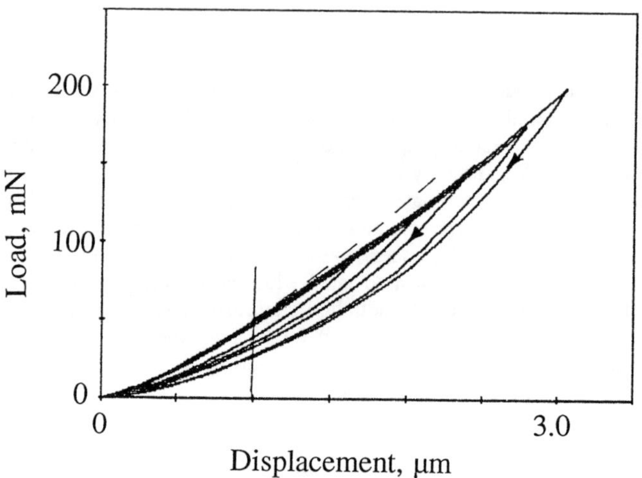

Figure 5. Influence of maximum load on the hysteretic response of glassy carbon indented with a 10 micron radius tipped spherical indenter. This figure is the superposition of a number of tests made at increasing maximum loads.

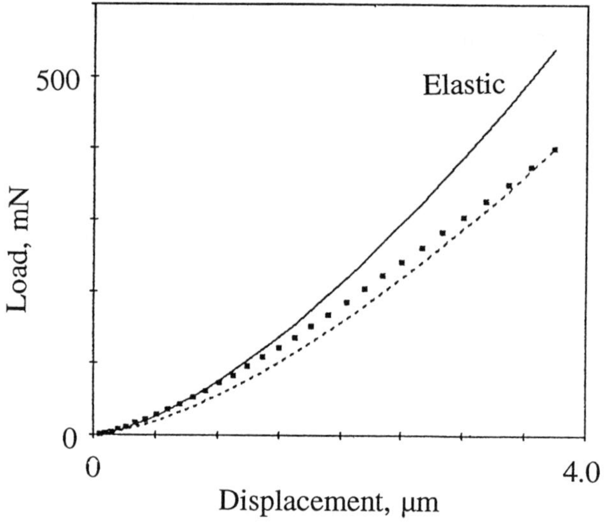

Figure 6. Comparison of the elastic and measured force-displacement response for the glassy carbon material indented with a spherical tipped indenter.

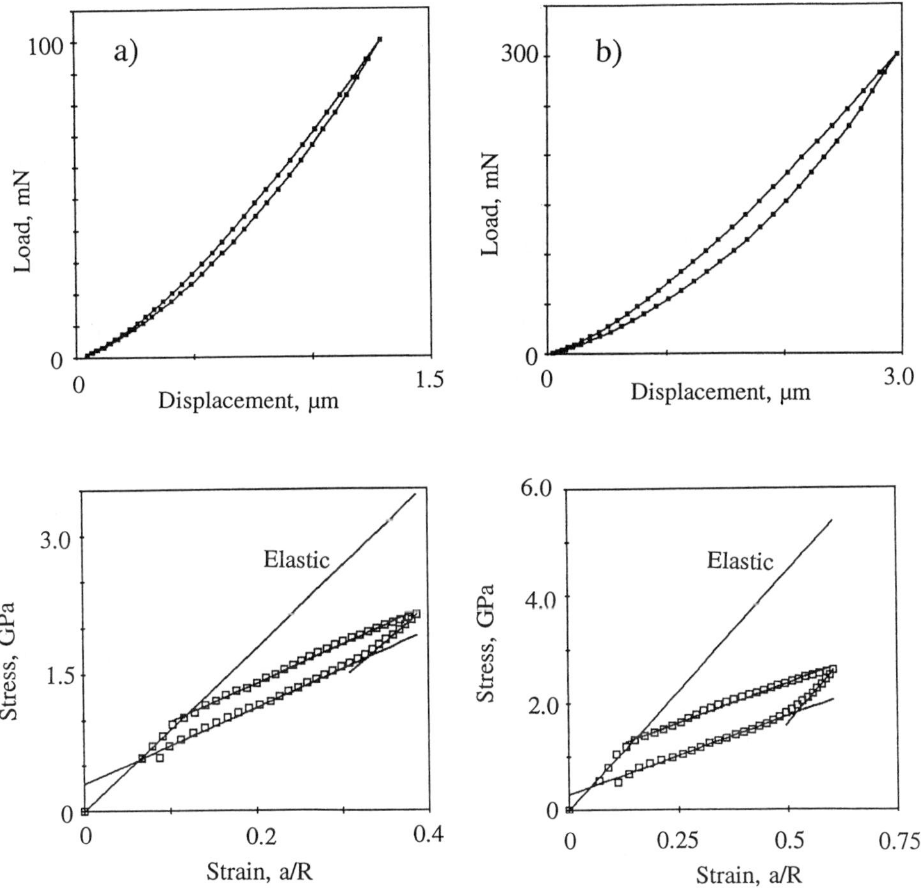

Figure 7. Two examples of force-displacement curves and the resultant calculated stress-strain response of glassy carbon during spherical indentation.

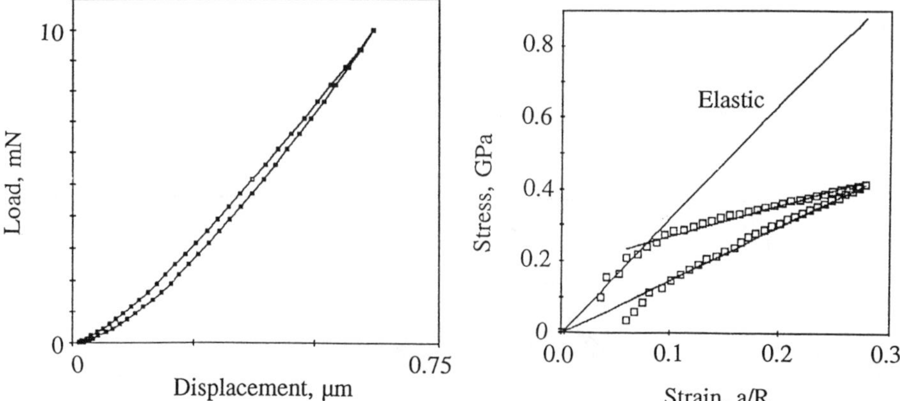

Figure 8. Force-displacement response and calculated stress-strain curve for a pyrolytic graphite indented with a spherical tipped indenter.

loading and unloading response indicating that no "damage" or modulus reduction had been achieved by these external pressures.

A mechanism is now proposed that enables us to rationalise the observed hysteretic response of these materials without recourse to a changing modulus. Firstly, we consider a simple model for pyrolytic graphite that is extendable to glassy carbon if one considers the latter to be a (nano) polycrystalline graphite material. Consider a pyrolytic graphite material indented normal to the graphitic sheet structure. Usually, graphite readily shears parallel to the basal plane by easy dislocation motion. However, with the superposition of a normal force to the graphite layers the shear resistance to slip is enhanced by a frictional component $\tau_y = \tau_{yo} + \mu F$ where μ is a "frictional" resistance term. The situation beneath a spherical indenter is that shear will attempt to occur by sliding of layers with dislocations propagating between the planes. With increasing normal load the above simple relationship implies that the yield or contact pressure should increase. Upon unloading, the frictional resistance to dislocation motion would be reversed, the rigid clamped sheets would attempt to return to their initial pre-loaded position. Initially this would mean that the permanent set ("plastic" strain) would be retained (although with elastic response decreasing) until the restraining frictional force (normal force) was sufficiently reduced.

A simple means of appreciating this response is to decompose the force-displacement data as shown in Figure 6 into two components, elastic plus "plastic". The displacement at each force step maybe analysed following the procedure outlined by Field and Swain [15] to determine the radius of contact and contact pressure (equations 14, 15 and 9). An effective indentation strain may also be estimated from equation 20. Typical results for this procedure, carried out for the complete loading and unloading cycle for the glassy carbon, are shown in Figure 7.

This figure shows two examples of increasing hysteretic response and the resultant stress-strain curve. The initial curve is elastic until a stress of 1.2 to 1.4 GPa, whereupon a steep linear hardening region develops to the maximum load. Upon unloading the response is elastic with virtually identical slope to the initial loading curve until a "reverse yield" occurs again followed by a linear region parallel to the hardening response until almost the original elastic loading line is met. There is some suggestion that the curves do not exactly coincide at the lowest load but this may be a consequence of noise limits of the instrument noise or surface asperity deformation.

Analysis of a pyrolytic graphite curve shows a similar trend (Figure 8), although in this case the unloading curve does not show the initial elastic unloading response with a "reverse yield" point. A more critical appraisal of this behaviour will be published elsewhere [18].

The observations for the pyrolytic and glassy carbon lend support for the deformation model proposed by Kotlensky and Martens [4]. In their model they proposed that basal slip was the primary mechanism of deformation and that even in poorly aligned crystal the deformation led to a preferential realignment of the graphitic planes. It is suggested that a similar mechanism is occurring for the glassy carbon materials but at a much finer scale than proposed by Kotlensky and Martens.

In the case of the DLC films where there was less evidence of a reversible response recourse is made to the load partial unload technique [15] for analysis of the spherical indentation results. The contact pressure versus depth and modulus versus depth for the films are shown in Figure 9. These observations suggest that the maximum contact pressure (hardness) lies between 15.7 and GPa with both the DLC films on silicon and germanium having the lower value. The composite modulus E* for the films varies from

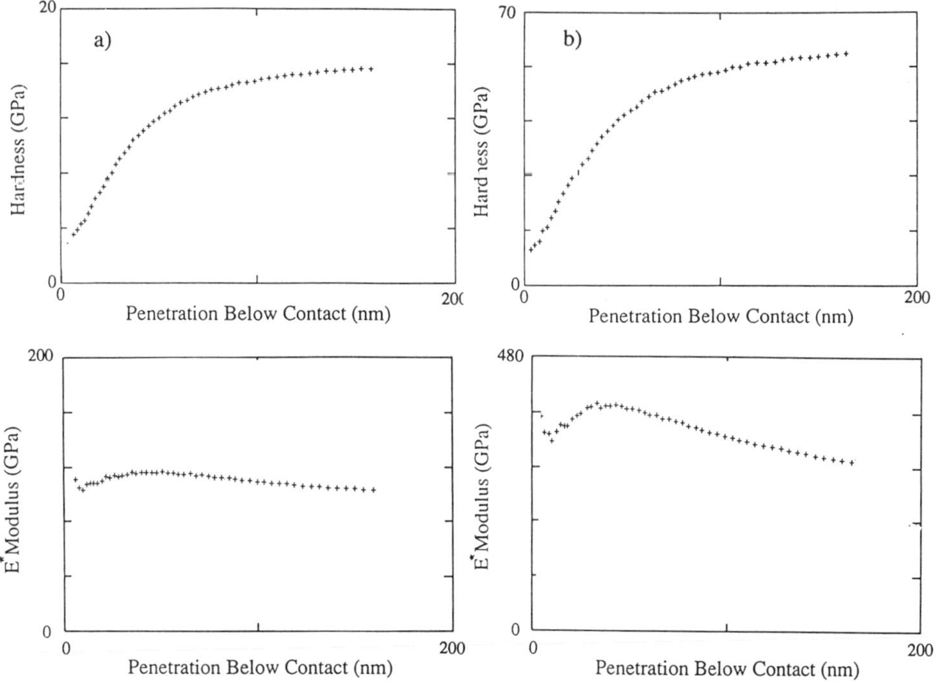

Figure 9. Estimates of the contact pressure and elastic modulus versus depth for the DLC materials, a) 1 μm film on Ge and b) 8 μm film on WC-Co hardmetal.

118 to 405 GPa with the film on the germanium substrate having the lower value. A critical appraisal of the hardness and modulus indicates that the contact dimension to the film thickness is an important consideration if a "true" indication of the film properties are to be measured [19].

CONCLUSIONS

The present observations have shown that a range of carbon materials exhibit interesting indentation force-displacement responses with both pointed and spherical tipped indenters. The common feature of all these materials is the substantial elastic recovery upon unloading, a feature that poses limitations for the usual analysis for pointed indenters. Of particular interest was the behaviour of glassy carbon which showed almost complete recovery upon unloading despite a considerable hysteretic response. The analysis of spherical indentations of this material and the pyrolytic graphite enabled a rational explanation of such behaviour and enabled a complete stress-strain curve to be generated. This curve showed an initial elastic response followed by a steep linear hardening response to the maximum load, and upon unloading there was an initial elastic response followed by a reverse yield point and a reversal of the steep hardening curve to the initial elastic loading line. These observations are in essential agreement with a model proposed by Kotlensky and Martens [4] for the deformation of pyrolytic graphite. Analysis of the spherical indentation data for the DLC materials suggests that the substrate plays an important role and that the hardness and modulus measurements using this approach are more realistic than many claimed after using pointed indentation analysis.

REFERENCES

1. O.L. Blackslee, D.G. Proctor, E.J. Seldin, G.B. Spence and T. Weng, J.Appl. Phys., 41, 3373 (1970).
2. B.T. Kelly, "The Physics of Graphite", Appl. Sci. Publ., Lond. (1981).
3. G.M. Jenkins, Brit. J. Appl. Phys.,13, 30 (1962).
4. W.V. Kotlensky and H.E. Martens, J. Amer. Ceram. Soc., 48, 35 (1965).
5. J. Skinner and N. Gane, Phil. Mag., 28, 827 (1973)
6. G. M. Jenkins and K. Kawamura, "Polymeric Carbons- Carbon Fibre, Glass and Char", Camb. Uni. Press (1976).
7. M. Shiraishi, Kaitei Tansozairyo Nyuma. Carbon Soc. Japan Tokyo 29 (1984)
8. M. Sakai, Acta. Metall. & Mater., 41, 1751 (1993).
9. M. Sakai, H. Hanyu and M. Inagaki, Proc. Int Symp. on Carbon, Tsukuba, Japan (1990).
10. J. Mencik and M. V. Swain, Materials Forum 18, 277 (1994).
11. W. C. Oliver and G. M. Pharr, J. Mater. Res., 7, 1564 (1992).
12. J-L. Loubet, J.M. Georges, J. Marchesini and G. Mielle, J. Tribol. 106, 43 (1984).
13. I.N. Sneddon, Int. J. Engng. Sci., 3, 47 (1965).
14. K.L. Johnson, "Contact Mechanics", Camb. Uni. Press (199?).
15. J.S. Field and M.V. Swain, J. Mater. Res., 8, 297 (1993).
16. J.S. Field and M.V. Swain, J. Mater. Res., 10, 101 (1995).
17. D. Tabor, "Hardness of Materials", Oxford Uni. Press (1951).
18. J.S. Field and M.V. Swain, in preparation.
19. H. Gao, C.H. Chiu and J. Lee, Int. J. Solids and Struct., 29, 2471 (1992).

THE MECHANICAL PROPERTIES OF CVD DIAMOND FILMS, AND DIAMOND COATED FIBRES AND WIRES

E D Nicholson*, J E Field**, P G Partridge***, MNR Ashfold*
*School of Chemistry, University of Bristol, Bristol BS81TS UK. **Cavendish Laboratory, University of Cambridge, Cambridge CB3OHE UK. ***Interface Analysis Centre, University of Bristol BS28BS UK.

ABSTRACT

Two areas of thin film property measurement are addressed. The first is that of flat films, either on a substrate or free-standing. The film properties only are of interest. Therefore, when the film remains attached to a substrate during testing, an appropriate analysis is used to subtract the effect of the substrate. The films under test are prospective protective coatings and 'window' materials for infrared applications, namely CVD diamond (Hot filament Assisted, HFACVD and Microwave plasma assisted, MPACVD) and Germanium carbide (Ge:C). The mechanical properties under investigation are the Young's modulus and the internal film stress.

In the second case the substrates are small diameter fibres and wires coated with CVD diamond. The mechanical properties measured were composite, containing contributions from both the substrate and the film. These coated fibres and wires, have possible applications as reinforcement phases in the production of composites. They are silicon carbide (SiC) and Tungsten (W) of diameters varying between 10 and 125μm. A technique has been developed to measure the Young's modulus of individual coated fibres.

MECHANICAL PROPERTIES OF THIN FILMS

The environments to which infrared (3-5μm and 8-12μm range) transparent 'window' materials used in airborne applications may be subjected include rain and solid particle bombardment and temperatures up to and possibly exceeding 900°C. The use of many bulk infrared materials (e.g. Ge, ZnS) is limited, as they cannot survive such conditions. In order to insure survival one of two routes can be taken; coating existing infrared materials with a resistant film, or replacing the window material with another of superior mechanical properties. Infrared materials tend to lack mechanical strength. There are few exceptions, these include diamond and Ge:C. Young's modulus and internal film stress are two of the properties of most interest.

Young's modulus

For flat films a variety of methods have emerged for measuring the Young's modulus, some of which are reviewed briefly below. A detailed review of techniques is given elsewhere[1].

The Young's modulus can be determined from stress-strain properties of the material. An example is the uniaxial tensile test[2], which has been used widely for measuring the modulus of metal foils, but not for very thin films produced by vapour deposition. Beam bending methods[3]

Mat. Res. Soc. Symp. Proc. Vol. 383 © 1995 Materials Research Society

have also been applied to thin film modulus measurement, the basis of which is the measurement of the deflection, or change in radius of curvature, of a sample by application a known load. This information, together with the sample dimensions is used to calculate the modulus. The sample can either be deflected centrally whilst being supported at both ends (3-point bend), or loaded at one end with the other rigidly held (2-point bend). The modulus of thin films deposited on substrates can be measured using this technique from the difference in bending behaviour of the composite (film + substrate) and the uncoated substrate.

In the bulge test the stress-strain properties of the film are measured by applying pressure to one side of a free-standing membrane (usually circular but can be square) and measuring the resultant deflection as a function of the applied pressure. This technique is applicable to many film types including diamond[4]. For very thin films on substrates, the substrate can be etched away over a circular region producing a hole through which the film can be bulged. Alternatively, thicker free-standing films can be pressurised through a die of known radius[5]. Pressure can be applied in various ways depending on the modulus, thickness, size of membrane and the degree of deformation required. Techniques include use of a syringe, compressed air, or applying a vacuum to one side of the membrane. Methods of measuring the applied pressure include a calibrated water manometer and a silicon pressure transducer. The height of the pressurised membrane can be measured using techniques such as interferometry, microscopy, displacement probe and surface profilometry. The bulge test has the advantage of simultaneously measuring the internal films stress and, if the film is bulged to failure, the fracture strength[5].

In recent years indentation methods have been extended to the measurement of both hardness and Young's modulus of thin films, through the development of low load indenters known as nano-indenters. The indents produced are nano-sized, thus their depth is only a small proportion of the film thickness (typically < 10%). This should allow the properties of the thin film to sampled without influence from the substrate material. Nano-indentation has been used to characterise natural diamond and diamond films[6-8].

Resonance methods of measuring modulus are a popular choice. Their basis lies in the measurement of the natural or resonant frequency of a sample which together with knowledge of the sample density and dimensions, yields the Young's modulus. Such methods have been widely used for bulk materials, and have more recently been applied to thin films including diamond.

The most accurate resonance methods impose the minimum constraint on the sample. The simplest method of excitation is by dropping the sample onto a surface. This principle was used by Hunt[9], who placed rectangular samples in a horizontally mounted drum which was then rotated. When airborne, after hitting the side of the drum, the sample vibrates freely emanating a tone. The vibrational modes of the sample can be detected using a microphone. Knowing the frequencies of these modes the Young's modulus can then be determined using the appropriate analysis.

Another resonance method involves supporting the sample at the nodes for a particular mode of vibration. This can be achieved by suspending the sample (rectangular of cylindrical) by a material that will not restrict its vibration (cotton thread or silica fibre)[10,11]. Resonance can then be produced in a variety of ways e.g. electrostatically or by acoustic waves. Alternatively, the sample can rest on supports positioned at the nodes[10]. These supports may be knife edges, or foam blocks for larger samples. Resonance can be determined, for example, using a record styli or by piezo-electric, capacitive or optical detection. This technique has been used to measure the modulus of diamond coated films as a function of temperature[12]. Samples can also be clamped at one or both ends and excited into vibration, as in the vibrating reed test[1]. Excitation and detection of resonance can be achieved in various ways, as described above. The modulus of very thin films, that cannot survive without a supporting substrate, can be measured by any one of the above methods. This is achieved by first measuring the resonant frequency of the composite, then repeating the experiment for the substrate alone. The shift in resonance frequency that results can be used to calculate the Young's modulus[13]. The measurement of the resonant frequency and hence the Young's modulus of circular membranes is also possible, using a similar technique[14]. The bulge test, the vibrating reed test and nano-indentation were chosen for further investigation.

The bulge test

Membranes were produced by a cold anisotropic etching technique, at room temperature. The etchant was a 1:1 mixture of HF (40%) and HNO_3. Five membranes were produced from MPACVD diamond (with thickness varying between 1-9μm); and one from HFACVD diamond of thickness 5.8μm. The samples were 16mm x 16mm squares of silicon with the film deposited on one side. Each had a central membrane of ~6mm in diameter. The samples were clamped within a chamber (figure 1), and pressurised (up to 80 mbar) using a syringe. The pressure was measured using a water manometer. Interferometry was used to monitor the membrane deflection.

The appropriate mathematical interpretation relating the central bulge height to the modulus of the film, will depend on the shape of the bulged membrane. In this investigation it was approximately spherical and therefore the spherical membrane model, originated by Beams,[15] was used. The film stress, σ, and strain, ε, are expressed by the following equations

$$\sigma = \frac{Pa^2}{4tY_{max}}, \tag{1}$$

$$\varepsilon = \frac{2Y_{max}^2}{3a^2}, \tag{2}$$

where P is pressure, a is membrane radius, t is film thickness and Y_{max} is the central membrane deflection. Stress over strain yields the biaxial modulus = $E/(1-\nu)$, from which the Young's modulus E can be calculated, if the Poisson's ratio ν is known. A further refinement by Cabrera (referred to by Beams) takes into account internal tensile stress in the film

$$P = \frac{8tY_{max}^3 E}{3a^4(1-\nu)} + \frac{4tY_{max}\sigma_0}{a^2}, \tag{3}$$

where σ_0 is the internal film stress. Plotting pressure versus deflection, and fitting the curve to a polynomial with first and third- order terms, yields Young's modulus and the internal stress.

During this investigation it was discovered that four of the six films tested were compressively stressed. This technique is not applicable to such films. The results for the remaining two samples are shown in table I. Unfortunately the pressure capabilities did not allow the thicker film to be bulged sufficiently for the modulus to be measured[1], however it was possible to measure the internal stress. The errors in measurement of pressure, deflection and variation in membrane thickness, amounted to a total uncertainty of ± 8%.

Table I. Sample details and moduli measured by the bulge test

Sample type	Film thickness (μm)	Young's Modulus (GPa)	Internal stress (MPa)
HFACVD diamond	5.8	1040	100
MPACVD diamond	9	------	365

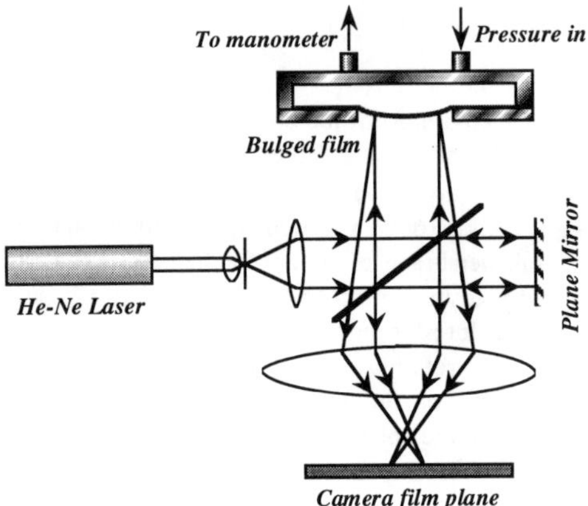

Figure 1. Schematic of the bulge test.

The vibrating reed test

The samples tested were CVD diamond films on substrates, free-standing diamond, single crystal silicon and Ge:C. The samples were rectangular, of lengths varying between 15-50mm, width 3-5mm, and were clamped at one end to form a capacitor with a reference electrode. Vibration was produced electrostatically and detected by means of a Michelson interferometer (figure 2). Resonance was detected by placing a photodiode behind a pinhole in a screen on to which the interference pattern falls; the pinhole is positioned at the edge of a bright fringe. When the frequency of the excited signal equals that of the resonant frequency of the beam, the amplitude of vibration passes through a maximum. This is observed as a blurring out of the interference pattern and corresponds to a maximum in the amplitude of the signal and a phase shift, both of which can be seen on an oscilloscope. The actual resonant frequency is, due to the nature of the test, twice that of the applied electrical frequency as the beam is electrostatically attracted to the reference electrode with either voltage polarity. This frequency is then used to calculate the modulus. To achieve an accurate determination of modulus the samples were successively shortened, and the resonant frequency measured as a function of length, l. A plot of frequency versus l^{-2} was drawn. The slope of this plot is related to the Young's modulus by equations, the precise form of which depend upon whether the film is free-standing or composite[16]. The equation for a monolithic beam is

$$E = f_{res}^2 l^4 \frac{48\pi^2 \rho}{t^2 (1.84)^4} \qquad (4)$$

where f_{res} is the resonant frequency and ρ, l and t are, respectively the density, length and thickness of the beam.

The modulus of single crystal silicon of two thicknesses was measured, as a standard. The results (table II) agree well with literature values. Details of other samples under test and their moduli are also given in table II. The major source of error, for all the samples tested, was due to uncertainty in the clamped length and poor film quality. Perfect sample clamping is difficult and, as the length appears as the fourth power in the equation for modulus, small errors in the measured length result in much larger errors in the modulus. Variation in sample thickness along their length also produced errors. The worst case was for the free-standing diamond films, which is reflected in the total predicted error (15-20%). The maximum error in the measured modulus for the single crystal materials was ~5%, with the error for the coated substrates slightly higher at 7-10%.

Table II. Sample details and Young's modulus measured by the vibrating reed test.

Sample type	Film thickness (μm)	Dimensions (mm)	Young's Modulus (GPa)
Free-standing polycrystalline MPACVD diamond			
D1 (3 samples)	150-180	30 x 3	802 ± 20
D2 (3 samples)	230-270	30 x 3	795 ± 15
D3 (4 samples)	230	12 x 4	760 ± 20
D4 (2 samples)	770-775	30 x 4	860 ± 50
Composite beams, polycrystalline MPACVD diamond on silicon (3 samples)	1-8	50 x 5	785 ± 15
Single crystal silicon			
(100) <100>,385μm (3 samples)		50 x 5	129.35 ± 4
(100) <100>, 532μm (3 samples)		50 x 5	129.65 ± 5
RF-Plasma sputtered Ge:C on Si (6 samples)	13.8 - 15	50 x 4	290 ± 30

errors are in the mean

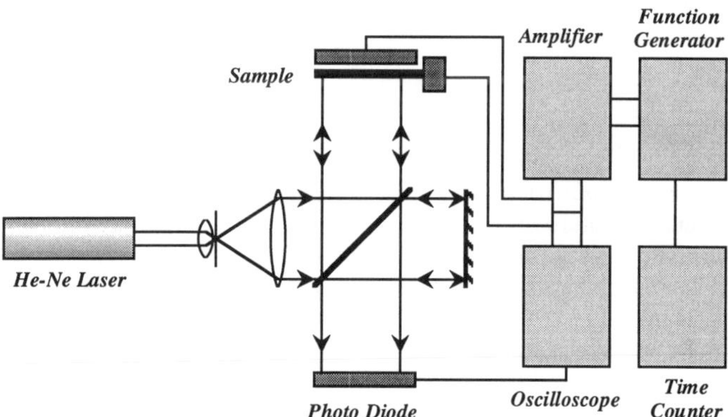

Figure 2. Schematic of the vibrating reed test.

Nano-indentation

The measurement of modulus of thin films by nano-indentation involves continuously recording the depth of the indentation as a function of load, during both loading and unloading[17]. The Young's modulus is extracted from the elastic region of the curve produced, whilst the plastic region yields the indentation hardness (H). Tests were carried out using a Nano-Test instrument (Micromaterials Ltd.). The indenter has a 90° trihedral diamond tipped stylus. 15 indents were carried out on each sample, from which a continuous plot of the load versus depth during loading and unloading was produced. These appear as a hysteresis curve, from which the hardness, plastic depth and the elastic recovery parameter can be determined. These values together with knowledge of the indenter shape, can be used to calculate the Young's modulus using the following equation[17],

$$E = \frac{H(1-\nu^2)\sqrt{\pi K}}{2R}, \tag{5}$$

where R is the elastic recovery parameter and K is the indenter shape parameter which, for a 90^{0} indenter, is 2.6 x $(\text{depth})^2$. Analysis of polycrystalline diamond films by indentation proved unsuccessful, due to indenter blunting and fracturing. Only the first few indents yielded values. Several samples showed two distinct values of modulus or hardness, suggesting separate regions of differing properties .Samples details, the Young's modulus and hardness are given in table III

Table III. Sample details, Young's modulus and Hardness measured by indentation.

Sample Type	Film thickness (μm)	Young's Modulus (GPa)	Hardness (GPa)
MPACVD diamond on Si	11.0	1240	83.1
HFACVD diamond on Si	5.8	1070, 830	80.3, 52.5
RF-sputtered Ge:C on Si	14.6	205 ± 4 300 ± 6	16.4 ± 0.2
RF-sputtered Ge:C on Ge	9.4	325 ± 13	19.8 ± 0.4

Internal stress

Curvature methods have proved popular for determining the internal stress of thin films on substrates. If the initial curvature of the substrate before the film is deposited is known, the film stress can be determined from the distortion of the substrate due to the presence of the film on its surface. Analyses for relating both the curvature of the samples in the form of long thin beams and a disc have been described. Methods of measuring curvature include interferometry and surface profilometry. Curvature methods are applicable to both compressively stressed and tensile films.

In the calculation of stress, an improved theory was used to determine the internal stress of beam shaped samples. The theory behind this analysis is presented elsewhere[1]. In brief, the analysis assumes that the internal stress is defined by the directions and magnitudes of the principle components of stress in the plane of the specimen, and that the principle stress axes coincide with the principle axes of curvature. Thus by measuring the principle curvatures of the film the principle stresses and their directions can be determined. The equation for stress in the x direction, where x is a direction of principle curvature of the beam, is given by,

$$\sigma_x = \frac{E_s}{6(1-\nu_s^2)} \frac{\phi_x + \nu_s \phi_y}{t_f} [6(t_f + \frac{t_s}{2}) - t_s^2], \quad (6)$$

where ϕ_x and ϕ_y are the measured curvatures of the substrate in the x and y directions respectively and t_s and t_f are the thicknesses of the substrate and film respectively. The stress in the y direction can be obtained by interchanging the subscripts x and y.

The curvature of coated and uncoated substrates was measured by both profilometry and interferometry. Sample details and measured stress are shown in table IV. Where stress in the x and y directions are approximately equal, the film stress was considered isotropic. A Sloan Dektak II profilometer was used to traverse the central regions of the samples. The length of the traversed segment was between 3 and 20mm depending on the sample size.

A phase-stepping interferometer was also used to determine curvature. In brief it is an automated method of analysing interference patterns, which is more accurate and which dramatically reduces the time required to analyse data[18]. Using this method the curvature, and hence the stress, in any direction across the surface can be determined. The phase stepped apparatus can be seen in figure 3. A beam splitter divides the light from a He-Ne laser. One half travels directly along the optical fibre to the beam collimator and the other half is phase-stepped. The light then continues down the optical fibre to the beam collimator. Half this light travels to and is reflected from the sample to the beam splitter, where it is reflected in the direction of the camera. The unshifted beam and the sample beam combine producing an interference pattern which is focused on to a video camera plane. Only Ge:C films were tested by this method. Stress was found to be isotropic, and of similar magnitude as measured by profilometry (table IV).

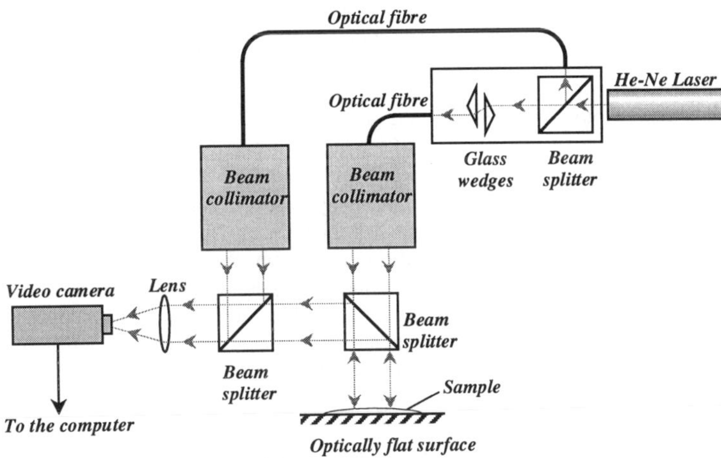

Figure 3. Schematic of the phase-stepped interferometry apparatus

Table IV. Sample details and stress values measured by profilometry and interferometry.

Sample Type	Substrate Dimensions (mm^3)	Film Thickness (μm)	Internal Stress by profilometry (GPa)	Internal Stress by Interferometry (GPa)
MPACVD diamond on (100) silicon sample S5	**50x50x 0.385**	**2.4**	**0.353 ± 0.003 (x)** **-0.082 ± 0.001 (y)**	
S4		**7**	**0.375 ± 0.015 (x)** **0.078 ± 0.001 (y)**	
RF-sputtered Ge:C on (100) silicon	**50 x 4 x 0.532**	**14.1-15**	**0.050 - 0.058 (x,y)**	**0.049 - 0.056 (x,y)**

MECHANICAL PROPERTIES OF DIAMOND COATED FIBRES AND WIRES

The continuous effort to develop new materials with improved properties has sparked an interest in diamond coated fibres and wires[19-21], which is ever increasing due to many prospective applications. One such application is in the production of composites, for aerospace applications. In the aerospace industry, rigidity and weight savings are of concern, consequently lighter, stronger, stiffer materials are always in demand. A substantial increase in fibre Young's modulus can be expected by diamond coating existing reinforcement fibres.[21]. These coated fibres, once incorporated into a composite, should transfer their stiffness to the composite. This suggests that fewer fibres would be required in the composite in order to produce significant increases in stiffness compared with uncoated fibre reinforcements. This, together with diamond's fairly low density, may lead to significant weight savings. In addition, its extremely high thermal conductivity and chemical inertness widen the scope of applications for such a product.

Reinforcement phases can be particles, whiskers, chopped fibres or continuous fibres, while the matrix can be polymeric, metallic, intermetallic or ceramic. In this investigation, metal matrix composites, reinforced with unidirectional, continuous fibres are of interest. Such composites offer outstanding unidirectional mechanical properties, and are particularly attractive for components that exploit these anisotropic properties. Candidate reinforcement fibres include tungsten (W) wire and silicon carbide (SiC) fibre, of diameters ranging from 10μm to 125μm. Tungsten fibres have been used to reinforce superalloys for heat engines and offer the potential for significantly raising hot component operating temperatures and therefore improving heat engine performance[22]. SiC fibres incorporated in Ti alloy, or Ti aluminide matrices, already in use in the aerospace industry, can lead to significant improvements in specific strength and stiffness of components at ambient and elevated temperatures, with predicted weight savings up to 75%[23].

Matrix materials of interest include metals such as aluminium, magnesium, copper and titanium. Al and Mg are favoured due to their high strength to density ratio, while a Cu matrix composite may be suitable for production of high strength thermal conductors.

The mechanical properties of most interest in this article are the Young's modulus and strength, both of individual reinforcement fibres and of the resulting composite material. Due to the small size of the coated fibres (10-270 μm diameter) mechanical property measurement is difficult. This investigation concentrated on the measurement of Young's modulus of individual fibres.

Production of diamond coated fibres and wires: The fibres and wires were coated by HFACVD (Thomas Swan reactor). The sample wires were positioned parallel to and surrounding a vertically held tantalum filament, approximately 5-6mm distant from the filament. The filament was made up of 5 individual filaments, of total length 12-13 cm. Samples were heated by the filament only, the temperature of which was measured using an optical pyrometer. A maximum of 14 substrate fibres/wires may be coated in any one run. The workable sample length is 10-11cm. The samples were tensioned during coating. Deposition was carried out under the following conditions: 20 Torr pressure, 0.75 CH_4 in H_2, Filament temperature 2150-2180^oC. Deposition times to date have been ≤72 hours, producing coatings of up to 70μm in thickness.

Young's Modulus

Many of the tests mentioned above for flat films are also applicable to long, small diameter fibres. The Young's modulus of high modulus single fibres and wires (including diamond coated fibres), has most often been measured using the tensile test[24], were E is calculated from the load-elongation of a sample of known cross-sectional area. By stressing the fibres/wires to failure the tensile strength can also be measured. However, difficulties have been experienced in gripping the sample successfully and measuring the very small strains experienced by very stiff fibres.

Bend testing to measure modulus of rod shaped samples, such as coated fibres and wire should also be feasible. Two-point and three-point bend tests have been used to measure stiffness of nickel-titanium alloy wire, of both rectangular and circular cross section[25,26].

From the resonance tests previously mentioned, we have chosen the technique of supporting the sample at nodes for a particular mode of vibration. As constraints on the sample are less, it was thought more accurate than the vibrating reed test. This test has previously been applied to much larger samples, therefore its applicability to smaller samples is under test here. One advantage of using rod shaped samples is that the mathematical relationship between resonant frequency and Young's modulus is more exact. The Young's modulus can be determined from either the flexural or the longitudinal vibrations. Generally it is easier to excite the flexural vibration than the longitudinal vibration and so this was employed. Knife edges were positioned at the nodes (0.224 of the length from each end) to excite the fundamental flexural vibration. For cylindrical rods, the Young's modulus and the fundamental frequency of the flexural vibration are related by:

$$E = 1.261886\frac{\rho l^4}{d^2}f_{res}^2 \qquad (7)$$

Where d is the diameter.

The resonance apparatus is shown in figure 4. The sample mount consists of micrometer calipers attached to an X-Y translation stage. The translation stage is mounted on a ridged platform bolted to a vibration isolated table. Metal shim formed in to knife edges is mounted on the inside edges of the micrometer calipers. Each shim has a central V of material removed, in to which the sample sits. This was thought necessary to prevent lateral movement of the sample during excitation. The distance between the knife edges (15-75mm) is set to that required for fundamental flexural resonance. Sample excitation is achieved using an audio range speaker, located directly beneath the centre of the sample. It is acoustically isolated from the table to prevent coupling to other parts of the apparatus. The frequency of excitation is tuned using a variable frequency function generator which is connected to the loud speaker via an audio amplifier. Resonance can be detected in two different ways depending on the nature of the sample surface. For specimens with a fairly smooth reflecting surface, resonance is detected by interferometery, using a similar method to that used for the vibrating reed test. In brief, the sample beam is reflected from the sample surface and interferes with the reference beam. The resultant beam passes through a pinhole and then is incident on the face of a photodiode. When the excitation frequency approaches the resonant frequency of the sample, the sample will start to vibrate. The changes in intensity at the photodiode are measured by an oscilloscope and can be seen as a maximum in the amplitude of the measured signal. Alternatively, for specimens with a rough surface, such as the diamond coated wires and fibres produced in this investigation, resonance was detected using position detection. This works using the same apparatus as above but omits the need for the reference arm of the interferometer. In this case, the intensity of the light detected at the photodiode is attenuated as the fibre moves. At resonance, the attenuation is greatest and again was seen as a maximum in the amplitude of the signal on the oscilloscope. A lens in front of the sample was required to focus the light to a spot on the sample surface, and to collimate the reflected light.

The resonant frequency was measured as a function of length (l~8-12cm). The slope of frequency versus l^{-2}, together with the sample density, diameter etc., is used with equation 7 to determine the Young's modulus. The frequency range for the fibre samples tested was 100-500 Hz. Density was determined both by mass and dimension measurements and, theoretically, using the rule of mixtures[21]. Calibration of the equipment was carried out using Pyrex (borosilicate glass) rods of know properties, and of different lengths (8-12cm). The value of modulus measured, 61.4 GPa, compares well with the data book value of 61 GPa.

Three samples of each CVD diamond coated SiC fibre and of CVD diamond coated W wire were tested. Sample details and measured moduli are given in table V. SEM photographs of these samples can be seen in figure 5 and 6.

Table V Sample details and modulus of coated wire and fibre samples measured by resonance.

Sample type		Average Diameter (μm)	Film Thickness (μm)	Length (cm)	Young's Modulus (GPa)
W/Diamond	**1**	**202.5**	**38-40**	**8.3-12.2**	**970±95**
	2	**197.8**	**38-40**	**8.4-11.7**	**1015± 100**
	3	**197.0**	**38-40**	**9.9-12.1**	**980± 100**
SiC/Diamond	**1**	**195.0**	**48-52**	**8.5-11.1**	**560±78**
	2	**197.4**	**48-52**	**8.8-10.1**	**690± 95**
	3	**193.2**	**48-52**	**8.5-10.0**	**522± 73**

As the coated fibres under test are not isotropic (which is assumed by equation 7), equation 7 is not exact, and was merely used here as a guide to modulus until a more exact solution is developed. Use of equation 7 gives an overestimation of modulus. Theoretical calculations using the rule of mixtures predicts a modulus of 800 GPa and 897 GPa for the W/diamond and the SiC/diamond samples respectively. The low modulus value for the SiC/diamond fibres was thought to be due to longitudinal cracks in the film. The measured modulus of W/diamond samples may also have been increased by inaccuracies in the density measurements. The measured density of these samples was greater than the theoretical value. The measured density would be expected to be lower as the theoretical value was calculated using the density of single crystal diamond. The diamond film density should be lower than its single crystal form. A 10% uncertainty is also present, due to errors in length measurement and variation of the fibre diameter. These results are preliminary; future experiments are required for verification.

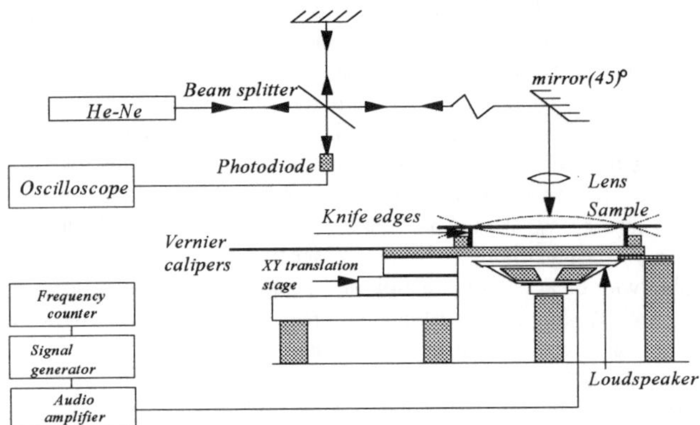

Figure 4. Schematic of the simply supported fibre resonance test.

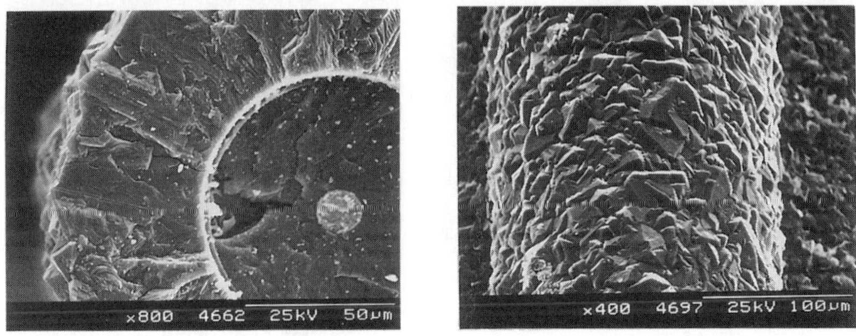

Figure 5. Diamond coated silicon carbide fibre, (a) cross section (b) surface.

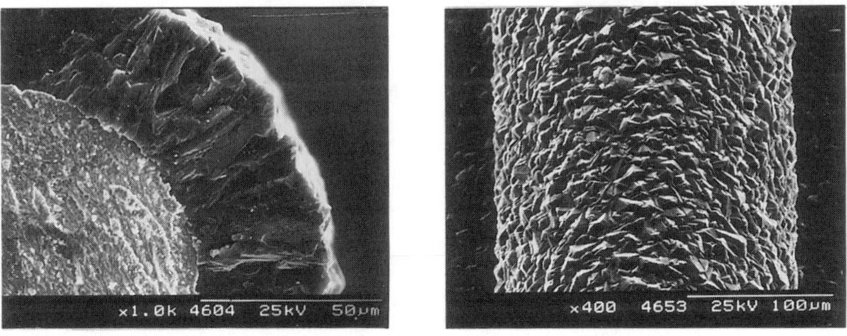

Figure 6. Diamond coated Tungsten wire, (a) cross section (b) surface.

Diamond Composite Work To Date: Diamond coated fibres produced during this investigation have been incorporated into a Ti-6Al-4V alloy matrix. They were individually coated with the titanium alloy matrix material, using physical vapour deposition, after which they were consolidated by hot vacuum pressing, either unidirectionally or isostatically at 900°C, producing a diamond coated fibre reinforced composite, in which the fibre spacing was controlled primarily by the metal matrix coating thickness and the diamond volume fraction by the diamond deposit thickness. A Ti-alloy composite microstructure containing fibres, with various diamond deposit thickness on SiC fibre cores, is shown in figure 7. The thin and thick diamond deposits were undamaged during consolidation, and Raman spectroscopy confirmed that they were still diamond[27]. A fibre with 85% volume fraction of diamond is shown in figure 8. Such fibres, when coated with 10μm of Ti-alloy and consolidated into a composite will produce a composite with 75% volume fraction diamond, and an elastic modulus of ~720 GPa (based on the rule of mixtures). This value is ~ 3.5 times greater than the modulus of current Ti-alloy/SiC fibre composites (~206 GPa) with about 30% fibre volume fraction[21].

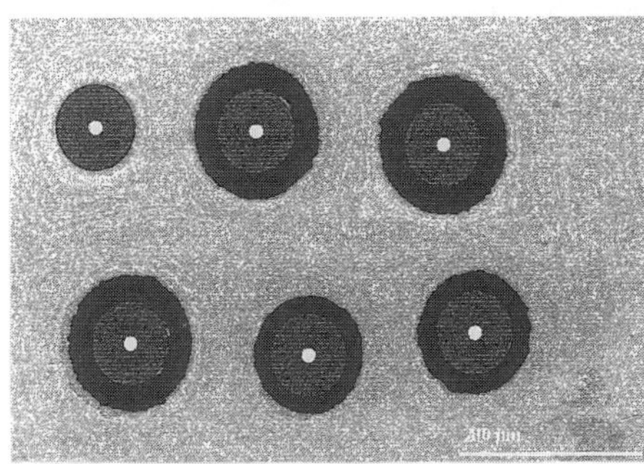

Figure 7.Diamond/SiC fibre-Ti alloy Composite.

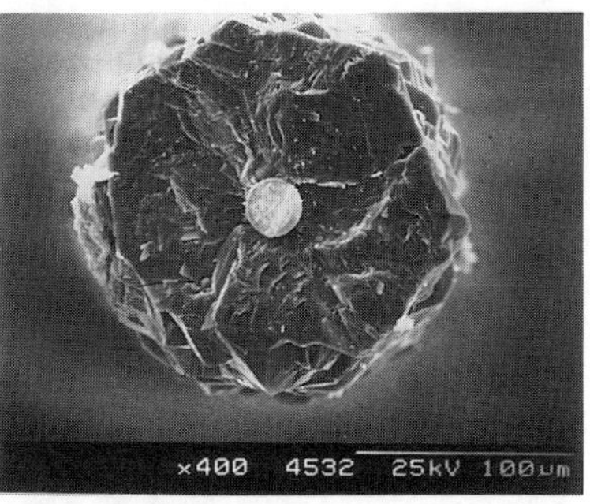

Figure 8. A SiC fibre with 85% volume fraction of diamond.

ACKNOWLEDGEMENTS

Many thanks to; S A Redman, T Baker and K N Rosser for their work on the fibre modulus measurement technique; H T Goldrein for the use of the phase-stepped interferometer and the EPSRC for financial support.

REFERENCES

1. E D Nicholson, J E Field, J. Hard. Mater. **5**, 89 (1994).
2. J M Blakely, J.Appl. Phys. **35** 1756 (1964).

3. J L Davidson, R Ramesham, C Ellis, J. Electrochem. Soc. **137** 3202, (1990).
4. G F Cardinal, R W Tustison, Proc. SPIE, **1325**, 90, (1990); J. Vac. Sci. Technol. **A9** 2204 (1991)
5. R S Sussmann, J R Brandon, G A Scarsbrook, C G Sweeney, T J Valentine, A J Whitehead, C J H Wort, Diamond and Related Materials **3** 303 (1994).
6. C J McHargue, Applications of Diamond and Related Materials eds Y Tzeng et al. (Amsterdam: Elsevier) 113 (1991).
7. N Savvides, TJ Bell, J. Appl. Phys. **72**, 2791 (1992).
8. C P Beetz, C V Cooper, T A Perry, J. Mater. Res. **5**, 2555 (1990).
9. H E M Hunt, Trans. Inst. Chem. Eng. **71**(A), 257 (1993).
10. S Spinner, W E Tefft, Am. Soc. Test. Mater. Proc. **61**, 1221 (1961).
11. ASTM Designation C623-71, 229 (reapproved 1985).
12 Y Seino, S Nagai, J. Mater. Sci. Lett. **12**, 324 (1993).
13. E D Nicholson, Ph.D Thesis, University of Cambridge, (1993).
14. B S Berry, W C Pritchet, J J Cuomo, C R Guarnieri, S Whitehair, Appl. Phys. Lett. **57** 302 (1990).
15. J W Beams, Structure and Properties of thin films, ed CA Neugebaur et al.(New York:Wiley) 183, (1959).
16. S Chandrasekar, Private communication (1990).
17. H M Polock, M Maugis M Barquins, Microindentation Techniques in Materials Science and Engineering ed P Blau, BR Lawn (Philadelphia:ASTM) 72 (1985);ASM Handbook **18** 419 (1992)
18. H T Goldrein, Ph.D Thesis, University of Cambridge, (1995).
19.A A Morrish, J W Glesener, M Fehrenbacher, P E Pehrsson, B Maruyama, P M Natishan, Diamond and Related Materials. **3** 173 (1993).
20. J Ling, M Lake, J. Mater. Res. **19** (3) (1994).
21. P G Partridge, P W May, C A Rego, M N R Ashfold, Mater. Sci. Technol. **10,** 505 (1994).
22. D W Petrasek, R A Signorelli, NASA Lewis Research Centre, Tech. Memo. 82590, (1981).
23. P G Partridge, C M Ward-Close. International Materials Reviews **38**(1) 1, (1993).
24. E Kalaugher, N M Everitt, University of Bristol, Private communication.
25. C J Burstone, A J Goldberg, Am. Orthod. **84** 95 (1983)
26. F Miura, M Mogi, Y Ohura, H Hamanaka, Am. J. Orthod. Dento. Orthop. **90** 1 (1986).
27. P G Partridge, M N R Ashfold, P W May, E D Nicholson, G Meaden, A Wisbey, in press.

ELASTIC PROPERTIES OF CVD DIAMOND VIA DYNAMIC RESONANCE MEASUREMENTS

MARK P. D'EVELYN,* DAVID E. SLUTZ,** AND BRADLEY E. WILLIAMS**
*General Electric Corporate Research and Development, P.O. Box 8, Schenectady, NY 12301
**General Electric Superabrasives, P.O. Box 568, Worthington, OH 43085

ABSTRACT

Control of the mechanical properties of CVD diamond is essential to achieve optimal performance in various applications. While several methods have been applied to the measurement of the Young's modulus of thick-film CVD diamond, in general these methods are not suitable for diamond characterization on a production scale. In addition, many of these methods cannot determine the shear modulus (or Poisson's ratio), which is necessary for a complete description of the elastic properties. We have developed a simple dynamic resonance method for determining both the Young's and shear modulus of free-standing CVD diamond in the shape of rectangular plates or round disks. The specimen is supported along nodal lines of flexural or torsional modes. Oscillations induced by impact from a falling ceramic bead are sensed by a microphone, and the resonant frequencies are determined by a signal analyzer. The Young's and shear modulus are calculated from the frequencies of the fundamental flexural and torsional modes, respectively, using quasi-analytic formulas. CVD diamond grown by several methods routinely achieves Young's and shear modulus values above 1000 GPa and 500 GPa, respectively, in good accord with theoretical values for pure polycrystalline diamond.

INTRODUCTION

The extreme properties of diamond, together with the capability for growth of thick, high-quality polycrystalline diamond films by chemical vapor deposition (CVD) developed over the past decade-and-a-half, has motivated the development of a number of CVD diamond products. Many of these products, including heat sinks, optical windows, machine tool inserts, and wire dies, involve thick, free-standing polycrystalline diamond plates rather than a thin film of diamond on a substrate. Achievement of optimal performance in commercial products, however, obviously hinges on the material properties of the diamond and their reproducibility. In the case of heat sinks for microelectronic thermal management applications, for example, it was necessary to develop suitable methods for routine determination of the thermal conductivity.[1]

The mechanical properties of CVD diamond, including the elastic constants and fracture strength, are critical for many other applications. For determination of the elastic modulus of CVD diamond, the bulge test method is the most widely applied technique.[2-5] In this technique the displacement of the center of a disk is measured as function of the differential pressure applied across it. The bulge test method has the advantages of requiring only inexpensive equipment and permitting fracture strength measurements in the same apparatus. However, it is arguably too tedious for routine use in production. Only the biaxial modulus can be determined by this method, requiring an assumed value for the Poisson's ratio. In addition, the boundary conditions of the disk (fixed versus free) must be determined, either directly or by empirical correction,[4-6] which complicates extraction of the modulus from the pressure-displacement data.

Several other methods have been applied to the determination of elastic constants for CVD diamond, but arguably are also unsuitable for routine measurements. Nanoindentation measurements can determine both the Young's modulus and hardness,[7-9] but require expensive,

Mat. Res. Soc. Symp. Proc. Vol. 383

specialized equipment and are rather time-consuming. Speed-of-sound measurements[10,11] are significantly simpler to apply and also have the capability for determining both the Young's and shear modulus, but the cost of the equipment and time required to characterize each specimen are still unattractive. Moreover, the accuracy of the measurements is limited by the extremely high speed of sound in diamond.

In this paper we report the first application of another technique, dynamic resonance, to the determination of the elastic constants of free-standing CVD diamond. This method requires only inexpensive equipment and is very fast, yet can accurately determine both the Young's and shear modulus. A similar technique has been applied to determination of the Young's modulus of thin-film CVD diamond.[12]

EXPERIMENTAL: IMPULSE DYNAMIC RESONANCE METHOD

The dynamic resonance method is a well-established technique for the determination of elastic constants.[13-15] As developed originally the technique involved driving the specimen with a variable-frequency external oscillator and measuring the response as a function of frequency. With the ready availability of digital signal processing techniques, an impulse technique, where oscillations are induced in the part by tapping and the transient "ringing" signal is detected by a microphone, digitized, and Fourier-analyzed, is much simpler experimentally and yields the same information. Although only approximate equations relating the measured resonant frequencies to the elastic constants are available, the accuracy of these expressions has been demonstrated to be approximately 1% or better for a wide range of specimen geometries and elastic properties, more than adequate for our purposes.

In the case of rectangular bars or plates, the Young's modulus may be determined from the frequency of the one-dimensional flexural mode.[13-15] The fundamental flexural mode has nodes at positions $0.224l$ and $0.776l$, where l is the sample length, and can be excited by supporting the specimen along the nodes and tapping in the middle, as illustrated in Fig. 1(a). The shear modulus may be determined from the frequency of the torsional mode. The fundamental torsional vibration has nodal lines bisecting the face in each direction, as illustrated in Fig. 1(b), and can be excited by

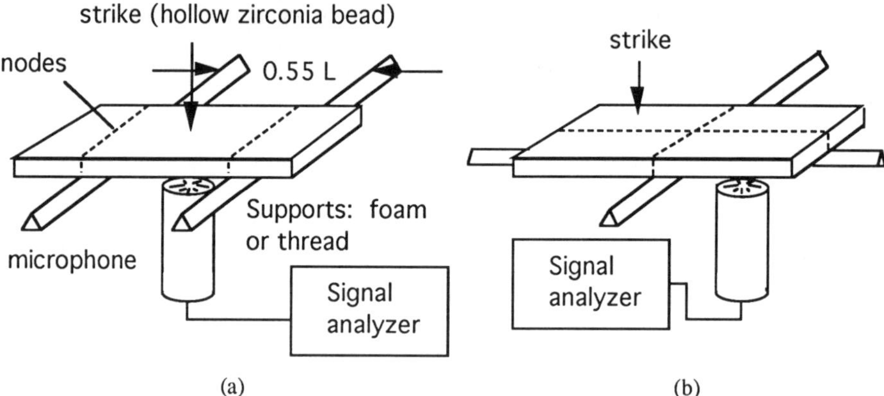

FIG. 1. Schematic illustration of ringing measurements on the (a) flexural mode or (b) torsional mode of rectangular plates. For disks the torsional mode is equivalent to (b), whereas the flexural vibration is a biaxial drumhead mode with a nodal circle at 0.681 of the diameter of the part.

supporting the specimen along the intersecting nodes and tapping near one corner. In the case of round disks, the torsional mode is essentially equivalent to that shown in Fig. 1(b). The flexural vibration of a round disk is a biaxial drumhead mode, however, with a nodal circle of diameter $0.681d$, where d is the diameter of the disk.[16,17]

We have analyzed free-standing rectangular plates and disks of GE CVD diamond produced by two different growth techniques by the impulse dynamic resonance method. 25-mm square plates were laser cut on one side to produce rectangular plates, and 10-mm diameter disks were laser-cut from larger wafers. The specimens were then supported in a fixture on either thin foam strips or on stretched cotton thread along the nodal lines for flexural or torsional vibrations, as appropriate (see Fig. 1). Oscillations were initiated by dropping a hollow zirconia bead from a height of 3-8 in. through a tube onto the center or one corner of the specimen. The ringing sound was detected by a microphone and preamplifier (Bruel & Kjaer, models 4165 and 2639, respectively), and the resonant frequencies were determined using a digital oscilloscope with Fast Fourier Transform capability.

Placement of the specimen on the fixture, tapping, and acquisition of several ringing measurements requires less than one minute per sample. For the initial experiments we performed, described here, the most time-consuming part of the operation was manual extraction of the resonant frequencies from the digital oscilloscope. For routine measurement in production, however, we use a commercial instrument[18] that performs the signal analysis automatically and reports the resonant frequency on a digital readout.

The approximate expressions used to extract the Young's and shear modulus from the measured flexural and torsional frequencies, f_F and f_T, respectively, and the weights and dimensions of the parts are summarized below. In these equations the length and width (transverse to flexure) of the rectangular plates are denoted l and w, respectively, d is the diameter of the disks, t is the thickness of either type of specimen, and ρ is the density.

The Young's modulus of the rectangular plates was determined from[13-15]

$$E = 0.9465\rho(l^2 f_F / t)^2 T_3 \tag{1}$$

$$T_3 = 1 + 6.585[1 + 0.0752\nu + 0.8109\nu^2](t/l)^2 - 0.868(t/l)^4 - \frac{8.340[1 + 0.2023\nu + 2.173\nu^2](t/l)^4}{1 + 6.338[1 + 0.1408\nu + 1.536\nu^2](t/l)^2} \tag{2}$$

These equations are due to an approximate solution by Pickett[19] and have been shown to be accurate to better than 1%.[15,20,21] The correction term T_3 contains the Poisson's ratio ν and so in principle requires iterative solution along with the determination of the shear modulus. However, for the aspect ratio ($t / l << 1$) of the CVD diamond specimens examined here T_3 is nearly equal to one and insensitive to the assumed value of ν, and iteration of the solution is unnecessary.

The shear modulus of the rectangular plates was determined from the equation[13-15]

$$G = 4\rho l^2 f_T^2 R \tag{3}$$

where, for the shape factor R we have used the equation recommended by Spinner and Tefft:[14]

$$R = \left[\frac{1 + \left(\frac{w}{t}\right)^2}{4 - 2.521\left(\frac{t}{w}\right)\left(1 - \frac{1.991}{e^{\pi w/t} + 1}\right)}\right]\left[1 + 0.00851\left(\frac{w}{l}\right)^2\right] - 0.060\left(\frac{w}{l}\right)^{3/2}\left(\frac{w}{t} - 1\right)^2 \tag{4}$$

For relatively thin plates, such as employed here,

$$R \approx (w/t)^2/4$$

The Poisson's ratio then follows from the usual relation

$$\nu = E/(2G) - 1 \tag{5}$$

With disks the flexural and torsional frequencies are each strongly influenced by both the Young's and shear modulus. Conveniently, the Poisson's ratio can be derived directly from the ratio of the two frequencies.[16,17] In general ν also depends on the t/d ratio, however, for relatively thin disks ($t/d \lesssim 0.05$) the latter dependence is negligible. In this case a quadratic fit to tabulated values[16,17] gives, to an excellent approximation,

$$\nu = -3.3382 + 3.834(f_F/f_T) - (f_F/f_T)^2 \tag{6}$$

Given the determined value of ν, the Young's modulus is determined from

$$E = \tfrac{3}{2}\pi^2\rho(1-\nu^2)(d^2/t)^2[(f_F/K_F)^2 + (f_T/K_T)^2] \tag{7}$$

The correction factors K_F and K_T depend weakly on ν and the t/d ratio and are tabulated.[16,17] For the parts analyzed here K_F and K_T are approximately equal to 8.4 and 5.9, respectively. The value of the shear modulus then follows from

$$G = E/[2(1+\nu)] \tag{8}$$

RESULTS AND DISCUSSION

The flexural and torsional frequencies, f_F and f_T, respectively, measured on representative samples are summarized in Table 1, along with the weights and dimensions of the samples. The rectangular bars were lapped on both faces and polished on one; the first disk was polished on both faces, and the second was as-grown (after removal from the substrate). These samples are all of high quality, as indicated by thermal conductivities between 9 and 20 W /cm-K in the case of the rectangular bars and excellent transparency in the case of the disks.

TABLE 1. Summary of dimensions, weights, resonant frequencies, elastic moduli, and Poisson's ratio of free-standing CVD diamond plates in rectangular or circular form.

Rectangular bars:

Sample #	l (mm)	w (mm)	t (mm)	weight (g)	f_F (kHz)	f_T (kHz)	E (GPa)	G (GPa)	ν
R-1	26.19	10.11	0.316	0.291	8.61	15.01	1145	530	0.08
R-2	25.68	12.52	0.302	0.339	8.76	12.11	1212	547	0.11
R-3	25.68	17.37	0.398	0.613	10.76	11.00	1037	467	0.11
R-4	25.68	9.06	0.312	0.248	8.24	15.78	980	461	0.06
						mean:	1093	501	0.09

Circular disks:

Sample #	d (mm)	t (mm)	weight (g)	f_F (kHz)	f_T (kHz)	E (GPa)	G (GPa)	ν
C-1	10.24	0.260	0.075	69.81	49.41	1161	536	0.08
C-2	10.08	0.365	0.102	100.41	71.00	1154	532	0.08

The accuracy of the elastic constants of CVD diamond determined by the dynamic resonance method is typically limited mainly by the accuracy of the dimensional measurements, particularly the thickness. The weight and frequencies can easily be determined to 0.1% accuracy or better, and the formulas are believed to be accurate to approximately 1% or better. The inferred modulus values are approximately proportional to t^{-3} ($\rho \propto$ weight/t) and therefore an accurate thickness is particularly important. For unpolished plates of negligible porosity an approximate average thickness can be determined by assuming the ideal density of 3.515 g cm^{-3} for diamond. This assumption, while only approximate for plates with a rough surface, is superior to measuring the thickness with a micrometer.

The measured values of the Young's and shear modulus and Poisson's ratio are in good agreement with theoretical values for pure polycrystalline diamond. Two groups have shown independently that the elastic constants for single-crystal natural diamond, when rotationally averaged as appropriate for polycrystalline diamond with no preferred orientation, yield a Young's modulus of 1143 GPa, a shear modulus of 534 GPa, and a Poisson's ratio of 0.07.[4,22] Thick-film, free-standing CVD diamond normally exhibits a significant fiber texture, and all of the samples examined in this study have a preferred (110) orientation. The effective elastic constants of polycrystalline diamond, based on an assumption of isotropic elastic behavior, depend on the texture. The Young's modulus exhibits only modest orientational dependence, and both Klein and Cardinale[4] and Werner *et al.*[22] calculate a value of 1151 GPa for (110)-oriented diamond. These authors differ somewhat on the effective shear modulus for (110)-textured diamond, obtaining 536 and 524 GPa, respectively.[4,22] These values correspond to Poisson's ratios of 0.07 and 0.10, respectively. Not all the grains in textured diamond have a fiber axis precisely aligned along the sample normal, of course, and a more complex rotational averaging procedure is probably required for precise specification of the effective elastic constants of textured CVD diamond. In any case, we have not investigated the relationship between the fiber texture of CVD diamond and the elastic constants.

In conclusion, we have demonstrated that the impulse dynamic resonance method as applied to free-standing CVD diamond is simple, fast, and apparently yields accurate values for the Young's and shear modulus and the Poisson's ratio, and is therefore suitable for routine characterization of commercial material. In addition, we have shown that GE CVD diamond produced by two different growth methods routinely achieves values of the Young's and shear modulus in good agreement with the corresponding values for pure polycrystalline diamond.

ACKNOWLEDGMENTS

MPD thanks Dr. Curt Johnson (GE CR&D) for many helpful discussions on the dynamic resonance method and for the use of his equipment, Dr. Andre Van Leuven (J. W. Lemmens, Inc.) for suggestions on sample fixturing and for providing the formulas and references 16 and 17 for extraction of the elastic constants from the flexural and torsional frequencies of disks, and Drs. Jim Ruud and Donna Hurley for bringing references 12 and 13 to the authors' attention.

REFERENCES

1. For example, P. G. Kosky, Rev. Sci. Instrum. **64**, 1071 (1993).
2. G. F. Cardinale and R. W. Tustison, J. Vac. Sci. Technol. A **9**, 2204 (1991).
3. H. Windischmann and G. F. Epps, Diamond Relat. Mater. **1**, 656 (1992).
4. C. A. Klein and G. F. Cardinale, Diamond Relat. Mater. **2**, 918 (1993).
5. T. J. Valentine, A. J. Whitehead, R. S. Sussman, C. J. H. Wort, and G. A. Scarsbrook, Diamond Relat. Mater. **3**, 1168 (1994).
6. W. C. Young, *Roark's Formulas for Stress and Strain*, 6th Edition (McGraw-Hill, New York, 1989) p. 429.
7. C. P. Beetz, Jr., C. V. Cooper, and T. A. Perry, J. Mater. Res. **5**, 2555 (1990).
8. C. McHargue, in *Applications of Diamond Films and Related Materials*, ed. Y. Tzeng, M. Yoshikawa, M. Murakawa, and A. Feldman (Elsevier, Amsterdam, 1991) p. 113.
9. N. Savvides and T. J. Bell, J. Appl. Phys. **72**, 2791 (1992).
10. K. Dunn and F. Bundy, J. Appl. Phys. **49**, 5865 (1978).
11. K. J. Gray, SPIE Diamond Optics V, **1759**, 203 (1992).
12. L. Chandra and T. W. Clyne, J. Mater. Sci. Lett. **12**, 191 (1993).
13. E. Schrieber, O. L. Anderson, and N. Soga, *Elastic Constants and Their Measurement*, (McGraw-Hill, New York, 1974), pp. 82-125.
14. S. Spinner and W. E. Tefft, Am. Soc. Test. Mater. Proc. **61**, 1221 (1961).
15. "Standard Test Method for Dynamic Young's Modulus, Shear Modulus, and Poisson's Ratio for Advanced Ceramics by Impulse Excitation of Vibration," ASTM Standard C 1259-94 (ASTM, Philadelphia, PA, 1994). .
16. J. C. Glandus, "Rupture fragile et résistance aux chocs thermiques de céramiques à usages mécaniques," Ph.D. Dissertation, University of Limoges (1981, unpublished).
17. "Synthesis of the elastic formulas for discs by J. C. Glandus," Private communication to MPD by J. W. Lemmens, Inc. (St. Louis, Missouri, USA).
18. GrindoSonic, J. W. Lemmens, Inc. (St. Louis, Missouri, USA).
19. G. Pickett, Am. Soc. Test. Mater. Proc. **45**, 846 (1945).
20. S. Spinner, T. W. Reichard, and W. E. Tefft, J. Res. Natl. Bur. Stand. **64B**, 237 (1960).
21. J. S. Smith, M. D. Wyrick, and J. M. Poole, *Dynamic Elastic Modulus Measurement in Materials*, ed. by A. Wolfenden, ASTM STP 1045 (ASTM, Philadelphia, PA, 1990).
22. M. Werner, S. Hein, and E. Obermeier, Diamond Relat. Mater. **2**, 939 (1993).

Study of Mechanical and Elastic Properties of Diamond Films by Surface Acoustic Wave Spectroscopy (SAWS)

R. KUSCHNEREIT AND P. HESS
University of Heidelberg, Institute of Physical Chemistry,
Im Neuenheimer Feld 253, D-69120 Heidelberg, Germany

ABSTRACT

The density and elastic properties of a 1.9 μm thick polycrystalline diamond film deposited on a silicon substrate were measured by surface acoustic wave (SAW) spectroscopy. A density of 3.45 ± 0.05 g/cm^3, Young's modulus of 940 ± 20 GPa and Poisson's ratio below 0.12 were determined from the dispersion of a broadband coherent surface acoustic wave pulse propagating in the layered system. The surface wave pulses were generated using a ns UV laser pulse and detected with a piezoelectric foil transducer.

INTRODUCTION

Diamond has always been of great scientific and technical interest because its atom density, and thermal, mechanical and elastic properties reach extraordinary values compared to all other materials [1]. Therefore diamond films are of interest as hard and resistant coatings, as thermal conductors and in other thin film applications. Since all diamond films available are polycrystalline, the characterization of the average value of the density and elastic constants is important for quality control and also of basic scientific interest. The laser method described in this paper allows the simultaneous determination of several mechanical and elastic properties for films with a thickness in the micrometer range.

EXPRIMENTAL

To excite broadband coherent surface acoustic wave (SAW) pulses, the laser pulses of a frequency-tripled Nd:YAG laser with a wavelength of 355 nm and a pulse duration of 7 ns FWHM were focused onto the sample surface by a system of three cylindrical lenses (Fig. 1). The size of the line focus was about 7 μm x 12 mm.

The expansion of the illuminated volume leads to the emission of nearly plane SAW pulses with a broad frequency distribution, due to the thermoelastic effect.

A piezoelectric foil transducer, consisting of a polyvinylidenedifluoride (PVDF) foil fixed on a metal wedge with a curvature of about 5 μm, was used to detect the acoustic pulses [2]. The foil detector was positioned on the sample surface at distances between 15 and 30 mm from the excitation source. Compression of the PVDF foil due to the passing SAW pulses causes detectable electrical signals, which were recorded by a digital oscilloscope.

The transducer signals detected at different distances from the excitation source were Fourier transformed and from the phase information of these pulses the dispersion curve could be extracted. The dispersion of SAWs in a layer/substrate system is determined by the density, the elastic properties of the substrate and the film material and the film thickness [3]. To evaluate

Mat. Res. Soc. Symp. Proc. Vol. 383 © 1995 Materials Research Society

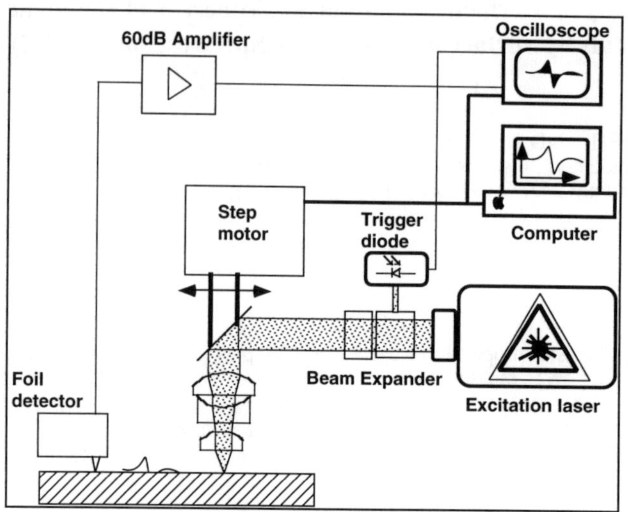

Fig. 1 Experimental setup with excitation laser, focusing optics, foil transducer, and detection electronics

the elastic constants of the film, a theoretical dispersion curve was fitted to the experimental one. This fit is based on a numerical solution of the exact model by taking into account the bulk acoustic equations for the displacements and the boundary conditions for the film surface and the interface between the film and the substrate. The fitting procedure was performed with a computer program developed for an isotropic film on either an isotropic or an anisotropic substrate by the variation of four properties of the film, namely the density, Young's modulus, Poisson's ratio and film thickness. The minimization of the variance between the experimental and the theoretical dispersion curves provided the set of film parameters.

The investigated sample was a 1.9 μm thick polycrystalline diamond film which was deposited on the (111)-plane of a silicon substrate. By scanning force microscopy (SFM) measurements we determined the surface roughness of the diamond film as 15 nm rms.

RESULTS

A typical acoustic pulse, detected at a distance of about 20 mm from the excitation line on the diamond film, is shown in Fig. 2a. The oscillating pulse shape indicates the interference of the different coherent partial waves of the frequency spectrum of the SAW pulse traveling with different phase velocities. Fig. 2b shows the corresponding amplitude spectrum of the Fourier transformation. This amplitude spectrum has a maximum at about 30 MHz. For higher frequencies the amplitude decreases and reaches zero at about 160 MHz. Therefore the dispersion curve could also be determined only up to a frequency of 160 MHz. Fig. 3 shows the experimental dispersion curve and the theoretical fit evaluated for the best set of parameters. The amount of information on the film material contained in the dispersion curve increases with the high frequency limit and the film thickness. The frequency limit reached in these measurements is rather low, compared to the limits of the setup, which was proved to detect frequencies up to 300 MHz. Nevertheless we were able to extract all four film properties, i.e. thickness, density, Young's modulus and Poisson's ratio. The mechanical and elastic constants determined for the

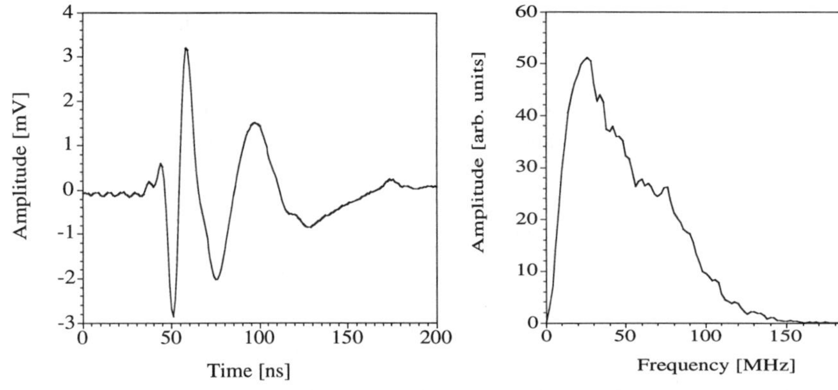

Fig. 2a SAW pulse shape

Fig. 2b Amplitude spectrum

polycrystalline diamond film are listed in Table I, where they are compared with the literature values of single crystal diamond [1,4] and the results recently obtained for state-of-the-art diamondlike carbon films investigated by SAWS [5]. The density of 3.45 ± 0.05 g/cm^3 is about 98% of the single crystal diamond value. The Young's modulus of 940 ± 20 GPa is about 82% of the mean single crystal value. A comparison with Poisson's ratio for different directions in the crystalline diamond is more difficult. In anisotropic media we have to define two Poisson's ratios for each direction, and especially in diamond the variation of Poisson's ratio with the direction is relatively great [4]. The average value given in [4] for crystalline diamond is 0.0691. We could only determine an upper limit of 0.12 from our measurements, which is much higher than this average value but already much lower than for most other materials.

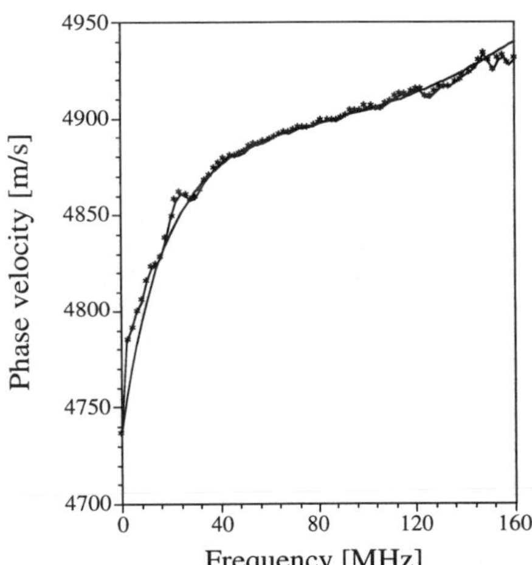

Fig. 3 Experimental (points) and theoretical dispersion curve (solid line)

DISCUSSION

The strong decrease of the amplitude spectrum for frequencies higher than 30 MHz is thought to be caused by scattering of the SAW pulse by the surface roughness and the grain boundaries. The size of the crystallites in the polycrysrystalline material is much smaller than 1 μm. This can be predicted from the deposition parameters. Thus the structure is much smaller than the shortest acoustic wavelength of 15 μm at 300 MHz which we expect to be efficiently excited by the excitation laser pulses. For this relation between wavelength and microstructure it is clear that scattering effects increase for higher frequencies.

Since the SAW pulse travels a distance of at least 15 mm from the excitation line to the nearest detection point the higher frequency components between 150 MHz and 300 MHz contained in the initial acoustic pulse may be already eliminated

Tab. I Density and elastic constants of crystalline diamond, polycrystalline diamond films, and diamondlike films.

	Single Crystal Diamond	Polycrystalline Diamond Films	Diamondlike Films
Density $[kg/m^3]$	3515	3450 ± 50	2900
Average Young's modulus [GPa]	1143	940 ± 20	400
Average Poisson's ratio	0.0691	< 0.12	

by scattering effects and do not appear in the detected SAW pulse. Quantitative predictions about the grain size or the surface roughness may be possible if the frequency dependence of the attenuation can be measured more accurately.

The limitation given by this lower frequency range was partly compensated by the large film thickness, which is the second parameter influencing the information content by increasing the nonlinearity of the dispersion curve. This can be easily understood as higher frequencies, i.e. shorter wavelengths, and also thicker films provide a stronger interaction between the wave pulse and the film, which is always much thinner than the shortest wavelength. In fact the ratio between film thickness and smallest wavelength realized in the pulse is the quantity appearing in theory. Additionally, because of the high sound velocity in the diamond material the dispersion effect of the SAWs is rather strong (Fig. 3), and therefore the fitting procedure yields results with relatively small error bars.

Since single crystal diamond is transparent for a wavelength of 355 nm, only the grain boundaries are expected to absorb the light pulses. Most of the energy will be absorbed by the silicon substrate, which has an extremely high absorption coefficient for this wavelength. However, the particular excitation mechanism is not important for our method and will not affect the accuracy of the measurement. The physical information comes only from the change in the SAW pulse propagating through the film/substrate system.

Table I shows the intermediate position of the properties of the polycrystalline diamond film between the single crystal diamond and the amorphous diamondlike carbon films. However, density and Young's modulus are much nearer the single crystal values. Valentine et al. determined even higher Young's moduli of up to 1079 GPa for bulk (free standing films of about 300 μm thickness) polycrystalline CVD diamond using a differential pressure technique [6]. This can be easily understood since for rather thin films of a few micrometers the influence of the silicon - carbon interface cannot be neglected. This interface is known to have properties very different from the single crystal diamond material [7], and therefore the values we measure have to be regarded as average values over the whole film thickness. In addition, the diamond film investigated here was not ideal for the SAW method, because the best diamond films that can be produced at the moment have enormous surface roughnesses and vary in thickness over the whole film area. These films are not accessible to our technique because it uses acoustic waves that travel macroscopic distances. Therefore it is important that the film is homogenous in thickness and quality over a relatively large area. Otherwise it will be difficult to extract the parameters and the error bars will increase enormously. The film we investigated was therefore a compromise between best material properties and satisfactory homogeneity and smoothness of the film.

Jiang et al. determined the elastic properties of up to 400 μm thick diamond films using Brillouin scattering. Since the lateral dimensions of the crystallites at the surface were between 20 and 40 μm the properties of these crystallites have been measured and the elastic constants determined were in agreement with the values of single crystalline diamond [8].

We point out that our technique measures properties of the whole film. This gives information on the polycrystalline material and the overall quality of the film. As far as we know this is the first measurement of the density and elastic properties of thin diamond films averaging over a macroscopic area.

Several measurements were published about the hardness of diamond films, for example, Savvides et al. measured the microhardness of a 3 μm thick polycrystalline diamond film with a ultralow-load microhardness instrument. From this data it was possible to estimate a Young's modulus of about 500 GPa [9].

Table I also shows a comparison between the amorphous and crystalline networks for carbon. Amorphous silicon networks nearly reach the density of the crystalline material, and also the Young's modulus is comparable to the crystalline value, for example for high quality amorphous hydrogenated silicon films [10]. Compared to this, state-of-the-art amorphous carbon films differ substantially from single crystals. The reason for this different behavior can be found in the ability of carbon atoms to form not only sp(3) but also sp(2) and sp hybrid bonds. Therefore carbon has the possibility of diminishing stress in the amorphous network not only by forming dangling bonds, voids, columnar structures, etc., but also by the formation of sp(2) bonded clusters, for example [11], which lower the density and Young's modulus to a significant extent. If the formation of diamond crystallites is induced, these effects can be avoided and the mechanical and elastic quality of the film material increases significantly.

ACKNOWLEDGEMENTS

The authors would like to thank Dr. B. Schreck for the preparation of the diamond film.

Financial support of this work from the Bundesministerium für Bildung, Wissenschaft, Forschung und Technologie under contract No. 13N6005, the European Union and the Fonds der Chemischen Industrie is gratefully acknowledged.

REFERENCES

1. J.C. Angus, Thin Solid Films **216**, 126 (1992).
2. H. Coufal, R. Grygier, P. Hess and A.Neubrand, J. Acoust. Soc. Am. **92**, 2980 (1992).
3. G.W. Farnell, Acoustic Surface Waves, edited by A.A. Oliner (Springer, Berlin, 1978), pp. 42.
4. C.A. Klein and G.F. Cardinale, Diamond Relat. Mater. **2**, 918 (1993).
5. H. Fath, Dissertation, University of Heidelberg, Institute of Physical Chemistry (1993).
6. T.J. Valentine, A.J. Whitehead, R.S. Sussmann, C.J.H. Wort and G.A. Scarsbrook, Diamond Relat. Mater. **3**, 1168, (1994).
7. K. Plamann, D. Fournier, E. Anger and A. Gicquel, Diamond Relat. Mater. **3**, 752 (1994).
8. X. Jiang, J.V. Harzer, B. Hillebrands, Ch. Wild and P. Koidl, Appl. Phys. Lett. **59**, 1055 (1991).
9. N. Savvides and T.J. Bell, J Appl. Phys. **72**, 2791 (1992).
10. R. Kuschnereit, H. Fath, A. Kolomenskii, M. Szabadi and P. Hess, Appl. Phys. A, accepted for publication (1995).
11. J. Robertson, Phys. Rev. Lett. **68**, 220 (1992).

MICROHARDNESS, STRUCTURE, AND COMPOSITION STUDY OF AMORPHOUS HYDROGENATED BORON CARBIDE

SHU-HAN LIN,* DONG LI** AND BERNARD J. FELDMAN*
*Department of Physics and Center for Molecular Electronics, University of Missouri, St. Louis, MO 63121
**Department of Materials Science and Engineering and Center for Engineering Tribology, Northwestern University, Evanston, IL 60208

ABSTRACT

We have grown amorphous hydrogenated boron carbide thin films by rf plasma decomposition of diborane and methane. The chemical composition, infrared absorption, optical absorption, and microhardness of these thin films were measured. As a function of increasing diborane concentration in the feedstock, we observe increasing boron and hydrogen concentrations, increasing infrared absorption at 1330 cm^{-1} due to boron icosahedra, increasing optical bandgaps, and an unchanging microhardness in the grown films. The microhardness should have decreased due to the increasing hydrogen concentration; this expected decrease may have been balanced by an increased microhardness due to the boron icosahedra.

INTRODUCTION

Amorphous hydrogenated carbon (a-C:H) has become an important industrial wear-resistant coating, presently being widely used to protect computer hard disks from damage by crashing magnetic heads. Because of its hardness and wear-resistance, a-C:H has been commonly called diamond-like carbon. Its physical, structural, optical, tribological and compositional properties have been extensively studied, and References 1 and 2 are excellent review articles of this work. Last year we extended our microhardness study from a-C:H to amorphous hydrogenated carbon nitride (a-C:N:H), noting that the microhardness of a-C:N:H was identical to that of a-C:H with the same optical bandgap.[3] In this report we extend our microhardness study to amorphous hydrogenated boron carbide (a-B:C:H).

Amorphous hydrogenated boron carbide was first grown as a p-type doped material of a-C:H.[4,5] Since then, it has been characterized[6,7] and used as an inner wall coating for fusion reactors,[8] as a wear-resistant coating for mechanical systems,[9] and as a diode material.[10] We recently investigated the transition from doping to alloying in the electrical properties of a-B:C:H.[11]

Koidl and co-workers first measured the hardness of a-C:H as a function of self-bias voltage.[2] They observed that with increasing self-bias voltage, the hardness increased, and the optical

Mat. Res. Soc. Symp. Proc. Vol. 383

bandgap and hydrogen concentration decreased. They concluded that "the hardness appears to be correlated with the hydrogen concentration and thus with the degree of three dimensional cross-linking of the a-C:H network."[2] Last year, we confirmed the experimental results of Koidl et al and the critical role of hydrogen in determining the hardness.[3] We also pointed out that the hydrogen is primaily found in the clusters, not in cross-links, because the only way to cross-link two clusters is by carbon-carbon bonds. Hydrogen can only terminate a carbon bond, and thus most of the hydrogen atoms are terminating the carbon clusters. Also, the clusters due to their high hydrogen concentrations are the weak links in this material's microhardness. The experimental evidence for this comes from our NMR studies observing large concentrations of rotating methyl groups, a striking example of the flexibility and compressibility of this material.[12] It is well known that boron tends to form small icosahedra. The question posed in this work is what effect would such boron icosahedra inside the amorphous clusters have on the microhardness of the material.

EXPERIMENTAL PROCEDURES

We grew both a-C:H and a-B:C:H from a feedstock of methane (CH_4) and diborane (B_2H_6) in a capacitively coupled rf plasma reactor with the conditions described in Reference 11. We varied the boron concentration in our films by varying the fraction of diborane in the feedstock, as listed in the Table. The thin films were grown on the anode, using glass, aluminum foil, and silicon substrates. The silicon substrate samples were mounted in an ultramicro-indentation system (UMIS-2000). Mutiple indentations were made at different locations of the film surface at different loads. At each load, the load versus displacement curve was recorded, from which the effective modulus and hardness can be calculated using standard formulae.[13] The films on aluminum foil were immersed in dilute HCl, dissolving the aluminum; the free standing films were sent to Galbraith Laboratories for chemical analysis. The films grown on silicon substrates were also used for infrared absorption measurements using a Perkin Elmer Model 1610 FTIR spectrophotometer. The films grown on quartz substrates were used for visible and ultraviolet absorption measurements on a Hitachi Model U3100 spectrophotometer; the optical bandgaps were determined by fitting the optical absorption to the Tauc equation.

RESULTS

The chemical compositions, microhardnesses, and optical bandgaps of our various a-B:C:H samples are listed in the Table. For comparison, we have also included the results on an anode-grown a-C:H sample. Figure 1a shows the infrared absorption spectra of a-B:C:H (sample 3); again for comparison, Figure 1b shows the infrared spectra of anode-grown a-C:H (sample 1). Finally, Figure 2 is a plot of the microhardnesses of various amorphous samples as function of their optical bandgaps; the solid circles are from a-B:C:H and the open circles, from a-C:H and a-C:N:H. We have fitted

TABLE

Chemical Composition, Optical Bandgap and Microhardness

Sample	B_2H_6/CH_4	Boron (at.%)	Carbon (at.%)	Hydrogen (at.%)	E_0 (eV)	Hardness (GPa)
1	0.0	0	47	53	2.1	
2	1.0	1	40	59	2.5	0.69
3	5.0	16	23	61	3.6	0.58
4	10.0				4.2	0.77

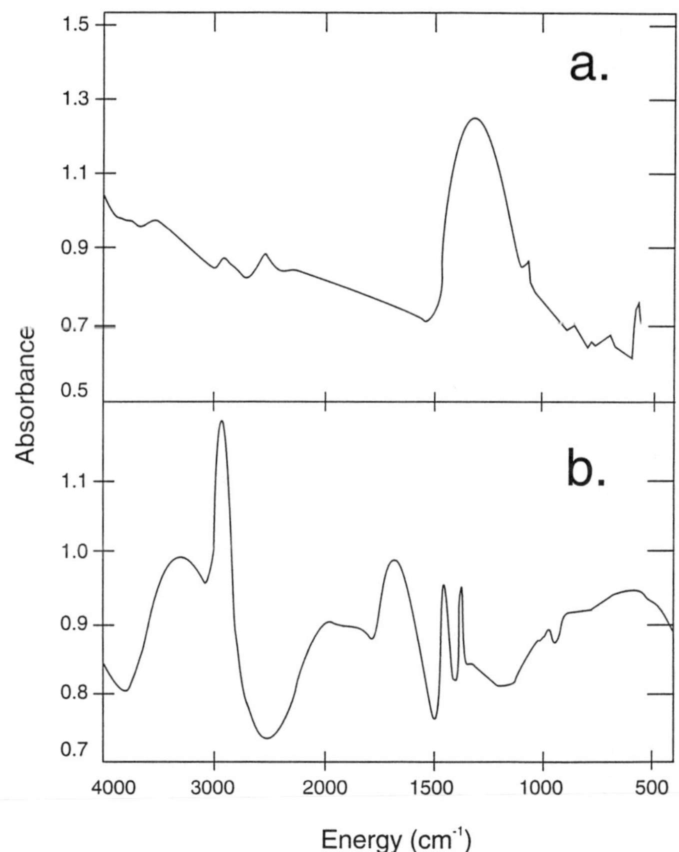

Figure 1. The infrared spectra of (a) a-B:C:H (Sample 3) and (b) a-C:H (Sample 1).

all the data to the function

$$H(E_0) = 3.3\ E_0^{-1.7}$$

where H is the microhardness, E_0 is the optical bandgap, and the linear correlation coefficient for this fit is 0.94.

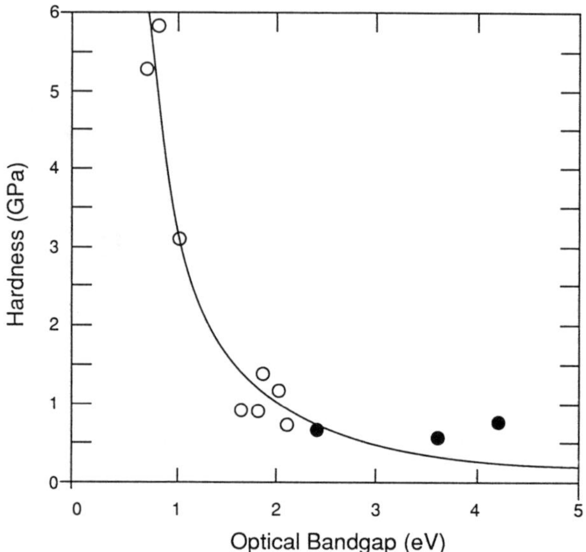

Figure 2. The microhardnesses of various a-B:C:H (solid circles), a-C:H and a-C:N:H (open circles) thin films as a function of their optical bandgaps. The curve is the least-squares fit to all the data points, $H(E_0)$.

DISCUSSION

With the addition of diborane to the feedstock, the following changes are observed: (1) the appearance of a very strong infrared line around 1330 cm^{-1}, as well as the disappearance of the C-H stretching and bending modes at 2950 and 1300 cm^{-1}; (2) an optical bandgap that increases from 2.1 eV to 4.5 eV; (3) an increase in the amount of boron and hydrogen in the thin films; and (4) a microhardness that is small but constant. The intense infrared line at 1330 cm^{-1} is most likely due to boron icosahedra.[14] The increase in absorption from boron icosahedra in the infrared spectra is due to (1) the increasing boron concentration creating more icosahedra and (2) the large oscillator strength of the icosahedra structure due to its large size and tight binding. The increasing optical bandgap probably has a number of causes, including the increase in hydrogen concentration and the increased number of boron icosahedra.

In regard to the microhardnesses of the a-B:C:H thin films, from Figure 2 they appear to be unchanged as a function of optical bandgap and deviate from the trend of the a-C:H and a-C:N:H thin films. One possible interpretation is that the presence of the boron icosahedra provides increased hardness that balances the expected decrease in hardness due to the increased hydrogen concentration. If this is the case, then cathode-mounted a-B:C:H, which would contain much less hydrogen and have smaller optical bandgaps than anode-mounted a-B:C:H, should be harder than a-C:H or a-C:N:H material of the same optical bandgap. We are investigating this posssibilty presently.

CONCLUSIONS

In conclusion, we have grown amorphous hydrogenated boron carbide by plasma-assisted chemical vapor deposition of diborane and methane. As a function of increasing diborane in the feedstock, we observed increasing concentrations of boron and hydrogen, increasing strength of the 1330 cm^{-1} line in the infrared spectra due to boron icosahedra, increasing optical bandgap, but unchanged microhardness. We suggest that the unchanging microhardness is due to a balance between the increased hardness due to the boron icosahedra and the decreased hardness due to the increasing hydrogen concentration.

ACKNOWLEDGEMENTS

This research is partly supported by a University of Missouri Center for Molecular Electronics grant and the National Science Foundation Surface Engineering and Tribology Program Contract No. MSS-9203239.

REFERENCES

[1]. J. C. Angus and C. C. Hayman, Science **24**, 913 (1988)

[2]. P. Koidl, Ch. Wild, B. Dischler, J. Wagner and M. Ramsteiner, Mat. Sci. Forum **52 & 53**, 41 (1989).

[3]. S.-H. Lin, B. J. Feldman, D. Li, Y.-W. Chung, M.-S. Wong and W. D. Sproul in Novel Forms of Carbon II, edited by C. L. Renschler, D. M. Cox, J. J. Pouch and Y. Achiba (Mater. Res. Soc. Proc. **349**, Pittsburgh, PA, 1994) pp. 519-524.

[4]. D. I. Jones and A. D. Stewart, Phil. Mag. **46**, 423 (1982).

[5]. B. Meyerson and F. W. Smith, Solid State Commun. **41**, 23 (1982).

[6]. S. Lee, J. Mazurowski, G. Ramsmeyer and P. A. Dowben, J. Appl. Phys. **72**, 4925 (1992).

[7]. B. M. Way, J. R. Dahn, T. Tiedje, K. Myrtle and M. Kasrai, Phys. Rev. B **46**, 1697 (1992).

[8]. S. Veprek, S. Rambert, M. Heintze, F. Mattenberger, M. Jurick-Rajman and W. Portmann, J. Nuclear Materials **162-164** 724 (1989).

[9]. J. Onate, A. Garcia, V. Bellido and J. Viviente, Surface and Coating Tech. **49**, 548 (1991).

[10]. S. Lee, Appl. Phys. A **A58**, 223 (1994).

[11]. B. Sylvester, S.-H. Lin and B. J. Feldman, Solid State Commun. **93**, 969 (1995).

[12]. S.-H. Lin, B. J. Feldman, J. R. Bodart, V. P. Bork, M. J. Kerman, P. A. Fedders, and R. E. Norberg in Novel Forms of Carbon II, edited by C. L. Renschler, D. M. Cox, J. J. Pouch and Y. Achiba (Mater. Res. Soc. Proc. **349**, Pittsburgh, PA, 1994) pp. 489-494.

[13]. M. F. Doermer and W. D. Nix, J. Mater. Res. **1**, 78 (1986).

[14]. H. Stein, T. Aselage and D. Emin in Boron-Rich Solids, (Amer. Inst. Phys. Conf. Proc. **231**, New York, NY, 1991) pp. 322-325.

Part III

Residual Stresses

MEASUREMENT OF STRESS IN CVD DIAMOND FILMS

K.J. GRAY, J.M. OLSON, AND H. WINDISCHMANN
Norton Diamond Films, Goddard Road, Northboro, MA 01532

ABSTRACT

Measurement of stress in CVD diamond by Raman spectroscopy and by the substrate curvature method is discussed. A correction to the commonly applied Stoney thin-film equation for the substrate curvature technique is presented for coatings with a large stiffness mismatch with the substrate. Stress measurement by Raman spectroscopy is complicated by factors such as temperature, domain size, non-hydrostatic stress, and degeneracy lifting. With proper consideration of these complicating factors, substrate curvature and Raman spectroscopy stress measurement results can be reconciled with the predictions based on thermal modeling.

INTRODUCTION

Since the first western reports[1] confirming the growth of diamond by CVD processes, interest in CVD diamond film synthesis has grown rapidly because of the potential for a wide range of products not possible with naturally occurring or high pressure, high temperature synthesized diamond. With the ongoing development of the CVD diamond industry, it has become clear that to insure product reliability and integrity it is necessary to characterize and quantify stresses and stress distribution within diamond films. A comprehensive knowledge of the magnitude and sign of the thermal and intrinsic stress is generally desirable for any thin film application; however, it is particularly important for diamond, since the combination of a rigid atomic structure, high Debye temperature (1800 K), and virtually immobile dislocations produce an extremely brittle material which can fail catastrophically.

There has been significant disagreement among various researchers regarding the magnitude of the stress observed under seemingly similar deposition conditions. The lack of agreement may by associated with the stress measurement techniques. For example, the thin film approximations and underlying assumptions inherent in the common stress measurement techniques such as substrate curvature (Stoney Equation) and $\sin^2 \Psi$ XRD technique are readily violated by diamond with its high elastic modulus and textured growth of columnar grains. The interpretation of Raman spectroscopy results can also be problematic. A film on a substrate is in a biaxial stress state, and under non-hydrostatic stresses the cubic symmetry of the lattice is destroyed, resulting in splitting of the triply degenerate Raman peak. Multiple peaks may also appear as a result of stress inhomogeneity.

In this paper we discuss stress measurement by Raman spectroscopy and by the substrate curvature technique. We report corrections to the Stoney equation frequently required for diamond films. We present several corrections which are necessary for proper interpretation of Raman spectra. The results of stress measurement by both Raman spectroscopy and the substrate curvature technique are compared.

Mat. Res. Soc. Symp. Proc. Vol. 383 © 1995 Materials Research Society

2.0 STRESSES IN CVD DIAMOND FILMS

The total stress in a film is composed of the intrinsic stress and in the case of a films on a substrate, the stress induced by the thermal mismatch with the substrate. The thermal stress is readily computed based on the well known thermal properties of most substrate materials. Intrinsic stresses, however, are not so well known. Many stress studies have been undertaken with various deposition techniques and conditions.[2–14] In general, the results show that the macroscopic intrinsic stress is dependent on methane fraction and deposition temperature[9] and crystallographic orientation[12]. The origin of the stress is attributed to porosity, hydrogen incorporation, and the presence of non-diamond phases at grain boundaries. Both compressive and tensile intrinsic stress are observed, depending on deposition conditions, but independent of deposition technique. Both in situ[6] and ex situ[15] stress analysis show that the assumption of uniform stress throughout the film thickness is valid for thin films but not for thicker films, where the transition from a fine grained equiaxed microstructure to a columnar microstructure is complete. Furthermore, strong strain anisotropy has been observed on a microscopic scale using high resolution Raman spectroscopy[7].

Thermal Mismatch Induced Stress

As the diamond/substrate combination cools following deposition, stress is generated due to differing linear thermal expansions (contractions). This thermal stress is given by

$$\sigma_{\mathrm{th}} = \frac{E}{1-\nu}\Big[\big(\frac{\Delta \mathrm{L}}{\mathrm{L}_0}\big)_{\mathrm{d}} - \big(\frac{\Delta \mathrm{L}}{\mathrm{L}_0}\big)_{\mathrm{s}}\Big], \tag{1}$$

where $\left(\frac{E}{1-\nu}\right)_s$ is the diamond biaxial modulus of 1180 GPa [16] and $\left(\frac{\Delta \mathrm{L}}{\mathrm{L}_0}\right)_{\mathrm{d,s}}$ are obtained from the diamond and substrate linear thermal expansions from room temperature to deposition temperature. As an example, Table 1 lists the stress resulting from deposition of diamond on reaction bonded silicon nitride (RBSN) at various deposition temperatures.

Table 1. Stress vs. deposition temperature for diamond on RBSN

Temperature	ΔL/L$_0$ Diamond	ΔL/L$_0$ RBSN	Strain in Diamond	Tensile Stress in Diamond
900 °C	-0.289%	-0.253%	3.6×10^{-4}	425 MPa
800 °C	-0.241%	-0.217%	2.4×10^{-4}	280 MPa
700 °C	-0.196%	-0.182%	1.4×10^{-4}	165 MPa

Intrinsic Stress

Figure 1 presents the variation of the bulk[17] and elastic[18] modulus of microwave plasma deposited diamond with methane fraction, normalized to the value corresponding to deposition at the lowest methane fraction. As is shown, the modulus of a film may decrease as much as 50% when deposited with a methane fraction of 2.5%. The decrease

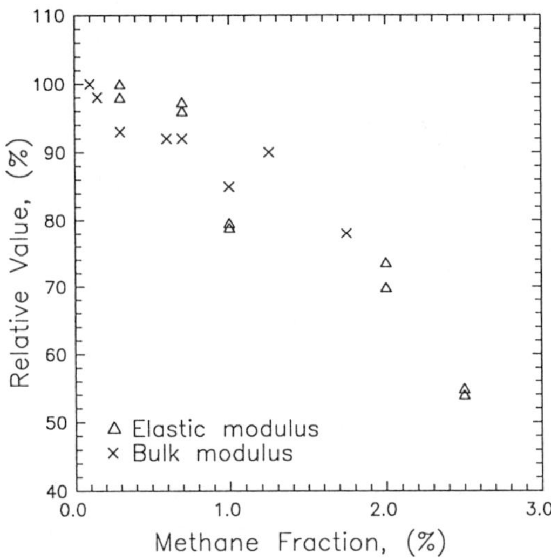

Fig. 1. Variation of the bulk and Young's modulus of CVD diamond with methane fraction during deposition. The moduli are normalized to those resulting from the lowest methane fraction deposition condition.

in modulus with increasing methane fraction is attributed to reduction in the ideal bond coordination number. As the methane fraction increases, the non-diamond component of the film increases and the grain boundary area increases due to decreasing grain size[9].

Within a diamond film produced at a given deposition condition, non-uniform stresses may exist throughout the thickness of the film. Many applications require diamond film thicknesses greater than 50 μm, and such films often exhibit significant chemical and structural heterogeneity due to the transition to a columnar microstructure. Departure from stress isotropy can therefore be expected. Direct evidence of departure from this assumption has been reported using in situ Raman spectroscopy[6]. Similar evidence that this assumption is not always valid has been reported[15] based on measurements using the substrate curvature technique applying the relationship developed by Brotherton, et.al.[19], to calculate the stress variation with distance from the interface.

STRESS MEASUREMENT

Two common techniques employed to characterize the stress state of a diamond film/substrate combination or a free-standing film are substrate curvature and Raman spectroscopy. Since many diamond applications require thickness extending beyond the "thin-film" definition and because of the high elastic modulus of diamond, stress measurement by the substrate curvature technique may require corrections to the approximate relations used, while in other cases the basic assumptions underlying the theory are violated. In the application of Raman spectroscopy, the straightforward model of stress-induced, linear shift in the peak position has been widely applied to estimate stress in diamond films despite the several sources of error lurk beneath this simple model.

Substrate Curvature

The substrate curvature method is a convenient method for measuring the global stress in thin films because it involves only the physical properties of the film and substrate and does not require knowledge of the elastic properties of the film. This is true for the case where the film thickness is small compared to the substrate thickness so that the bending stresses associated with the deformation can be ignored. Furthermore, the film and substrate biaxial elastic moduli must be nearly equivalent.[20] With these restrictions, the stress can be calculated using

$$S_0 = \frac{1}{6}\left(\frac{E}{1-\nu}\right)_s \frac{H^2}{hr}, \tag{2}$$

known as the Stoney equation[21], where H and h are the substrate and film thickness, $\left(\frac{E}{1-\nu}\right)_s$ is the substrate biaxial modulus, and r is the net change in the radius of curvature of the substrate.

As the thickness of the film increases, the film bending moment and resultant flexure stress are no longer negligible, and the film/substrate structure must be treated as a composite beam. The relationship derived by Brenner and Senderoff[22] and Roll[23] must be used to avoid substantial errors in the calculated stress. By ignoring terms of order higher than $\left(\frac{h}{H}\right)$ the modified Brenner-Senderoff equation is given by

$$S = S_0\left(1 + 4R\frac{h}{H} - \frac{h}{H}\right), \tag{3}$$

where R is the ratio of the film and substrate biaxial moduli and S_0 is the Stoney thin-film approximation given by Eq.(2). The elastic properties of the films are included in the second term in Eq.(3), which represents the composite beam correction. This term can be viewed as the spring constant ratio of film and substrate. The net effect is that the deflection of the substrate for a given film stress is reduced by the stiffening effect of the film. Were this term to be ignored, significant errors in the calculated stress might result. Figure 2 presents the Brenner-Senderoff relationship as a function of the thickness ratio h/H, with the stiffness ratio R as a parameter. The resultant stress is normalized to the stress S_0 given by the Stoney equation, Eq.(2). As can be seen, the application of the thin film approximation can produce substantial errors as the film thickness increases above 5% of the substrate thickness for $R > 2$. For example, consider the case of a 25 μm diamond film deposited on a 500 μm thick silicon substrate. Since the biaxial modulus ratio for polycrystalline diamond to (100) silicon is 6.08 the error in using the Stoney equation is greater than 100%.

Raman Spectroscopy

Knight and White[2] first reported the use of Raman spectroscopy as a stress measurement tool for CVD diamond. As discussed above, several factors must be considered

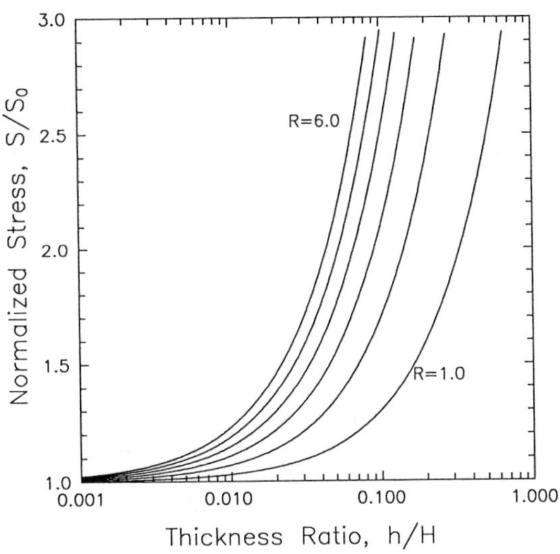

Fig. 2. Stress calculated from Eq. (2) normalized to the stress calculated from Eq. (1) as a function of film to substrate thickness ratio. The parameter R is the ratio of the film biaxial modulus to the substrate biaxial modulus.

when interpreting Raman spectra for stress information: the shift of the Raman peak with temperature, non-hydrostatic stress (which leads to lifting of the triple degeneracy of the line), and the effect of domain size.

Temperature Effect

The diamond Raman peak shifts to lower values with increasing temperature. Excessive laser beam power density applied to a diamond film with a highly absorbing non-diamond phase can produce a measurable peak shift. Henchen[24] has measured a peak position temperature sensitivity of -0.2 $cm^{-1}/10$ °C temperature rise, which can represent a 80 MPa error. In most cases, if care is taken, no temperature correction will be necessary.

Domain Size Effect

To correctly calculate the stress in diamond films, domain size correction may be required depending on the grain size. The domain-size peak shift is obtained by measurement of the diamond Raman peak FWHM. In the absence of domain-size correction data for diamond, LeGrice, et.al.[8], applied an analysis based on a silicon model to correct the Raman peak position and calculate the stress in diamond deposited on Si and Mo by RF plasma CVD. They argue that since the shapes of the phonon dispersion curves near the center of the Brillouin zone along (011) are similar for diamond and silicon, the silicon data may be applied to diamond.

Hydrostatic vs. Biaxial Stress

Several groups have measured the shift in Raman frequency of single crystal diamond as a function of hydrostatic stress. The average hydrostatic stress dependence is 2.58 cm^{-1}/GPa compressive stress (387 MPa/ cm^{-1}). Care must be taken in application of this data, however since films deposited on a substrate are not in a hydrostatic but a biaxial stress state.

In general, a stress tensor can be expressed as a superposition of a hydrostatic stress component and a shear stress component. For the specific case of biaxial stress, the stress tensor is given by

$$\underbrace{\begin{pmatrix} \sigma & 0 & 0 \\ 0 & \sigma & 0 \\ 0 & 0 & 0 \end{pmatrix}}_{\text{biaxial}} = \frac{2}{3}\underbrace{\begin{pmatrix} \sigma & 0 & 0 \\ 0 & \sigma & 0 \\ 0 & 0 & \sigma \end{pmatrix}}_{\text{hydrostatic}} + \frac{1}{3}\underbrace{\begin{pmatrix} \sigma & 0 & 0 \\ 0 & \sigma & 0 \\ 0 & 0 & -2\sigma \end{pmatrix}}_{\text{shear}} . \tag{4}$$

Since the hydrostatic stress component is only 2/3 of the biaxial stress, the calculated stress based on the hydrostatic stress dependence, must be multiplied by 3/2 to obtain the biaxial stress.

As noted above, the stress dependence of both the shift and splitting of the peak varies with crystallographic orientation. Grimsditch, et.al.[24], measured the stress sensitivity of the singlet and doublet for uniaxial compressive stress applied along two crystallographic directions of single crystal diamond. Ager and Drory[25] developed a general model for shift and splitting of the three degenerate components for the more common biaxial plane stress state as a function of crystallographic direction.

Application of Raman measurement corrections

All of the factors listed above are accounted for via

$$\Delta\nu_{\text{tot}} = \frac{3}{2}\Delta\nu_{\text{obs}} - \Delta\nu_{\text{dom}} - \Delta\nu_{\text{temp}} \tag{5}$$

where $\Delta\nu_{\text{tot}}$ is the total (corrected shift), $\Delta\nu_{\text{obs}}$ is the observed or measured shift, $\Delta\nu_{\text{dom}}$ is the domain size correction obtained from reference 8, and $\Delta\nu_{\text{temp}}$ is the measurement temperature correction described above. Equation 5, with the $\frac{3}{2}$ factor, applies only in the case of a film in a biaxial stress state. Note that the biaxial correction factor applies only to the first term in the equation. This is because defects limit domain size to a scale much smaller than the crystal size. The domain size effect is a microscopic phenomenon, and is independent of applied, macroscopic stresses. Similarly, the effect of sample temperature serves only to shift the reference.

EXPERIMENTAL RESULTS

The results of stress measurements by the techniques described above on diamond films deposited on molybdenum, tungsten carbide (cobalt cemented), and RBSN are summarized in Table 2.

We measured the stress in a 70 μm diamond film deposited on a 3 mm thick molybdenum substrate by both the substrate curvature method and by Raman spectroscopy. Application of the 3/2 correction factor to the Raman-shift derived stress value of 1350 MPa, based on the hydrostatic stress dependence, yields a measured stress of 2030 MPa. The stress measured by substrate curvature was 2200 MPa. The agreement between the models is well within the measurement precision.

In a second experiment, two RBSN disks were coated with diamond at a deposition temperature of 700 °C. Upon cooling, an extensive network of tensile failure cracks developed on one of the disks. Raman measurements of the uncracked disk produced a stress measurement of -155 MPa, in reasonable agreement with thermal predictions of -165 MPa. Raman measurements of the cracked disk, taken adjacent to a crack junction (where the cracks have relieved the diamond of any thermal stresses) yielded an intrinsic stress measurement of 60 MPa ± 100 MPa.

Finally, in the case of a diamond coated tungsten carbide cutting tool insert, these measurements were used to estimate the magnitude of the intrinsic stress of the diamond coating using $\sigma_{int} = \sigma_{total} - \sigma_{th}$. With a calculated thermal stress of 1.3 GPa and a Raman measured total stress of 1.46 GPa, the intrinsic stress is estimated at approximately 150 MPa, which is only somewhat greater than the measurement error of 100 MPa.

Table 2. Stress measurements of diamond films on a substrate

	Molybdenum	WC(6% Co)	RBSN
Thermal (calculated)	2120 MPa	1300 MPa	-165 MPa
Curvature method	2200 MPa		
Raman components			
Hydrostatic shift	1350 MPa	540 MPa	-560 MPa
Biaxial correction	680 MPa	270 MPa	-280 MPa
Domain correction	0 MPa	650 MPa	685 MPa
Total Raman result	2030 MPa	1460 MPa *	-155 MPa

Raman measurements ± 100 MPa

* Diamond film known to exhibit high intrinsic stress. Measurement of intrinsic stress could be obtained by curvature method once the substrate is thinned or removed

Conclusion

Given the unique nature of diamond from a materials science viewpoint, it is not surprising that many of the straightforward and simple techniques used to characterize stress in other materials become more complicated when applied to diamond. The substrate curvature technique can yield accurate results if corrections are made for the film thickness and for the biaxial modulus of diamond. Consideration of the nature of the stress in the sample is essential in order to achieve accurate stress measurements by Raman spectroscopy. Experiments demonstrate that if these factors are accounted for, good agreement can be reached between the results of these stress measurement techniques.

ACKNOWLEDGEMENTS

This work was supported by ARPA through ONR/NRL contract number N00014-91-C-266.

REFERENCES

1. S. Matsumoto, Y. Sato, M. Tsutsumi, N. Setaka, J. Mater. Sci., **17** 3106 (1982).
2. D.S. Knight and W.B. White, J. Mater. Res., **2** 385 (1989).
3. Y.H. Lee, K.J. Bachmann, J.T. Glass, Y.M. LeGrice, and R.J. Nemanich Appl. Phys. Lett. **57** 1916 (1990).
4. E.D. Specht, R.E. Clausing, and L. Heatherly J. Mater. Res. **5** 2351 (1990).
5. M. Yoshikawa, G. Katagri, H. Ishida, A. Ishitani, M. Ono, and K. Matsumura, Appl. Phys. Lett. **55** 2608 (1989).
6. L.J. Bernardez, and K.F. McCarty, Diamond and Relat. Mater. **3** 22 (1993).
7. D.J. Vestyck, D. Shechtman, and J.E. Butler, *Workshop on Characterization of Diamond Films, Washington, D.C., Feb. 1994,* NIST, 1994.
8. Y.M. LeGrice, R.J. Nemanich, J.T. Glass, Y.H. Lee, R.A. Rudder, and R.J. Markunas, *Materials Research Society Symp. Proc. 162,* 219 (1990).
9. H. Windischmann, G.F. Epps, Y. Cong, and R.W. Collins, J. Appl. Phys. **69** 2231 (1991).
10. S.A. Stuart, S. Prawer, and P.S. Weiser, Appl. Phys. Lett. **62** 1227 (1993).
11. P.K. Bachmann, H. Lade, D. Leers, D.U. Wiechert, Diamond and Relat. Mater. **3** 799 (1994).
12. S. Sails, M. Bowden, D.J. Gardiner, S. Haq, and J. Savage, Appl. Phys. Lett., in press.
13. W. Wanlu, L. Kejun, and L. Aimin, Thin Solid Films **215** 174 (1992).
14. M. Yoshikawa, H. Ishida, A. Ishitani, T. Murikami, S. Koizumi, and T. Inuzuka, Appl. Phys. Lett. **57** (1990).
15. N.S. Van Damme, D.C. Nagel and S.R. Winzer, Appl. Phys. Lett. **58** 2919 (1991).
16. G.F. Cardinale and R.W. Tustison, J.Vac.Sci.Tech.A, **9**, 2204 (1991).
17. Y. Sato and M. Kamo, in J.E. Fields (ed.),*Properties of Natural and Synthetic Diamond*, Academic Press, San Diego, 1992, p. 444.
18. H. Windischmann and G.F. Epps, Diamond and Relat. Mater. **1** 656 (1992).
19. D. Brotherton, T.G. Read, D.R. Lamb and A.F.W. Willoughby, Solid State Electr. **16** 1367 (1973).
20. F.J. Von Preissing, J. Appl. Phys. **66** 4262 (1989).
21. G.G. Stoney, Proc. Royal Soc. London **A 82** 172 (1909).
22. A. Brenner and S. Senderoff, J. Res. Natl. Bur. Stand. **42** 105 (1949).
23. K. Roll, J. Appl. Phys. **47** 3224 (1976).
24. H. Henchen, Thin Solid Films **212** 206 (1990).
24. J.W. Ager III, D.K. Veirs. and G.M. Rosenblatt, Phys. Rev. B **43** 6491 (1991).
25. J.W. Ager III, and M.D. Drory, Phys. Rev. B **48** (1993) 2601.

RESIDUAL STRESS IN DIAMOND AND AMORPHOUS CARBON FILMS

JOEL W. AGER III
Center for Advanced Materials, Materials Sciences Division, Lawrence Berkeley Laboratory, Berkeley, CA 94720

ABSTRACT

Both diamond and amorphous carbon films can contain substantial residual stresses (up to 10 GPa and higher) due to a combination of thermal and intrinsic effects. Residual stresses in the films limit the maximum thickness to which films can be grown and can lead to weakening and/or failure of the film/substrate interface. In addition, a quantitative measurement of the residual stress in the film is required for an accurate determination of the film/substrate interface fracture toughness. Spectroscopic measurements of residual stresses in diamond and amorphous carbon films based on frequency shifts of Raman-active phonon modes are presented. The spatial resolution of the technique is 1 μm in the lateral direction. In transparent materials such as single crystal diamond, stress profiles with 10 μm resolution in the axial direction can be measured. Examples of such stress profiles are presented for natural and CVD-grown single crystals. Raman measurements of thermal stress in thin polycrystalline diamond films on metal substrates are presented. Initial, spatially resolved measurements of large, compressive intrinsic growth stresses in high-hardness amorphous carbon films grown by ion beam techniques are presented. The accuracy of the Raman residual stress measurement depends on the film structure and is ±60 MPa for single crystal diamond, ±120 MPa for polycrystalline diamond, and ±500 MPa for hard amorphous carbon.

INTRODUCTION

There is considerable interest in depositing hard thin films on ductile substrates, e.g. metals, in order to reduce friction in wear in applications such as tooling for high speed machining. Both polycrystalline diamond and hard amorphous carbon coating processes have been developed for this purpose. Coating delamination is a common failure mode during coating testing, particularly for the hardest coatings. To address this issue, quantitative methods have been developed to assess the film's adhesion by measurements of the interface fracture toughness [1,2].

The value of the film's residual stress enters as squared term in the analysis of both indentation [1] and blister test measurements [2] of the interface fracture toughness. Notably, both diamond and amorphous carbon films can contain substantial residual stresses (up to 15 GPa) due to a combination of thermal and intrinsic effects. Therefore, a quantitative measurement of the film residual state is required for an accurate determination of the interface fracture toughness. Traditionally, the residual stress, σ, of a thin film is measured by depositing the film on a thin substrate and relating the observed curvature to the film stress using the Stoney equation:

Mat. Res. Soc. Symp. Proc. Vol. 383

$$\sigma = \frac{E z_s^2}{6(1-\upsilon) z_f R} \quad (1)$$

where E and υ are the Young's modulus and Poisson ratio of the substrate and z_s and z_f are the thicknesses of the substrate and film, respectively. The substrate bend method is adequate for estimating the bulk stress of the film, but for (1) ensuring that the expected stress is actually present when films are deposited on real (i.e., not thin) substrates and (2) obtaining spatial resolution, a local method is desirable.

It will be shown that for both diamond and amorphous carbon films, Raman microscopy can be used as a local probe of residual stress. In the case of diamond, the measurement of stress by Raman spectroscopy is relatively straightforward and example of stress measurements from polycrystalline and single crystal material are presented. Measurements in amorphous carbon are more difficult, because the phonon frequencies in stress-free material cannot be known *a priori*. Initial work in a highly stressed, hard, amorphous carbon film is presented that illustrates the advantages and limitations of the Raman technique.

EXPERIMENTAL

CVD-grown diamond

Both single crystal and polycrystalline diamond films were used in this study. Single-crystal diamond films were grown homoepitaxially on high temperature high pressure (HTHP) type IIa substrates provided by GE using a microwave plasma enhanced CVD process described elsewhere [3]. Stress analyses will be presented here from two epitaxial films: (1) a 400 μm layer and (2) a 200 μm layer, both of which were grown on (100)-oriented substrates that were nominally 400 μm thick. Polycrystalline diamond films (nominally 1 μm thick) were grown on 60 x 20 x 3 mm rectangular substrates of Ti-6Al-4V alloy by microwave plasma CVD processes described previously [4,5].

Vacuum-arc carbon

A detailed description of the film growth method is given elsewhere [6]; only a brief description will be given here. The vacuum arc plasma source produces a stream of fully ionized carbon plasma with a directed energy of the ions of 20 eV. The substrate which is exposed to the plasma is repetitively pulse biased at a negative voltage (1 μs pulse duration, variable pulse off time) to combine low energy deposition (pulse off, C^+ ion energy 20 eV) with implantation (pulse on, C^+ ion energy = pulse bias value + 20 eV). The structure of the film can be controlled by varying the bias voltage and the duty cycle (ratio of pulse on to pulse off time). It has been found previously that the conditions that produce the maximum hardness films are -100 V bias and 50% duty cycle. For substrate curvature stress measurements, 100 nm films were grown on 200 μm thick Si wafers (25 mm diameter). The substrate curvature was measured over the central 10 mm of the wafer before and after growth with a profilometer and the film stress was calculated with the Stoney equation. The maximum hardness film (-100 V bias) also has the maximum biaxial compressive stress, 10.5 GPa, occurs at a bias voltage of

-100 V [7]. Preliminary nanoindentation, electron energy loss spectroscopy (EELS), and hydrogen forward scattering characterization of films deposited with the same process at -100 V yield a peak hardness of 60 GPa, approximate Young's modulus of 600 GPa, an sp^3 content of 85%, and a hydrogen concentration of less than 5% [8].

Raman microscopy

The optical stress measurements are based on obtaining Raman spectra using a homebuilt Raman microprobe based on a Olympus metallurgical microscope. Using either a 50x (NA 0.8) or 80x (NA 0.90) objective, a ca. 1 μm diffraction-limited spot size is obtained in the focal plane. Excitation at 488 nm was used for all measurements reported here. For measurements with spatial resolution in the axial (lateral) direction the z (x) position of the sample stage was controlled to ±2 μm by a DC motor with a closed-looped optical encoder (Microkinetics). For axially resolved measurements in the transparent single crystal diamond samples, the top and bottom surfaces of the sample were located precisely by the specular reflection of the laser light. The sample volume is estimated to be ca. 2 μm in lateral extent and ca. 10 μm in vertical extent. In all cases the laser power was kept below 10 mW (1 mW for the amorphous films) to avoid heating of the sample volume.

The analysis of the Raman spectra depended on the sample. For the single crystal diamond samples, Raman spectra were obtained as a function of depth through the crystals. The diamond phonon was fit to a Lorentzian lineshape by a non-linear least-squares fit. The precision of the line center determinations is ±0.1 cm^{-1}. In all cases at least two separate experiments at different lateral position were performed with opposite faces of the diamond facing the microscope objective. It all cases it was assumed that the shift of the Raman phonon from the unstressed value of 1332.5 cm^{-1} is due entirely to biaxial stress. The frequency shift of the allowed singlet phonon is -610 MPa/cm^{-1} for biaxial stress in the (001) plane [4].

For the polycrystalline diamond films grown on metals, the diamond Raman peak is split and shifted to higher frequency by the biaxial compressive growth stress. The split peak is fit to two overlapping Gaussians, the "doublet," at higher frequency and the "singlet" at lower frequency. In this case, the precision of the line center determination is ±0.2 cm^{-1}. The Raman spectrum of vacuum-arc deposited carbon is a broad, asymmetric peak whose maximum ranges from ca. 1490 to 1560 cm^{-1}, depending on deposition conditions. The peak center of the observed asymmetric phonon feature was determined by fitting a Gaussian line shape to the symmetric, central region. The precision of the peak center determination is ±1 cm^{-1} by this method.

Stress analysis of Raman spectra of diamond and amorphous carbon

The shift and splitting of the Raman phonon into a singlet and doublet mode caused by biaxial stress for diamond in different crystal orientation has been published previously [4,9]. A summary of the results will be given here. For biaxial stress in the indicated plane, the shift and splitting of the Raman phonon can be summarized as follows.

Biaxial stress orientation	Singlet shift (cm^{-1}/GPa)	Doublet shift (cm^{-1}/GPa)
(100)	-1.64	-2.37
(111)	-0.67	-2.86
(110)	-0.90	-2.25
$(11\bar{2})$	-0.90	-2.80
average	-0.93	-2.60

In cases where the biaxial stress produces three peaks (i.e. the double is split slightly in the (110) and $(11\bar{2})$ cases), the values have been averaged. The "average value" is averaged over crystallographic orientation and is intended for use in polycrystalline films. When the stresses are small in the film, the doublet and singlet may not be resolved. In this case, the weighted average value of -2.05 cm^{-1}/GPa is recommended. Note that this value for the shift of the Raman phonon caused by biaxial stress is approximately 2/3 of the value for hydrostatic stress in diamond measured in diamond anvil cell experiments, -2.9 cm^{-1}/GPa [10,11]

It is not possible to calculate the shift of the Raman peak of amorphous carbon by using parameters derived from measurements on crystalline materials. However, in what follows, it will be instructive to compare values observed in amorphous carbon to those expected for disordered graphites. The shift of the Raman phonon in graphite with applied biaxial stress in the graphite plane can be calculated following the analysis of Sakata *et al.* [12]. Using a coordinate system (x, y, z) with x and y in the plane and z perpendicular to the graphite plane, the strains produced by a biaxial stress $\tau = \sigma_{xx} = \sigma_{yy}$ are:

$$\begin{bmatrix} \varepsilon_{xx} \\ \varepsilon_{yy} \\ \varepsilon_{zz} \\ \varepsilon_{yz} \\ \varepsilon_{zx} \\ \varepsilon_{xy} \end{bmatrix} = \begin{bmatrix} S_{11} & S_{12} & S_{13} & & & \\ S_{12} & S_{11} & S_{13} & & & \\ S_{13} & S_{13} & S_{33} & & & \\ & & & S_{44} & & \\ & & & & S_{44} & \\ & & & & & 2(S_{11}-S_{12}) \end{bmatrix} \begin{bmatrix} \tau \\ \tau \\ 0 \\ 0 \\ 0 \\ 0 \end{bmatrix} \quad (2)$$

such that

$$\varepsilon_{xx} = (S_{11} + S_{12})\tau \quad (3a)$$
$$\varepsilon_{yy} = (S_{11} + S_{12})\tau \quad (3b)$$
$$\varepsilon_{zz} = 2S_{13}\tau \quad (3c)$$

with all shears equal to zero. The secular equation is

$$\begin{vmatrix} A(\varepsilon_{xx} + \varepsilon_{yy}) - \lambda & B(\varepsilon_{xx} - \varepsilon_{yy} + 2i\varepsilon_{xy}) \\ B(\varepsilon_{xx} - \varepsilon_{yy} - 2i\varepsilon_{xy}) & A(\varepsilon_{xx} + \varepsilon_{yy}) - \lambda \end{vmatrix} = 0 \quad (4)$$

where

$$\lambda = \omega_\tau^2 - \omega_0^2 \quad (5)$$

and the subscript indicates the frequencies of the stressed and unstressed Raman active E_{2g2} phonon. For biaxial stress the solutions are degenerate (no splitting of the phonon) and are

$$\omega_\tau - \omega_0 = \frac{1}{2\omega_0}(2A\varepsilon_{xx}) = \frac{A}{\omega_0}(S_{11} + S_{12})\tau. \quad (6)$$

Using A = -1.44x10^7 cm^{-2} from Sakata *et al.*, the graphite elastic constants S_{11} = 0.98x10^{-12} Pa^{-1} and S_{12} = -0.16x10^{-12} Pa^{-1} [13], and ω_0 = 1540 cm^{-1} yields a stress shift of -7.7 cm^{-1}/GPa. Due to the very weak interplane bonding, very little effect is expected from plane stress perpendicular to the graphite planes. Assuming a random orientation of the graphite planes in the disordered system with respect to the stress plane reduces this value by a factor of $2/\pi$ to yield -4.9 cm^{-1}/GPa for a disordered graphite system.

RESULTS AND DISCUSSION

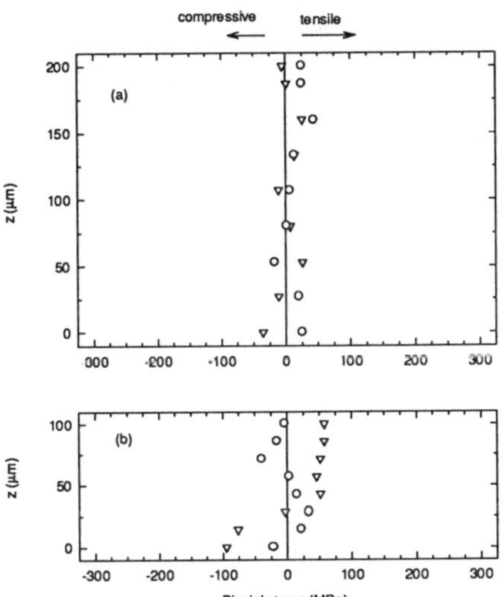

Fig. 1 Axial profiles of stress (as biaxial) in (a) type Ia and (b) type IIa natural diamond measured by depth profiling Raman spectroscopy. In both cases, the circles and squares are measurements from different lateral positions on the sample.

Single crystal diamond films

To test the stress measurement technique, stress profiles were measured through a 200 μm thick natural Ia and a 100 μm thick natural IIa diamond crystal. The results are shown in Fig. 1. The Ia sample is uniform (with respect to stress) in both the z and lateral directions. The spread in the measured stress values, ±30 MPa, is within the expected precision, ±60 MPa, of the measurement. The IIa profile shows more variation as a function of z; in addition, the two stress profiles at different lateral positions on the crystal are not the same. The difference is attributed to the presence of inhomogeneous stress in the IIa crystal. This is to be expected; it is well-known from X-ray topography and other measurements that type II diamond are more plastically deformed than type I diamonds, because dislocation movement is hindered by the platelets that occur in many type Ia damonds [14].

Raman spectra from sample volumes inside the CVD-grown epilayer have a higher background than spectra from the HTHP substrate; otherwise, except for the peak shifts due to stress, the Raman spectra from the substrate and epilayer were identical. Figures 2 and 3 show the results of depth profiling scans through the as-grown 400 µm and 200 µm homoepitaxial film and HTHP substrates. In the crystal with the 400 µm film, the two stress profiles obtained in the substrate at different lateral positions agree to within experimental error. The stress is slightly compressive at the surface and becomes more tensile near the interface. The stress

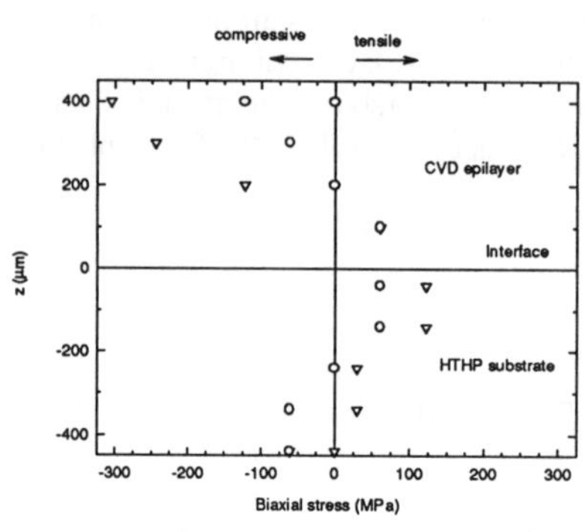

Fig. 2. Biaxial stress as a function of vertical position, z, in a 400 µm homoepitaxial diamond film grown on a HTHP substrate. The circles and squares are the results of separate profiles at different lateral positions.

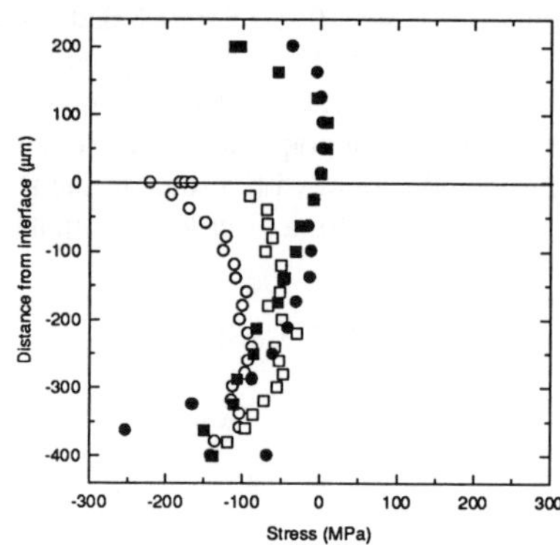

Fig. 3. Biaxial stress as a function of vertical position, z, in a 200 µm homoepitaxial diamond film grown on a HTHP substrate. The circles and squares are the results of separate profiles at different lateral positions. The open circles are values measured in the substrate before epilayer growth.

profiles in the epilayer are tensile near the interface and compressive near the surface. Profiles obtained at different lateral points do not agree, which indicates the presence of inhomogeneous stress. In the crystal with the 200 µm film, the stress profile obtained in the substrate at before and after growth agree to within experimental error. The epilayer has an overall lower stress compared to the film shown in Fig. 2, although there is still compressive stress near the surface. These results are consistent with preliminary X-ray topography measurements [15], in which the 400 µm epitaxial film was shown to have an much higher density of extended defects compared to the 200 µm epitaxial film. In both cases, the substrate has a lower concentration of individual dislocation and stacking faults.

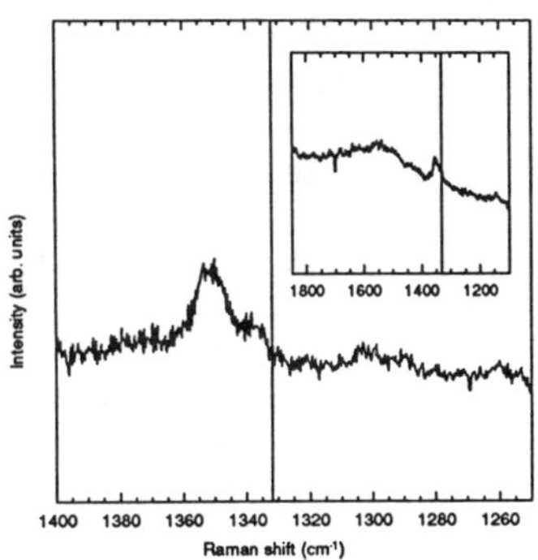

Fig. 4 Raman spectrum of CVD diamond grown on Ti-6Al-4V. The inset shows a lower resolution spectrum (with a broad feature at ca. 1500 cm^{-1} which is attributed to a small amount of amorphous carbon. The vertical lines indicate the Raman frequency of stress-free diamond, 1332 cm^{-1}.

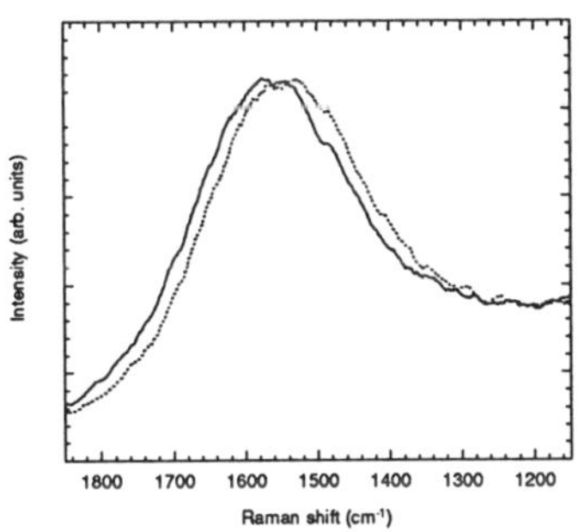

Fig. 5 Raman spectra of a 100 nm film deposited at -100 V bias voltage: solid line, adhering film, residual stress = -10.5 GPa (compressive); dotted line, delaminated film, residual stress relaxed. The shift of the peak position to lower frequency in the delaminated film is 20 cm^{-1}.

Polycrystalline diamond films

The Raman spectrum of a 1 μm polycrystalline diamond film deposited on Ti-6Al-4V is shown in Fig. 4 [4]. Using the analysis described above, a compressive stress value of -7.1 GPa (compressive) was measured. This is in good agreement with the value of -6.7 GPa predicted from the thermal expansion coefficient data. The agreement of the measured value and the value predicted from the thermal expansion coefficient mismatch indicates that the diamond film has accomodated all of the thermal stress and that plastic deformation of the interface and/or buckling of the film has not occured.

Amorphous carbon films

Films deposited at ground and -100 V sample bias by vacuum-deposition are have both the highest sp^3 content (81% and 85%, respectively, as measured by K-edge EELS [8]) and the highest biaxial, compressive, residual stress, 9.0 and 11.0 GPa, respectively [7]. Due to the high stress, at a nominal film thickness of 50 nm, some portions of these film delaminated from the substrate after a period of time. Raman spectra were measured from the adhering regions and from the delaminated regions of the ground potential and -100 V films. In each case, the Raman peak of delaminated material is at lower frequency than adhering material. The shift is substantial (10 - 20 cm^{-1}) and is well above the peak determination uncertainty. This is shown in Fig. 5 for the -100 V film (maximum stress).

The shift of the Raman peak to lower frequency in delaminated material is attributed to relaxation of the compressive growth stress in the film. The shift is in expected direction and is consistent between repeated measurements. We have observed previously that CVD-grown diamond films with substantial compressive thermal growth stresses (up to 7 GPa) relax completely (less than 0.3 GPa) upon delamination [4]; we assume that the hard amorphous carbon films in this study behave in the same way. Averaging the shifts of the ground (14 cm^{-1}) and -100 V (20 cm^{-1}) measurements, we obtain a stress shift of -1.9 cm^{-1}/GPa for biaxial plane stress in hard, vacuum-arc-deposited a-C films. It should be noted that to compare this value to what would be expected for applied hydrostatic stress, as in a diamond anvil cell, it should be multiplied by 3/2.

It is interesting to compare the observed value to what would be expected for (1) a disordered diamond structure (100% sp^3) and (2) a disordered graphite structure (100% sp^2). The latter analysis is outlined above. The observed value (-1.9 cm^{-1}/GPa) is much closer to the disordered diamond value (-2.05 cm^{-1}/GPa) than it is to estimated disordered graphite model (-4.9 cm^{-1}/GPa). Although the agreement between the observed value and the disordered diamond model is good and is nominally consistent with the observed sp^3 fraction in the hard amorphous films, some caution is warranted. The Raman spectrum of amorphous carbon films is known to probe the sp^2 bonds of the material rather than the sp^3 bond. Furthermore, due to resonance Raman effects, the Raman spectrum at a given laser energy (2.54 eV in this work) probes only some fraction of the sp^2 bonds in the material. Finally, the actual stress distribution in the film may be inhomogeneous on the length scale of the sp^2 clustering — the present experiments probe only the stress state in those domains with resonance enhancement at the laser energy. Measurements at different laser energies may yield different values. Nevertheless, the observed value can be used as an empirical calibration for Raman stress in films deposited by vacuum-arc and similar (i.e. ion beam and laser ablation) techniques. The method can be used to measure stress variations with the lateral resolution of the Raman microprobe, 1 μm. Combining the stress shift value and the line center precision yields a stress measurement precision of ±500 MPa.

CONCLUSIONS

Several examples have been given of using Raman microscopy to measure residual stress in diamond and amorphous carbon films. An accurate, non-destructive, and rapid method for measuring vertical stress profiles in transparent crystals is demonstrated. The method is accurate to ±60 MPa in diamond and has a 10 μm spatial resolution. It is shown that intrinsic growth stresses in homoepitaxial diamond films can be measured using this method. Thermal growth stresses with 1 μm lateral resolution are measured in diamond grown on metals. The precision of the stress measurement is ±120 MPa. Finally, a Raman stress measurement technique is developed for amorphous carbon films. The calibration is based on comparing the Raman spectra of adhering and delaminated material. For hard, amorphous carbon films (ca. 80% sp^3 content) the calibration factor is -1.9 cm^{-1}/GPa and the overall precision of the stress measurement is ±500 MPa.

ACKNOWLEDGMENTS

This work was by the LBL was supported by DMS/DOE Center for Excellence in Synthesis and Processing project, "Processing for Surface Hardness," and by a Office of Energy Research Laboratory Technology Transfer (ER-LTT) Cooperative Research and Development Agreement [CRADA BG 94-120(00)], through the Director, Office of Energy Research, U.S. Department of Energy, under contract No. DE-AC03-76SF00098. M. D.

Drory and M. A. Plano of Crystallume provided the diamond films and S. Anders, A. Anders, and I. G. Brown of LBL provided the amorphous carbon films used in this work.

REFERENCES

1. M. D. Drory and J. W. Hutchinson, Science **263**, 1753 (1994); M. D. Drory and J. W. Hutchinson, this volume.

2. W. D. Nix, J. J. Vlassak, R. J. Holdfelder, and J. T. Sizemore, this volume.

3. M. I. Landstrass, M. A. Plano, M. A. Moreno, S. McWilliams, L. S. Pan, D. R. Kania, and S. Han, in *Proceedings of Diamond Films '92/ICNDST-3*.

4. J. W. Ager III and M. D. Drory, Phys. Rev. **B 48**, 2601 (1993).

5. S. S. Perry, J. W. Ager III, G. A. Somorjai, R. J. McClelland, and M. D. Drory, J. Appl. Phys. **74**, 7542 (1994).

6. S. Anders, A. Anders, I. G. Brown, B. Wei, K. Komvopoulos, J. W. Ager III, and K. M. Yu, Surf. Coatings Technol. **68/69**, 388 (1994).

7. J. W. Ager III, S. Anders, A. Anders, I. G. Brown, Appl. Phys. Lett., in press.

8. G. M. Pharr, D. L. Callahan, S. D. McAdams, T. Y. Tsui, S. Anders, A. Anders, J. W. Ager III, I. G. Brown, C. S. Bhatia, S. R. P. Silva, and J. Roberton, Appl. Phys. Lett., submitted.

9. M. H. Grimsditch, E. Anastassakis, and M. Cardona, Phys. Rev. **B 18**, 901 (1978).

10. H. Boppart, J. van Straaten, and I. F. Silvera, Phys. Rev. **B 32**, 1423 (1985).

11. M. Hanfland, K. Syassen, S. Fahy, S. G. Louie, and M. L. Cohen, Physica **139/140B**, 516 (1986).

12. H. Sakata, G. Dresselhaus, M. S. Dresselhaus, and M. Endo, J. Appl. Phys. **63**, 2769 (1988).

13. M. S. Dresselhaus, G. Dresselhaus, K. Sugihara, I. L. Spain, and H. A. Goldberg, *Graphite Fibers and Filaments*, (Springer Verlag, Heidelberg, 1988), pp. 85 - 95, 120-123.

14. W. J. P. Van Enckevort, "Physical, Chemical, and Microstructural Characterization and Properties of Diamond, in *Synthetic Diamond*, ed. K. E. Spear and J. P. Dismukes, (Wiley, New York, 1994) pp. 307-353.

15. M. A. Plano, M. D. Moyer, M. A. Moreno, D. Black, H. Burdette, J. W. Ager III, and A. L. Chen, presented at the 1995 Spring Electrochemistry Society Meeting (unpublished).

[illegible] and [illegible] for the [illegible] films and [illegible] provided [illegible] film used in this work.

REFERENCES

[illegible]

STRESS IN THIN DIAMOND FILMS ON VARIOUS MATERIALS MEASURED BY MICRORAMAN SPECTROSCOPY

V.G. RALCHENKO, E.D. OBRAZTSOVA, K.G. KOROTUSHENKO, A.A. SMOLIN, S.M. PIMENOV AND V.G. PEREVERZEV
Institute of General Physics, Russian Academy of Sciences, ul. Vavilova 38, Moscow 117942, Russia

ABSTRACT

MicroRaman spectroscopy has been used to determine residual stress in polycrystalline diamond films grown by CVD process in a DC arc plasma from CH_4/H_2 gas mixtures. A variety of substrate materials, including silica glass, Si, SiC, Mo, Cu, Ni, Fe-Ni alloy, WC-Co and steel, were selected in order to extend as much as possible the range of the substrate thermal expansion coefficients (CTE). Diamond adhesion to catalytically active materials (Ni, Fe, Co) and Cu was improved with a thin buffer layer of CVD tungsten. The observed Raman peak shifts were converted to stress values σ_{meas} in the framework of biaxial stress model. Compressive stress has been found in diamond films at all substrates, except SiO_2, which provided a tensile stress up to +2.3 GPa. The maximum compressive stress has been detected for Ni substrate σ_{meas} =-11.4 GPa at room temperature and σ_{meas} =-14.3 GPa at T=78 K. A linear dependence of stress on temperature was found at Ni in the temperature range 78-520 K, in good agreement with prediction of thermal stress model. The splitting of diamond peak to singlet and doublet becomes observable at σ>8-9 GPa. The measured stress correlates in general with calculated thermal stress, increasing with CTE value of respective substrate materials, copper being the only exclusion because of its high plasticity.

INTRODUCTION

Now more than 10 different methods of diamond growth by CVD processes are utilized to obtain polycrystalline diamond coatings on various materials. Typically the deposition is performed at high temperatures 700-900°C, and after cooling down the sample to room temperature a residual thermal stress is generated in the film because of a mismatch in thermal expansion coefficients between diamond and substrate. Almost in all cases the thermal stress appeared to be compressive due to very low CTE for diamond, $\alpha=0.8\times10^{-6}\,K^{-1}$ at room temperature. The maximum values of residual stress of about 7 GPa were reported for diamond films deposited on Ti alloy [1] and steel [2]. There is also a tensile intrinsic stress caused, presumably by defects and nondiamond carbon inclusions, but its value is relatively low, less than 1 GPa [3 ,4]. An information about thermal stress is important for many applications of diamond films, including fabrication of thin membranes, optical protective coatings, cutting tools and antifriction coatings, which may undergo high mechanical load upon performance. Too high residual stress may cause a cracking or delamination of the coating. Besides, the knowledge of maximum stress, which can be accommodated without film peeling off, could serve as a quantitative measure of diamond adhesion to substrate.

Raman spectroscopy is a very convenient technique for stress measurements both in single crystal diamonds [5 ,6] and polycrystalline diamond films [1,2,7 ,8 ,9]. The narrow Raman peak at 1332 cm^{-1} corresponds to stress-free diamond, but it shifts to lower frequencies upon tensile stress, or to higher ones upon compressive stress. The peak shifts measured for thin film can be converted to the stress using hydrostatic [7], uniaxial [8] or biaxial [1] models. In the present paper we used Raman spectroscopy to determine the residual stress in diamond films on following materials: insulator (SiO_2), semiconductors (Si, SiC) and metals (Mo, Ni, Fe–Ni alloy, steel, WC–Co, Cu). For Raman shift-stress linear relation we used the biaxial stress calculations made by Ager and Drory [1], which are adequate for thermal stress induced in film plane.

Mat. Res. Soc. Symp. Proc. Vol. 383 © 1995 Materials Research Society

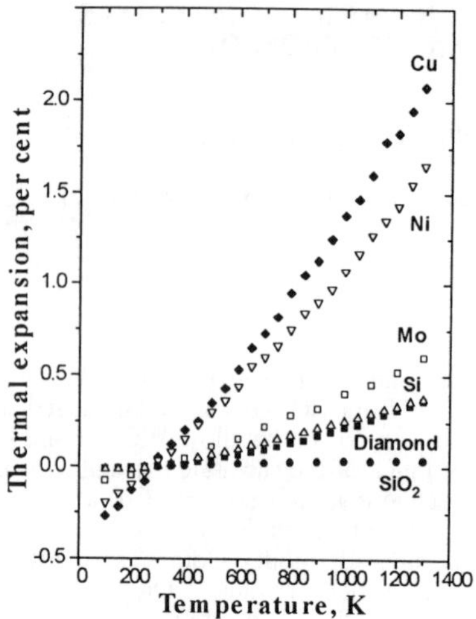

Fig.1. Thermal expansion, ε, as a function of temperature for diamond and some substrates. Zero expansion corresponds to T=293 K.

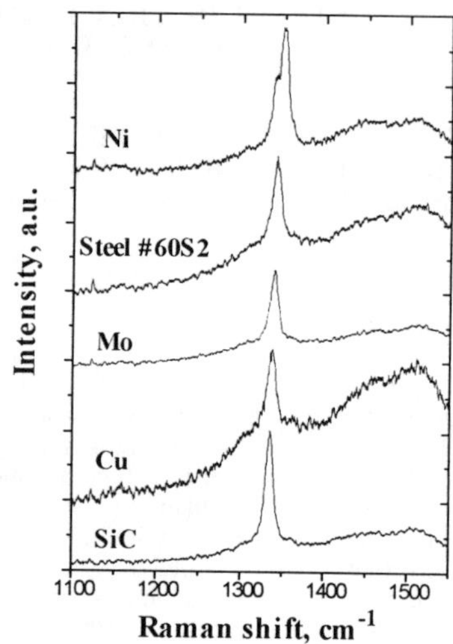

Fig.2. Raman spectra in wide spectral range for diamond films at five substrates. A broad peak at 1500 cm^{-1} is from amorphous carbon.

EXPERIMENTAL

The list of used substrates includes silica glass, Si and 6H–SiC single crystals, polycrystalline Mo, steels #60S2 and #R18 (Fe–1.7W–4Cr), WC–Co (6%), Ni, Fe–Ni alloy (Fe–38Ni–1.7V), and Cu. A part of these materials are known to have pore adhesion to CVD diamond because of either a primary graphite layer formation due to catalytic activity (Fe, Ni, Co), or absence of respective carbide interlayer (Cu). To improve the adhesion we used a thin buffer layer of tungsten to pre-coat Ni, Fe-Ni alloy, steel, WC-Co and Cu. The W layer of typical thickness 5-7 μm was deposited by a CVD process from WF_6/H_2 mixtures at 550°C [2].

Diamond films were grown on square 9x9 mm substrates of 1-3 mm thickness in a DC arc discharge using CH_4/H_2 gas mixtures as described elsewhere [10]. Briefly, the plasma was ignited between a tantalum carbide rode (cathode) and graphite sample holder (anode), on which the substrate was placed. The total gas pressure was 100 Torr, methane content was 4% or 5%, gas flow rate was 100 sccm, substrate temperature T_d was kept between 900-950°C for all samples except SiO_2 (690°C). The diamond film thickness was about 10 μm for majority of samples, and about 1 μm for Si and SiO_2 substrates.

The CTE value of the substrate materials tested spans a wide range from $\alpha=0.5\times10^{-6}K^{-1}$ for SiO_2 to $\alpha=17\times10^{-6}K^{-1}$ for Cu. The thermal expansion ε, is determined as

$$\varepsilon = \int_{293K}^{T} \alpha(T)dT \quad (1)$$

and plotted in Fig.1 as function of temperature for some of the materials as tabulated in [11]. The thermal stress can be calculated as

$$\sigma_{th}=E(\varepsilon_f-\varepsilon_s)/(1-\nu) \quad (2)$$

where E=1143 GPa and ν=0.07 are Young's modulus and Poisson coefficient for diamond, ε_f and ε_s are temperature depended thermal expansions for diamond film and substrate, respectively. The copper substrate is expected to generate the maximum stress in a diamond film.

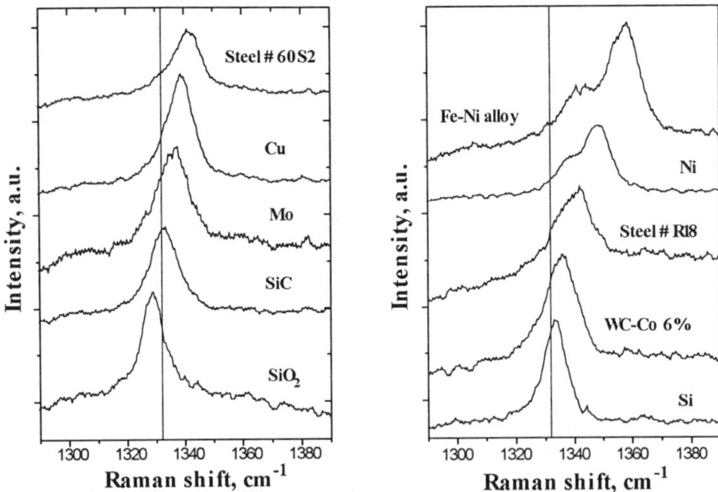

Fig.3. High resolution Raman spectra of diamond films on different substrate materials at room temperature. The vertical line at 1332cm^{-1} indicates the peak position for stress-free diamond. Note a peak splitting for Ni and Fe-Ni alloy.

Raman spectra were taken using a Jobin-Yvon S-3000 micro-Raman spectrometer (triple monocromator, 640 mm focal length, gratings 1800 lines/mm) with x50 objective which focuses the laser beam in a spot of 2 μm. All spectra were recorded by CCD matrix with 2 cm^{-1} spectral resolution. The excitation of the Raman scattering was performed with an Ar^+ ion laser at 514.5 nm wavelength. To avoid local heating of the sample we worked at a low laser power ≈40 mW. The temperature upon spectroscopic measurements could be varied in the range from 76 K to 600 K using a small vacuum chamber with a substrate holder cooled by liquid nitrogen or heated by furnace. The substrate temperature was measured by a thermocouple and controlled by Eurotherm SMC BT100 attachment. The spectra have been taken at least at five different points at the film surface.

RESULTS AND DISCUSSIONS

Raman spectra of the films on SiC, Cu, Mo, steel and Ni in frequency range 1100 to 1600 cm^{-1} are shown in Fig. 2. The narrow peak around 1333-1360 cm^{-1} characterizes diamond. A small amount of amorphous carbon is also present as revealed by a broad peak at 1500 cm^{-1}.

Fig. 3 shows Raman spectra of diamond films on ten different materials with a higher resolution. SiO_2 substrate is the only material which causes a negative peak shift relative to 1332 cm^{-1} (tensile stress), while other substrates induce positive peak shifts (compressive stress). A splitting of the peak into two lines is clearly seen for Ni and Fe-Ni alloy substrates, for which the shift of the high frequency component reaches 18-27 cm^{-1}.

In case of biaxial stress σ the Raman line of polycrystalline diamond splits for singlet and doublet and shifts from normal position at 1332 cm^{-1} on $\Delta\nu_s$ and $\Delta\nu_d$, respectively, according to relationships derived by Ager and Drory [1]:

$$\Delta\nu[cm^{-1}] = -0.93\ \sigma[GPa] \qquad \text{for } \textit{singlet} \qquad (3)$$

$$\Delta\nu[cm^{-1}] = -2.60\ \sigma[GPa] \qquad \text{for } \textit{doublet} \qquad (4)$$

These linear expressions have been obtained by averaging the "stress-peak shift" relations different for stress applied to different crystal planes over all crystallites orientations, which were

Table I. Raman peak position shift $\Delta\nu$; measured stress σ_{meas}; and thermal stress σ_{th}, calculated from eq.(2) for diamond films on various substrate materials. Coefficient of thermal expansion is given for T=293K.

Substrate	α, $10^{-6}K^{-1}$	$\Delta\nu$, cm^{-1}	σ_{meas}, GPa	σ_{th}, GPa
SiO_2**	0.5	-1.0÷-3.7	+2.3	+2.1
Si**	2.5	1.2÷1.6	-1.0	-0.25
SiC	2.8	0.9÷2.6	-1.6	-1.0
WC-Co (6%)	5.2	3.8÷6.2	-3.8	-2.0
Mo	5.3	6.5÷6.7	-4.1	-2.6
steel #R18	11.2	6÷12	-7.4	-10.4
steel #60S2	12	4.5÷8.9	-5.5	-11.1
Ni	13	17.5÷18.0*	-6.9	-13.4
Ni**	13	29.5*	-11.4	-13.4
Fe–Ni alloy	4.5	26.0÷27.5*	-10.6	-
Cu	16.7	4.5÷4.7	-2.9	-18.1

* position for doublet peak
** diamond film thickness is ≈1÷2μm

assumed to be random. If the film has a texture the coefficients in eqs. (3,4) may change, but we do not take into account this possibility in the present study. At low stresses (typically at σ<8 GPa) the peak splitting is not resolved, and we have to use a "weighted" shift $\Delta\nu=(1/3)\Delta\nu_s+(2/3)\Delta\nu_d$ to calculate the stress:

$$\Delta\nu\ [cm^{-1}] = -1.62\ \sigma\ [GPa] \qquad \text{for } \textit{unsplitted peak} \qquad (5)$$

The Raman peak parameters, measured stress σ_{meas} found from peak shifts using eqs.(3-5), and thermal stress are summarized for substrate materials in Table I. The thermal stress σ_{th} was calculated from eq. (2) for deposition temperature T_d=950°C (T_d=690°C for SiO_2).The substrates are arranged in ascending order of predicted thermal stresses. Silica glass with tensile stress $\sigma_{th}\approx2$ GPa and copper with compressive stress $\sigma_{th}=-18$ GPa are the extreme samples on this stress scale. Generally, the measured stress is in a reasonable correlation with thermal stress, however a noticeable reduction in σ_{meas} is found for materials with very high thermal stress $\sigma_{th}>10$ GPa beginning from steel (we remind that these materials were covered with tungsten). The measured stress for 3 mm thick copper substrate was in six times lower than the predicted one. This is the evidence of strong stress relaxation due to plastic deformation of copper. Moreover, the diamond–coated 1 mm thick Cu substrates were even severely curved after cooling, but such samples were not analyzed by us.

The large dispersion in peak shifts measured is mainly due to rough surface relief in our diamond films. Actually, the tungsten buffer layer is crystallized in faceted grains during its deposition, forming non-flat relief similar to that known for CVD diamond. This may cause the enhanced roughness of successively grown diamond film, the stress on top of protrusions being strongly relaxed compared to ones at the film "valleys" [2]. In addition, the Raman spectra reflect a stress averaged over the film thickness, while the real stress at the film/substrate interface may be higher. Therefore we preferred using maximum measured peak shifts to obtain σ_{meas} in Table I.

The splitting of diamond Raman peak at Ni and Fe-Ni alloy substrates is in agreement with biaxial stress model [1]. However, for other substrates the splitting was not observed because of large width of singlet and doublet lines relative to shift value (full width at half maximum of 10–15 cm^{-1}). The thermal nature of stress allows to perform a controllable stress variation at the same film simply by changing the sample temperature. To observe this effect we prepared one diamond film of 2 μm thickness on Ni with a thin W interlayer (≈2-3 μm). Fig. 4 illustrates the

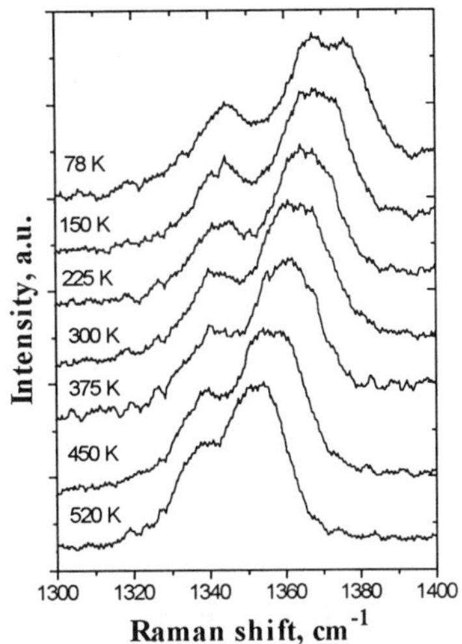

Fig.4. Raman peak evolution for diamond on Ni substrate with temperature in the range from T=78 K (top) to 520 K (bottom). Note a doublet splitting (right peak) at T=78 K.

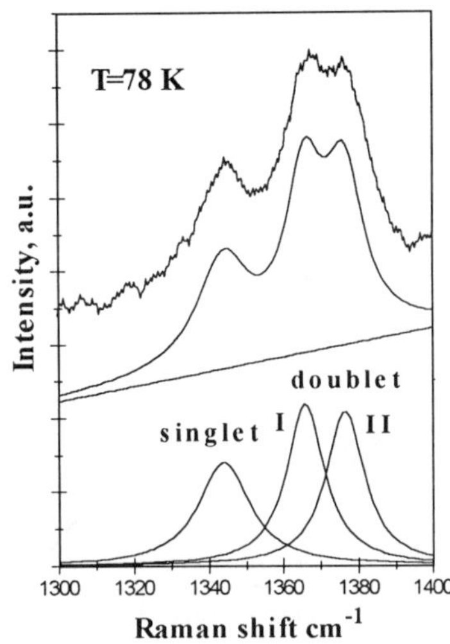

Fig. 5. Raman peak for diamond on Ni measured at T=78 K (top), its deconvolution to singlet and doublet peaks (bottom) and synthesized peak (center).

changes in Raman peak shape and position upon temperature scan from 78 K to 520 K. Peak splitting becomes well-resolved at low temperatures. The less intensive left peak associated with singlet mode shows only a slight shift to higher frequencies with temperature decrease, while the right peak associated with doublet shifts significantly, its position approaching ≈ 1370 cm^{-1} at T=78 K. Moreover, the tendency to doublet splitting is observed at low temperatures.

We performed a computer deconvolution of the spectra presented in Fig. 4, using Lorentzians to fit the three observed components. An example of the deconvolution for the peak taken at T=78 K is presented in Fig. 5. The singlet and two lines from doublet denoted (I) and (II) have width (FWHM) of 12 cm^{-1}. The three components have nearly same integral intensity, thus the doublet/singlet intensity ratio is close to 2:1.

It should be taken into account that the shift in Raman peak position to lower frequencies at elevated temperatures (mainly at T>300 K) occurs not only due to thermal stress relaxation, but also because the fundamental ν=1332 cm^{-1} mode of stress-free diamond shifts to the same direction. To separate these two effects we selected a stress-related peak shift by subtracting from the measured peak position the temperature dependent data $\nu(T)$ for diamond reported by Herchen and Cappelli [12]. We used their fit expression $\nu=a_1T^2+a_2T+a_3$; where $a_1=-1.075\times10^{-5}$ cm^{-1}K^{-2}, $a_2=-0.00777$ cm^{-1}K^{-1}, $a_3=1334.5$ cm^{-1}. The temperature dependence of thus determined peak shifts for singlet and doublet are shown in Fig. 6. All three lines shift linearly to lower frequencies with temperature in agreement with thermal stress behavior. Doublet lines I and II intersect the horizontal line of zero shift at around 1000°C, which is in a reasonable agreement with deposition temperature T=950°C. However, it is still unclear why the singlet is so insensitive to temperature variation.

To calculate the temperature-depended stress from doublet peak shift we used eq.(4) for biaxial stress which reefer to position of unresolved doublet. Therefore we averaged the shifts for doublet lines (I) and (II). The plot of stress versus temperature is shown in Fig. 7. The stress linearly decreases with T, that again confirms its thermal nature. The maximum stress σ_{meas}=14.3 Pa was achieved at T=78 K. No film delamination was observed after the sample undergone so high stress, indicating a very good adhesion both at diamond/tungsten and tungsten/nickel interface. The experimental data are in a good agreement with predicted temperature dependence of thermal stress (straight line in Fig.7).

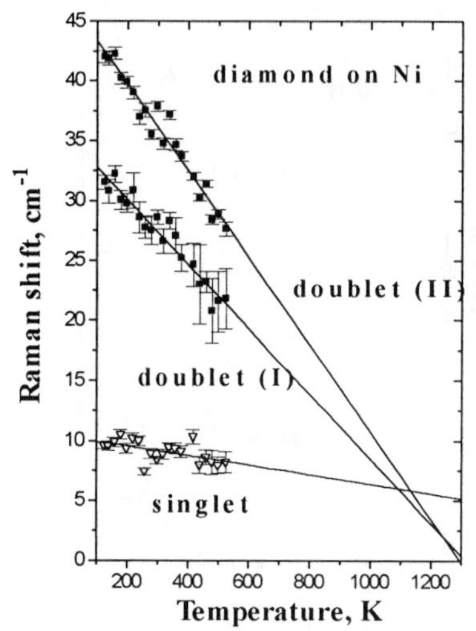

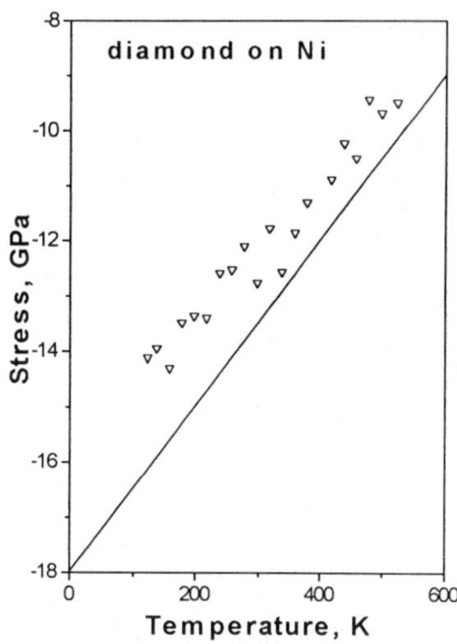

Fig.6. Dependence of doublet (squares) and singlet (triangles) Raman peak shifts on temperature. Linear extrapolation of doublet to zero shift yields "deposition temperature" ~1000°C.

Fig.7. Biaxial stress for 2 μm diamond film on Ni vs temperature (triangles), as deduced from doublet peak position and eq. (4). Solid line is thermal stress calculated from eq. (2) for T_d=950°C.

CONCLUSIONS

The stress in polycrystalline diamond films deposited on ten different substrate materials (SiO_2, Si, SiC, Mo, Cu, Ni, Fe-Ni alloy, WC-Co and steel) was measured using micro Raman spectroscopy. The stress determined in the framework of biaxial stress model correlates with expected thermal stress originated from CTE mismatch between diamond and substrate. The highest stress at room temperature σ_{meas} =11.4 GPa has been found at Ni substrate, and it increased to 14.3 GPa at T=78 K. The measured stress linearly depends on temperature in accordance with thermal stress model.

REFERENCES

1. J.W. Ager III and M. Drory, Phys.Rev. B, **48**, 2601 (1993)
2. V.G. Ralchenko, A.A. Smolin, V.G. Pereverzev, E.D, Obraztsova, K.G. Korotoushenko and V.I. Konov, presented at the Diamond Films'94, Il Ciocco, Italy, Sept.25-30, 1994.
3. H. Windishmann and G.F. Epps, J. Appl. Phys. **69**, 2231 (1991).
4. D. Schwarzbach, R. Haubner and B. Lux, Diamond and Related Materials, **3**, 757 (1994).
5. A. Tardieu, F. Cansell and J.P. Petitet, J. Appl. Phys. **68**, 3243 (1990).
6. M. Hanfland and K. Syassen, J. Appl. Phys. **57**, 2752 (1985).
7. D.S. Knight and W.B. White, J. Mater. Res. **4**, 385 (1989).
8. M. Yoshikawa, G. Katagiri, H. Ishida, A. Ishitani, M. Ono and K. Matsumura, Appl. Phys. Lett. **55**, 2608 (1989).
9. C. Johnston, A. Crossley, P.R. Chalker, I.M. Buckley-Golder and K. Kobashi, Diamond and Related Materials, **1**, 450 (1992).
10. N.I. Chapliev, V.I. Konov, S.M.Pimenov, A.M. Prokhorov and A.A. Smolin, in Y.Tzeng, M Yoshikawa, M Murakawa and A. Feldman (eds), *Applications of Diamond Films and Related Materials*, Elsevier, Amsterdam, 1991, p.417
11. *Handbook of Thermophysical Properties of Solid Materials*, ed. by A. Goldsmith et al., Macmillan, New York, 1961.
12. Herchen and M.A. Cappelli, Phys. Rev. B, **43**, 11740 (1991).

RESIDUAL STRESSES ANALYSIS IN DIAMOND LAYERS DEPOSITED ON VARIOUS SUBSTRATES

D. RATS*, L. BIMBAULT**, L. VANDENBULCKE*, R. HERBIN*, and K. F. BADAWI**
*LCSR, CNRS, 1C av. de la Recherche Scientifique, 45071 Orléans Cedex 2 (France).
**LMP, URA 131, 40 av. du Recteur Pineau, 86022 Poitiers (France).

ABSTRACT

A major problem for diamond coating applications is that diamond films tend to exhibit poor adherence on many substrates and typically disbond at thicknesses of the order of few micrometers due especially to residual stresses. Residual stresses in diamond are composed of thermal expansion mismatch stresses and intrinsic stresses induced during film growth. Diamond films were deposited in a classical microwave plasma reactor from hydrocarbon-hydrogen-oxygen gas mixtures. Thermal stresses were directly calculated from Hook's law. On silicon substrate, intrinsic stresses were deduced by difference from measurements of total stresses either by the curvature method or by X-ray diffraction using the $\sin^2\psi$ method. These investigations allow us to discuss the origin of the intrinsic stresses. The residual stress level was also investigated by Raman spectroscopy as a function of the deposition conditions and substrate materials (SiO_2, Si_3N_4, Si, SiC, WC-Co, Mo and Ti-6Al-4V). We show that the thermal stresses are often preponderant.

INTRODUCTION

Considerable efforts have been made in recent years to understand and predict the residual stresses that develop during diamond film growth. The total residual stresses of a diamond film on a substrate are composed of two types : intrinsic stresses (due almost to microstructure and impurities in the layer) that always occur during the layer growth, and thermal stresses. The later are due to the difference between the thermal expansion of diamond and the substrate and they are generated after deposition during cooling to room temperature.

In the case of low-pressure diamond synthesis, total residual stresses are generally determined by the radius of curvature method [1], by a vibrating-membrane method [2], by the Raman diamond peak shift [3-5] or by X-ray diffraction (XRD) using the "$\sin^2\psi$" method [6,7].

Thermal stresses were directly calculated from Hook's law. On silicon substrate, intrinsic stresses were deduced here by difference from measurements of the total stresses by the curvature method and by XRD using the "$\sin^2\psi$" method. The intrinsic stresses were investigated as a function of the gaseous composition and the deposition temperature.The stress-free lattice parameter and the microdistorsions are also reported. The origin of the intrinsic stresses are discussed in terms of grain size, non-diamond incorporation, or diamond grain microstructure. For other substrate materials, Raman spectroscopy was used to estimate the total stresses with regard to the theoretical models.

EXPERIMENTAL PROCEDURE

Diamond deposition was carried out in a classical plasma tubular reactor that crossed a wave-guide connected to a microwave generator (2.45 GHz). Various $CH_4/H_2/O_2$ and $CH_4/H_2/CO$ gas mixture were used. The pressure was set between 5 and 30 Torr and the

Mat. Res. Soc. Symp. Proc. Vol. 383 © 1995 Materials Research Society

deposition temperature was in the range 500-900°C. It was measured by an optical pyrometer and by an axial thermocouple. An enhancement of the nucleation was obtained by seeding the surface with a 1 μm diamond powder in an ultrasonic bath.

The quality of the films was deduced from Micro-Raman spectra obtained with a Dilor XY spectrometer. The measurements were performed in back-scattering configuration, using a 514.5 nm Ar laser beam light. The average grain size was deduced from SEM observation.

The total residual stress on silicon have been determined by the Stoney equation with Brenner-Senderoff correction [8] :

$$\sigma = E_s' \cdot \frac{t_s^2}{6t_f} \cdot \left(\frac{1}{R} - \frac{1}{R_0}\right) \cdot \left[1 + 4 \cdot \frac{t_f}{t_s} \cdot \left(\frac{E'_f}{E'_s} - 1\right)\right]$$

where E_s' and E_f' are the substrate and the film biaxial Young's modulus (180 GPa for silicon and 1230 GPa for CVD diamond), t_f and t_s are the film and substrate thickness, R_0 is the initial curvature of the substrate and R is the final one, measured by a Dektak II profilometer.

XRD analysis of stresses and strains in thin films is a precise and non-destructive technique [9]. When the film is stressed, the lattice spacing $d_{\phi\psi}$ of the (hkl) planes perpendicular to a $\phi\psi$ (Euler angles) direction is different from the stress-free lattice spacing d_0. The whole method and the principal equations are reported elsewhere [7,9]. The diffraction measurements were performed with a synchrotron radiation.

CALCULATED THERMAL STRESSES

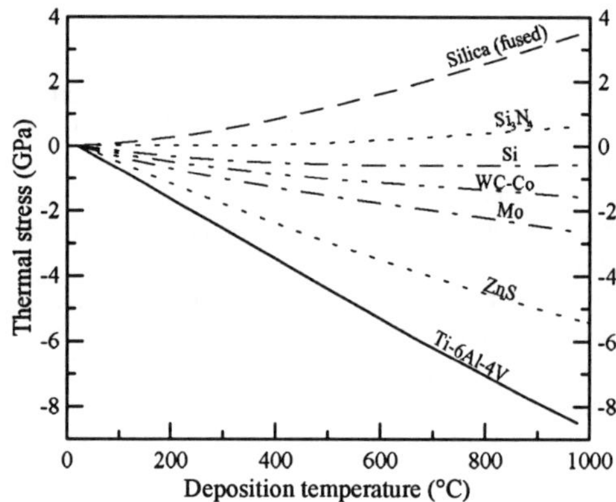

Fig. 1: Variation of thermal stress with temperature for various substrate materials.

During cooling of a diamond coated substrate to room temperature, a thermal stress component developed which can be calculated from Hook's law using the biaxial Young's modulus E_f' for polycristalline diamond (1230 GPa). For substrate materials, the temperature dependence of thermal expansion [10] is already known and Pickrell et al [11] have recently determined the thermal expansion of CVD diamond. Thermal stress (Fig. 1) is almost compressive (except for oxides like silica, or nitrides like Si_3N_4). Their absolute values are very high (>3 GPa) principally for metallurgical substrates. This is one reason that explain the important difficulties to obtain a well-adherent diamond coating on these substrates.

SILICON, A CONVENIENT SUBSTRATE

Because of its low thermal stress component (-0.6 GPa), silicon is a convenient substrate to study the intrinsic stresses. They were calculated for films with an average thickness of 2 µm as a function of the deposition temperature and the C-atomic fraction, defined by $X_C=C/(C+O)$, in the inlet C-O binary mixture. The intrinsic stresses are tensile and increase with increasing X_C (or methane concentration) or the deposition temperature (fig. 2). Raman spectroscopy analysis of these samples shows that an increase of the methane concentration or the deposition temperature tends to favor the deposition of non-diamond carbon (graphite, disordered sp^2 and amorphous). These results are in good agreement with those already published. From direct in-situ measurements Shäfer et al [12] and Schwarbach et al [13] found increasing tensile stresses with increasing deposition temperature or decreasing grain size. Windischmann et al [1] found increasing tensile stresses with increasing deposition temperature but decreasing tensile stresses with decreasing methane fraction. This last result is different from ours, but the deposition conditions were slighty different. Normal stress measurements obtained by the "$\sin^2\psi$" method are close to those obtained by the curvature method. The XRD value at 900°C is much lower than the value obtained by the curvature method; this result might be attributed to a higher bending of the silicon wafer which was already observed at temperature above 850°C, in conditions of plastic deformation of the sample [12]. The shear stress remains constant and its value is equal to about 120 ± 40 MPa. The genesis of such shear component is a consequence of the polyphase structure of the film [7,14].

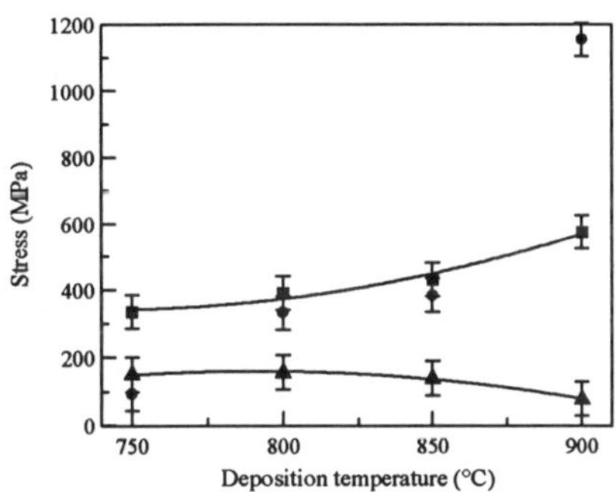

Fig. 2a: Variation with the temperature of shear stresses (▲) and intrinsic stresses deduced from the $\sin^2\psi$ method (■) and the curvature method (●).

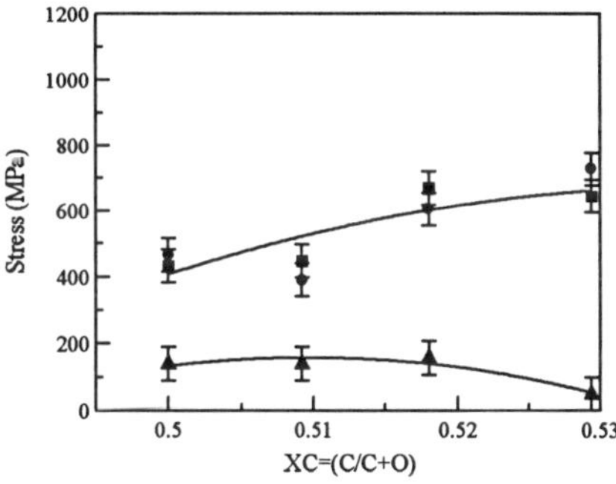

Fig. 2b: Variation with the deposition X_C of shear stresses (▲) and intrinsic stresses deduced from the $\sin^2\psi$ method (■) and the curvature method (●).

The normal stresses were determined for an isotropic material from the slope of $a^+=½.(\varepsilon^++\varepsilon^-)$ as a function of $\sin^2\psi$ and the shear stress from the slope of $a^-=½.(\varepsilon^+-\varepsilon^-)$ versus $\sin2\psi$ where ε^+ and ε^- are the strain in the direction $\phi\psi$ for $\psi>0$ and $\psi<0$ respectively [7,9]. Fig. 3 presents an example of a plot of a^+ versus $\sin^2\psi$ which shows that the linear law, representative of a biaxial stress state of the material, is fairly well observed.

Previous studies [7] have shown that the stress-free lattice parameter a_0 can be deduced. It is about 3.566 ± 0.001 Å, a value very close to the bulk one (3.5671 Å) that might indicate a good crystalline quality or a small amount of impurities within the grains. This result could also be obtained if the vacant sites or voids compensate the impurities influence. However insignificant microdistorsions within the grains were found previously [7] and the average size of the coherently diffracting domains (or subgrains) was higher than 1000 Å (a monocrystalline bulk material like silicon would have a value of about 2000 Å), indicating a very good crystallinity which did not vary a lot as a function of the deposition conditions employed here. Then the diamond phase microstructure would not contribute to a variation of the stress level.

The main difference between the two methods of stress measurements is the probed volume. XRD gives only the stresses in the diffracting diamond phase, while the curvature method gives the weighted means of stresses in the film. The whole film is composed of the diamond grains observed by SEM, which include many subgrains, and the grain boundaries which incorporate principally graphitic and amorphous carbon. This non-diamond carbon can constitute a real second phase at sufficiently high volume fraction, the mechanical properties of which are very different from the diffracting one. Because the results of the XRD and the curvature methods are similar, the stress state in both phases is similar. The two methods give also the same variation of the intrinsic stresses as a function of the grain size (Fig. 4). It appears that processes at grain boundaries resulting from the nucleation density, and then grain growth by elimination of grain boundaries, annihilation of excess vacancies and void shrinkage by diffusional process could be esssential in the absence of impurities. They lead to different volume shrinkage of the already deposited films thus causing different tensile stresses as a function of the grain size and in accordance of the amount of grain boundaries [1,12]. Finally, intrinsic growth stresses variations appear to depend on the grain boundary density, without any important variation due to the diamond phase itself. It can be noted that this result is obtained in conditions where a steady-state growth resulting in a columnar structure were generally not reached (grain size greater than the film thickness).

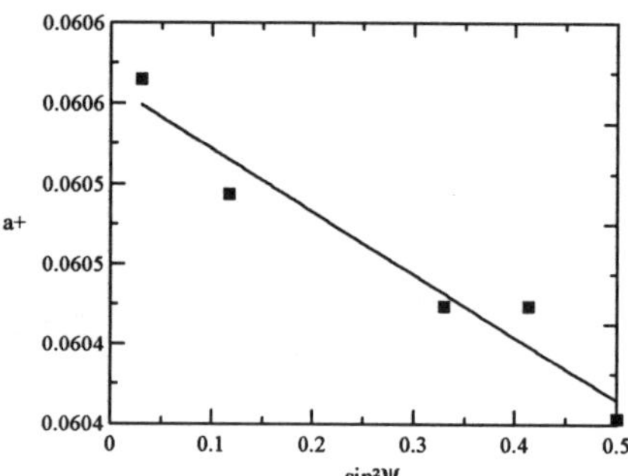

Fig. 3: Example of a plot of a^+ as a function of $\sin^2\psi$.

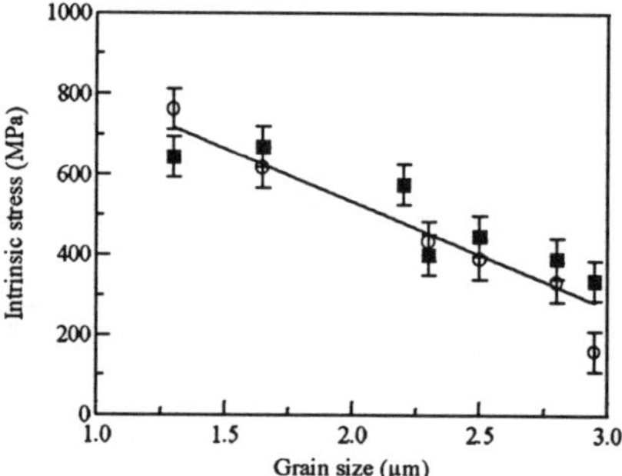

Fig. 4: Intrinsic stress variation with the average grain size measured by SEM analysis. "$\sin^2\psi$" method (■) and curvature method (O).

OTHER SUBSTRATE MATERIALS

The curvature method permit to measure easily the whole film stress when the deposition conditions do not lead to a plastic deformation of the substrate. The "$\sin^2\psi$" method is a powerful technique, but the high X-ray transparency of carbon often requires a synchrotron source or low-angled beam especially when diffraction peaks overlap. Raman spectroscopy has thus been used till now to estimate the residual stresses for other materials than silicon. For a strained lattice, the interatomic potentials change, and the lattice vibrations may deviate from that of the unstrained lattice; this implies an effect on the Raman active mode of diamond. Residual stress can be evaluated from the shift of the Raman diamond peak with respect to natural diamond value at 1332 cm^{-1}. The peak shift does not change a lot as a function of the inlet gaseous composition (see Fig. 5, at 850°C). The results reported in figures 5 and 6 show that for a given temperature, the residual stresses are principally related to the substrate material and are less sensitive to coating quality. However, the intrinsic stress component should be taken into account to relate the shift with the total stress.

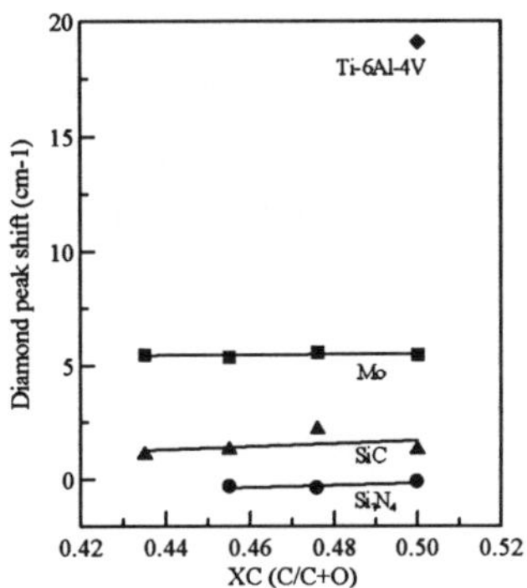

Fig. 5: Diamond peak shift variation with X_C for several substrate materials (at 850°C).

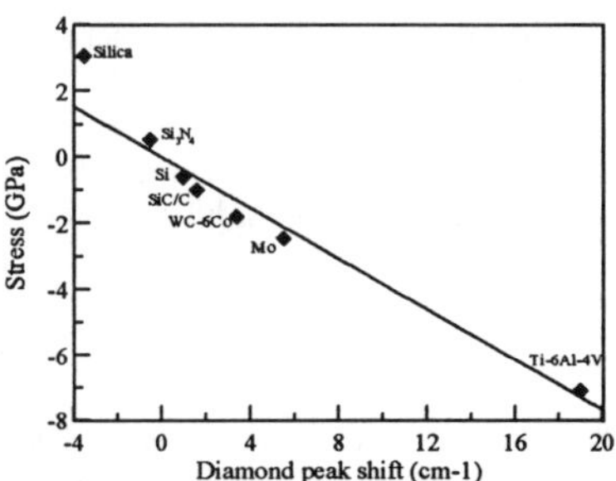

Fig. 6: Calculated thermal stress versus experimental diamond peak shift for various substrates (T=850°C). The Ager model, which gives a residual stress gage factor of -0.384 GPa/cm^{-1}, is represented by the continuous line on the same stress scale.

Raman spectra of diamond-coated Ti-6Al-4V also exhibit a partial splitting of the diamond feature into two peaks which is attributed to the compressive residual stress [5]. One must remind that a dispersion of the crystallite (subgrains) size or orientation could also produce multiple Raman peaks [15,16]. However weakly-stressed coatings on other substrates did not exhibit obvious multiple peaks. For the same deposition conditions (especially T=850°C), the variation of the diamond peak shift obtained for various substrate materials shows a strong correlation with the thermal stress (Fig. 6). The film deposited on silica exhibits some cracks, which could explain the low shift value due to some stress relaxation. Ager and Drory [5] have recently developed a general model to measure biaxial stress in polycrystalline diamond films. The shift and splitting of the Raman phonon into a singlet and a doublet are shown as a function of the crystallographic direction. The average stress gage factor for the principal peak (corresponding to the doublet phonon) is used to calculate the variation of the residual stress as a function of the diamond peak shift (Fig. 6). A fairly good agreement is obtained between our results and the Ager et al. model, which allows therefore to attribute the residual stresses principally to the thermal component.

CONCLUSION

The XRD $\sin^2\psi$ method is a powerful tool to analyze residual stress in diamond films and the results are fairly well compared to those obtained by the classical curvature method. The intrinsic stresses in diamond films deposited on silicon are sensitive to the deposition conditions and increase with the deposition temperature and X_C (C/C+O). They are tensile, indicating either a process of grain boundaries attraction or a diffusion process in the film decreasing the volume of the grain in favor to the non-diamond carbon part of the film. However the incorporation of impurities or the presence of vacancies and voids could also occur. In all deposition conditions reported here, the microstructure of the diamond diffracting phase is of good quality, particularly the dislocations density and the point defects one, which are responsible of microdistorsions, are very low and the size of the coherently diffracting domains is relatively important. On other substrates, the technical difficulties of the XRD $\sin^2\psi$ method have led us to use Raman spectroscopy for stress estimation. Diamond films deposited on metallurgical materials present dominant thermal stress in comparison with the intrinsic one. The residual stress evaluation from the shift of the diamond peak can be particularly employed for such high-stressed coatings, because this shift is not related to the residual stress only. Other phenomena are known to influence the Raman results like the dispersion of the crystallite size and orientation, or the partial lifting of the triple degeneracy of the diamond peak [5,6,15,16].

ACKNOWLEDGMENTS

The authors would like to thank for his collaboration: C. Beny (BRGM) for Raman analysis. This work was supported in part by DGA/DRET.

REFERENCES:

1. H. Windishmann and G. F. Epps, J. Appl. Phys. 69 (4), 2231 (1991).
2. B. S. Berry, W. C. Pritchet, J. J. Cuomo, C. R. Guarnieri and S. J. Whitehair, Appl. Phys. Lett. 57 (3), 302 (1990).
3. R.J. Nemanich, S.A. Solin and R. M. Martin, Phys. Rev. B 23, 6348 (1981).
4. D.S Knight and W.B White, J. Mat. Res. 4 (2), 385 (1989).
5. J. W. Ager III and M. D. Drory, Phys. Rev. B 48 (4), 2601 (1993).
6. P. R. Chalker, A.M. Jones C. Johnston and I. M. Buckley-Golder, Surf. Coat. Tech. 47, 365 (1991).
7. D. Rats, L. Bimbault, L. Vandenbulcke, R. Herbin, and K. F. Badawi, (submitted to J. Appl. Phys.).
8. A. Brenner and S. Senderoff, J. Res. Natl. Bur. Stand. 42, 105 (1949).
9. Ph Goudeau, K. F. Badawi, A. Naudon and G. Gladyszewski, Appl. Phys. Lett 62 (3), 246 (1993).
10. Y.S. Touloukian, R.K. Kirby, R. E. Taylor and T. Y. Lee, Thermophysical Properties of Matter: Vol. 13, Thermal Expansion of Nonmetallic Solids (Plenum Press, New York, 1977).
11. D. J. Pickrell, K. A. Kline and R. E. Taylor, Appl. Phys. Lett., 64 (18), 2353 (1994).
12. L. Shäfer, X. Jiang and C.P. Klages, in Applications of Diamond and Related Materials, edited by Y. Tzeng, M. Yoshikawa, M. Murakawa, A. Feldmann (Elsevier Science Publishers B.V., Amsterdam, 1991), p. 121.
13. D. Schwarzbach, R. Haubner and B. Lux, Diamond and Related Materials 3, 757 (1994).
14. L. Castex and K. F. Badawi, Matériaux et Structures, (Editions Hermés, Paris, 1987) p. 277.
15. J. W. Ager III, D. K. Veirs, and G. M. Rosenblatt, Phys. Rev. B 43 (8), 6491 (1991).
16. S. Prawer, K. W. Nugent, and P. S. Weiser, Appl. Phys. Lett. 65 (18) 2248 (1994).

LOCAL STRESS ASSESSMENT OF CVD DIAMOND WITH MICRO-RAMAN SPECTROSCOPY UP TO 1200K

LI-CHYONG CHEN+, KUEI-HSIEN CHEN*, YEN-LIANG LAI*, and JIN-YU WU*
+Center for Condensed Matter Sciences, National Taiwan University, Taipei, Taiwan.
*Institute of Atomic and Molecular Sciences, Academia Sinica, Taipei, Taiwan.

ABSTRACT

Raman spectra in CVD diamond films were measured at room temeprature and at high temperatures up to 1200 K. With micron spatial resolution, variations of Raman line from different crystals in the same sample were observed. Both single-peak and multiple-peak analyses were used to assess the stress distribution within the CVD diamond film. In addition, the evolution of Raman line position, line width, and line intensity was monitored as a function of annealing temperature and isothermal holding time. Both (100) and (111) CVD diamond crystals were studied. The detailed evolution of the Raman line width was found to depend on the crystallographic orientation. The temperature dependence of the Raman spectra for CVD and natural diamonds are compared.

INTRODUCTION

The structural stability and mechanical strength of CVD diamond film are of fundamental importance in its applications. These properties are in turn related to the phonon behavior, especially, the anharmonic effects in the bonding between carbon atoms in the diamond lattice. Raman spectroscopy is a powerful method to probe atomic arrangement, and has been rapidly emerging as the technique to characterize the quality of diamond films. Recently, several authors have used the Raman technique to assess the residual stress in the diamond film [1-5]. In the present report, we'd like to address the residual stress from two aspects, namely, the variations within crystals of the same sample and the temperature dependence of the Raman spectra.

The variation of the Raman spectrum with temperature in diamond was first reported by Nayar in 1941 [6]. Since then a number of extensive studies have been made of the frequency shift and linewidth in diamond over a wide range of temperatures [7-12]. All of the measurements mentioned above are made in natural diamond. The temperature dependence of Raman spectrum of the CVD diamond film has never been investigated. Due to potential applications of CVD diamond films in situations where they are subjected to high temperatures, it is of considerable interest to address this temperature dependence.

EXPERIMENTAL

The CVD diamond samples were prepared by an AsTex 5 kW microwave system. The process conditions employed to produce diamond film with a mixture of both (100) and (111) orientations as well as epitaxial films are described elsewhere [5, 13]. A Renishaw system 2000 micro-Raman spectrometer, equipped with a 125 mW Ar^+ laser operating at 514.5 nm as the light source, was used to measure the Raman spectra of diamond films. The line shift was calibrated by comparing the spectrum with that of natural diamond. The incident laser beam was normal to the surface of thin films. Scattered light of all polarization components was collected. With a 5 micron slit, the spectral resolution is 1 cm^{-1}.

For high-temperature Raman studies, a heating stage capable of reaching 1200 K was used. The temperature was measured with a type-S Pt-Pt/10%Rh thermocouple placed underneath the sample holder. Measurements can be conducted in air or in inert gas. For the latter operation, a

Mat. Res. Soc. Symp. Proc. Vol. 383

chamber enclosing the heating stage is connected to a pumping system. High purity He gas was introduced into the chamber after a base pressure of 10^{-6} Torr was reached.

RESULTS AND DISCUSSION

Stress Distribution at Room Temperature

Figure 1 shows the SEM micrograph of a high quality (100) CVD diamond film. The room temperature Raman spectrum for one grain in this film together with the same for a natural diamond are shown in Figure 2. For natural diamond, the peak width (full width at half maximum) of the 1332.5 cm^{-1} line is 2.7 cm^{-1}. For CVD diamond, some line broadening and line shift with respect to that of natural diamond were observed. The magnitude of the shift and broadening varied from grain to grain in the same film. However, the histograms made from 118 grains in the same film suggested that characteristic peak position and peak width can still be determined (Figure 3a and 3b). From Gaussian distribution fit to the histogram, the line position for this specific diamond film was 0.35 ± 0.60 cm^{-1} higher than that for natural diamond. The corresponding peak width was 7.2 ± 1.6 cm^{-1}.

It should be mentioned that the above single-peak approximation of the line shape is valid only when the residual stress is hydrostatic in nature. That is to say, the diamond cubic symmetry is preserved and, therefore, the 1332.5 cm^{-1} peak remains triplet degenerate. A Raman line of diamond shifts to higher and lower frequency under compressive and tensile stresses, respectively. Using the formula established by Grimsditch *et al.*, the diamond film shown above was under an average of -0.1 GPa compressive stress [14].

Considering that the growth process of CVD diamond is far away from equilibrium, the single-peak analysis is apparently oversimplified. The residual stress results from the sum of thermal strains, epitaxial strains, and the growth strains. The first two are due to the thermal expansion and lattice mismatch between the substrate and the film. Since the direction perpendicular to the substrate is less constrained than in-plane directions, the stress induced by the substrate is biaxial (therefore, anisotropic). Under heteroepitaxial growth conditions, lattice parameter mismatch between Si and diamond produce tensile stress. On the other hand, the thermal expansion mismatch induced stress is compressive. Under our experimental conditions, this thermal stress was estimated to be about -0.2 GPa [15]. In contrast to the former two origins of stress, the stress induced by growth strain is often called "intrinsic" stress [2, 15].

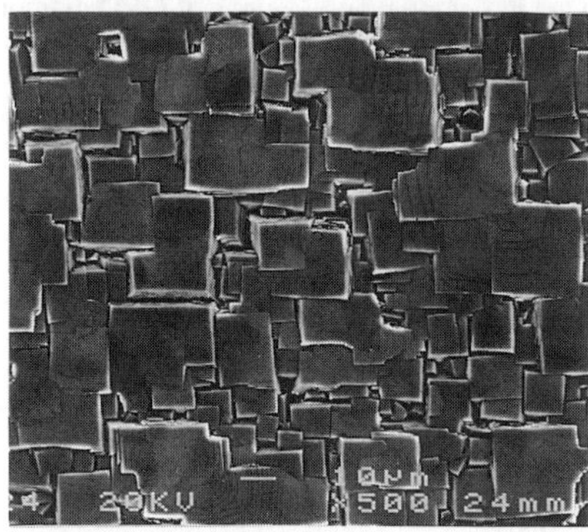

Figure 1 SEM micrograph for a continuous and well oriented (100) CVD diamond.

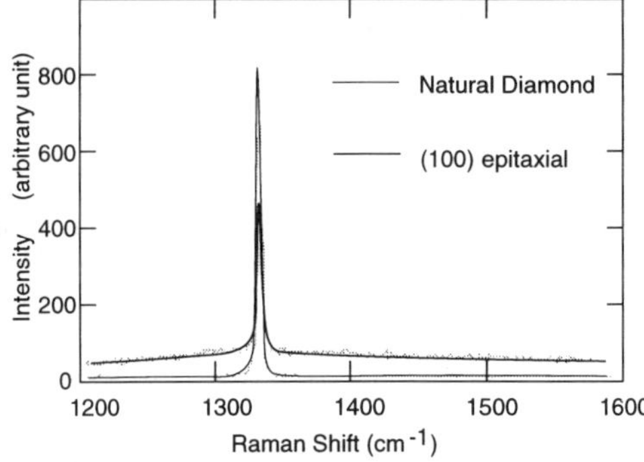

Figure 2 Raman spectra of a (100) CVD diamond and a natural diamond.

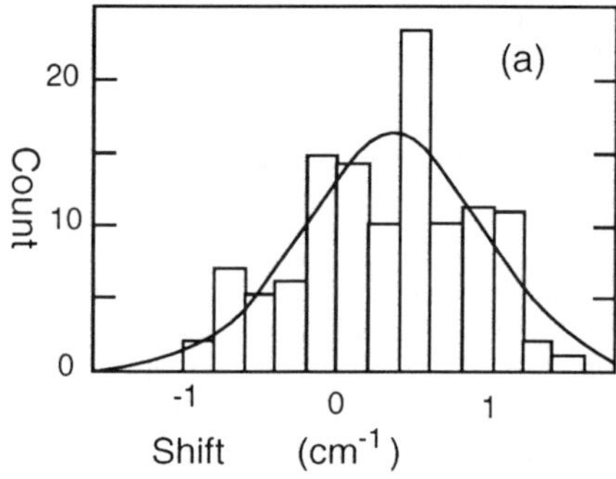

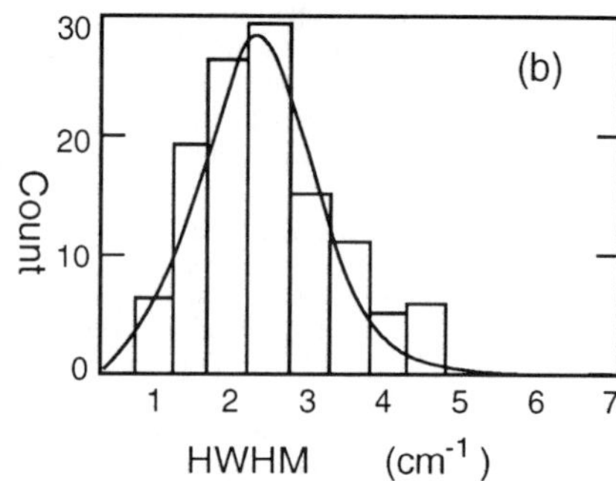

Figure 3 The histograms for (a) Raman peak shift and (b) Raman line broadening with respect to that of natural diamond. Half width at half maximum was used in (b).

Growth strains arise from a density change during film growth. For diamond grown by CVD method, common growth strains may include graphite inclusions, hydrogen clusters, and voids. The stress induced by growth defects is mostly inhomogeneous. Plane defects such as stacking faults and line defects such as screw dislocations were observed in some crystals. While difficult to determine, it is speculated that the point-like defects such as voids and hydrogen clusters were also incorporated during crystal growth. The plane defects and the line defects are anisotropic in nature, while the point defects can be isotropic. Voids and dislocations produce tensile stress, whereas hydrogen clusters and impurity phases produce compressive stress [2].

Practically, multiple mechanisms are operating simultaneously. Since some are tensile and some are compressive, the sum can be either compressive or tensile. Furthermore, since both isotropic and anisotropic components are present, the sum is anisotropic. The single-peak approximation does not reflect the anisotropic nature of the residual stress.

Most growth defects distort the atomic arrangement in the diamond lattice. The distortion in crystallographic symmetry removes the three-fold degeneracy of the zone-center Raman peak, either completely or partially. Therefore, the width of an unresolved split peak reflects the degree of distortion. Indeed, detailed examination of the Raman spectra reveals asymmetric line shape.

In most cases, the asymmetric spectra can be adequately fitted with two or three Lorentzian curves. The magnitude of the split and shift varies from grain to grain within the same CVD diamond film, suggesting an inhomogeneous and anisotropic stress distribution. For each grain, a Raman spectrum fitted with two-peak line shape gave a pair of numbers, $\Delta\omega_1$ and $\Delta\omega_2$, the wave number shifts (with respect to 1332.5 cm^{-1} for natural diamond) of the first and the second peaks, respectively. The notation is chosen as such that $\Delta\omega_2 > \Delta\omega_1$. Figure 4 shows the loci of peak shifts for over 100 grains in the well oriented (100) film. A phenomenological model, whereby the residual stress was separated into an isotropic and an anisotropic component has been established to describe the loci of Raman line shifts and splits [5].

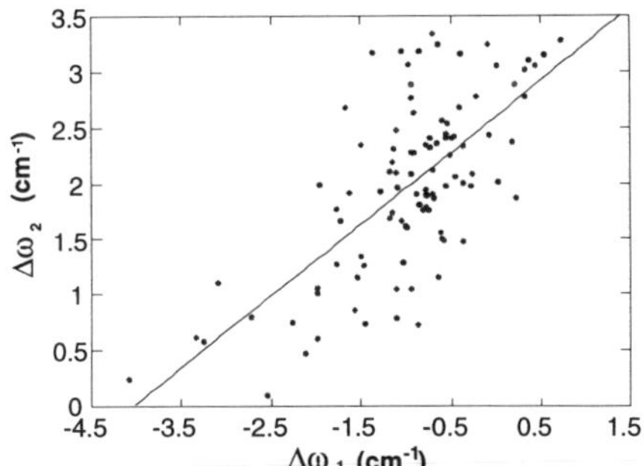

Figure 4. The Raman shift of the higher frequency versus the Raman shift of the lower one for (100) CVD diamonds.

From the slope and the intercept of a linear fit to the loci of $\Delta\omega_1$ and $\Delta\omega_2$, a compressive stress of about 2.3 GPa was determined for the isotropic component. The corresponding anisotropic component of the stress was tensile and was estimated to range between 2.8 - 6.3 GPa. The predominant point-like defects can be inferred from the sign of the isotropic component of the stress. The type of the predominant defect is a strong function of the growth parameters such as methane concentration. A dependence on the orientation of growth was also observed [5].

Raman Studies at High Temperature

A series of high temperature Raman spectra for CVD diamond were recorded at various point of temperature and time during annealing. For simplicity, we assumed single-peak approximation for the Raman line. For (111) CVD diamond, the Raman line position, line width, and intensity evolution as a function of heating time are shown in Figure 5a, 5b, and 5c, respectively. A complete annealing cycle consists of three segments: heating at a constant rate, holding at a constant temperature, and cooling also at a constant rate same as heating. These three segments are denoted as regions I, II, and III, respectively. Typical ramp rate was 50 °C/min. A marked downward shift of the Raman spectra as well as line width broadening with temperatures were observed (region I and III). During 973 K isothermal holding (region II), despite that there was no noticeable change of Raman peak position, the line width decreased initially with isothermal annealing time, and finally settled at a constant level. Whereas the line intensity increased initially, reached a peak, and then decreased with isothermal annealing time. SEM imaging suggested that the final decrease of intensity in region II was caused by reaction of CVD diamond with air. When high purity He gas was used during heating, such reaction was suppressed.

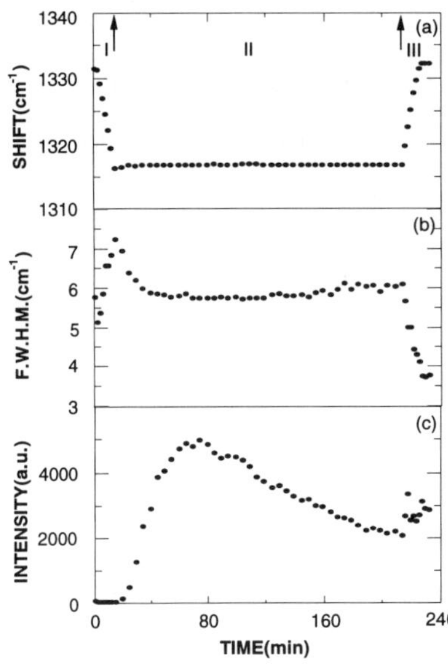

Figure 5. The evolution of Raman spectra for a (111) CVD diamond upon heating, holding at 973 K, and cooling, which are denoted as region I, II, and III, respectively, are described by their (a) peak position, (b) line width, and (c) intensity as a function of heating time. Experiments were performed in air.

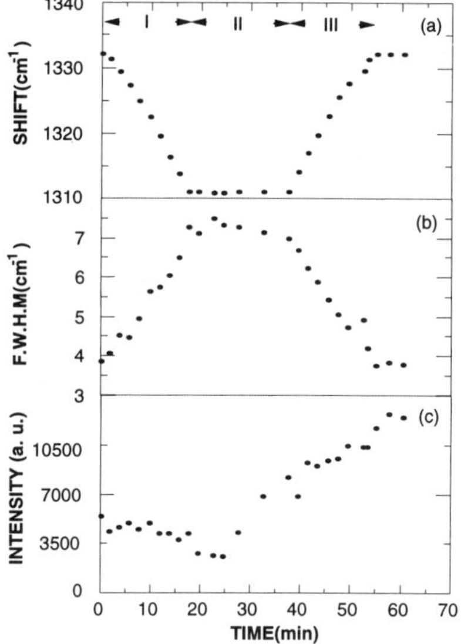

Figure 6. The evolution of Raman spectra for a (100) CVD diamond upon heating, holding at 1173 K, and cooling, which are denoted as region I, II, and III, respectively, are described by their (a) peak position, (b) line width, and (c) intensity as a function of heating time. Experiments were performed in air.

The evolution of Raman line position, line width, and intensity for (100) oriented CVD diamond are shown in Figure 6a, 6b, and 6c, respectively. The Raman spectra exhibited a downward shift with temperatures in a manner similar to that for (111) oriented CVD diamond. However, in contrast to the case for (111) oriented diamond, there is hardly any change of the line width during isothermal holding at 1173 K, which is 200 K higher than the isothermal annealing temperature for (111). Furthermore, the variation in the intensity for (100) is not as significant as that for (111). During the isothermal holding at 1173 K, the intensity increased only by a factor of 3.

It is intriguing to observe that (111) exhibited some appreciable reduction of Raman line width upon annealing in air, while (100) did not. This may be in part due to a less pronounced carbon-oxygen reaction, as reported in the literature that the {111} diamond surfaces were found to have higher oxidation rate than the {100} surfaces in the temperature range of 970 and 1270 K [16].

The overall temperature dependence of the Raman line position for (100) is shown in Figure 7 (solid square). The result for (111) was identical to (100) data within experimental error. For comparison, Figure 7 also contains the results for natural diamond reported in the literature: Anastassakis, Hwang, and Perry, open circles [7]; Borer, Mitra, and Namjoshi, triangles [8]; Krishnan, crosses [9]; Solin and Ramdas, plusses [10]; and Zouboulis and Grimsditch, open squares [12]. The solid line in the figure is the thermal-expansion contribution to the frequency shift [12]. Our results agree well with those of Zouboulis and Grimsditch [12]. Like the case for natural diamond, the downward shift of Raman spectra up to 1200 K for CVD diamond can not be accounted for by thermal expansion effects only. Whatever the additional effects that contribute to the shift, no crystallographic orientation dependence is to be considered.

The temperature dependence of the line width, on the other hand, strongly depends on the orientation as well as the ambient gas environment. Figure 8 shows the temperature dependence of the linewidth. The (100) data measured in air upon cooling are shown here as solid square. The corresponding (111) data obtained in air were identical to (100) data within experimental error, and threfore are not shown here. Instead, the (111) data obtained in high purity He gas are shown here as solid triangle for comparison. Literature data of Refs. 7-10, and 12 are again shown in Figure 8 as open circles, triangles, crosses, plusses, and open squares, respectively. The solid line is based on Klemen's model [17]. Our instrumental line width was approximately 1 cm^{-1} and no attemp was made to deconvolute our FWHM data with this instrumental line width. The instrumental resolutions in Refs. 7, 9, 10, 12 were 1, 1.5, 0.4, and 1.6 cm^{-1}, respectively. Only the data from Ref. 8 (triangles) have been corrected to zero slit width. It is quite clear that there is a systematic difference between the data of CVD diamond and those of natural diamond.

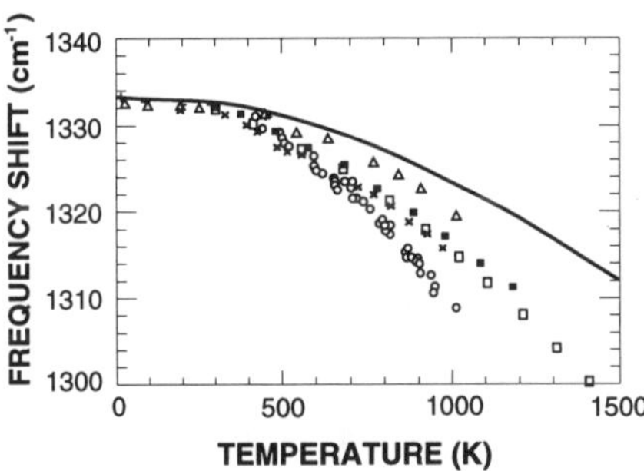

Figure 7. The temperature dependence of Raman line position for CVD diamond (solid square) is compared with those for natural diamond reported in literature. The open circles, triangles, crosses, plusses, and open squares are for Refs. 7-10, and 12, respectively.

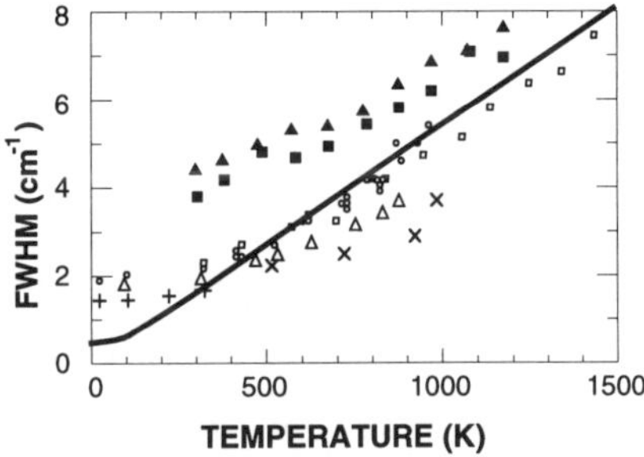

Figure 8. The temperature dependence of Raman line width for (100) CVD diamond (solid square) and (111) CVD diamond (solid triangle) are compared with those for natural diamond reported in literature. The open circles, triangles, crosses, plusses, and open squares are for Refs. 7-10, and 12, respectively.

SUMMARY

Raman spectra measured for CVD diamond often exhibited some broadening and shift with respect to that for natural diamond. In addition, Raman spectra varied from grain to grain even for a seemingly high quality (continuous and well oriented) film. By analyzing the Raman line shape, stress distribution can be conveniently assessed. Single-peak analyses are useful for estimating the net stress, provided that the stress is hydrostatic in nature. Multiple-peak analyses are useful to address the inhomogeneous and anisotropic aspects of the stress. For the (100) oriented film, the localized anisotropic stress ranged between 2.8 - 6.3 GPa, with a characteristic isotropic stress of 2.3 GPa.

The Raman peak frequency decreases with temperatures in a manner quite similar to that of natural diamond for both (100) and (111) oriented CVD diamonds. The corresponding temperature dependence of Raman line width somewhat depends on the crystal orientation. For (111) CVD diamond, appreciable decrease of Raman line width was observed at 973 K, whereas the Raman line width hardly varied for (100) oriented crystal subjected to isothermal heat treatment at 1173 K. Furthermore, the Raman peak intensity of (111) CVD diamond was found to vary more significantly after high temperature annealing than the same of (100) CVD diamond. Presumably, the Raman spectra variations can be accounted for by the different origins and annealing effects of the residual stress in different crystallographic orientation.

ACKNOWLEDGMENT

The authors would like to thank Dr. T.J. Chuang, Dr. Y.L. Wang, Dr. K. J. Song, and Dr. J. C. Lin for helpful discussions. The project is supported by the fund from the National Science Council, Taiwan, ROC under contract No. NSC-84-2113-M-002-031.

REFERENCES

1. M. Yoshikawa, G. Katagiri, H. Ishida, *et al.*, Appl. Phys. Lett. **55**, 2608 (1989).
2. W. Wang, K. Liao, J. Gao, and A. Liu, Thin Soild Films **215**, 174 (1992).
3. S. A. Stuart, S. Prawer, and P. S. Weiser, Diamond and Related Materials **2**, 753 (1993).
4. J. E. Butler, D. J. Vestyck, Jr., B. Mackey, *et al.*, Proceedings of 1994 Workshop on Processing & Applications of Diamond Films, Hsin-Chu, Taiwan, pp. 133 (1994).
5. K. H. Chen, Y. L. Lai, J. C. Lin, K. J. Song, L, C. Chen, and C. Y. Huang (accepted for publication in Diamond and Related Materials).
6. R. G. N. Nayar, Proc. Indian Acad. Sci., Sect. A **13**, 284 (1941).
7. E. Anastassakis, H. C. Hwang, and C. H. Perry, Phys. Rev. B **4**, 2493 (1971).
8. W. J. Borer, S. S. Mitra, and K. V. Namjoshi, Solid State Commun. **9**, 1377 (1971).
9. R. S. Krishnan, Proc. Indian Acad. Sci. **24**, 45 (1946).
10. S. A. Solin and A. K. Ramdas, Phys. Rev. B **1**, 1687 (1970).
11. H. Herchen and M. A. Cappelli, Phys. Rev. B **43**, 11740 (1991).
12. E. S. Zouboulis and M. Grimsditch, Phys. Rev. B **43**, 12490 (1991).
13. J. Y. Wu, C. C. Juan, K H. Chen, and L. C. Chen (in preparation).
14. M. H. Grimsditch, E. Anastassakis, and M. Cardona, Phys. Rev. B **18**, 901 (1978).
15. See, for example, H. Windischmann, G. F. Epps, Y. Gong, and R. W. Collins, J. Appl. Phys. **69**, 2231 (1991).
16. T. Evans and C. Phaal, Proc. 5th Biennial Conf. on Carbon (Penn. State Univ.) 147 (1962).
17. G. Klemens, Phys. Rev. **148**, 845 (1966).

Part IV

Fracture and Adhesion

AN INDENTATION TEST FOR MEASURING ADHESION TOUGHNESS OF THIN FILMS UNDER HIGH RESIDUAL COMPRESSION WITH APPLICATION TO DIAMOND FILMS

M. D. Drory[1] and J. W. Hutchinson[2];
[1]Crystallume, 3506 Bassett Street, Santa Clara, CA 95054
[2]Division of Applied Sciences, Harvard University, Cambridge, MA 02138.

Abstract

Adhesion measurement of brittle coatings on ductile substrates is examined with emphasis on diamond–coated metals. The experimental aspects of determining film adhesion is discussed through measurement of the interface toughness by a brale indentation method.

Introduction

Adhesion measurement has many scientific and engineering aspects which are of current interest with important applications in developing thin film technologies. For example, present efforts to develop diamond–coated tooling illustrate the need to advancing current testing methods for adhesion, and represent several challenges as summarized in Table 1. An adhesion method is needed to obtain an engineering design parameter, the interface toughness, G_c. This test is ideally performed without the use of special coupons since the chemistry of the substrate and interface can be size (or area) dependent. This is of concern in comparing the interface properties of diamond–coated cemented carbide tool inserts with test coupons which are much larger. A small probe of the interface toughness is needed to examine variations in G_c for a given sample, which is best performed as a non–destructive test. Finally, it is desirable to rely on low–cost testing equipment to allow for wide–spread use of the test procedure from the laboratory to production. Several tests are routinely used for examining adhesion, including the scratch, tape, and tensile (pull) tests [1-3]. These tests do not provide a measure of the interface toughness. In addition, the scratch and tape tests are not appropriate for testing hard brittle coatings. Alternatively, indentation tests of adhesion

Mat. Res. Soc. Symp. Proc. Vol. 383

provide a measurement of G_c for brittle coatings under certain conditions.

Indentation tests involve the use of diamond–tipped penetrators in hardness testing machines. One test uses the 'macro–hardness' testing machine with a Vickers diamond indenter[4-6]. Alternatively, a Rockwell hardness tester is used with the diamond brale indenter[7-11]. The brale indentation test has the potential as a rapid test of adhesion with wide-spread use, since the Rockwell hardness tester is routinely used for measuring the hardness of metals and is of low–cost.

Table 1 – Practical Goals for Adhesion Measurement

ATTRIBUTE	OTHER RELEVANT ISSUES
(1) Engineering design parameter for adhesion (Interface Toughness, G_c)	• needed for scientific studies • reproducibility • development tool
(2) Measurement of interface of interest	• Minimize size (area) dependence of substrate (interface)
(3) Small probe of adhesion	• examine variation in G_c • non–destructive test
(4) Simplified testing procedure	• robust test for various film & substrate combinations: – ductile/brittle – tensile/compression film stresses • routine testing procedure in production environment
(5) Low–cost testing equipment	• important for implementing in production environment

The brale indentation test utilizes the Rockwell hardness test with the diamond brale "C" indenter, producing delamination in the vicinity of the indentation for *brittle films* on a *ductile substrate* under the large loads of 60, 100 & 150 kg provided by the testing machine. An important additional requirement for this test in providing a valid measurement of G_c is that the films be in *residual compression.* Details of the mechanics analysis for obtaining the interface toughness is given elsewhere [11]. Further aspects of the adhesion testing procedure are given in the next section for a diamond–coated titanium alloy.

Experimental Procedure

A coupon of a common titanium alloy, Ti-6Al-4V, was diamond coated by a chemical vapor deposition method at low pressures. Diamond growth was achieved on a coupon of approximately 60 x 20 x 3mm which was cleaned in solvents and scratched with a fine diamond powder for enhancing diamond nucleation. A diamond film of about 1μm thickness was obtained in a microwave reactor at 2.45Ghz excitation in a predominately hydrogen gas flow with one percent methane. The total gas flow was about 300 cm^3min^{-1} at a pressure of 50–90 torr. The sample temperature was estimated by a thermocouple as 700–900°C.

The presence of diamond was confirmed by scanning electron microscopy (SEM) where a faceted microstructure is evident (Fig. 1) and by Raman spectroscopy (Fig. 2). A characteristic peak for diamond is observed which is greatly shifted from the 1332cm^{-1} as a result of very large compressive film stress. This peak shift was analyzed in a separate study to give a value of the residual stress as about 7GPa (compression) [12]. This value of the compressive stress has found to be accounted for by elastic strains arising from the large difference of the thermal expansion coefficients between diamond and titanium.

The adhesion between the diamond coating and the Ti-6Al-4V substrate was determined by brale indentation using a Rockwell hardness tester with the standard "Brale C" diamond indenter. Several indentations were made at the major loads of 60, 100, and 150kg supplied by the hardness tester. The indentation produced considerable regions of annular film delamination and radial film splitting at all three loads. As described in the next section, these effects were more pronounced with increasing load. The indentation and delamination radii were measured with an optical microscope at 50X magnification using side illumination. In addition, an "in-situ" photograph

Figure 1: Scanning electron micrograph of diamond microstructure deposited on Ti-6Al-4V coupon.

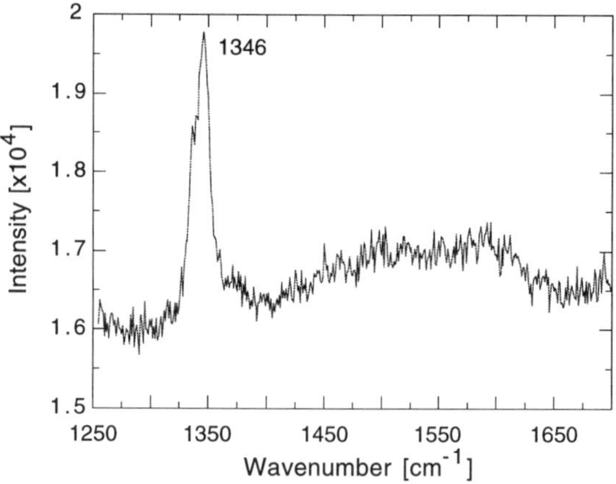

Figure 2: Raman spectrum of diamond–coated titanium coupon

was taken to trace the formation of film delamination during the loading cycle provided by the Rockwell hardness tester. This involves first manually applying a 'minor' 10 kg load, follow by the 'major' load with a loading rate predetermined by the testing machine. The minor load is manually released following the unloading of the major load (Fig. 3).

Results and Discussion

Adhesion testing of the diamond–coated Ti-6Al-4V coupon was performed by the brale indentation test where considerable splitting of the film was observed along with an axisymmetric delamination crack. In most cases, debris is formed exposing the underlying substrate. Several tests were performed at each of the major loads supplied by the Rockwell hardness tester, 60, 100 and 150 kg (Fig. 4). The size of the hardness impression and film delamination crack was measured for each indentation by an optical microscope (Fig. 5). The film delamination and cracking (and film spall debris) at each indentation allowed for a film thickness measurement to be made by SEM examination. The film thickness is an important geometric parameter for determining the interface toughness necessitating a thickness measurement at each indentation. The impression size and delamination radius increased with increasing applied load. After several indentations, the diamond tip of the brale indenter was inspected in a low–magnification stereomicroscope for fracture and wear. No wear was observed for indenting the diamond-coated titanium alloy, however erosion and damage of the penetrator is an issue for testing harder substrates, such as cemented carbide [13].

Several indentations were made at 150 kg while observing the formation of the delamination crack during the loading cycle (Fig. 6). The majority (~80%) of the delamination process occurs during the application of the major load. The axisymmetric delamination crack appears to advance to its final value while removing the major load. This behavior differs considerably from the Vickers indentation test on *ductile* films in which delamination occurs on unloading.

Values of the interface toughness were obtained by using the mechanics solution to a particular case of the diamond–coated titanium substrate shown in Fig. 7 for given values of the substrate Young's modulus, yield stress, and work hardening coefficient. In addition, the residual film stress is a critical parameter for obtaining G_c since it strongly influencing the value of G_o given by,

$$G_o = \frac{(1-\nu^2)t}{2E}{\sigma_o}^2$$

LOADING STEP	TOTAL LOAD
I. No Applied Load	0
II. Apply Minor-load, P_0	P_0
III. Apply Major-Load, P_1	$P_0 + P_1$
IV. Release Major Load	P_0
V. Release Minor Load	0

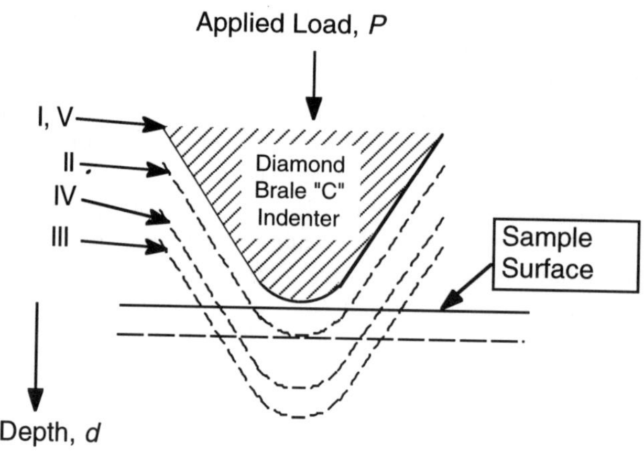

Figure 3: Schematic of loading cycle for the Rockwell hardness test.

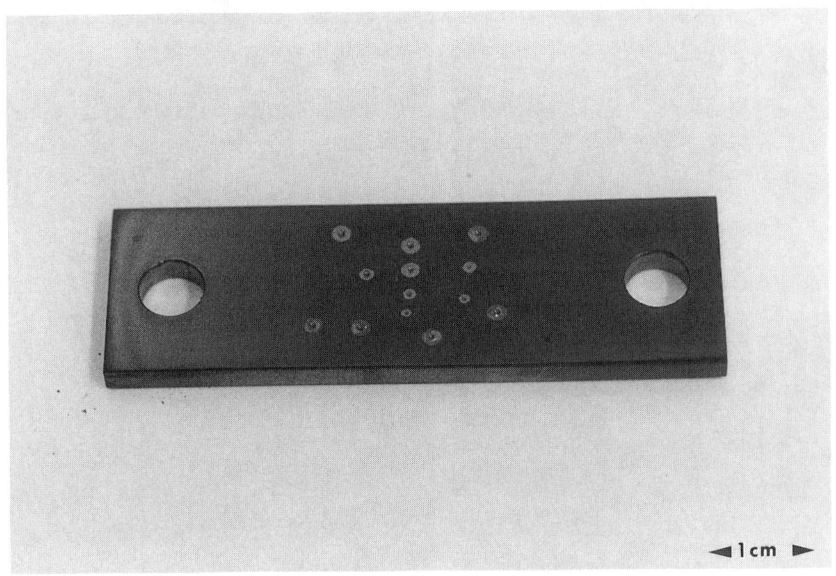

Figure 4: Diamond–coated titanium coupon with multiple indentations.

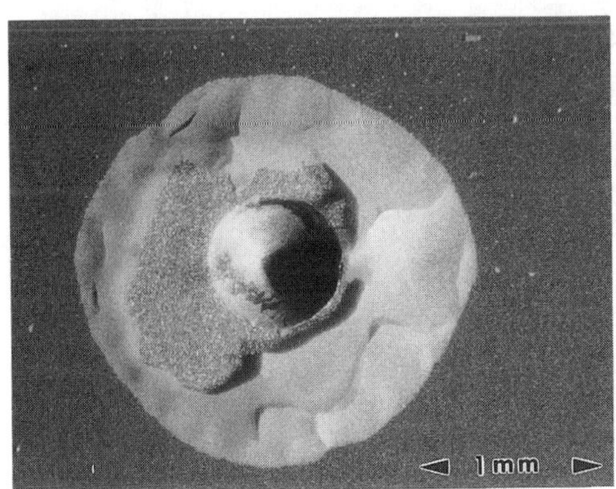

Figure 5: Indentation at 1568N load.

Figure 6: "In–situ" photograph of the brale indentation test during the loading cycle on a diamond–coated titanium coupon.

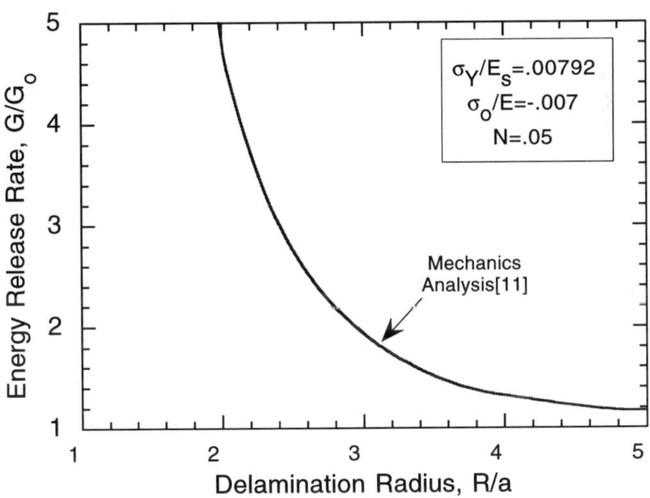

Figure 7: Trends in energy release rate with delamination radius.

Using nominal values for the elastic constants of diamond, as E = 1000GPa, $\nu = 0.07$ [14], and the residual film (compressive) stress σ_o= 7GPa, gives an average value of G_c of ~50J/m^2. These are very high values of the interface toughness for a brittle film, but may explain the extreme difficulty in delaminating the film by (other) mechanical means [13]. The origins of this strong interface are under study.

References

1. A. J. Perry, Thin Solid Films, **107**, 167 (1983).
2. P. A. Steinmann and H. E. Hintermann, J. Vac. Sci. Technol. A **7** [3] 2267 (1989).
3. P. C. Jindal, D. T. Quinto and G. J. Wolfe, Thin Solid Films, **154**, 361 (1987).
4. C. Rossington, A. G. Evans, D. B. Marshall and B. T. Khuri-Yakub, J. Appl. Phys. **56**, 2639 (1984).
5. M. R. Lin, J. E. Ritter, L. Rosenfeld and T. J. Lardner, J. Mat. Res. **5** [5] 1110 (1990).
6. C. A. Gamlin, E. D. Case, D. K. Reinhard and B. Huang, Appl. Phys. Lett. **59** [20] 2529 (1991).
7. P. K. Mehrotra and D. T. Quinto, J. Vac. Sci. Technol. A **3** [6] 2401 (1985).
8. M. R. Hilton and P. D. Fleischauer, Thin Solid Films, **172**, L81 (1989).
9. T–Y Yen, C–T Kuo and S. E. Hsu, Mat. Res. Soc. Symp. Proc., **168**, 207 (1990).
10. M. D. Drory and J. W. Hutchinson, Science **263**, 1753 (1994).
11. M. D. Drory and J. W. Hutchinson, Proc. Roy. Soc. To be published.
12. J. W. Ager and M. D. Drory, Phys. Rev. B, **48** 2601 (1993).
13. M. D. Drory, unpublished work.
14. J. E. Field, ed. The Properties of Natural and Synthetic Diamond, Academic Press, 1992.

THIN FILM DECOHESION AND ITS MEASUREMENT

A. Bagchi and A.G. Evans
Division of Applied Sciences, Harvard University, Cambridge, MA 02138

ABSTRACT

The mechanics of thin films are used to define quantitative procedures for predicting interface decohesion motivated by residual stress. The emphasis is on the role of the interface debond energy, especially methods for its accurate and reliable measurement. Experimental results are reviewed and possible mechanisms are discussed.

1. INTRODUCTION

The number of applications for thin films and multilayers that take advantage of their special mechanical, thermal, electronic and optical characteristics has steadily increased. The incidence of interface decohesion, motivated by residual stresses, has limited more widespread use. Such stresses are inevitable in vapor deposited layers and are exacerbated when the constituent materials have vastly differing thermomechanical properties. The stresses arise for two reasons. (1) Intrinsic stresses develop during deposition. These stresses persist, unless they are relaxed by plastic deformation or annealing. (2) The mismatch in thermal expansion induces stresses when the temperature is changed.

Controlling the stress in order to inhibit decohesion without compromising the functional characteristics of the system is not usually an option. Instead, thermomechanical design to resist these failure modes is required. This goal is crucially dependent upon the attainment of an adequate interface debond toughness, Γ_i. The role of the toughness is manifest through the fail-safe design requirement [1, 2],

$$\Gamma_i \geq h\sigma_R^2/\bar{E}\lambda \tag{1}$$

where h is the film thickness, $\bar{E}$ is the appropriate Young's modulus (plane strain or biaxial plane stress), σ_R is the residual stress and λ is a cracking number (of the order unity). When Eqn. (1) is satisfied, there is insufficient energy stored in the film to permit an interface crack to propagate and the film *must* remain attached to the substrate.

In order to implement this fail-safe criterion, methods for the accurate measurement of Γ_i on actual interfaces of relevance must exist. Most thin film adhesion tests empirically infer the adhesive strength by subjecting the film to some external loading and measuring the critical value at which decohesion occurs. These tests are simple and effective for routine ranking of bond quality. However, they do not measure Γ_i, because the strain energy release rate cannot usually be deconvoluted from the work done by the external load. An ideal test should *duplicate* the practical situation as closely as possible, while also being able to modulate the available strain energy. Moreover, the method must explicitly incorporate the contribution to decohesion from the residual stress.

2. MECHANICS OF FILM DECOHESION

Films decohere by relaxing the residual stress above interface cracks. For a thin, homogeneous film subject to uniform residual stress on a thick substrate, the steady-state energy release rate, $\mathcal{G}_{ss}$, for an interface crack is given by *the strain energy in the film.* The non-

Mat. Res. Soc. Symp. Proc. Vol. 383

dimensional form is,

$$\Pi = \bar{E}\mathcal{G}_{ss}/\sigma_R^2 h \tag{2}$$

where Π is a non-dimensional quantity of the order unity. The same form arises for other problems, but its numerical magnitude differs, as elaborated below. Decohesion takes place when $\mathcal{G}_{ss}$ exceeds the interface debond energy, Γ_i [3],

$$\mathcal{G}_{ss} \geq \Gamma_i . \tag{3}$$

However, Γ_i may be a strong function of the mode mixity angle ψ. Hence, it is *not sufficient* to know $\mathcal{G}_{ss}$; ψ must also be calculated. Moreover, to design against decohesion, the interfacial fracture toughness must be measured as a function of ψ. Solutions are presented below for monolayer and bilayer films.

2.1 Monolayer Films

For a semi-infinite interface crack between two isotropic elastic layers, under generalized edge loading conditions (Fig. 1) with forces P_i and moments M_i, equilibria dictate that [4, 5],

$$\begin{aligned} P_1 - P_2 - P_3 &= 0, \\ M_1 - M_2 + P_1\left(\frac{h}{2} + H - \delta\right) + P_2\left(\delta - \frac{H}{2}\right) - M_3 &= 0 \end{aligned} \tag{4}$$

where h is the film thickness and H is the substrate thickness. The quantity δ is the height of the neutral axis from the bottom surface. *Only four* among these six loading parameters are actually independent. These are, P_1, P_3, M_1 and M_3. The number of independent load parameters can be further reduced to only *two*, through superposition (Fig. 1). These parameters are P and M, given by,

$$\begin{aligned} P &= P_1 - C_1 P_3 - C_2 \frac{M_3}{h} \\ M &= M_1 - C_3 M_3 , \end{aligned} \tag{5}$$

where the C's are dimensionless numbers that must be calculated in accordance with the following five steps: obtain the position of the neutral axis, evaluate the sectional modulus, obtain the section area, calculate the stresses, determine the C's. These requirements give

$$C_1 = \frac{\Sigma}{A_o}, \quad C_2 = \frac{\Sigma}{I_o}\left(\frac{1}{\eta} - \Delta + \frac{1}{2}\right), \quad C_3 = \frac{\Sigma}{12 I_o}, \tag{6}$$

where $\Sigma = E_1/E_2$, $\eta = h/H$, $A_o = \Sigma + 1/\eta$ and I_o is given by

$$I_o = \frac{1}{3}\left[\Sigma\left\{3\left(\Delta - \frac{1}{\eta}\right)^2 - 3\left(\Delta - \frac{1}{\eta}\right) + 1\right\} + 3\frac{\Delta}{\eta}\left(\Delta - \frac{1}{\eta}\right) + \frac{1}{\eta^3}\right]. \tag{7}$$

The steady-state strain energy release rate, computed from P and M and using the difference in stored energy *far ahead* and *far behind* the crack tip, becomes

$$\mathcal{G}_{ss} = \frac{1}{2\bar{E}_1}\left[\frac{P^2}{Ah} + \frac{M^2}{Ih^3} + 2\frac{PM}{\sqrt{AI}h^2}\sin\gamma\right], \tag{8}$$

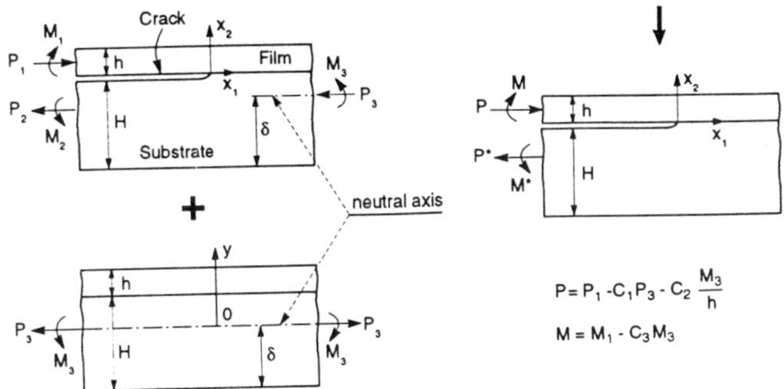

Figure 1. Forces and moments that simulate a residually stressed bimaterial system.

where γ is obtained from the expression $\sin\gamma = 6\Sigma\eta^2(1+\eta)\sqrt{AI}$ and the quantities A and I are

$$A = \frac{1}{1+\Sigma\left(4\eta+6\eta^2+3\eta^3\right)}, \quad I = \frac{1}{12\left(1+\Sigma\eta^3\right)}. \tag{9}$$

The mode mixity is given by

$$\psi \ = \ \tan^{-1}\left[\frac{\xi\sin\omega - \cos(\omega+\gamma)}{\xi\cos\omega + \sin(\omega+\gamma)}\right], \tag{10}$$

where ξ measures the loading combination:

$$\xi \ = \ \sqrt{\frac{I}{A}}\frac{Ph}{M}. \tag{11}$$

For *thin* monolayers subject to a *uniform stress* σ_R, these results reduce to [4]

$$\begin{aligned} P_1 &= P_3 = \sigma_R h, \\ M_1 &= 0, \\ M_3 &= \sigma_R h\left(H - \delta + \frac{h}{2}\right) \end{aligned} \tag{12}$$

such that

$$P \ = \ \sigma_R h, \qquad M \ = \ 0. \tag{13}$$

In this limit, $\mathcal{G}_{SS}$ is given by Eqn. (2) and the mode mixity is

$$\psi \ = \ \lim_{\xi\to\infty}\left\{\tan^{-1}\left[\frac{\xi\sin\omega - \cos(\omega+\gamma)}{\xi\cos\omega + \sin(\omega+\gamma)}\right]\right\} \ \equiv \ \omega. \tag{14}$$

Moreover, since the film stress *diminishes* as the interface decoheres, this energy release behavior is entirely controlled by *elasticity, even when the film has yielded*.

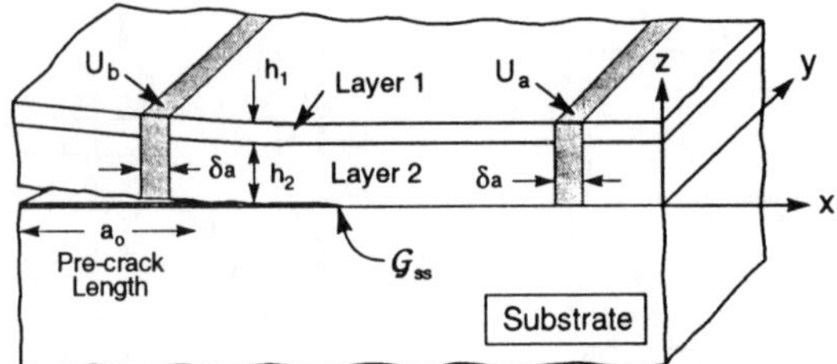

Figure 2. The effect of bending on the strain energy release rate: the energy left far behind the crack (due to bending) is to be subtracted from the energy stored far ahead of the crack to find $\mathcal{G}_{SS}$.

2.2 Bilayer Films

Decohesion may occur at an interface below the surface film. Additional considerations are then involved. The key new feature is that the decohered films bend, such that the stresses above the crack are not fully relieved, causing the energy release rate to be diminished (Fig. 2). The *redistributed stresses* must be determined before obtaining $\mathcal{G}_{SS}$ and ψ. Explicit results are given for the bilayer [6, 7]. The retained strain energy results in a diminished energy release rate,

$$\Delta\mathcal{G}_{ss} = -\partial U_c / \partial a \equiv \sum_i \frac{1}{E_i}\left[\frac{P_i^2}{h_i} + \frac{12M_i^2}{h_i}\right] \tag{15}$$

For a thin bilayer on a substrate, the bending forces are related to the curvature after decohesion κ by

$$P = P_1 = P_2 \equiv \left[\frac{\bar{E}_1 h_1^3 + \bar{E}_2 h_2^3}{6(h_1 + h_2)}\right]\kappa \tag{16a}$$

with

$$\kappa = \frac{6(h_1 + h_2)(\varepsilon_1 - \varepsilon_2)}{\left[h_1^2 + \bar{E}_2 h_2^3/\bar{E}_1 h_1 + \bar{E}_1 h_1^3/\bar{E}_2 h_2 + h_2^2 + 3(h_1 + h_2)^2\right]} . \tag{16b}$$

The ε_i are the misfit strains, which are related to the residual stresses $\sigma_{R,i}$ by

$$\varepsilon_i = \sigma_{R,i}(1 - \nu_i)/E_i . \tag{17}$$

The moments are related to the curvature by

$$M_i = E_i h_i^3 \kappa/12 . \tag{18}$$

A numerical example for a Cr/Cu bilayer, decohering from a substrate, illustrates the trends (Fig. 3). For such structures, the strain energy in the top Cr layer (the *superlayer*) drives the decohesion at the Cu/substrate interface. The trends show that variation in the Cr thickness produces a spectrum of energy release rates, $\mathcal{G}_{ss}$. Also note that $\mathcal{G}_{ss}$ *decreases as the Cu thickness increases*, which can be explained as follows. The contribution of a detached Cu layer to the energy release rate (without bending) is marginal, since it scales with the square of the film stress, and $\sigma_{R,Cr} >> \sigma_{R,Cu}$. However, during stress redistribution by bending, following decohesion, the Cu layer acts as a *sink* by storing elastic strain energy, which increases $\Delta\mathcal{G}_{ss}$ (Eqn. 15). The same phenomenon causes the phase angle to be small, compared with the single layer result ($\psi \approx 50°$) (Fig. 4).

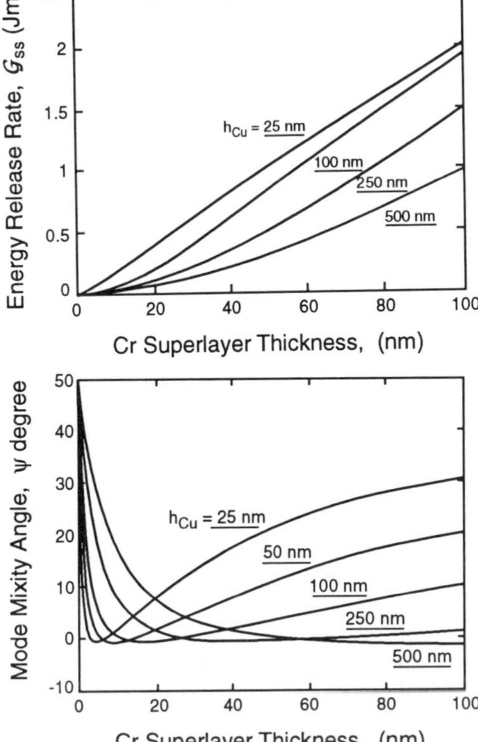

Figure 3. The effect of Cu layer thickness on the energy release rate for a Cu/Cr bilayer.

Figure 4. The effect of Cu layer thickness on the mode mixity angle for a Cu/Cr bilayer.

3. MECHANISMS OF INTERFACE CRACK GROWTH

Cracks at interfaces extend in accordance with several different mechanisms. In some cases, the interfaces have sufficient bond strength, relative to the metal yield strength, that the cracks extend in the *metal*, by a ductile mechanism. Such "strong" interfaces are not considered in this discussion. The emphasis is on "brittle" interfaces, devoid of reaction layers, in which the interface crack causes *complete separation of the constituent materials.*

Decohesion is fundamentally controlled by atomic bonding across the interface, coupled with bridging and frictional effects induced by roughness. The bonding is characterised by the maximum stress needed to separate the bonds, designated the bond strength $\hat{\sigma}$, and the energy dissipated during the bond rupture process, designated the work of adhesion, W_{ad} (Fig. 5). As debonding proceeds from an interface crack, the displacements induce large stresses. These stresses, in turn, activate inelastic mechanisms in the adjoining materials. The associated inelastic dissipation , magnitude Γ_p, may substantially exceed W_{ad}. The interface fracture energy Γ_i is the sum of these two contributions. Consequently, an understanding of the magnitude of Γ_i at various interfaces requires a coupling of the atomistics of bond rupture with the mechanics of inelastic dissipation as well as friction.

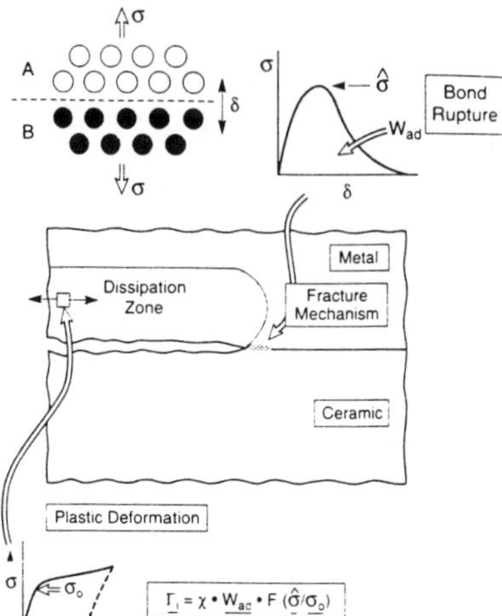

Figure 5. A schematic of the energy dissipated upon interface crack propagation.

3.1 Atomically Sharp Cracks

When the crack propagation mechanism allows the crack front to remain atomically sharp as it extends, an energy based fracture criterion is governing, because the *singularity necessarily permits the stress to attain the bond strength.* This criterion enables the energy release rate at the crack tip to be equated to the work of adhesion in order to simulate the energy dissipation upon interface crack extension. For such simulations, matching is required between the continuum inelastic zone and the bond rupture zone. For instance, when one of the constituents is a metal, matching to the plastic zone is required. This has been achieved by using a dislocation exclusion zone, width D, around the crack [8]. Upon enforcing this consistency, numerical analysis and a subsequent fit yields (Fig. 6)

$$\Gamma_i/W_{ad} = \left[EW_{ad}/qD\sigma_o^2\right]^c \tag{19}$$

where σ_o is the tensile yield strength, $q{\sim}40$ and $c \approx 3$. This model is known as the Suo-Shih-Varias (SSV) model. A major effect of the yield strength on the interface toughness is evident (to the power 6), as well as the expected scaling with W_{ad}. To use this solution and predict trends, the magnitude of D is first inferred by matching one set of results. Then, D is considered to remain fixed and other values of Γ_i determined. An example for the Al_2O_3/Nb interface is shown in Fig.6 using experimental data [9]. The fit requires that D~10nm.

3.2 Blunt Cracks

When dislocations interact with the crack front and induce blunting, crack extension cannot be simulated by using W_{ad}, because this criterion does not ensure that the peak stress reaches the bond strength. Instead, the criterion must be *stress-based* and must also satisfy basic energy requirements. As slip progresses from the crack, the blunting displacement increases and the stress ahead of the crack redistributes. If the slip is unrestricted, the redistribution process is efficient and the stress only attains values up to a few times the yield strength.

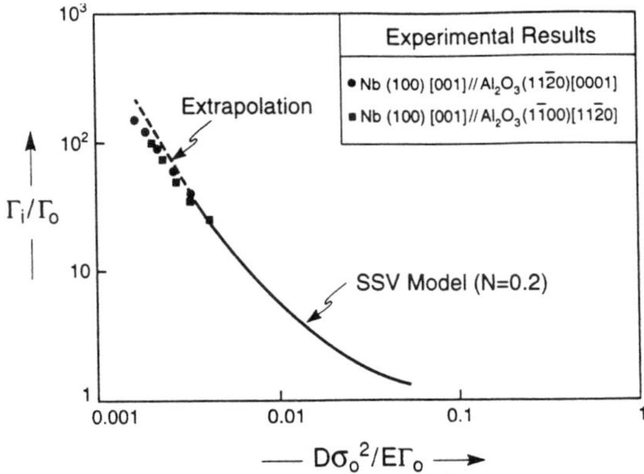

Figure 6. The SSV model for propagation of an atomically sharp interface crack. D is the width of the plasticity-free zone. Also plotted are experimental results for the Al_2O_3/Nb interface: these have been fit to the model at one Γ_i in order to specify D. Here Γ_o should be interpreted as the work of adhesion, W_{ad}, and N as the metal strain hardening exponent.

For the stress to build up as the load is increased, *a barrier to slip is required.* When such a barrier exists, one approach for simulating crack growth is to equate the peak magnitude of this stress to the bond strength. Calculations of this type give solutions that can be expressed in the form [10]

$$\Gamma_i/\mu b \;=\; \mathcal{F}\left[z/b,\, \hat{\sigma}/\mu,\, \gamma_m/\mu b\right] \tag{20}$$

where μ is the shear modulus, b is the Burgers' vector, z is the slip impediment length (for a very thin layer z = h), γ_m is the surface energy of the metal and $\mathcal{F}$ is the function plotted on Fig. 7. This approach does not explicitly consider the energy and W_{ad} does not appear in the solution. But, it is implicit that the energy available to the crack appreciably exceeds that needed to break the bonds.

When the precrack blunts and there are no barriers to slip, debonding may still occur within a *cohesive zone* on the interface ahead of the crack [11]. This zone ruptures according to a dissipation Γ_0, subject to a peak stress, σ^*. These quantities replace W_{ad} and $\hat{\sigma}$, respectively, in the sharp crack model and differ in magnitude ($\Gamma_0 > W_{ad}$ and $\sigma^* < \hat{\sigma}$). Various results have been calculated subject to this rupture mode [12, 13]. Among these is the effect of metal layer thickness h on Γ_i, when the layer is thin relative to the plastic zone size, R_0. The result is given by:

$$\Gamma_i/\Gamma_0 \;\approx\; 1+(h/R_o)\left[\sigma^*/\sigma_0 - a\right]^b \tag{21}$$

for $\hat{\sigma}/\sigma_0 > a$, where $a \approx 2.5$ and $b \approx 3$ (Fig. 8).

4. MEASUREMENT METHODS

There are three challenges to be met when designing a test for obtaining accurate and reliable measurements of $\Gamma_i(\psi)$ pertinent to thin films. (1) One arises primarily from the *thinness* of the

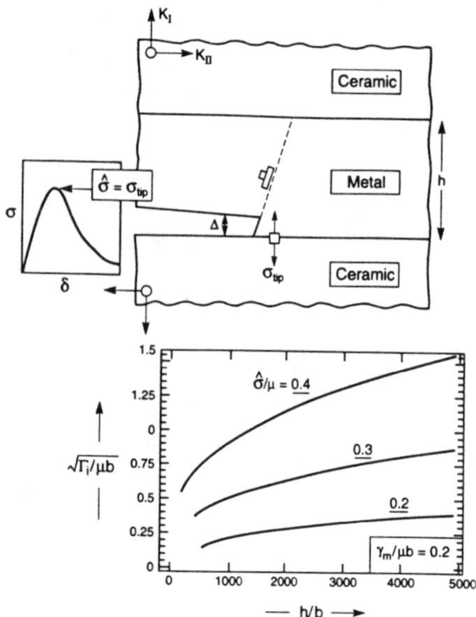

Figure 7. The interface toughness for very thin layers, as a function of the layer thickness, calculated using a restricted slip model.

films. When an external load is applied, extensive plastic deformation occurs. In extreme cases, the film may fail before decohesion initiates. De-coupling of the fracture energy Γ_i from the work done by the external load remains a difficult task. (2) It is necessary to characterize for the mode mixity. Ideally, *it should be possible to vary the mixity over a wide range, covering all values of interest.* (3) In general, as the decohesion grows, the strain energy release rate $\mathcal{G}$ varies. In such cases the debond radius must be measured to obtain Γ_i. This requirement presents problems for opaque films. In some configurations, the debond attains a steady-state value, $\mathcal{G}_{SS}$. *The preferred test method should exploit the crack length independence of steady-state configurations.*

4.1 Microscratch Test

The Microscratch test has been applied to a wide assortment of metallic and ceramic films [14]. Its primary advantage includes quickness, reproducibility and ease in implementation. However, the *critical load* is influenced by many factors. Deconvolution to evaluate Γ_i and ψ from the load is impeded by the complex nature of the deformation fields, which arise through interactions between the film and the probing indenter. The models developed to estimate Γ_i have been approximate [15, 16]. They use a point contact to approximate film stresses around the indenter using the elastic field. The elastic strain energy contained in the film above the delamination is obtained from the stress. This energy is considered to be available for interface decohesion, plus a contribution from the substrate *assumed* to be equal to that from the film. Residual stresses present in the film are difficult to take into account because the sign of the stress relative to that from the point force varies spatially around the scratch. Practical implementation requires measurement of the delamination geometry from scanning electron microscopy observations of the scratch track.

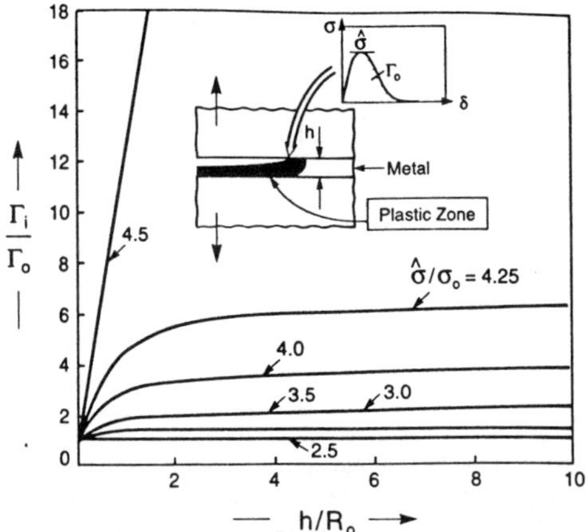

Figure 8. The effect of layer thickness on the interface toughness calculated using a cohesive zone model. R_0 is the plastic zone size.

4.2 Peel Test

Peel tests have been embraced by the electronic packaging industry to assess the adhesion of metallic and polymeric thick films, deposited on various dielectric substrates [17]. The test has the attribute that the peeling force is measured in *steady-state*, when the shape of the strip remains invariant. *This force is used as an interfacial quality measure.* A detailed analysis of the test has been developed [18]. The principal result may be expressed through the parameter,

$$\eta' = 6EP/\sigma_o^2 h. \tag{22}$$

where P denotes the applied force per unit width of the strip and σ_0 is the yield strength of the film. When $\eta' \leq 1$, the film deforms *elastically* and the *peel force* P *becomes a direct measure of the interfacial fracture resistance* Γ_i:

$$\Gamma_i = P. \tag{23}$$

However, elastic behavior is atypical. The *minimum* film thickness h^* for elastic peeling, obtained from Eqn. (22) by equating η' to unity, is

$$h^* = 6E\Gamma_i/\sigma_o^2 \tag{24}$$

For example, Cu films (yield strength $\sigma_0 \approx 100$ MPa) exhibiting typical peel energies [19], h^* is as large as 1 cm . All thinner films yield.

When $\eta' > 1$, plastic deformation occurs around the base of the film, where a *plastic hinge* forms. The normalized debond energy Γ_i/P has a strong dependence on the base angle (Fig. 9), θ_B, a parameter that is difficult to either measure experimentally or model theoretically. The analysis reveals that for thin films, most of the work done by the applied load is dissipated as heat [20] and it has not been possible to obtain reliable measurements of Γ_i.

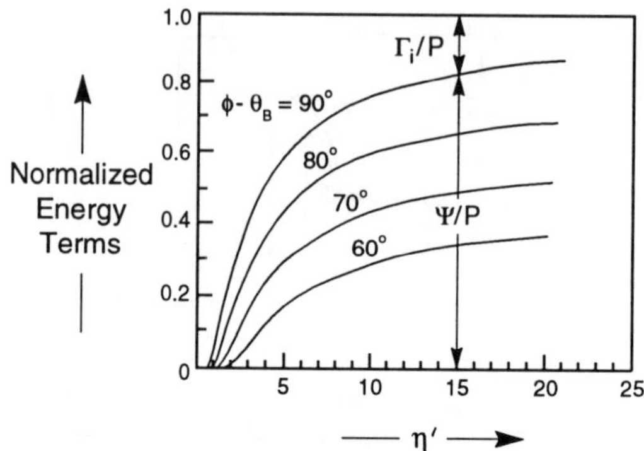

Figure 9. The effect of the base angle θ_B on the normalized energy terms. The contribution to the interfacial fracture (Γ_i) and the extraneous dissipation (Ψ) are normalized by the total peel energy P. Here ϕ denotes the peeling angle.

4.3 Blister Test

In the Blister Test, a hydrostatic load is imposed at the interface by introducing a fluid through a perforation, leading to progressive interfacial debonding. The critical pressure needed to initiate the debond is related to Γ_i through the mechanics of a pressurized elastic, circular blister [21]. The effects of residual stress can be readily included [22]. The only limitations of this test are the complex sample preparation and the narrow range of accessible mode mixities. The method *cannot* be used if the film yields before it decoheres. Other articles in this volume disscuss this approach.

4.4 Superlayer Test

There are few options for the steady-state loading of a thin film system at the mode mixities relevant to thin film and multilayer decohesion. One approach uses a *residual stress* which *duplicates* the problem of interest. For typical thin films (h < 1 μm), and representative residual stress levels ($\sigma_R \sim 100$ MPa), the energy release rate is small (Eqn. 2): of order, $\mathcal{G}_{ss} \approx 0.1$ Jm^{-2}. Such values are below the fracture toughness of interfaces having practical interest. A procedure that substantially increases $\mathcal{G}_{ss}$, is required. Such a procedure involves the deposition of a *superlayer* that increases the effective film thickness and also elevates the residual stress. The superlayer is selected in accordance with four characteristics. The deposition should be conducted at ambient temperature. The layer should not react with the existing film. It should have good adhesion and be subject to a large residual tension upon deposition. A Cr film, deposited by evaporation, meets all four criteria [6].

An optical micrograph of a superlayer test specimen is shown in Fig. 10. The test configuration has be analyzed for $\mathcal{G}_{ss}$ and ψ, using the solutions presented in Section 2. The Cr superlayer thickness is varied in order to produce a range of energy release rates.When the strips decohere, the energy release rate exceeds the debond energy, $\mathcal{G}_{ss} > \Gamma_i$. Conversely, when the film remains attached, $\mathcal{G}_{ss} < \Gamma_i$. Consequently, Γ_i is determined from the *critical Cr layer thickness* above which decohesion always occurs, designated h_c. The method entails tedious specimen preparation.

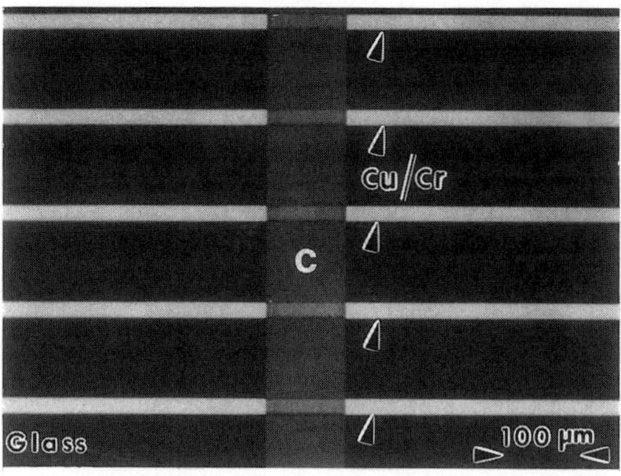

Figure 10. A micrograph of the superlayer test structure.

4.5 Multistrain Test

Multistrain Tests have many of the attributes of the Superlayer Test, but they avoid some tedious aspects of specimen preparation (Fig.11) [23]. Such tests require a ductile template in the form of a beam that can be deformed after the films have been deposited. Stainless steel has been used for this purpose. One surface is polished to an optical finish and a thin layer of polyimide is spun onto this surface. Various films and multilayers are then deposited onto the polyimide and patterned into strips, parallel to the long axis of the beam, with a gap at the center. The beam is subsequently subjected to bending, *with the coated surface on the side.* As bending occurs, each strip experiences a different strain : zero at the neutral axis and a maximum adjacent to the tensile surface. There is a corresponding variation in the strain energy release rate at interface cracks that originate at the gaps along each strip.

The obvious advantage of the test is *the ability to obtain a wide range of energy release rates on a single specimen.* Upon testing, those strips located near the tensile surface decohere, but the others remain attached, enabling identification of a critical strain at which decohesion occurs, designated ε_c. Then, if the film behaves elastically and its Young's modulus is known, the steady-state energy release rate may be obtained using the results from Section 2. An important limitation of this test is that it cannot be used with brittle substrates prone to cracking before interfacial decohesion commences.

5. EXPERIMENTAL RESULTS

5.1 High-Temperature Bonds

One particularly important factor governing the decohesion energy between metal and non-metal interfaces is the temperature experienced by the interface relative to the melting temperature of the constituents. Interfaces made at *high relative temperature* T/T_m (where T_m is the melting temperature) typically differ in their decohesion properties from those made at low temperature. At "brittle-clean" interfaces, produced by using a *high temperature* annealing step, the crack progresses by the growth and coalescence of debond patches *ahead of the crack front.* This yields

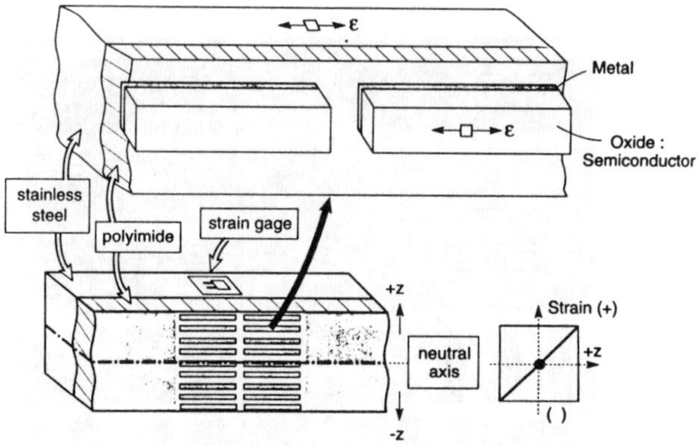

Figure 11. A schematic of the multistrain test.

a relatively large nominal toughness, $\Gamma_i \sim 100 Jm^{-2}$ [11, 24]. The debonds usually nucleate at defects on the interface, such as small pores, precipitates and grain boundaries. However, Γ_i can be considerably diminished when certain impurities are present that segregate to the interface. The two best documented examples are Ag at the Al_2O_3/Nb interface [9] and C at the Al_2O_3/Au interface [25]. The general trends appear to be in accordance with the SSV model (Fig.6). In particular, the major effect on Γ_i of a small change in W_{ad} is explicable. This happens because the toughening ratio is in a range where there is a substantial leveraging effect of the tip toughness on the plastic zone size.

5.2 Low Temperature Bonds

There have been few reliable measurements of Γ_i made on interfaces formed at *low temperature*. All of the results obtain for Cu (Fig.12), either as a thin film deposited onto dielectric/semiconductor substrates or as a substrate layer in contact with a polymer film [6, 7]. There are two substantial differences from the behavior found for bonds formed at elevated temperature. (1) The Γ_i are quite small and of the same order as W_{ad}. (2) There is a large elevation of Γ_i caused by a thin interlayer (5 nm) of either Cr or Ti. These results can be qualitatively interpreted using the cohesive zone model [13]. A negligibly small plastic dissipation, consistent with $\Gamma_i \Rightarrow W_{ad}$, arise because of the high yield strength of the thin film [26, 27]. That is, thin films require high $\hat{\sigma}$ to promote plastic dissipation. Hence, adhesion promoters (Cr or Ti) which increase $\hat{\sigma}$, results in higher Γ_i.

6. CONCLUDING REMARKS

The mechanics of thin films are now well established and provide a framework for the quantitative prediction of interface decohesion. A central parameter is the interface debond energy. The limitation on the application of the approach has been the lack of reliable experimental procedures for measuring Γ_i and of models that relate Γ_i to structural parameters for the film and the interface. The experimentally straightforward measurement methods involve complex loading paths, rendering unreliable measures of Γ_i. Test methods amenable to rigorous analysis require more extensive specimen preparation. Some of these tests involve multiple steps and are tedious. Others involve fewer steps and are more readily used. In practice, the most straightforward method that best satisfies the multilayer system being investigated, would be chosen.

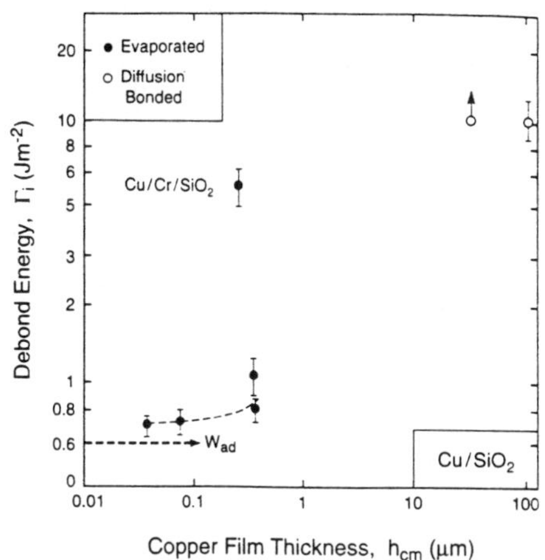

Figure 12. The effect of layer thickness on the toughness of the interface between thin Cu layers and SiO_2.

Because reliable tests have only recently been devised and calibrated, $\Gamma_i(\psi)$ data are sparse. One of the major findings to date is that films deposited at low homologous temperatures have much lower Γ_i than nominally identical interfaces produced at high homologous temperature. These results suggest that atom rearrangements by diffusion and, perhaps the dissolution of surface contaminants, has a crucial influence on Γ_i. The effect of high yield strength of thin films on plastic dissipation is also noteworthy. Further research on these phenomena is clearly of fundamental interest.

REFERENCES

1. M. D. Drory, M. D. Thouless, A. G. Evans, *Acta Metall. Mater.* **36**, 2019–2028 (1988).
2. M. D. Thouless, *J. Vac. Sci. Technol. A* **9**, 2501–2515 (1991).
3. Z. Suo, *J. Vac. Sci. Technol. A* **11**, 1367–1372 (1993).
4. Z. Suo, J. W. Hutchinson, *Intl. J. Frac.* **43**, 1–18 (1990).
5. J. W. Hutchinson, Z. Suo, *Adv. Appl. Mech.* **29**, 63–191 (1992).
6. A. Bagchi, G. E. Lucas, Z. Suo, A. G. Evans, *J. Mater. Res.* **9**, 1734–1741 (1994).
7. A. Bagchi, A. G. Evans, *submitted to Thin Solid Films.*
8. Z. Suo, C. F. Shih, A. G. Varias, *Acta Metall. Mater.* **41**, 1551–1557 (1993).
9. G. Elssner, D. Korn, M. Rühle, *Scripta Metall. Mater.* **31**, 1037–1042 (1994).
10. K. J. Hsia, Z. Suo, W. Yang, *J. Mech. Phys. Solids.* **42**, 877—896 (1994).
11. M. R. Turner, A. G. Evans, *Technical Report MECH-250*, Division of Applied Sciences, Harvard University (1995).
12. V. Tvergaard, J. W. Hutchinson, *J. Mech. Phys. Solids* **41**, 1119–1135 (1993).
13. V. Tvergaard, J. W. Hutchinson, *Phil. Mag. A* **70**, 641–56 (1994).
14. T. W. Wu, *J. Mater. Res.* **6**, 407–426 (1991).
15. S. K. Venkatraman, D. L. Kohlstedt, W. W. Gerberich, *J. Mater. Res.* **7**, 1126–1132 (1992).
16. S. K. Venkatraman, D. L. Kohlstedt, W. W. Gerberich, *Thin Solid Films* **223**, 269–275 (1993).

17. W. T. Chen, T. F. Flavin, *IBM J. Res. Develop.* **16**, 203–213 (1972).
18. K.-S. Kim, N. Aravas, *Intl. J. Solids Struct.* **24**, 417–435 (1988).
19. K.-S. Kim, J. Kim, *J. Engr. Mater. Tech.* **110**, 266–273 (1988).
20. J. F. Goldfarb, R. J. Farris, Z. Chai, and F. E. Karasz in *Materials Science of High Temperature Polymers for Microelectronics*, edited by D. T. Grubb, I. Mita, D. Y. Yoons (Mater. Res. Soc. Proc. **227**, Pittsburgh, PA, 1991) pp. 335–340.
21. H. M. Jensen, *Engr. Frac. Mech.* **40**, 475–86 (1991).
22. H. M. Jensen, M. D. Thouless, *Intl. J. Solids Struct.* **30**, 779–95 (1993).
23. D. K. Leung, N. T. Zhang, R. M. McMeeking, A. G. Evans, submitted to *J. Mater. Res.* .
24. A. G. Evans, B. J. Dalgleish, *Acta Metall. Mater.* **41**, S295–S306 (1992).
25. D. Lipkin, D. R. Clarke, University of California, Santa Barbara (private communication),
26. W. D. Nix, *Met. Trans. A* **20**, 2217–2245 (1989).
27. A. Bagchi, PhD thesis, University of California, Santa Barbara (1994).

MEASURING THE ADHESION OF DIAMOND THIN FILMS TO SUBSTRATES USING THE BLISTER TEST

Jim Sizemore, R. J. Hohlfelder, J. J. Vlassak, and W. D. Nix
Department of Materials Science & Engineering, Stanford University, Stanford CA 94305

ABSTRACT

It is shown that the blister testing technique can be used to measure the adhesion of thin films to their substrates. A brief discussion of blister test mechanics is presented here, leading to a simple equation relating adhesion to the height of the blister and the pressure causing it to grow. Blister test data for plasma-enhanced CVD diamond films on Si substrates have been analyzed using this relation. The tests show adhesion energies of 1.8- 2.6 J/m^2.

INTRODUCTION

Due to their very high hardness, low coefficient of friction and low wear rate, CVD diamond coatings are very attractive candidates for cutting tool and bearing applications. For such mechanical applications good adhesion of the film to the substrate is needed. However, it is known that diamond adheres poorly to most substrates[1]. Thus, much research is currently being focused on measuring and improving the adhesion of diamond thin films to various substrates.

The stresses that develop during growth of diamond and on cooling from the growth temperature can create high driving forces for diamond film delamination. Accordingly, the resistance of the film-substrate interface to debonding frequently controls the manufacturability and reliability of diamond films. Quantitative measurements of adhesion are, however, not straightforward.

Adhesion of thin films to substrates can be measured using the *blister test*, in which a pressurized free-standing film window is forced to debond from its substrate. The adhesion energy of the film is determined by measuring the relation between applied pressure and deflection of the pressurized window during debonding. Large-scale deformation of the film is avoided, and most of the energy dissipation occurs at the film-substrate interface. The blister test therefore provides a very direct measurement of the interfacial work of adhesion.

ANALYSIS

The *bulge test* and the *blister test* are closely-related techniques for measuring the mechanical properties and adhesion of thin films. In the *bulge test*,

Mat. Res. Soc. Symp. Proc. Vol. 383 © 1995 Materials Research Society

pressure applied to one side of a free-standing thin-film "window" causes the window to deflect outward (figure 1a). As stress and strain in the film are determined by the applied pressure and the resulting deflection, measurement of these quantities allows one to extract information about the residual stress, in-plane elastic modulus, and plastic behavior of the film [2,3,4]. Windows are typically a few millimeters wide and are formed *in situ* by etching holes through the substrate from the backside (see *Sample Preparation*).

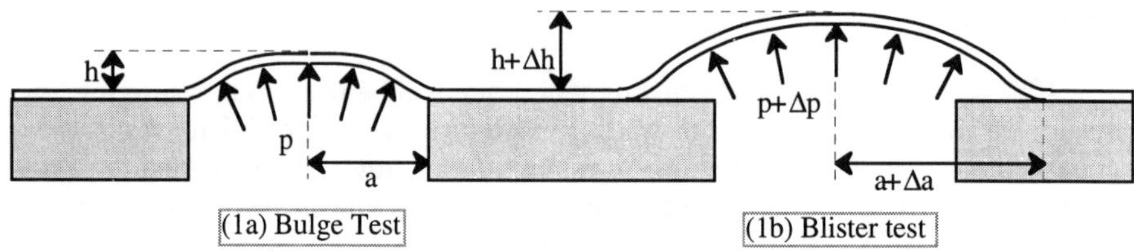

Figures (1a) and (1b): the bulge and blister tests

The *blister test* is identical to the bulge test except that pressures are high enough to cause the film to debond from its substrate, forming a circular "blister" which grows outward (figure 1b) [5,6,7,8]. The work required to separate a unit area of film from the substrate, which we term the *adhesion energy* (γ), can be obtained from measurements of pressure and deflection during periods of controlled blister growth.

The elastic deformation of a pressurized window can be modeled by treating the window as a piece of a thin-walled spherical pressure vessel, with uniform curvature and under a purely biaxial stress (the "spherical cap" model). It is straightforward to show that for such a geometry, pressure (p) and center deflection (h) are related by [9]:

$$p = \frac{8\,Mt}{3\,a^4}h^3 + \frac{4\,\sigma_0 t}{a^2}h \quad , \tag{1}$$

where σ_0 is the residual stress of the film, M is the in-plane biaxial modulus of the film (equal to E/(1-ν) for an isotropic material), t is the film thickness, and a is the radius of the window.

The spherical cap model is somewhat unrealistic because it neglects the fact that the film is fixed to the substrate at the outer edges of the window. In reality, plane-strain conditions exist at the edges of the window and biaxial strain is present near the center, and windows are accordingly more compliant than would be predicted by the spherical cap model. One can obtain an exact membrane solution to this problem by assuming series solutions for both the deflection and stresses in the film and imposing appropriate boundary conditions [10]. Another approach is to assume solutions for displacement which contain unknown

Table I: Geometric parameters c_1 and c_2 for the bulge equation (eq. 2)

Geometry	c_1	c_2
Circular Window	4	$2.667\,(1.014 - 0.244\,\nu)$
Square Window	3.393	$(0.792 + 0.085\,\nu)^{-3}$
Long Rectangular Window	2	$1.333\,/\,(1+\nu)$

parameters and to solve for those parameters by minimizing the total potential energy of the system [3].

In most cases, it can be shown that the basic form of the pressure-deflection behavior is similar to that of the spherical cap model. For purely elastic deformation, applied pressure and resulting deflection of a thin-film window can be related by a "bulge equation":

$$p = c_1 \frac{\sigma_0 t}{a^2} h + c_2 \frac{Mt}{a^4} h^3 \quad , \tag{2}$$

where c_1 and c_2 are functions of the shape of the window and the Poisson's ratio of the film (see table I) [10]. Residual stress and biaxial modulus may be determined by fitting eq. (2) to experimental data if the film thickness and window radius are known.

The volume underneath a bulged window is determined by the radius and height of the window, and a *shape parameter (κ_v)* defined in the following expression:

$$V = \kappa_v a^2 h \quad . \tag{3}$$

For an ideal spherical cap, $\kappa_v = \pi/2$. The parameter is slightly smaller than this because of the clamping effect at the window edges, which produces a less rounded shape. The true volume under the window may be calculated analytically or determined from finite-element simulations. It is found that κ_v changes very little throughout a test and may be considered to be constant.

For the blister test to be meaningful, there must be a way to relate the adhesion energy (γ) to measurable quantities. This can be done by considering the energy changes that occur when a blister grows by some small amount (da). When a blister grows, work is performed by the applied pressure ($dW_{applied}$) and there in an increase in the strain energy of the pressurized film (dU_{strain}). The difference between these two quantities, commonly called the *energy release rate (G)*, is the amount of energy available to separate the film from the substrate. G is defined per unit length along the interface, and for a circular blister is given by:

$$G \equiv \frac{1}{2\pi a} \left. \frac{\partial\left(W_{applied} - U_{strain}\right)}{\partial a} \right|_p \quad . \tag{4}$$

Growth will occur when the energy release rate is balanced by the adhesion energy. This may be expressed as:

$$\gamma = G \quad \Rightarrow \quad 2\pi a \, \gamma = \left(\frac{\partial W_{applied}}{\partial a}\right)_p - \left(\frac{\partial U_{strain}}{\partial a}\right)_p \quad . \tag{5}$$

The strain energy of the film can be calculated using eq. (2); it is a function of p, a, and h. The work done on the system is the applied pressure multiplied by the change in blister volume, and can be calculated using eqs. (2) and (3). By evaluating these quantities, eq. (5) may be reduced to a simple "blister equation":

$$C\left(M/\sigma_0 , h/a \right) p \, h = \gamma \quad , \tag{6}$$

where C is a coefficient in the range of 0.5 to 0.7, defined by the following expression:

$$C = \frac{\kappa_v}{2\pi} \left\{ \left(\frac{7}{6} + f(\xi)\right) - \frac{f(\xi)}{2\left(1 + \phi\,\xi^2\right)} \right\} \quad , \tag{7}$$

where: $\xi = \frac{h}{a}$, $\phi = \frac{c_2 \, M}{c_1 \, \sigma_0}$, $f(\xi) = \frac{4\phi\,\xi^2 + 2}{3\phi\,\xi^2 + 1}$.

It is straightforward, using eqs. (2, 6 &7) to show that C remains constant during blister growth. This is a convenient result, as it provides a simple and direct relation between adhesion, which is not directly measurable, and pressure and height, which are. Pressure may be easily and accurately monitored using an electronic transducer. Height can be measured directly, for example using laser interferometry [11], or determined indirectly from measurements of blister volume.

One can obtain an exact solution for the large-scale deflection of a pressurized circular membrane under residual stress [10]. Such an analysis is beyond the scope of this paper and will be presented in a future publication. The exact solution gives results which are similar to those obtained from the simpler model presented above, and allows one to evaluate some additional quantities which are not well described by the simpler model. The mode mixity at the crack tip of the debonding interface, which describes relative mode I (opening) and mode II (shear) stress intensities at the debonding interface [12], is found to remain constant during blister growth. This implies that it is not necessary to consider effects of changing mode mixity when analyzing blister test data.

Figure 2 shows a diagram of pressure vs. height data from a blister test. At first, pressure is too low to drive debonding, and the window simply stretches and deflects upward according to the bulge equation (eq. 2). This is followed by a transition region in which debonding has initiated but the blister is non-circular, growing unstably, or is otherwise poorly behaved. For example, one can conduct a test with square windows; debonding begins at the edge centers and proceeds until the blister is circular. Once the blister has assumed a circular shape and grows stably, the blister equation, (eq. 6), holds.

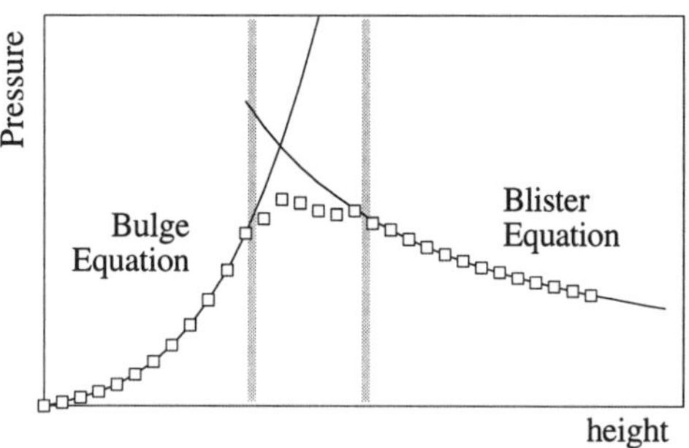

Figure 2: Schematic of blister test data

EXPERIMENTAL RESULTS

Sample Preparation:

Free-standing thin film windows can be manufactured *in situ* by etching openings through the substrate from the backside. Figure 3 depicts two suitable processes. For diamond films, one can simply etch a hole through the Si substrate without damaging the film. When necessary, more complex test structures can be made; figure 3b shows a process used to form free-standing windows of fragile polymer films [13]. First, a free-standing nitride window is fabricated. Polymer is deposited on top of the nitride, which is subsequently removed with a selective plasma etch, leaving a free-standing polymer film.

It is essential for bulge testing to have a film with residual tensile stress, so that the window remains taut, not wrinkled, when the substrate is etched away. The initial shape of the window is not critical for blister testing, as blisters tend to assume a circular shape, which is energetically favorable, as debonding proceeds; square or imperfect circular windows are therefore acceptable.

For these experiments, diamond was grown on (100) silicon substrates at 900°C using a DC plasma. The substrate was not scratched to enhance nucleation.

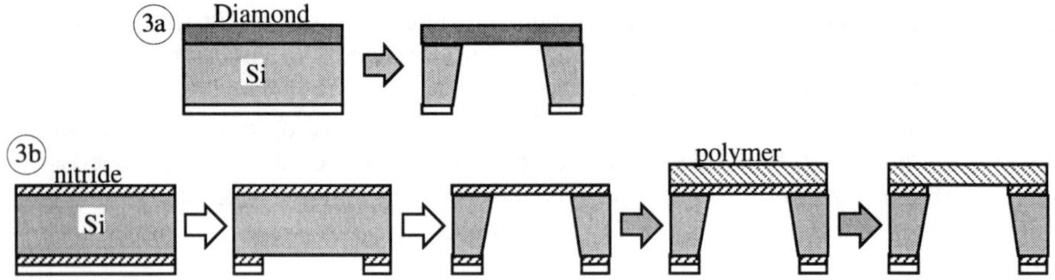

Figure 3: Sample Fabrication

A film thickness of 2 μm was measured in cross section using an SEM. Raman spectroscopy indicated the presence of diamond.

Experimental apparatus

As the product of pressure and blister height (p h) is constant during a test, the pressure required to drive film delamination continually decreases as the blister grows. If a test system is excessively compliant, pressure will not drop quickly enough when a blister starts to grow, and the blister will grow uncontrollably. This is undesirable, as the energy-balance of eq. (4) is valid only when a blister grows in a steady, controlled fashion. To minimize the compliance of the system, blisters should be pressurized with an incompressible liquid instead of with a compressible gas. An ideal apparatus would allow tests to be completely "displacement-controlled", the volume of the blister being directly controlled by the volume of fluid pumped into the system.

The blister test apparatus which was used for measurements of diamond film adhesion is depicted in figure 4. A rigid plate clamps the sample to the top of an oil-filled cell. A low-moisture mechanical pump oil was selected as a pressurizing fluid for its low vapor pressure and low viscosity. O-rings form a seal between the plate and the sample and between the sample and the cell. The volume between the plate and the sample is filled with oil, which rises up into a capillary tube. Using a microscope to observe the oil level in the tube, changes in the blister volume can be directly measured. This system of measuring volume displacement was developed in place of the usual laser interferometry for measuring the blister height because the rough surfaces of diamond precluded the formation of good interference fringes. The cell is pressurized with a syringe pump and pressure is monitored with an electronic transducer. Compression of the upper O-ring as the cell is pressurized produces an apparent increase in blister volume, but is easily measured and corrected for.

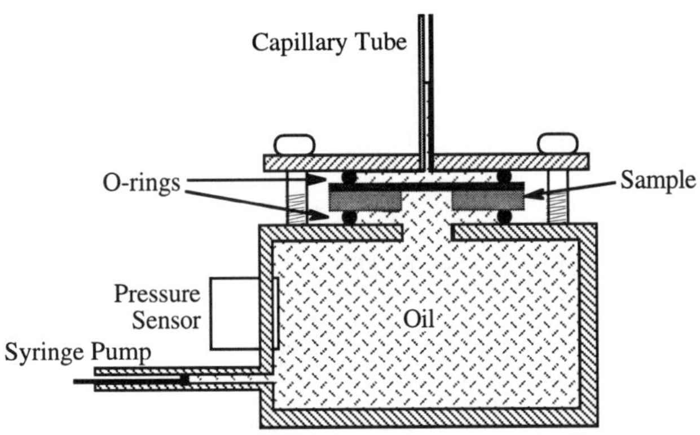

Figure 4: Blister Test Apparatus

Experiments:

Two experiments are presented for discussion. Data from the first are plotted in figure 5 as pressure vs. blister volume. Three distinct regions can be discerned. Before any debonding occurs (region I), the window is pressurized and "bulges" upward as described by eq. (2). In region II, pressure has reached a value high enough to cause the film to debond from the substrate. Compliance in the system causes the blister to grow rapidly and uncontrollably, as indicated by the sudden jump in the data. The unstable growth is followed by a regime of controlled, steady growth. The blister grows outward until it runs into the O-ring (see figure 4). Prevented by the O-ring from debonding any further, the membrane again stretches and deflects upward in accordance with the bulge equation (region III).

Since no debonding occurs in regions I and III, both can be analyzed to determine residual stress and biaxial modulus. It is appropriate to consider only region III, the larger and better defined of the two. Assuming a spherical-cap shape for the blister, region III was fitted to the bulge equation. Residual stress (σ_0) and biaxial modulus (M) were determined from the fitting parameters by knowing the film thickness ($t = 2$ μm) and the O-ring radius ($a = 7.35$ mm). Their values were:

$$\sigma_0 = -280 \text{ MPa} \quad , \quad M = 980 \text{ GPa}$$

These are reasonable figures; the biaxial modulus ($E/(1-\nu)$) of single crystal diamond is roughly 1100 GPa [14]. Even though the film was in compression, it was unwrinkled, implying that the bulge equation is still valid.

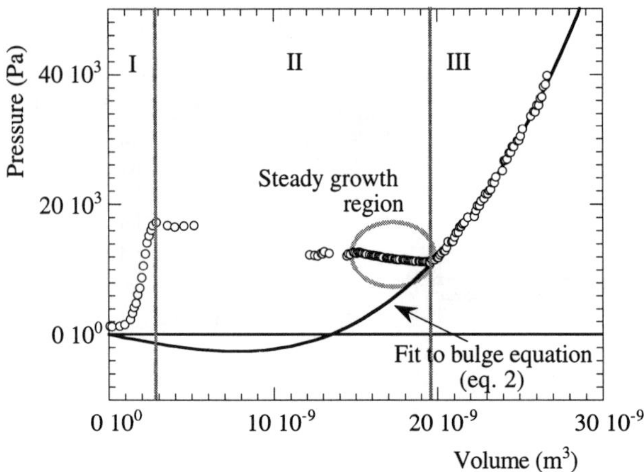

Figure 5: Blister Test Data

Blister height can be determined indirectly from measurements of blister volume by substituting eq. (3) into eq. (2) and solving for h:

$$h^5 + \frac{c_1 V \sigma_0}{c_2 \kappa_v M} h^2 - \frac{PV^2}{c_2 \kappa_v^2 Mt} = 0 \tag{8}$$

This polynomial has no analytic solutions but is easily solved numerically.

Figure 6 is a plot of pressure vs. calculated blister height over the steady-growth region. The best fit of the data to the eq. (6) is: p h = 2.63. C was calculated using eq. (7) and found to be 0.70, giving a value for adhesion:

$$\gamma = 1.8 \text{ J/m}^2$$

This result may not be very accurate because of the limited amount of crack extension in the steady growth regime. Additionally, the adhesion in this system may be affected by the use of pressurizing oil, as the oil wets both diamond and silicon and is likely to be present at the crack tip.

The second experiment was conducted with a smaller film window than the first. Stable blister growth could not be achieved because the blister was too small for growth to sufficiently reduce pressure in the compliant testing system. To circumvent this problem, the pressure at which debonding initiated was measured at different blister radii. Unstable blister growth was limited by testing with successively larger O-rings. Data from this experiment are plotted in figure 7; three successive loading curves are shown. On each curve the "critical point" at which debonding initiated can be discerned.

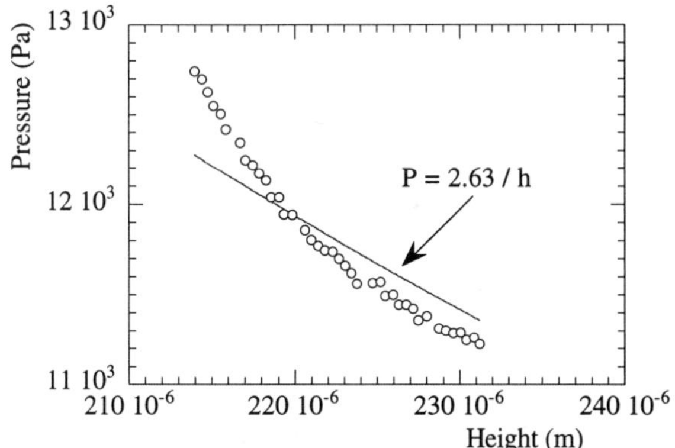

Figure 6: Pressure vs. Height

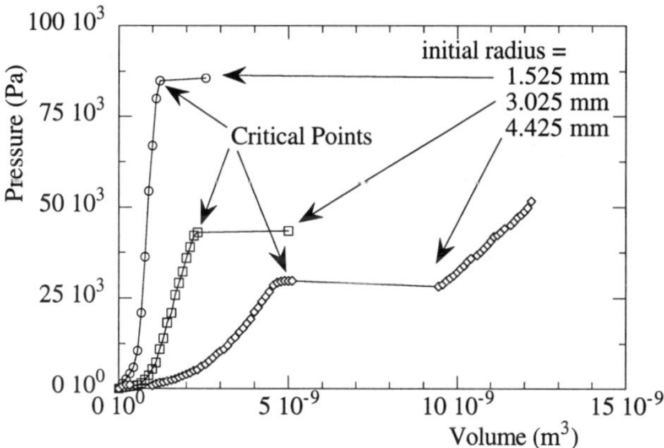

Figure 7: Blister Test Data

The residual stress and biaxial modulus of the film were determined by fitting the initial portion of the third loading curve to the bulge equation (eq. 2), again using the known film thickness and window radius. The following values were obtained:

$$\sigma_0 = 210 \text{ MPa} \quad , \quad M = 736 \text{ GPa}$$

The center deflection corresponding to each of the critical points was calculated using eq. (8). Figure 8 is a plot of pressure vs. calculated height at the three critical points. The best fit of the data to the bulge equation (eq. 6) is also

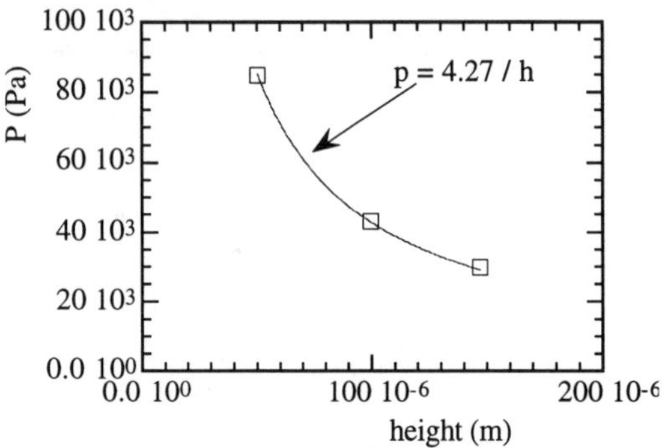

Figure 8: Pressure vs. Height

plotted: p h = 4.27. The coefficient C was calculated to be 0.61, giving for the adhesion energy:

$$\gamma = 2.6 \text{ J/m}^2$$

The adhesion energies we have measured are low, on the order that one would expect for free surface energies; this indicates that little energy is dissipated by plastic deformation at the crack tip for this brittle-brittle film-substrate system. In view of the high hardnesses of diamond and silicon, this result is not surprising. One would expect to measure much higher fracture energies for systems in which more opportunities for plastic deformation exist.

CONCLUSIONS

The blister test is a viable tool for quantifying the adhesion of diamond films to substrates. Work of adhesion can be directly determined from easily observed quantities (pressure, deflection). Sample preparation is not trivial, but the test provides a direct measurement of energy dissipation at the film-substrate interface. In tests of diamond films on Si substrates, low adhesion energies (approximately 2 J/m^2) were observed. Future work will be required to determine whether this technique will be applicable to the study of diamond film/substrate systems in which adhesion is much better.

ACKNOWLEDGMENTS

This research was supported by the Electric Power Research Institute, contract RP8042-06 (J. Sizemore), by the U.S. Air Force through an SBIR contract (AF93-066) to Crystallume (J. Vlassak and W.D. Nix) and by the U.S. Department of Energy, grant DE-FG03-89ER45387 (W.D. Nix). R.J. Hohlfelder wishes to gratefully acknowledge the support provided to him by Failure Analysis Inc. for the G. Marshall Pound Fellowship at Stanford.

REFERENCES

1) P.K. Bachmann and R. Messier, Chem. and Eng. News, **67** (20) 24-39 1989)

2) J. W. Beams, *Structure and properties of thin films*, C. A. Neugebauer, J. B. Newkirk, and D. A. Vermilyea, Eds., John Wiley and Sons, Inc., 1959.

3) J. J. Vlassak, W. D. Nix, *J. Materials Research*, **7**, 3242-3249 (1992).

4) V. M. Paviot, J. J. Vlassak, W. D. Nix, *Measuring the mechanical properties of thin metal films by bulge testing of micromachined windows*, Mat. Res. Soc. Symp. Proc., **356** 1995.

5) H. Dannenberg, *J. Appl. Polym. Sci.*, **5**, 125 (1961).

6) A. N. Gent, L. H. Lewandowski, *J. Appl. Polym. Sci.,* **33**, 1567-1577 (1987).

7) H.S. Jeong, Y. Z. Chu, C. J. Durning, R. C. White, *Surface and Interface Analysis*, **18** 289-292 (1992).

8) J. Sizemore, D. A. Stevenson, J. Stringer, *Modeling of the blister test to express adhesive strength in terms of measurable quantities*, Mat. Res. Soc. Symp. Proc., **308**, 165-170 1993.

9) M. K. Small, W. D. Nix, *J. Materials Research*, **7**, 1553-1563 (1992).

10) J. J. Vlassak, Ph. D. Dissertation, Stanford University, 1994.

11) M. K. Small, Ph. D. Dissertation, Stanford University, 1992.

12) Z Suo, J. W. Hutchinson, *Int. J. Fract.*, **43**, 1 (1990).

13) R. J. Hohlfelder, J. J. Vlassak, W. D. Nix, H Luo, C. E. D. Chidsey, *Blister Test Analysis Methods*, Mat. Res. Soc. Symp. Proc., **356**, 1995.

14) T. H. Courtney, Mechanical Behavior of Materials, McGraw-Hill, New York, 1990.

ADHESION EVALUATION OF CVD DIAMOND FILMS AND METAL REINFORCED COMPOSITE DIAMOND FILMS

D.F. BAHR*, J..C. NELSON*, D. ZHUANG**, E. PFENDER**, J. HEBERLEIN**, AND W.W. GERBERICH*
*Chemical Engineering and Materials Science, University of Minnesota, Minneapolis,MN 55455
**Mechanical Engineering, University of Minnesota, Minneapolis,MN, 55455

ABSTRACT

Poor adhesion of diamond films limits the use of CVD diamond films as coatings for cutting tools. The adhesion of these films is limited by stresses in the film caused by thermal expansion mismatch between the substrate and the film and by voids present at the interface due to the morphology of the crystal growth. A three step process of making diamond composite films has been developed, involving nucleation of individual diamonds on the substrate, electroplating a metal binder in the voids between the crystals, and lastly growing a complete film over the composite layer. The metal binder acts both to fill the voids at the interface and to absorb energy during fracture processes at the interface. Diamond growth was performed in a DC Triple Torch reactor using a mixture of methane and hydrogen with a molybdenum substrate. Measurements to determine the amount of improvement of the film adhesion have been performed. These tests include indentations using conventional hardness testing equipment and four point bend tests with the film in tension and compression. A correlation is shown between the plastic zone of the substrate and the area of the film which delaminated during indentation. Bend tests with the film in tension did not delaminate the film, instead the film underwent intergranular fracture. Bend tests in compression act similarly to pile up around an indentation, and cause film delamination. Residual stress measurements in the single step film show a compressive stress of 650 MPa.

INTRODUCTION

One of the limiting factors in applying diamond films as coatings for cutting tools is the adhesion of the films onto various substrates. The variety of methods presently used to deposit diamond films such as microwave plasmas, thermal plasmas, and combustion flame methods have led to great differences in growth conditions and film thicknesses between films grown by different methods[1,2]. An adhesion testing method for diamond films should be able to compare films of different thicknesses grown under various conditions on a variety of substrates.

A method presently used by several researchers involves indentations using a Brale indenter tip in a Rockwell hardness testing system[3-6]. The tip is driven into the sample at various loads, and after loading the indentation is examined. At some critical load diamond films tend to spall from the substrate, leaving behind a region from which the film has delaminated from the substrate. The slope of a plot of applied load versus the crack length (either the radius or diameter of the spalled region) is used to semi-quantitatively measure the adhesion of the diamond film. This method has been suggested to compare only films grown on the same substrate materials[4].

Nanoindentation techniques use a monitoring of load and depth during a light indentation of a film on a substrate to determine loads at which delamination occurs[7]. While nanoindentation has proven to be difficult to cause delamination of diamond films prior to tip fracture, the same idea can be applied to an indentation on a larger scale. By adapting a digitally controlled servo-hydraulic mechanical testing device to accept a Brale indenter tip, indentations can be made while monitoring the load and depth of the indentation. In addition, by adapting a mechanical testing device to perform indentations, any desired load can be applied, thereby allowing an accurate determination of the slope of the load versus crack length.

Mat. Res. Soc. Symp. Proc. Vol. 383

The indentation method was used to examine the effects of processing parameters on adhesion. In addition, a bend test of the diamond films was compared to the indentation test. A three step film developed[8,9] to improve adhesion was evaluated to determine the magnitude of adhesion enhancement of the three step process. This film consists of individual diamonds crystals grown in the first step, a metal binder electroplated in the voids between the crystals for the second step, and a continuous diamond film grown on the composite layer for the third step.

EXPERIMENTAL PROCEDURE

Diamond films were grown on molybdenum substrates in a previously described DC triple torch reactor[10] . This reactor uses 3 DC torches to create a plasma plume into which the reactant gasses are fed. The substrates were placed on a water cooled holder and raised into the plasma plume. Reactor pressure was kept at approximately 250 torr, and the substrate temperature was monitored via a Williamson optical pyrometer. Temperature measurements may be innacurate due to reflection from the plasma, but the measurements indicate two distinctly different growth regimes. Substrates were either 25 mm diameter discs 1 mm thick or the same disc with slots cut through half the thickness of the sample to form 6 small strips. These samples were polished on 240 grit SiC paper and then rinsed in methanol in an ultrasonic cleaner for 20 minutes. No pretreatment of diamond powder or paste was used. Growth times are shown in Table I. The high initial CH_4 concentration was used to enhance nucleation, and the concentration was then lowered to grow a better quality film. Therefore the times listed correspond to the methane concentrations listed during one continuous run. The three step film was grown at different conditions, and the conditions during the first and third step are separated by a colon. The second step was applied by electroplating nickel into the voids, as described by Tsai *et. al.* [8,9] Adhesion testing was carried out in a mechanical testing device adapted to accept a Brale indenter. Indentations were applied to the films at a rate of indenter travel of 20 μm per second to a predetermined load. When the load was reached the indentation was reversed and removed from the sample at the same rate. For bend tests the strips were sectioned apart by cutting the webs between the strips away. These strips were bent in four point bend tests in compression or tension, while monitoring the load and displacement, in a jig built to fit the hydraulic mechanical testing system. In both tests the indent was observed using a 20x optical microscope to determine if spalling occurred during loading or unloading.

Residual stress measurements were made by polishing the back of a strip until it was thin enough that the residual stress caused a deflection of the beam. The beam bowing was measured using optical microscopy. Characterization of the other films was accomplished using a JEOL 840 SEM or an Hitachi S800 SEM. Raman spectra of fims 4THMLI and 6THMLB are shown in Figure 1. The quality of the films varied dramatically with processing conditions, and showed variations when the laser was focused on various facets and positions within the film.

RESULTS AND DISCUSSION

A. INDENTATION

A typical indentation of a diamond film on a molybdenum substrate is shown in Figure 2. Observation of some of the films with an optical microscope revealed that the films spalled upon loading. The spalled region was not removed all at once, but spalled off in rings or sections as the load was increased, showing that the indenter was not transmitting any energy to the film directly after the ring of film contact with the indenter had spalled. Therefore it is suspected that the plasticity of the substrate leads to the energy transfer to the film, causing delamination. This plasticity can be either radial displacements or a pile up of substrate material around the indenter during the indentation. One of the indents at lower loads shows what is suspected to be the beginning of this process due to pile up of the substrate. Figure 3 shows an indent of 3TLMH, showing the initiation of film delamination. This type of mechanism, piling up the substrate, effectively causes beam bending of the film as the plastic zone increases, explaining why the films would delaminate during loading in ring shaped increments.

Table I
Growth Conditions in Triple Torch Reactor

Sample	Temp. (°C)	% CH_4	Run Time	Film Thickness
1TLMHI	1000	5.06 - 3.79 - 2.53	10 - 10 - 40 min.	13 µm
2TLMLI	1025	2.5 - 1.25	20 - 10 min.	8 µm
3THMLI	1200	2.5 - 1.25	10 - 27 min.	17 µm
4THMLI	1200	2.5 - 1.25	10 - 15 min.	7 µm
5THMHI	1300 - 1000	5.0 - 3.0 - 2.0	30 - 20 - 10 min.	21 µm
6THMLB	1150	2.5 - 1.25	10 - 90 min.	60 µm
3STEP	1100 : 1000	1.25 - 2.5 : 1.25	60 - 25 : 30 min.	10 µm

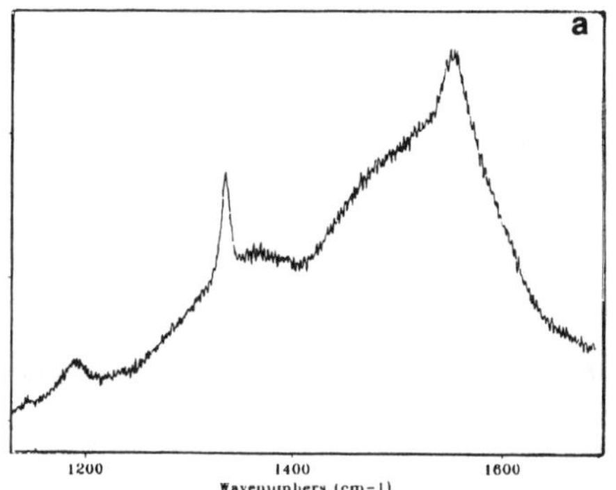

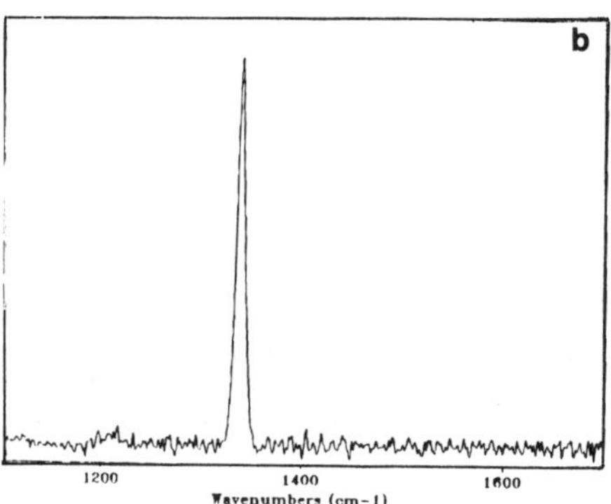

Figure 1
Raman spectra from samples grown at 2.5 to 1.25% CH_4 for a) 25 minutes ;b) 100 minutes

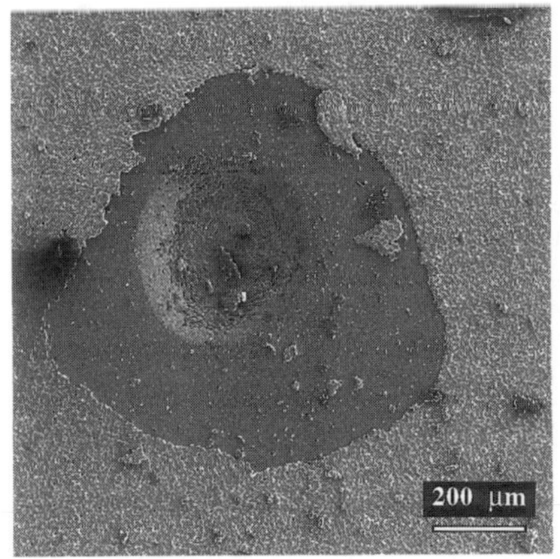

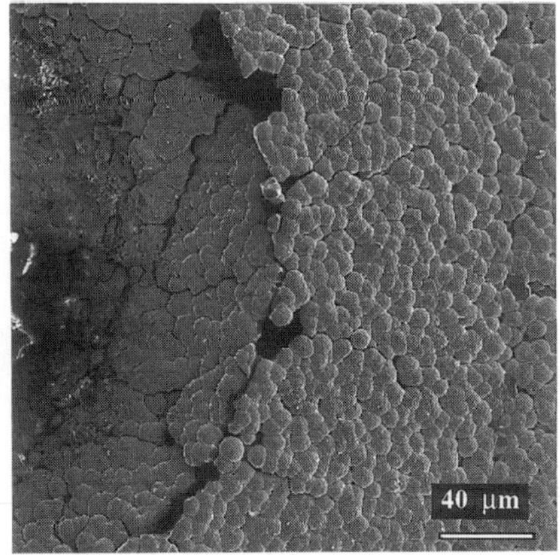

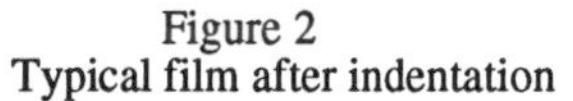

Figure 2
Typical film after indentation

Figure 3
Beginning of film delamination

To verify the extent of pile up a Dektek profilomiter was used across the indents. Profilometry results for and indent on sample 3THML are shown in Figure 4. The extent of the spalled film seems to approach the edge of the region of the substrate which underwent pile up. The pileup in this case was about 1/3 of the film thickness. This factor should not be neglected when examining the mechanism which caused the film to delaminate.

While visual observation of the indent shows delaminations during loading no specific load drop is clear when data taken during each indentation is examined, as shown in Figure 5. A slight slope change in the load versus depth data may be present. However, since this slope change occurs during the point where the indenter changes from hemispherical to conical, it is not conclusive that this change is due to film delamination. Load versus indenter depth monitoring of indentations of molybdenum before any plasma processing, several diamond films, and the three step film are shown in Figure 5. The first factor which should be noted is that for a given load the substrates with diamond films allow a further penetration of the indenter than the as received molybdenum. The three step film also exhibits this behavior. This is taken to mean that the substrates which have diamond films are softer than the as received molybdenum, and the three step film is even softer; suggesting a lower yield strength of the substrate. Since the molybdenum was heated to temperatures in excess of 1000 °C it is likely that annealing occurred, producing the observed behavior.

Since the films did not spall in exact circles the extent of the crack was traced using an image analysis program and the effective spallation area was determined and converted into an effective crack radius if that area had been circular. By doing a series of indents at increasing loads the maximum load versus crack length could be determined. These results for samples which spalled during indentation are presented in Figure 6. The proposed relationship between plastic zone size and the plastic zone as estimated by Harvey *et. al.* [11] is also presented in this plot. This estimation of the plastic zone is not exact but does match well with the plastic zones measured by finding the edge of the region which had piled up using the profilometer.

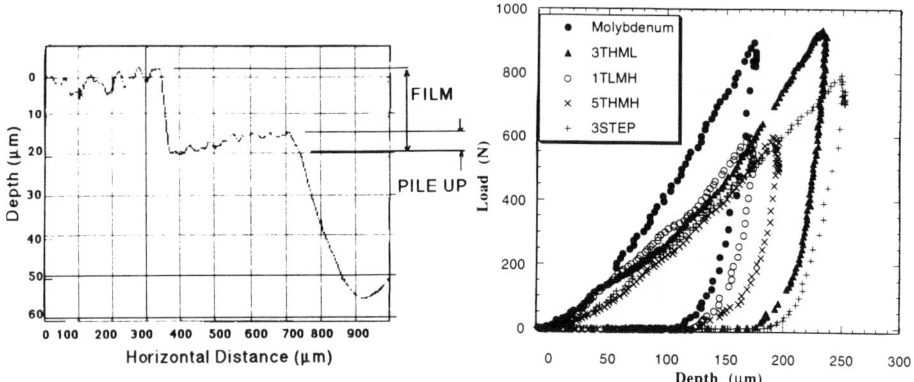

Figure 4
Height profile across indentation of 3TLMH

Figure 5
Load v. depth during indentation

This analysis suggests that the adhesion of a film is not directly related to just the slope of a given load versus crack length, but instead is actually related to the plastic zone size. If the piled up substrate acts effectively as a load at the edge of the film a treatment similar to that proposed by Cao and Evans [12] for film delamination by a wedge of a given displacent may be appropriate. This will be explored in a future publication.

The indentation data suggests that films grown at 1200 °C have significantly better adhesion than films grown at 1000 °C. Whether the data is examined just by comparing the slope of the load versus crack length data or by examination of the plastic zone, the conclusion is that films grown at the higher temperature are adhered better than the films grown at the lower temperatures.

Indentations were also carried out on a three step film. This film showed no observable delamination at loads of 800 N, as shown in Figure 7 when the indentation occured in the center of the test strip. Increased film bonding due to the presence of a metal binder which diffusion bonded to the substrate during the third step is credited to improving the adhesion of this film. Since this test was carried out on a sample which was cut into strips it was possible to get minute delaminations at 600 N when the indent was placed close to one edge of the test strip. In this case it was observed that the entire film thickness delaminated, not just the top layer.

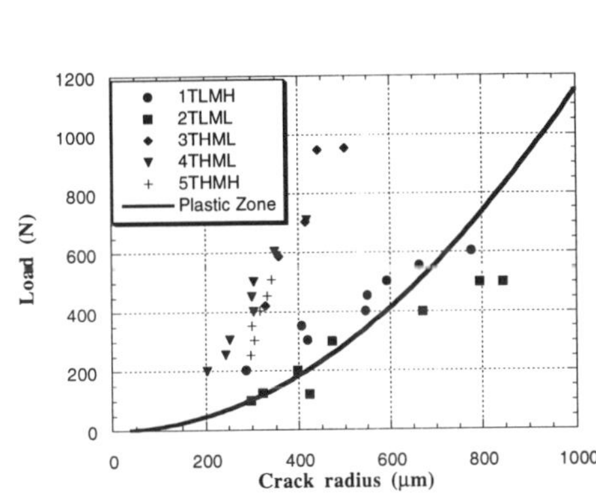

Figure 6
Maximum applied load v. crack radius
Calculated plastic zone size is also shown

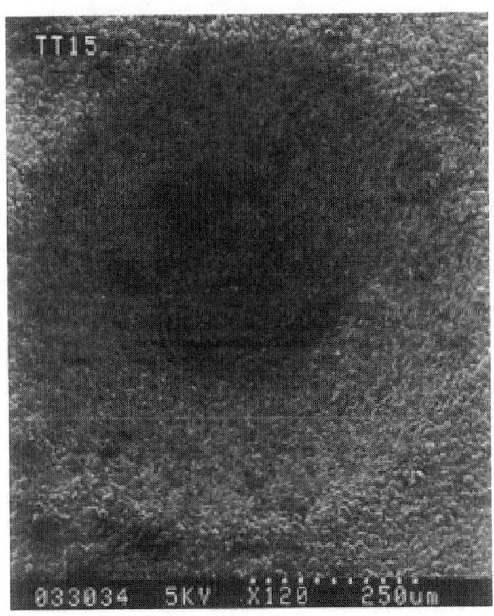

Figure 7
Indentation to 800 N on 3 step film

B: Bend Tests

Bend tests of small bars of molybdenum with diamond films tested in a four point bend test jig were done both in tension and compression. Placing the films in tension was attempted first to initiate a pre-crack in the film which would arrest at the film-substrate interface. However, the films did not crack only to the substrate in one place due to the polycrystalline nature of the films. Cracks formed along many grain boundaries in the diamond film and proceeded down to the interface, effectively destroying the film without separating the diamonds from the substrate.

By placing the film in compression during a four point bend test film delamination did occur. The film was observed with an optical microscope during the test while load versus displacement data was collected, and is shown in Figure 8. A section of the film delaminated at the same time that the load excursion was recorded. The same type of test was applied to the three step film, and no spalling of the film was observed, even with a significant amount of plastic work imparted to the specimen. This correlates well to the indentation tests, showing similar behavior for the diamond and composite films.

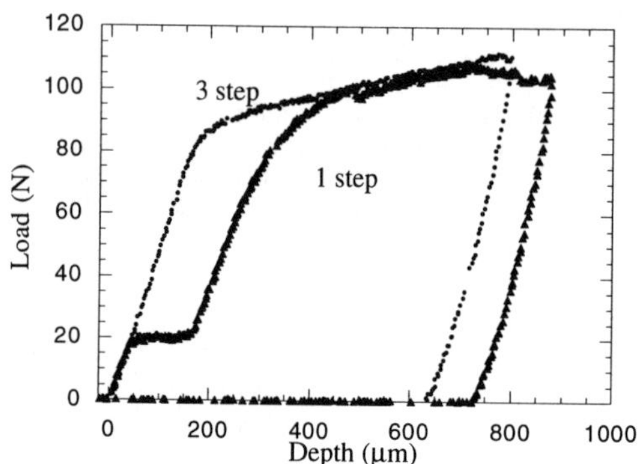

Figure 8
Bend testing load v. displacement for 1 and 3 step films

By cutting the strips of a diamond film apart and polishing the back of the substrate until it thinned enough to cause the beam to bow and deflect a mechanical measurement of the stress in the film can be calculated. Using the Stony equation and measuring the deflections allowed the residual stress in 6THML to be determined. This film was much thicker than previous films, and had grown for a longer time. The film was found to be in compression, and had a residual stress of 650 MPa[13], which is on the order of the yield strength of molybdenum.

SUMMARY

A need to test the adhesion of diamond films easily and reliably is needed to refine processing conditions of film growth to promote increased adhesion of the film. Experiments on diamond films on molybdenum and a novel three step method of film deposition have used both indentations using a Brale indenter tip and four point bend tests in compression to observe film delamination. The spalled region of a film indented with a Brale indenter has been shown to correlate to the plastic zone size in the substrate. Models previously used for indentations using the Brale indenter did not address the issue of the film cracking during the indentation, and therefore the energy transferred to the film causing delamination cannot be a function of indenter contact with the film but instead is a result of the substrate being raised up around the indenter during loading. The three step process has been tested both in bending and indentations and is found to have superior adhesion to the one step diamond films. Also, the adhesion of films has been shown to increase with increasing growth temperatures between 1000 and 1200 °C.

ACKNOWLEDGMENTS

The authors wish to thank David Bucci of Drexel University for the Raman spectroscopy. The assistance of Victor Indajang with diamond deposition is also greatly appreciated. This research was supported by the NSF under grant NSF/ECD-8721545, ERC for Plasma Aided Manufacturing and grant NSF/CDR-8721551, Center for Interfacial Engineering.

REFERENCES

1. K.E. Spear, J. Am. Cer. Soc., **72**, 171 (1989).
2. W.A. Yarbrough, J. Am. Cer. Soc,., **75**, 3179 (1992).
3. P.C. Jindal, D.T. Quinto, G.J. Wolfe, Thin Solid Films, **154**, 361 (1987).
4. K.J. Grannen, F. Xiong, R.P.H. Chang, Surf. Coat. Tech., **87**, 155 (1993).
5. C.T. Kuo, T.Y. Yen, T.H. Huang, S.E. Hsu, J. Mater. Res., **5**, 2515 (1990).
6. R.C. McCune, R.E. Chase, E.L. Cartwright, Surf. Coat. Tech., **53**, 189 (1992).
7. S.K. Venkataraman, J.C. Nelson, N.R. Moody, D.L. Kohlstedt, W.W. Gerberich, MRS Spring Meeting Symposium H, San Francisco, CA (1994).
8. C. Tsai, J.C. Nelson, W.W. Gerberich,J. Heberlein, E. Pfender, Dia. Rel. Matl., **2**, 617 (1993).
9. C. Tsai, J.C. Nelson, W.W. Gerberich,J. Heberlein, E. Pfender, J. Mater. Res., **7**, 1967 (1992).
10. Z. Lu, L. Stachowicz, P. Kong, J. Heberlein, E. Pfender, Plasma Chem. Plasma Proc.,**11**, 387, (1991).
11. S. Harvey, H. Huang, S. Venkataraman, W.W. Gerberich, J. Mater. Res., **8**, 1292 (1993).
12. H.C. Cao, A.G. Evans, Mech. Matl., **7**, 295 (1989).
13. A. Brenner, S. Senderoff, J. Res. Natl. Bur. Stnds., **42**, 105 (1949).

ADHESION AND YOUNG'S MODULUS OF CVD DIAMOND THIN FILMS GROWN OVER VARIOUS SUBSTRATES

R. RAMESHAM, R.F. ASKEW, AND M.F. ROSE
Space Power Institute, 231 Leach Center, Auburn University, Auburn, AL 36849-5320.

ABSTRACT

Diamond films were deposited by microwave plasma CVD using H_2 and CH_4 gas mixture over various substrate materials such as Si, Pd, Be, Cu, Mo, AlN, SiO_2, Si_3N_4, Al_2O_3, Sapphire, Quartz, Ni-base alloys, single crystal Ni, boron nitride, and Ti. We have used a Z-axis pull stud test to determine adhesion strength of diamond film to some of the substrates. Our observations on the adhesion of diamond films to the above substrates are reported. A method will be described to evaluate Young's modulus of CVD diamond films using fabricated diamond cantilever beams.

INTRODUCTION

Growth of polycrystalline diamond films using a CVD processes has received a considerable interest in recent years, since diamond has unique physical and chemical properties. Diamond films based on their unusual properties will have potential uses in electronics, optics, chemical resistance, protective mechanical coatings. These applications will mostly depend on stress in the films, adhesion strength to the substrate, smoothness of the films, and growth temperature, type of substrate, nature and quality of the films, patternability, and related properties.

Although diamond film has unique properties, the durability of diamond film grown on any substrate is largely dependent on the adhesion strength between the film and the desired substrate for eventual successful use in various applications. Adhesion is related to the nature and the strength of the binding forces at the interface of the film and the substrate where they are in physical contact with each other. Therefore, a measurement of the adhesion strength is primarily a practical interest but it also has fundamental importance. Adhesion of diamond film to any substrate is strongly dependent on several factors such as the chemical nature and cleanliness of the substrate, film growth temperature, lattice constant, thermal expansion coefficient, and surface energy of the film and the substrate, nucleation density of the film on the substrate, and substrate morphology. Acoustic emission scratch adhesion testing (Rockwell diamond indentor) is used to measure the critical load for failure of the diamond thin films grown on silicon substrate [1], tungsten substrate [2], and cemented carbide (94.3% WC + 5.7% Co) cutting tools [3]. Reports of the attempts to determine adhesion strength using the peel test are scanty. The peel test requires patterned films in a specified configuration on a chosen substrate. The films are patterned by selective growth [4,5]. Using selectively grown diamond on silicon substrates, infrared mask aligning, and standard silicon microetch techniques, polycrystalline diamond microstructures were fabricated and the Young's modulus has been measured.

Mat. Res. Soc. Symp. Proc. Vol. 383

EXPERIMENTAL DETAILS

Surface damaging is a necessary process in order to nucleate the CVD diamond on non diamond substrates. Ultrasonic agitation of the substrates is performed in methanol containing synthetic diamond particles of ~ 90 μm typical size for 10 - 60 minutes [6,7]. Microwave plasma (2.45 GHz) assisted CVD system (ASTeX, Woburn, MA) has been used to grow diamond films on damaged substrates like Mo and Si. Schematic diagram of the diamond deposition system is described earlier [4,5]. The substrate was placed at the center of the stage that is then loaded into the quartz bell jar reactor. The reactor was evacuated to a base pressure of 10^{-4} Torr. Plasma was obtained by adjusting pressure in the chamber, hydrogen flow rate, microwave power, and wave guide tuning. The substrate was *in-situ* heated by the microwave plasma to attain the desired substrate temperature and initiated the growth of diamond. UHP grade H_2 and research grade CH_4 were used in our experiments. The temperature was monitored remotely by an optical pyrometer. Diamond deposition was started by injecting methane into the system when the substrate reached the desired temperature (~ 925°C). The deposition rate under the typical conditions provided in Table 1 is normally ~ 1 μm/hour. A continuous film of diamond was usually obtained after 10 - 12 hours of growth. Typical deposition parameters are provided in Table 1

Table 1. Typical parameters for the growth of undoped diamond films.

Growth Parameter	**Undoped diamond**
Substrate temperature (°C)	925±25
Hydrogen flow rate(SCCM)	500
Chamber pressure(Torr)	45±5
Methane flow rate(SCCM)	3.6
Forward power(Watts)	1150±50
Reflected power(Watts)	<50
Substrate base	Si

RESULTS AND DISCUSSION

a. Z-axis pull stud test for adhesion strength measurement

Processes for the selective deposition of diamond films onto a variety substrates by ultrasonic agitation have been described earlier [6,7]. Z-axis pull stud method has been used to determine the adhesion strength of diamond thin films to the various substrates. A Sebastian V multipurpose tester was used to perform adhesion tests. Initially, the non-diamond side of the substrates (0.5 - 1 in^2) was bonded to a backing plate of black alumina (2 in x 2 in) using a fast set epoxy resin and hardener and cured at room temperature for 12 hours. backing plates are required in order to have a rigid test set-up. A test stud precoated with an epoxy is bonded to the diamond surface by holding it in contact using a spring mounting clip designed for that stud and cured at 150°C for 1 hour. The stud is inserted into the platen support of the equipment until the diamond or bonding material fails. The selectively grown diamond coating pattern on each substrate was 0.1 x 0.1 in^2. The length of the stud was 0.5 inch. Unfortunately, precoated studs were not bonded well to the diamond surface. This may be due to an insufficient amount of epoxy present with standard pull studs and non-uniformity of the surface of diamond thin films. Alternatively, we have mixed the epoxy resin and hardener before coating the studs and used the mix to bond the studs to cleaned diamond film pattern. Subsequently the test samples were cured at room temperature for 12 hours. Care was taken not to overflow the epoxy beyond the diamond film.

Table 2: Adhesion strength test results of diamond films on various substrates using z-axis pull test [7]

Test sample	t (μm)	Adhesion strength (lb inch^{-2})
Cu/epoxy/Al		8909
Diamond/Si[a]	19	3423
Diamond/Si[b]	10.5	3067
Diamond/alumina[b]	9	4492
Diamond/Si_3N_4[b]/Si	10	4709
Diamond/SiO_2[b]/Si	9	3843
Diamond/Mo[b]	10.5	poor adhesion of stud to diamond with epoxy

[t], thickness of diamond film
[a], damaged by diamond paste
[b], damaged ultrasonically by diamond particles

Table 2 lists adhesion strength results of selectively grown diamond films on various substrates using z-axis (perpendicular to the substrate) pull stud approach. The reference sample (Cu/epoxy/Al) has the highest adhesion strength when compared with all the diamond samples. In all the test samples, the diamond film is intact even after the adhesion test, however, failure of the epoxy has occurred during the test. Mixed epoxy did not adhere well to the rough diamond thin film on all the substrates. Thus, in practical application the adhesion strength might be higher than the values provided in the Table 1. The z-axis pull test measures the force required to peel the diamond film off the substrate, whereas the acoustic emission scratch adhesion test measured the critical load at which the diamond film fractures. It is difficult to compare the results obtained by these two different approaches at this stage since the studies are made with different thicknesses and areas of diamond film, and with different substrates. However, it should be noted that the z-axis pull test is made possible as a result of selective growth of diamond on various substrates.

Adhesion of CVD diamond to Si, Mo, AlN, BN is very good. Diamond films deposited over Pd, Be, Cu, Al_2O_3, sapphire, quartz, Ni-base alloys, single crystal Ni, and Ti contain several microcracks and in some cases diamond films peeled off the substrate.

b. Young's modulus evaluation

Davidson et al., [8] and Ramesham et al., [5] have described a procedure to deposit diamond selectively over the silicon substrate and also described a procedure to fabricate the diamond microstructures such as cantilever beams, membranes, and bridges. We have also determined the Young's modulus of thin film diamond using diamond's cantilever microstructures and load-deflection data. After the fabrication of the cantilever beams, the silicon wafers were mounted under the optical microscope. The deflection of the polycrystalline diamond "microbeams" by loading was used as a means of determining mechanical properties. The beams were deflected by applying loads manually and measuring the deflection using calibrated vertical focusing of an AO Scientific Instruments (Model Microstar) optical microscope. Beam lengths and widths were measured with a calibrated filar micrometer eyepiece on an optical microscope. Diamond film thickness was measured by a profilometer. Weights attached to fine wires were carefully hung, manually with the aid of tweezers, on the ends of the cantilever beam while in place under the microscope. After measuring the deflection, the weights and supporting wire measured for each case on an analytical balance. The amount of deflection was determined using the focusing of optical microscope with the calibrated vertical stage, of the loaded and unloaded diamond cantilever beam. The deformation of the cantilever beam can be expressed in terms of the deflection of the beam from its original zero-load position. The relationship between cantilever beam deflection, S, applied load, P, at the end of the cantilever beam deflection, and the length of the beam, L, is

$$S = (PL^3)/3EI$$

where E is the Young's modulus of thin film cantilever beam material (kg/cm^2), and I is the moment of inertia of the beam (cm^4) . Specifically

$$I = (bh^3)/12$$

where b and h are width and thickness of the cantilever beam, respectively [9]

From these formulas it is seen that the unknown, E, may be determined experimentally from deflection (S) vs., load (P) at a constant L or deflection (S) vs., length (L^3) at a constant load, P. Both approaches were used to estimate Young's modulus. The dimensions (L=3168 μm, t=13 μm, W=639 μm, clearance, L=2430.5 μm, clearance, W=639 μm, anchor=738 μm) of the fabricated cantilever of CVD diamond beam used to determine the Young's modulus by deflection vs load. Young's modulus determined using deflection vs. load at a constant length, width, and thickness was found to be 1.2×10^7 kg/cm^2. Similarly, we have also used the other approach such as deflection vs. length^3 at a constant load and the Young's modulus was found to be 1.25×10^7 kg/cm^2. These values compares interestingly with the reported value for the single crystal diamond of 1.15×10^7 kg/cm^2.

CONCLUSIONS

We have reported a method to determine the adhesion strength of CVD diamond thin films to the various substrates. A method has been qualitatively evaluated to determine the Young's modulus of thin films of diamond by fabricating the diamond microstructures and using deflection vs. load or length data.

ACKNOWLEDGMENTS

This work is supported by the NSWC, Crane, IN through Port Hueneme, CA and by the CCDS located at Auburn University, with funds from NASA Grant NAGW-1192-CCDS-AD,

Auburn University, Center's Industrial Partners, and in part by the Office of IST of the SDIO's Office through Navy Contract #N60921-91-C-0078 with the NSWC.

REFERENCES

1. D.E. Peebles and L.E. Pope, J. Mater. Res., **5,** 2589, 1990.
2. M. Alam, D. Peebles, and D. Tallant, Electrochemical Society Spring Meeting, Extended Abstracts, **91 (1)**, 151, 1991.
3. C.T. Kuo, T.Y. Yen, and T.H. Huang, J. Mater. Res., **5,** 2515, 1990.
4. J.L. Davidson, C. Ellis, and R. Ramesham, J. Electron. Mater., **18,** 711, 1989.
5. R. Ramesham, T. Roppel, C. Ellis, D.A. Jaworske, and W. Baugh, J. Mater. Res., **6,** 1278, 1991.
6. R. Ramesham and T. Roppel, J. Mater. Res., **7,** 1144, 1992.
7. R. Ramesham, T. Roppel, R.W. Johnson, J.M. Chang, Thin Solid Films, **212,** 96, 1992.
8. J.L. Davidson, R. Ramesham, C. Ellis, J. Electrochem. Soc., **137,** 3206, 1990.
9. W.A. Nash, in "Theory and problems of Strength of Materials," 2nd Edition, p. 169, McGraw Hill Book Company, New York (1972).

BRAZING CHARACTERISTICS OF DIAMOND ON VARIOUS SUBSTRATES WITH DIFFERIENT BRAZING ALLOYS

CHENG TZU KUO, CHII RUEY LIN AND HOW MIN LIEN
National Chiao Tung University, Institute of Materials Science and Engineering
Hsinchu 30050 Taiwan

ABSTRACT

The commercial diamond grits and the diamond films deposited by a microwave plasma chemical vapor deposition system were used to study the wetting characteristics of few commercial brazing alloys, e.g., Cu, Ni, Ag-Cu-Zn-Cd-Ni and Ag-Cu-In. The substrates for brazing experiments include AISI-304 stainless steel, AISI-1045 steel and cemented WC insert. The results shows that the wettability of the brazing alloys and the bonding strength between diamond and substrate are strongly influenced by the compositions of the brazing alloys and the surface conditions of diamond and substrate. The wettability of Cu with diamond grit or diamond film depends on the brazing substrate, and is in order of AISI-304 > WC > AISI-1045. The brazing alloys with Cu or Ag as the main component often show a poor brazing strength for diamond/plain carbon steel or diamond/WC.

INTRODUCTION

In recent years, great interest has been concentrated on various possible applications of diamond films [1~8]. This is mainly due to some of the fascinating properties of diamond, such as the extreme hardness, high thermal conductivity, high electrical resistivity, low dielectric constant and large band gap. However, the most important requirements for these applications are to solve the film adhesion problems and to find a cost effective and fast fabrication method. One way to solve the problems is to deposit a film on a proper substrate and then to etch off the substrate to obtain a free standing film , and finally to braze the film on a desired substrate.

Therefore, from the scientific point of view and practical applications, the brazing characteristics of diamond on various substrates with different brazing alloys are of significant interest. In this paper, the results of the investigations on the natural diamond grits and diamond film wettability on different substrates with different brazing alloys will be reported.

EXPERIMENTAL

The diamond films used in this research were deposited by a microwave plasma enhanced chemical vapor deposition system. The Si wafer and the cemented WC were used as the substrates for diamond film deposition. The details of the system and the deposition procedures have been reported alsewhere [5]. For comparison, the industrial natural diamond grits of 400 ~ 460 mesh were also used for brazing experiments. Few commercial available alloys, such as 99.99% Cu [liquidus temperature(l.t.)=1082℃], commercial Cu paste, 99.99% Ni (l.t..=1454℃), Ag-15.5%Cu-15.5%Zn-16.0%Cd-3.0%Ni (l.t. = 688℃), Ag-24%Cu-14.5%In (l.t.. = 630℃), were used as the brazing alloys. The diamonds were brazed to different substrates, such as

Mat. Res. Soc. Symp. Proc. Vol. 383

AISI-304 stainless steel, AISI-1045 steel and cemented carbides(WC).

The brazing experiments were carried out by two different processes: (1) an induction furnace under 100 torr argon atmosphere, (2) a commercial vacuum brazing furnace under 0.1 torr nitrogen atmosphere. In some experiments, Cu and Ni were deposited by a DC sputtering system. The sputtering conditions were: Ar flow rate = 15.4 sccm, pressure = 0.01 torr, 280 volts, 0.1 amp. for various times. The brazing experiments for diamond grits were carried out by sprinkling grits on a WC substrate. The wetting phenomena and the morphologies were examined by SEM. The Raman spectrometry and x-ray diffraction were used to identify the quality of diamond films.

RESULTS AND DISCUSSION

It was proposed that the wettability of a brazing alloy and the peeling strength of the bonding are strongly influenced by the chemical compositions of the brazing alloy and the surface conditions (e.g. oxide thickness, dense or porous oxide) of the substrate [7-10]. Figures 1 to 5 show the wettability of various brazing alloys with diamond films or diamond grits. Figure 1 shows a free-standing diamond film (~ 54 μm in thickness) brazed with pure Cu to the AISI-304 stainless steel substrate by an induction furnace under 100 torr argon atmosphere. The corresponding Cu, Ni, Fe, C and Cr mappings at the film/brazing alloy interface are also shown in Figs. 1(a) to 1(e), respectively. It indicates that a larger amount of elements, such as Fe, Cr, diffuses out from the steel substrate to the Cu brazing alloy. This may contribute to a good wetting and good bonding between diamond film and Cu. On the contrary, Fig. 2 shows a poor wetting and poor bonding between diamond grits and Cu, where grits were brazed to the WC substrate by an induction furnace. A similar poor bonding was also found between diamond film and Cu, as shown in Fig. 3, where diamond film was coated with pure Cu by a DC sputter. Therefore, Figs. 1, 2 and 3 seem to suggest the bonding between diamond and brazing alloy depending upon the surface conditions of diamond and substrate. Experiments were also conducted to barze diamond grits on three different substrates with commercial Cu paste by a commercial vacuum brazing furnace. The results are shown in Figs. 4(a), 4(b) and 4(c) for three different substrates: AISI-304 stainless steel, WC and AISI-1045 steel, respectively. Difference in bonding strength is obvious. It is known that carbon is so insoluble in most nontransition metals (e.g. Cu, Ag) that graphite-clay crucibles are frequently used for melting them. This explains a poorer bonding between diamond and AISI-1045 steel or WC substrates by brazing with Cu.

Figure 5 shows poor wetting between Ag-Cu-In and diamond grits, which were brazed to a WC substrate by an induction furnace. Figure 6 shows a fractograph between Ag-Cu-Zn-Cd-Ni brazing alloy and the diamond grits, which were also brazed to a WC substrate by an induction furnace. The bonding was not good enough.

The melting temperature of pure Ni is much higher. Therefore, instead of brazing, diamond films were deposited with a layer of Ni film by a DC sputter, as shown in Fig. 7. The sputtered Ni films were also deposited a layer of diamond film by a microwave plasma chemical vapor deposition system, as shown in Fig. 8. It indicates that Ni film can well covered on diamond film, but the reverse is not true.

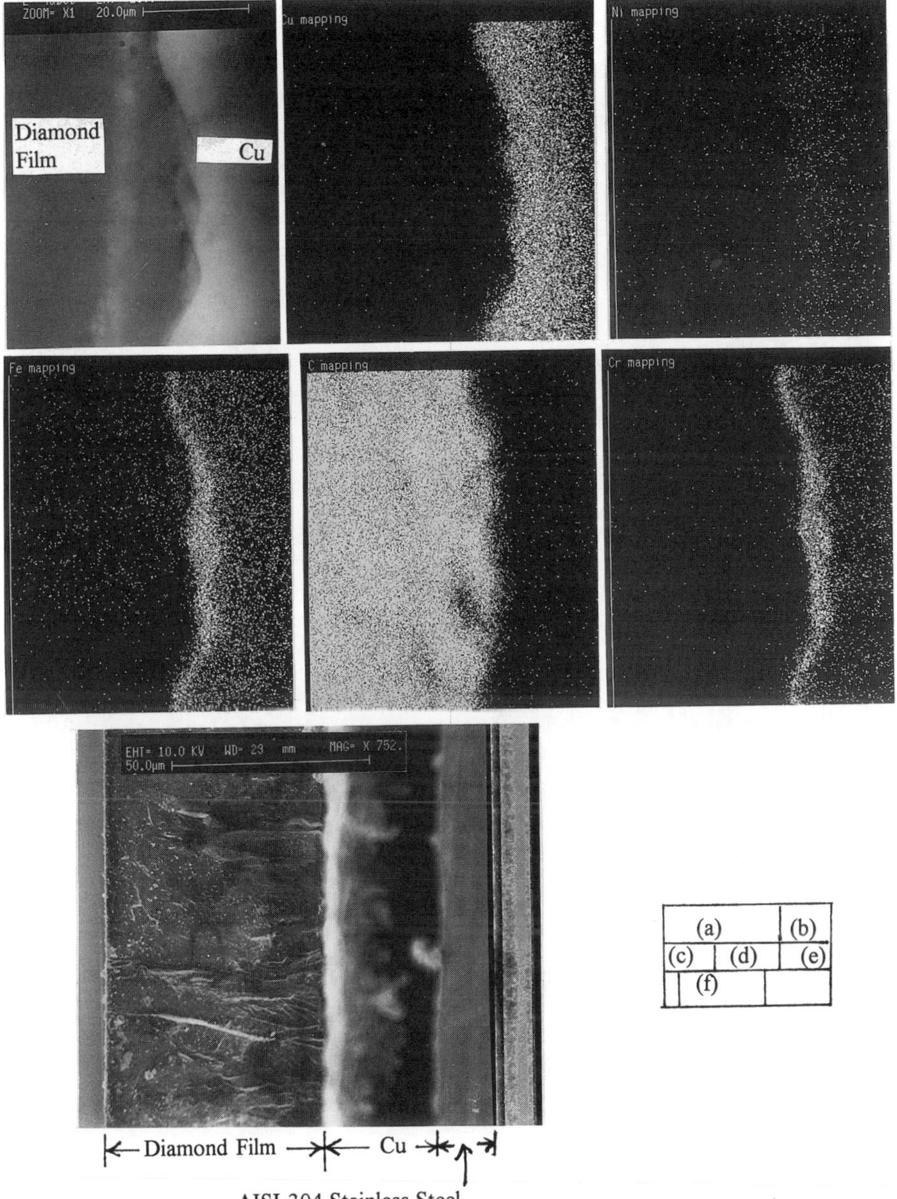

Fig. 1: Diamond Film brazed with pure Cu on the AISI-304 stainless steel substrate by an induction furnace: (a) to (e) are Cu, Ni, Fe, C and Cr mappings, respectively, (f) SEM micrograph at interfaces.

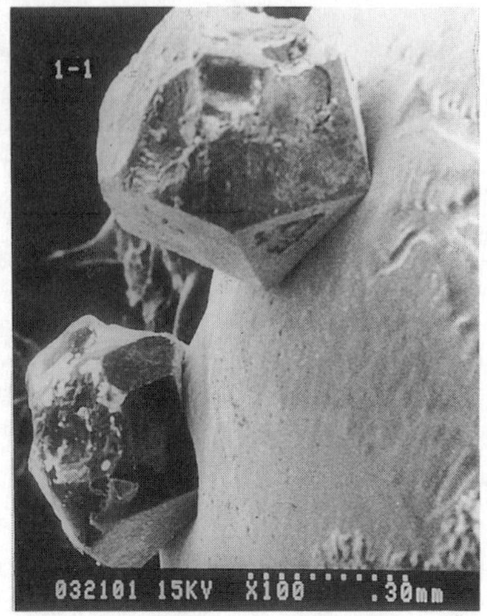

Fig. 2: Diamond grits brazed with Cu by an induction furnace.

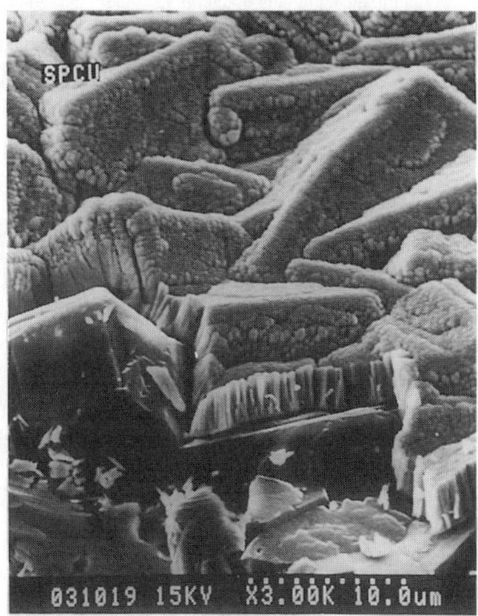

Fig. 3: Diamond films deposited with Cu by a DC sputter.

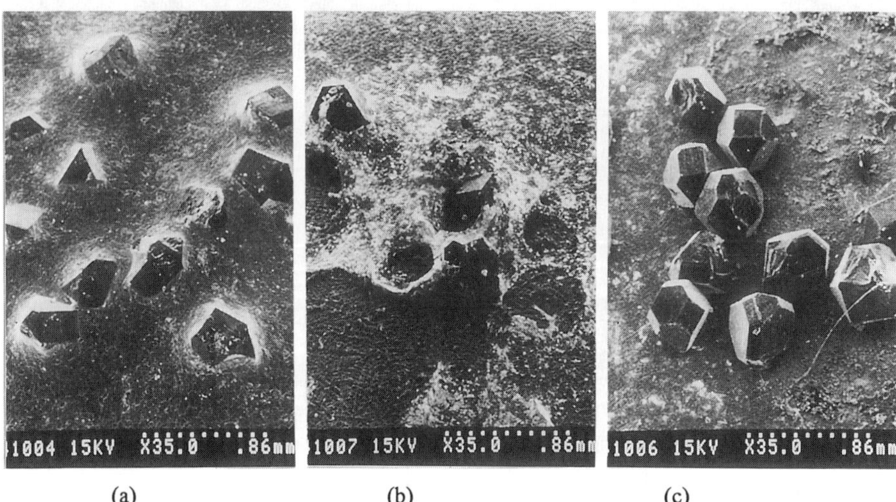

(a) (b) (c)

Fig. 4: Diamond grits brazed with commercial Cu paste to three different substrates by a commercial vacuum brazing furnace: (a) AISI-304 stainless steel substrate, (b) WC substrate, (c) AISI-1045 steel substrate.

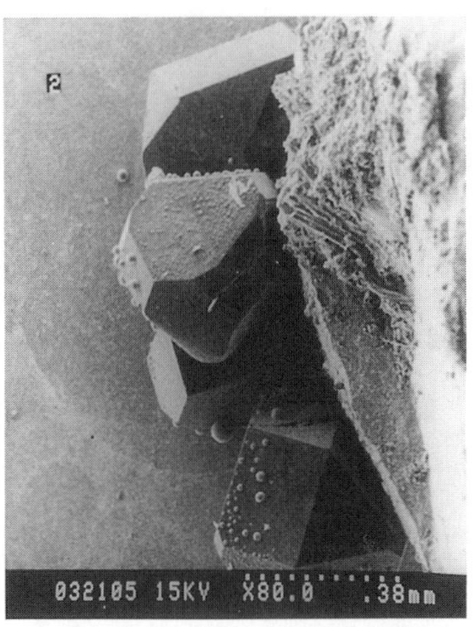

Fig. 5: Diamond grids brazed with Ag-Cu-In by an induction furnace.

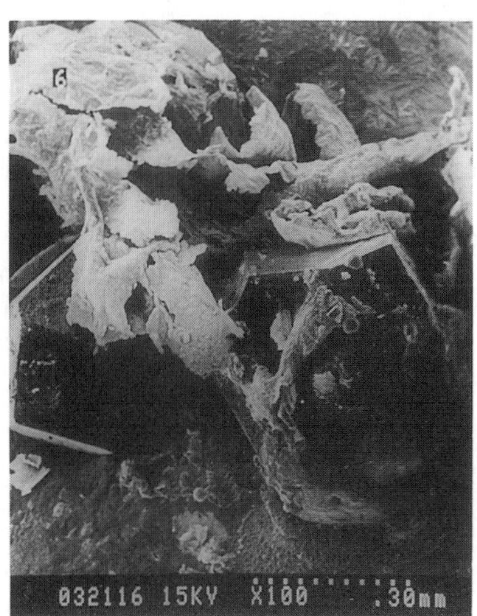

Fig. 6: Fractograph showing fracture area between Ag-Cu-Zn-Cd-Ni brazing alloy and diamond grits, which were brazed by an induction furnace.

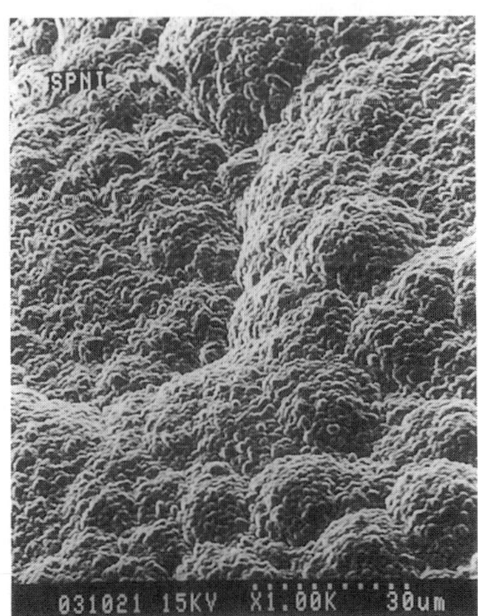

Fig. 7: Diamond films deposited by a layer of Ni film by a DC sputter.

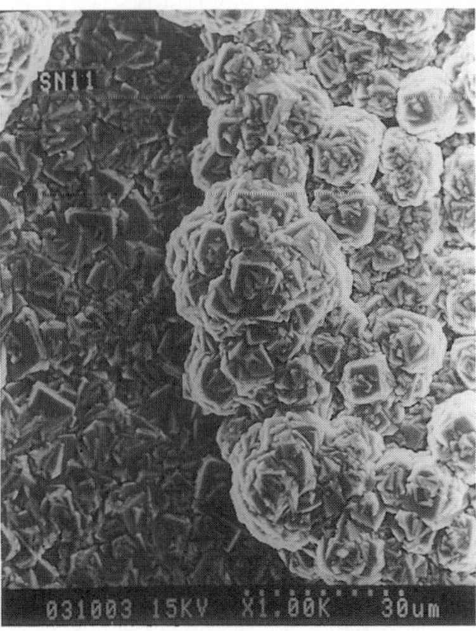

Fig.8: Ni films deposited by a layer of diamond film by a microwave CVD system.

CONCLUSIONS

Few commercial brazing alloys were used to investigate their wetting characteristics with diamond grits and diamond films. The results shows that the wettability of the brazing alloys and the bonding strength between diamond and substrate are strongly influenced by the compositions of the brazing alloys and the surface conditions of diamond and substrate. The wettability of Cu with diamond grit or diamond film depends on the brazing substrate, and is in order of AISI-304 > WC > AISI-1045 . Carbon is so insoluble in nontransition metals. Therefore, the brazing alloys with Cu or Ag as the main component often show a poor brazing strength for diamond/plain carbon steel or diamond/WC.

ACKNOMLEDGEMENTS

This research was sponsored by the National Sicence Council, R.O.C. under contract number: NSC84-2221-E009-037. The authors also thank Dr. K. H. Chen of Institute of Atomic and Molecular Science, Taipei, for assistance in Raman analysis.

REFERENCES

1. M. Murakawa and S. Takeuchi, in Applications of Diamond Films and Related Materials, edited by Y. Tzeng, M. Yoshikawa, M. Murakawa and A. Feldman (Elsevier Science Publishers B. V., New York, 1991), pp. 93-98.
2. R. A. Hay and C. D. Dean, ibid., pp. 53-60.
3. M. Murakawa and S. Takeuchi, Surf. Coat. Technol. 49, 359 (1991).
4. T. Yashiki, T. Nakamura, N. Fujimori and T. Nakai, Surf. Coat. Technol. 52, 81 (1992).
5. C. T. Kuo, T. Y. Yen and T. H. Huang, J. Mat. Res. 5, 11, 2515 (1990).
6. B. Lux, R. Haubner and P. Renard, Diamond and Related Materials 1, 1035 (1992).
7. A. K. Chattopadhyay and H. E. Hintermann, J. Mat. Sci. 28, 5887 (1993).
8. A. V. Andreyev, Diamond and Related Materials 3, 1262 (1994).
9. Ju. V. Naidich, Ju. N. Chuvashov, J. Mat. Sci. 18, 2071 (1983).
10. C. T. Kuo and W. K. Wang, in Proc. 1987 Annual Conf. Chinese Soc. for Mat. Sci., (May 23-24, Hsinchu, Taiwan, 1987) pp. 191-196.

FRACTURE, FATIGUE AND INDENTATION BEHAVIOR OF PYROLYTIC CARBON FOR BIOMEDICAL APPLICATIONS

R. O. RITCHIE,[1] R. H. DAUSKARDT,[2]
W. W. GERBERICH,[3] A. STROJNY,[3] AND E. LILLEODDEN[3]

[1]Department of Materials Science and Mineral Engineering, University of California, Berkeley, CA 94720-1760
[2]Department of Materials Science and Engineering, Stanford University, Stanford, CA 94305-2205
[3]Department of Chemical Engineering and Materials Science, University of Minnesota, Minneapolis, MN 55455

ABSTRACT

The fracture, fatigue and indentation properties of pyrolytic carbon, both as a monolithic material and as a coating on a graphite substrate, are described in light of its use for biomedical implant applications, specifically for the manufacture of mechanical heart valve prostheses. From the perspective of determining properties that are important for the prediction of safe structural lifetimes in such prostheses, it is found that by traditional engineering standards, pyrolytic carbon has low damage tolerance, i.e., fracture toughness values between 1 and 3 MPa√m and susceptibility to subcritical crack growth by both cyclic fatigue and stress-corrosion cracking (static fatigue). Subcritical crack-growth rates are evaluated in simulated physiological environments for both through-thickness "long" cracks, and for physically "small" surface cracks, the latter measurements being performed for cracks initiated at hardness indents. The unusual deformation characteristics of indentation in pyrolytic carbon are described based on instrumented microhardness indentation and scanning probe microscopy (AFM/STM) studies.

INTRODUCTION

Carbon is the most frequently found element in all organic molecules and compounds and as such performs a vital role in biological processes. As a crystalline material it can exist in numerous forms, most of which offer outstanding chemical inertness and biocompatibility. However, only three types of carbon are commonly used for biomedical devices [1-5]: the low-temperature isotropic (LTI) form of pyrolytic carbon, glassy (vitreous) carbon and the ultralow-temperature isotropic (ULTI) form of vapor-deposited carbon. These three forms of carbon have a disordered lattice structure and are collectively referred to as *turbostratic* carbons. Although pyrolytic carbons were developed originally for elevated-temperature applications (e.g., as coatings for nuclear fuel particles), LTI pyrolytic carbon has found wide appeal in the biomedical materials industry, in particular for mechanical cardiac-valve prosthetic devices, as it has been shown to be highly thromboresistant and to have inherent cellular biocompatibility with blood and soft tissue [3]; moreover, it displays good durability, strength and resistance to wear [6,7], and had been thought to be immune to cyclic fatigue failure [8,9]. In fact, the majority of modern mechanical heart-valve prostheses utilize components manufactured from silicon-alloyed

Mat. Res. Soc. Symp. Proc. Vol. 383 © 1995 Materials Research Society

LTI-pyrolytic carbon, either as a coating on a polycrystalline graphite substrate or as a monolithic material (where the substrate has been machined away). Up to 20 wt% silicon is added to improve mechanical properties without significant changes in biocompatibility.

STRUCTURE AND PROCESSING

Structurally, turbostratic carbon forms part of the disordered end of the wide range of crystalline states of carbon, from the perfect, three-dimensional graphite forms through the partially ordered graphite-like structures to the nearly amorphous state. Although disordered, its microstructure is closely related to that of graphite. Both structures consist of covalently bonded carbon atoms arranged in layers and stacked to produce a three-dimensional graphite with weak van der Waals bonding between the layers. In graphite, the layer planes are stacked in a regular ABAB sequence (Fig. 1a), crystallite sizes are typically 100 nm in diameter, and correspondingly single crystals have highly anisotropic properties. In pyrolytic carbon, conversely, the sequence is disrupted by stacking through random rotations or displacements of the layers relative to each other, termed *turbostratic* structures (Fig. 1b); moreover, as the randomly oriented crystallites are much smaller (~10 nm or less) in size (Fig. 1c), turbostratic carbons display isotropic mechanical and physical properties at the macro-scale [10].

The distortion and voiding associated with such turbostratic structures has a marked effect on both the density and strength of pyrolytic carbon. Missing carbon atoms lead to many vacant lattice sites in each of the turbostratic layer planes; imperfect matching of small segments of these planes also leads to wrinkles or distortions within each plane. Furthermore, there may also be a substantial fraction of disorganized carbon between crystallites in the bulk material. Densities, therefore, range from 1,400 $kg.m^{-3}$ to a theoretical limiting value of 2,200 $kg.m^{-3}$. High density Si-alloyed LTI carbons are the strongest bulk form of turbostratic carbon.

Dense and high strength LTI pyrolytic carbon components are typically made by co-depositing carbon and silicon carbide on a polycrystalline graphite substrate via a chemical vapor-deposition (CVD), fluidized-bed process using a gas mixture of a silicon-containing carrier gas with a hydrocarbon (e.g., propane, methyltrichlorosilane and helium gas mixtures) at elevated temperatures [11]. The resulting material contains typically 10 wt% silicon, often in the form of discrete sub-micron β-SiC particles, randomly dispersed in a matrix of roughly spherical micron-size subgrains of pyrolytic carbon; the carbon itself has a subcrystalline turbostratic structure, with a crystallite size typically less than 10 nm [12,13].

The CVD production of pyrolytic carbon is performed in a fluidized bed coater; this in principle consists of a vertical tube furnace (or reactor) containing a bed of granular particles, usually zirconium oxide. The reactant gas stream enters the reactor and fluidizes (i.e. supports and agitates) the bed of particles in which the components to be coated are suspended. Due to the high temperatures (typically in the range 1000 to 1500°C), the hydrocarbon gas "cracks" or pyrolyzes according to the reactions:

$$C_3H_8 \rightarrow 3C + 4H_2 ,$$

and

$$CH_3Cl_3Si \rightarrow SiC + 3HCl .$$

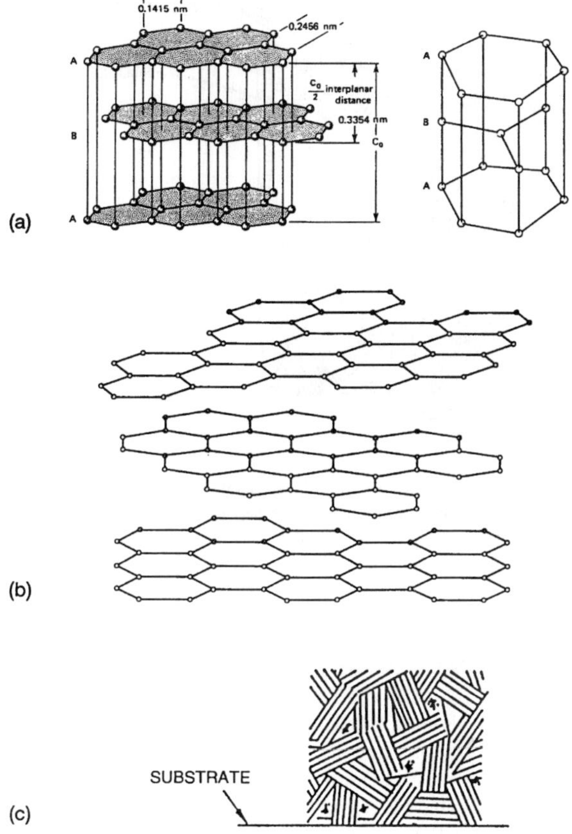

Figure 1. The crystallographic arrangement of carbon atoms in (a) hexagonal graphite where parallel layer planes are in a regular sequence and (b) a turbostratic carbon where the layers are arranged without order. Randomly oriented turbostratic crystallites are assembled to produce a bulk material in (c). Adapted from ref. [4].

The solid products of these pyrolysis reactions are carbon and silicon carbide which deposit as a coating on everything in the bed including the fluidized particles and the graphite components to be coated [11]. The silicon carbide, however, does not dissolve in the carbon matrix but instead forms discrete second-phase particles of silicon carbide with cubic crystal symmetry (Fig. 2). The quantity of silicon present has a marked influence on the mechanical properties of pyrolytic carbon. As the silicon carbide content increases, fracture strength, wear resistance, hardness and the elastic modulus increase significantly; its content and distribution must therefore be carefully controlled to achieve optimum properties.

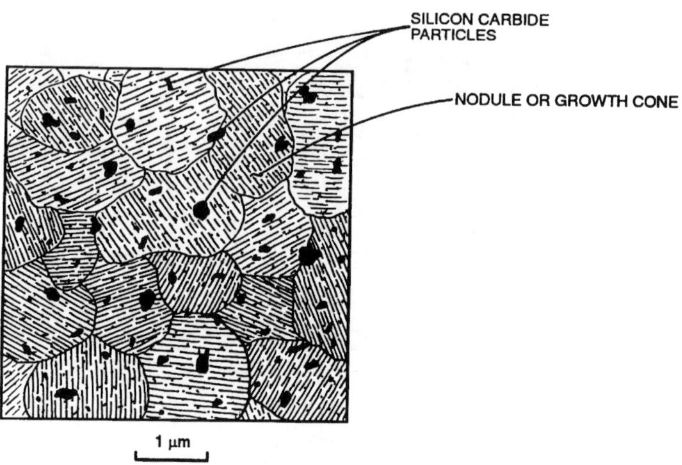

Figure 2. Schematic illustration of the microstructure of Si-alloyed LTI pyrolytic carbon. Nodules or growth cones, with sizes dependent on the processing parameters, are composed of millions of crystallites or turbostratic carbon; randomly dispersed throughout the matrix are particles of silicon carbide, which vary in size from ~1 nm to 1 μm.

APPLICATION TO BIOMEDICAL DEVICES

High-purity turbostratic carbons, like LTI pyrolytic carbon, have exceptionally good cellular biocompatibility and thromboresistance [2]. For cardiovascular applications, the compatibility with blood has received the most attention; unlike soft and hard tissue reactions to implanted materials which develop slowly, the rejective reactions in blood are dramatic and swift. Most materials in contact with blood quickly activate the tissue's clotting mechanism; LTI carbon has been shown to be equivalent to siliconized glass which causes little damage to blood. Theories to explain the excellent cellular biocompatibitily of LTI carbon with blood range from conditioning of the carbon surface with a passivating protein layer through selective adsorption to the complete inertness of the LTI carbon surface to proteins in general.

Accordingly, silicon-alloyed LTI pyrolytic carbon [1-5] has found extensive application in the manufacture of mechanical heart-valve prostheses, either as a ~250-μm-thick coating on a polycrystalline graphite substrate or as a monolithic material where the substrate is machined away. Such heart-valve prostheses are designed to regulate blood flow continuously in hostile physiological environments for periods in excess of patient lifetimes, with valve components subjected to cyclic loading, flexing and bending, wear at surfaces exposed to articulation, and cavitation erosion on surfaces exposed to blood flow. They are generally constructed with two semi-circular leaflets (bileaflets), or with a tilting disk (occluder), which open(s) and close(s) as the heart beats, thereby regulating blood flow under near-normal rheological conditions. The leaflets or disk are contained in a circular housing (or orifice), which in certain valves is stiffened by a metallic restraining ring; a cloth sewing ring is affixed to the outer diameter to facilitate attachment to cardiac tissue. In the U.S., the disk or leaflets are invariably manufactured from pyrolytic-carbon coated graphite; the housing can also be manufactured from this material, or alternatively from pure pyrolytic carbon, titanium or cobalt-chromium alloys. Unfortunately, the complex shapes of some of the metallic components have required the use of casting and welding technologies in their fabrication; as a result, serious structural problems have arisen with certain of these devices leading to many implant failures due to fatigue cracking in the valve housing [e.g., 14,15]. Accordingly, the majority of heart valves implanted in the U.S. today are made entirely from pyrolytic carbon and/or a pyrolytic carbon/graphite laminate; examples are shown in Fig. 3.

Figure 3. A selection of commercial pyrolytic-carbon bileaflet and tilting-disk prosthetic heart valves. A metallic tilting-disk valve (with a Co-Cr alloy housing and a pyrolytic carbon/graphite occluder) is shown for comparison (bottom left).

One of the primary restrictions on artificial heart valves is their potentially uncertain lifetime under complex physiological loading due to degradation by cyclic fatigue or stress corrosion. While hundreds of thousands of pyrocarbon valves have been successfully implanted, over forty recent structural failures, involving principally leaflet fractures, have been reported and in many cases attributed to cyclic fatigue [16]. Although the latter conclusion may be somewhat questionable due to the difficulty in distinguishing fractographic features produced under cyclic versus impact loading conditions [17], these failures do serve to illustrate that careful mechanical design and accurate life prediction are essential elements for the safe, reliable operation of heart-valve prostheses, which must endure sustained physiological loading and environmentally-induced degradation throughout their service lifetime.

Since the human heart beats some 40 million times per year, a primary mechanism of structural degradation can be considered to be cyclic fatigue. Accordingly, to maintain structural integrity, prosthetic heart valves must be designed to endure fatigue lifetimes in excess of 10^9 cycles in physiological environments. The principal material input to such considerations [18-20] involves information on subcritical crack-growth rate and fracture toughness behavior; this is reviewed below for pyrolytic carbon materials.

MECHANICAL PROPERTIES

The mechanical properties of turbostratic carbons are closely related to their coating density as the extent of porosity affects the internal area over which stress is distributed; properties also depend on specific composition and microstructure, including the volume fraction of SiC particles and the size of the crystallites and "growth features" [2,4,5]. Unlike certain forms of carbon with a high degree of preferred orientation and high density which are extremely strong and stiff in directions parallel to the preferred layer planes, because crystallites are randomly arranged and the material is not fully dense, LTI pyrolytic carbon has an uncharacteristically low modulus and comparatively high strength. Values of Young's modulus are typically between 27 and 31 GPa, which is close to the upper bound value for bone of 21 GPa and an order of magnitude less than that of surgical stainless steel or titanium alloys. The hardness, and hence wear resistance, of pyrolytic carbons is also a function of crystallite size and SiC content; in fact, Si-alloyed pyrolytic carbon has superior wear resistance than either graphite or unalloyed pyrolytic carbon [2].

The high strength and low moduli of turbostratic carbons results in large strains to failure when compared to other brittle ceramics, i.e., LTI pyrolytic carbon has a fracture strain of 1.5 - 2.0%, compared to ≥ 0.1% for alumina and 0.1 - 0.7% for polycrystalline graphite. This is thought to result in part from the network of strong covalent C-C bonds in the graphitic layers which must be broken before failure by shear or cleavage can occur. However, as discussed below, fracture toughness, K_c, values measured in the presence of a pre-existing crack, are low compared to many ceramics [5,21,22]. A list of structural and mechanical properties for polycrystalline graphite, Si-alloyed LTI pyrolytic carbon and pyrolytic-carbon coated graphite is given in Table I.

Table I: Structural and Mechanical Properties of Graphite, Si-Alloyed LTI Pyrolytic Carbon, and Graphite/Pyrolytic Carbon Laminate [5]

Property	Polycrystalline Graphite Substrate	Si-Alloyed LTI Pyrolytic Carbon	Si-Alloyed LTI Pyrolytic-Carbon Coated Graphite
Density ($kg.m^{-3}$)	1,500 - 1,800	1,700 - 2,200	1,500 - 2,200
Crystallite Size (nm)	15 - 250	3 - 5	-
Expansion Coefficient (10^{-6} K^{-1})	0.5 - 5	5 - 6	2 - 6
Hardness (DPH)	50 - 120	230 - 370	232 - 412
Young's Modulus (GPa)	4 - 12	27 - 31	34.5
Flexural Strength (MPa)	65 - 300	350 - 530	-
Fracture Strain (%)	0.1 - 0.7	1.5 - 2.0	1.5 - 2.0
Fracture Toughness, K_c (MPa√m)	~1.5	0.9 - 1.1	1.0 - 2.6

For the prediction of a safe lifetime in any structural component, the most relevant information for damage-tolerant calculations pertains to cyclic and static fatigue-crack growth and fracture toughness data. This "worst case" approach essentially assumes that all components already contain pre-existing defects and life is determined in terms of how long it takes these defects to extend to failure. Specifically, the concept of damage-tolerant design relies on the notion that the minimum safe structural life of a component can be determined from the time, or number of loading cycles, for incipient cracks to grow subcritically to instability or until catastrophic failure occurs. For pyrolytic carbon heart valve prosthesis subjected to roughly 40 million cycles a year, this translates into the time for the largest pre-existing defect to grow by cyclic fatigue (and/or stress corrosion) until final fracture ensues at K_c [18,20]. Accordingly, to estimate this life, a characterization of subcritical crack-growth rates for laboratory samples or actual components tested under realistic service (i.e., *in vitro*) conditions must be determined, and then integrated between the limits of the initial and final crack sizes. Such toughness and subcritical crack-growth results for pyrolytic carbon are discussed below.

Fracture Toughness

In general, the fracture toughness of pure Si-alloyed LTI pyrolytic carbon or pyrolytic-carbon coated graphite is of the order of 1 to 2 MPa√m [5,20-25], i.e., only slightly higher than that of soda lime glass and small compared to a K_c of ~3 to 6 MPa√m for alumina and ~3 to 15 MPa√m for zirconia [5]. However, due to imprecise measurements, higher values have been quoted in the literature. For example, it is vital that fracture initiates at an atomically sharp pre-crack for K_c determination; using machined notches instead can lead to unrealistically high K_c values as the toughness is a strong function of the root radius of the stress concentrator. More *et al.* [26], for example, used machined-notched double-torsion specimens to measure the toughness of pyrolytic carbon; their reported result of $K_c = 2.79 \pm 0.23$ MPa√m is some 50% higher than the accepted K_c value measured with pre-cracked [5,20-25] samples.

The first fracture toughness measurements on the Si-alloyed LTI pyrolytic-carbon coated graphite material using appropriate pre-cracked (compact-tension) samples yielded K_c values between 1.1 and 1.6 MPa√m for tests in room air and 37°C Ringer's solution* [21]. (There is no difference between results in air and Ringer's solution as little effect of environment would be expected during the brief duration of these tests). Subsequent studies [20-25] on both pure (monolithic) pyrolytic carbon and the pyrolytic carbon/graphite laminate generally show complex R-curve behavior with initiation and steady-state toughness values varying between 1 and 2 MPa√m. Results for pure pyrolytic carbon are invariably on the low end of this range [23,24]; for the laminate, the toughness appears to be increased as the ratio of the thickness of the pyrolytic carbon to graphite layers is made smaller [24]. It is interesting to note that with thickness ratios below about 2, the toughness of the composite can be higher than either the pyrolytic carbon or graphite layers. Although not completely understood, this has been attributed to variations in the stress intensity across the crack front due to the inhomogeneous (layered) nature of the material, which cause the crack in the pyrolytic-carbon coating to lead slightly the crack in the graphite substrate [24]. An example of fracture-toughness data for pyrolytic-carbon coated graphite is shown in Fig. 4; the average K_c value, measured in 37°C Ringer's solution at

* 37°C Ringer's solution is used here as a simulated physiological (blood analog) environment. It is prepared by dissolving 8.5 g NaCl, 250 mg KCl and 300 mg $CaCl_2$ in 1000 ml of distilled water.

crack initiation on compact specimens machined from actual heart-valve leaflets, is 1.64 MPa√m (std. dev. 0.53 MPa√m) [21].

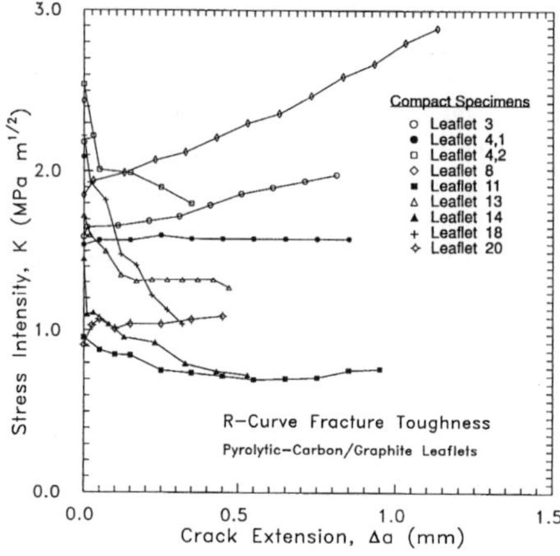

Figure 4. Resistance curves for half-round compact (leaflet) specimens of pyrolytic-carbon coated graphite containing "long" through-thickness (edge) cracks, tested in 37°C Ringer's solution. The fracture toughness, K_c, can be defined either at crack initiation, i.e., as $\Delta a \rightarrow 0$, or at nominal "steady state" on the plateau of the R-curve, if one exists [25].

An alternative method of estimating the fracture toughness in brittle materials is to use indentation methods [27-29]. Here, the K_c value is estimated from the size of the radial cracks emanating from Vicker's hardness indent at high applied indentation loads, typically in excess of 9 kg for pyrolytic carbon (Fig. 5). The radial cracking forms in response to the *residual* tensile stress field which surrounds the indent on unloading; the stress-intensity factor, K_R, from this field can be computed in terms of the peak indentation load P_{int}, and the half surface-crack length c, by [27]:

$$K_R = \chi_R P_{int} c^{-3/2} \quad . \tag{1}$$

χ_R is a material constant dependent upon the ratio of Young's modulus E, to hardness H_V, given by [28]:

$$\chi_R = \xi_V (E/H_V)^{1/2} \quad , \tag{2}$$

where ξ_V is a material-independent constant for Vicker's-produced radial cracks with a value of 0.016 [29]. The toughness is then simply computed from Eqs. 1,2 at $K_c = K_R$ by denoting c_0 as the equilibrium surface length of the post-indentation crack. An example of such radial cracking surrounding a Vicker's indent in pyrolytic carbon (for P_{int} = 9 kg) is shown in Fig. 5a; toughness measurements (Fig. 6) yielded an average K_c of 1.84 MPa√m (std. dev. 0.18 MPa√m) [21], which is consistent with the conventional "long-crack" values shown in Fig. 4.

In general, indentation toughness values must be considered to be approximate as they take no account of R-curve behavior; in essence, they are defined for crack arrest at some arbitrary point on the R-curve. Moreover, certain authors have questioned their validity for pyrolytic carbon because they claim that the residual stresses will be essentially zero after removal of the indenter [30]. This point is discussed in greater detail below.

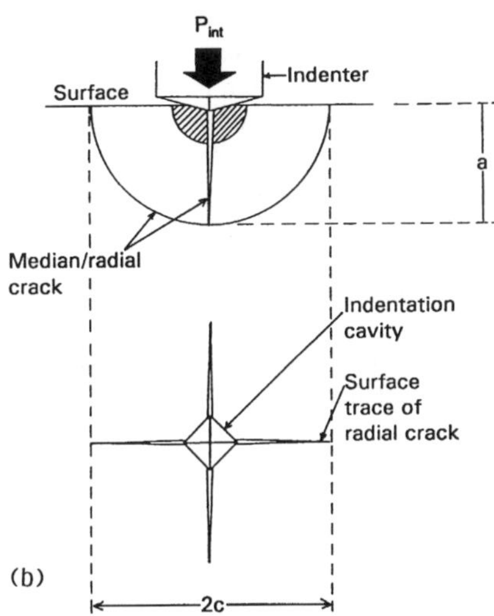

Figure 5. (a) Optical micrograph of a Vicker's indent (indentation load P_{int} = 10 kg) on pyrolytic-carbon coating, showing median/radial cracks emanating from the indent; (b) schematic illustration of the indentation system and crack configuration showing the characteristic crack-size dimensions a and c [25].

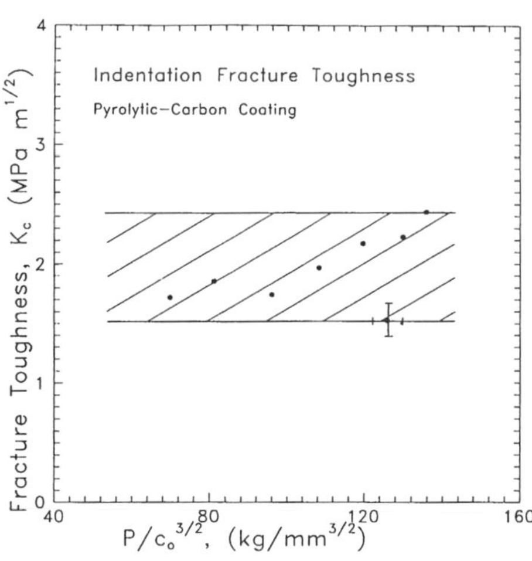

Figure 6. Variation in fracture toughness, K_c, in the presence of "small" cracks in the pyrolytic-carbon coating as a function of the indentation load and equilibrium radial crack depth, $P_{int}c_o^{-3/2}$, as measured using indentation-toughness techniques [25].

Whereas fracture toughness values between 1 and 2 MPa√m are now accepted for monolithic and composite pyrolytic carbon materials, it should be noted that the extreme scatter which characterizes the data is most probably the result of residual stresses, which are known to exist in the composite as a result of coating deposition and processing (e.g., from the thermal-expansion mismatch between the pyrolytic-carbon coating and the graphite substrate). Moreover, with respect to the pyrolytic carbon/graphite material, recent finite-element calculations [24], which take into account the composite elastic-property mismatch between the graphite and pyrolytic-carbon layers, indicate some degree of uncertainty in the stress-intensity solutions for this material. Although no studies to date have taken account of this effect, the numerical calculations do show that the K-solutions for homogeneous material may underestimate K in the pyrolytic carbon by up to 40% and overestimate K in the graphite layer by up to 60%. However, as shown below, comparisons of the fatigue-crack growth properties in the monolithic pyrolytic carbon and the pyrolytic carbon/graphite laminate, which are characterized in terms of K, do not show such major differences [21-25].

Subcritical Crack Growth

Although unstable fracture occurs at the fracture toughness value (under plane-strain conditions where $K = K_c$), subcritical crack growth can occur at lower K levels by mechanisms such as cyclic fatigue and stress-corrosion cracking (static fatigue); in ceramic materials, this typically occurs at stress intensities above ~50% of K_c [31]. Under cyclic loading conditions, the rate of crack growth per cycle, da/dN, can generally be described in terms of power-law relationships, which in their simplest form, can be written as:

$$da/dN = C \Delta K^m \quad , \tag{3a}$$

or

$$da/dN = C'(K_{max})^s (\Delta K)^p \quad , \tag{3b}$$

where ΔK is the applied stress-intensity range, given by the difference in the maximum and minimum stress-intensity values in the cycle, i.e., $\Delta K = K_{max} - K_{min}$, C and C' are experimentally-determined scaling constants ($C = C'/\{1 - R\}$, where the load ratio $R = K_{min}/K_{max}$), and $m = (s + p)$ are the crack-growth exponents. Typically, the exponent m takes values in metals between 2 and 4 (with $p > s$), whereas in ceramic materials, m values can be as high as 100 or more (with $p << s$) [31,32]. Similarly, under sustained loading, the corresponding relationships for stress-corrosion crack velocities, da/dt, are of the form:

$$da/dt = C''(K)^n \quad , \tag{4}$$

where is C'' and n are scaling constants; values of n are similar to m. At very low stress intensities, cracks (of a size larger than the dimensions of the microstructure or scale of local inelasticity) may appear to be dormant; these stress intensities are referred to as the fatigue and stress-corrosion thresholds, ΔK_{TH} and K_{TH}, respectively; the cyclic fatigue threshold, however, is a marked function of the load ratio R [e.g., 33].

Cyclic Fatigue of Pyrolytic Carbon: Until fairly recently, it had been assumed that turbostratic carbons such as pyrolytic carbon were insensitive to cyclic fatigue degradation. Early studies

[8,9] had claimed that since the fatigue endurance strength of these materials was similar to the single-cycle fracture stress, cyclic stresses less than this stress would do no microscopic damage. However, by specifically employing fracture-mechanics type testing procedures with pre-cracked (compact-tension) samples, Ritchie *et al.* [21] first demonstrated unequivocally that fatigue cracks can grow under alternating loads in the pyrolytic-carbon coated graphite laminate in both ambient temperature air and 37°C Ringer's solution. Subsequent studies have verified that the pyrocarbon/graphite material is indeed susceptible to cyclic fatigue [20,22-25], and further shown that fatigue-crack growth can similarly occur in monolithic pyrolytic carbon [23,24,34].

Typical crack-growth rate data, for leaflet specimens of the pyrolytic-carbon/graphite laminate tested in 37°C Ringer's solution, are shown in Fig. 7 as a function of the stress-intensity range, ΔK (= 0.9 K_{max}). It is apparent that similar to many ceramic materials [31], the slope of these plots, which is characterized by the exponent m in Eq. 3a, is extremely high; values for pyrolytic carbon materials are routinely well over 100 [22-25]. Moreover, similar to R-curve data for pyrolytic carbons (Fig. 4), the growth-rate behavior shows a very high degree of scatter. Based on a comparison of similar fatigue data for ceramic materials tested on the same mechanical-testing machine [e.g., 31-33], such a wide difference in growth rates (at a given stress intensity) is unusual and appears to result from inherent material differences attributable to variations in fabrication-induced residual stresses or local microstructure. Although measurements on the pyrolytic-carbon coating of leaflets used for manufacturing heart valves do in fact display large variations in residual stress from sample to sample, which would indeed affect properties, the precise origin of the scatter in toughness and crack-growth data for this material is still uncertain.

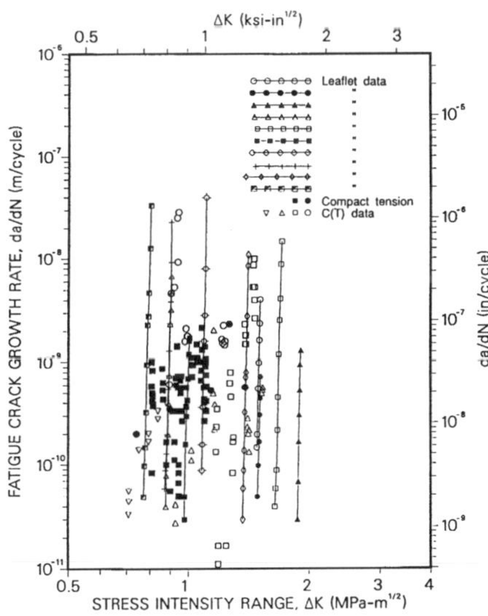

Figure 7. Experimental data showing cyclic fatigue-crack propagation rates, da/dN, as a function of the applied stress-intensity range, ΔK (= 0.9 K_{max}), for half round compact (leaflet) specimens of pyrolytic-carbon coated graphite containing "long" through-thickness (edge) cracks. Results from compact-tension C(T) specimens are included for comparison [21]. Tests were performed in a simulated physiological environment of Ringer's solution at 37°C [22].

Similar to behavior in many metals and ceramics [e.g., 31,35], an apparent threshold can be defined at low growth rates approaching 10^{-11} m/cycle; in most ceramics, this threshold is of the order of 50% of K_c [31]. Based on the results from several studies in 37°C Ringer's solution [20-25], measured values of ΔK_{TH} for the monolithic and composite pyrolytic carbon materials vary between ~0.7 and 2 MPa√m, with a typical average value being ~1 MPa√m. As with toughness data though, within this range, ΔK_{TH} values for pure pyrolytic carbon are at the low end (generally < 1 MPa√m) [23,24], whereas in the laminate, thresholds appear to increase with a decreasing ratio of the thickness of the pyrolytic carbon to graphite layers [24].

While cyclic fatigue in pyrolytic carbon is not longer in question, in view of past skepticism over cyclic fatigue in turbostratic carbons [8,9], it has been necessary to demonstrate unequivocally that the crack growth is cyclically induced and not simply a consequence of stress-corrosion cracking at maximum load. To achieve this, crack extension has been monitored with a) the stress intensity cyclically varied between K_{max} and K_{min} and b) the stress intensity held constant at the same value of K_{max}. A typical result, for K_{max} = 1.36 MPa√m and K_{min} = 0.14 MPa√m, is shown in Fig. 8 [21]. It is apparent that, whereas crack extension proceeds readily under cyclic loading conditions (region a), upon removal of the cyclic component by holding at the same K_{max} (region b), crack-growth rates are markedly reduced. Clearly, similar to behavior reported for other ceramic-like materials [e.g., 31-33], a true cyclic fatigue effect is apparent; furthermore, subcritical fatigue crack-growth rates under cyclic loading appear to be far in excess of those under equivalent sustained loading.

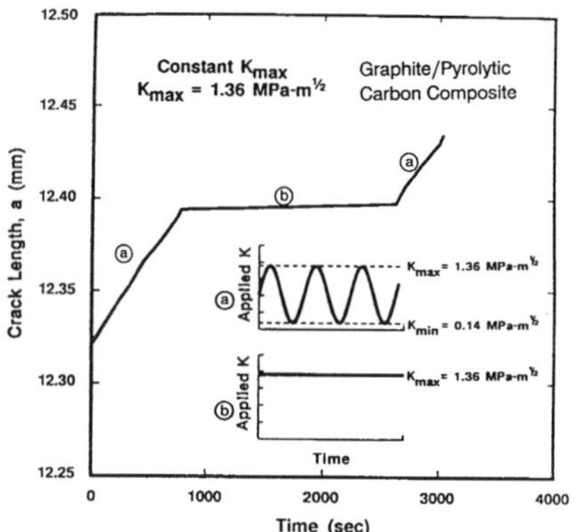

Figure 8. The effect of sustained-load vs. cyclic loading conditions, at a constant K_{max}, on the subcritical crack growth in pyrolytic carbon coated graphite tested in Ringer's solution at 37°C (blood analog). Note how crack-growth rates under cyclic loading (region a) are far in excess of those measured under sustained loading (region b) [21].

Stress-Corrosion Cracking in Pyrolytic Carbon: Similar to other ceramics, the various forms of carbon are prone to subcritical crack growth under the synergistic action of an applied load and a moist (or corrosive) environment [20-23,36-39]. Such stress-corrosion crack-growth behavior is plotted in terms of the crack velocity with respect to time, *da*/*dt*, as a function of the applied stress intensity, *K*, in Fig. 9 for several types of carbon [21]. The data include a medical-grade LTI pyrolytic carbon/graphite composite in an environment of 37°C Ringer's solution [21], glassy carbon in water [36] and in air [37], and for a pyrolytic graphite in air [38]. Crack velocities span many orders of magnitude from 10^{-9} to 10^{-1} m/sec and, similar to cyclic fatigue-crack growth, show a marked sensitivity to the stress intensity; values of the exponent n (in Eq. 4) typically exceed 100 for pyrolytic carbon. Similar to cyclic fatigue, a threshold can be defined for crack velocities approaching 10^{-10} m/sec; K_{TH} values range between ~0.8 and 1.8 MPa√m, with a typical average value of ~1.25 MPa√m [e.g., 21,23].

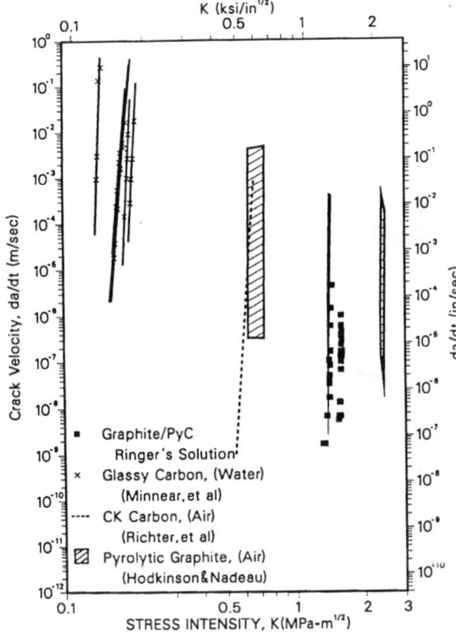

Figure 9. Stress-corrosion crack-growth behavior of LTI pyrolytic-carbon coated graphite composite tested in 37°C Ringer's solution [21], shown in comparison with data for glassy carbon tested in water [36] and air [37] and for pyrolytic graphite in ambient temperature air [38].

Fractography: Contrary to early reports in the literature [16], the morphologies of cyclic fatigue, stress-corrosion cracking and overload fractures in carbon materials are essentially indistinguishable [17,20-23]. Indeed, this identical morphology of fatigue and overload fracture surfaces is not unique to pyrolytic carbon (Fig. 10a,b); similar behavior is shown by the graphite core of the composite material (Fig. 10c,d), and by most ceramic materials where cyclic fatigue fractures have been documented [e.g., 31]. This is in marked contrast to metallic materials where the characteristic cyclic fatigue fracture mode, i.e., striations (Fig. 10e), is quite different to the transgranular cleavage (Fig. 10f), microvoid coalescence or intergranular fracture seen for failure under monotonic loads. There is, however, a clear change in the laminate material between fractures in the pyrocarbon coating and the graphite substrate as the latter surfaces are considerably rougher. Fracture surfaces in both the substrate and coating appear to be covered in fine-scale facets, the dimensions of which approach that of the grains or "growth features" in the microstructure. The implication of these results is that post-failure analysis of failed pyrolytic-carbon valves can be quite deceptive; unlike metallic components, it is clearly not feasible to rely

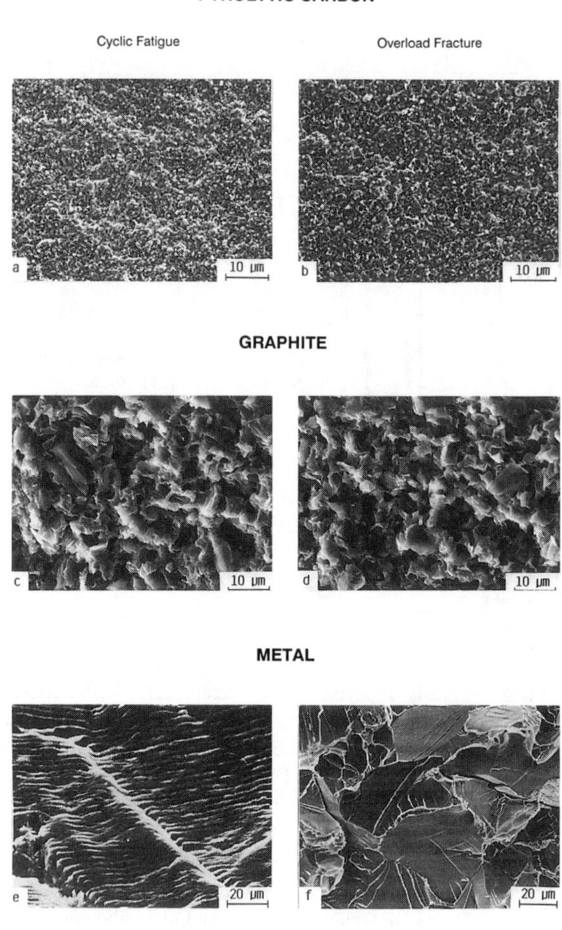

Figure 10. Scanning electron micrographs of cyclic fatigue and overload (fast) fracture in a,b) the pyrolytic-carbon coating, c,d) the polycrystalline graphite substrate, and e,f) a metallic material (low-strength ferritic steel), showing e) fatigue striations and f) transgranular cleavage fracture [17].

solely on fractography to conclude whether a specific failure was associated with a progressive time-dependent fracture or an instantaneous event.

Small Cracks: All of the subcritical crack-growth data described above are derived from conventional fracture-mechanics style samples containing "long" (larger than ~2 to 3 mm in length), pre-existing, through-thickness (edge) cracks [18-25]. Although life prediction based on such data is perfectly standard for numerous applications in the aerospace and nuclear industries for example, it has been increasingly recognized that a more complete analysis must include consideration of the apparently rapid growth of "small" surface cracks, particularly in smaller component cross-sections such as those found in heart-valve prostheses. This is particularly important since it is well known for metal fatigue [40-42], and has recently become apparent for the fatigue of ceramics [32,43], that where cracks are physically small (typically below ~500 μm), or small compared to microstructural dimensions or the extent of local crack-tip inelasticity, their growth rates can be far in excess of corresponding long cracks subjected to the same *applied K* levels; moreover, small cracks are known to propagate at applied stress intensities *less than* the fatigue threshold, below which long cracks are presumed dormant [32,40-43].

In metals and phase-transforming ceramics such as PSZ, fatigue cracks can form naturally; however, in the majority of ceramic materials, including pyrolytic carbon, crack initiation always occurs at pre-existing defects; in fact, analysis of *in vivo* failures of pyrolytic carbon valves has shown that small cracks initiate at material defects, or from regions suffering cavitation erosion [e.g., 16]. Such cracks are typically semi-elliptical, part-through, surface cracks with dimensions less than a few hundred microns over the vast majority of the device life. Since most life-prediction analyses [e.g., 18-20] currently utilize long-crack results (e.g., Fig. 7), it is important to show that these data adequately reflect the growth-rate properties of such small surface cracks.

To monitor the growth of such small surface cracks, several Vicker's hardness indentations are placed along the top (polished) surface of cantilever-bend specimens and cycled in 37°C Ringer's solution [22,25]; the residual tensile stress field surrounding the indent, together with half penny-shaped radial/median microcracks emanating from the indent corners (at $P_{int} > 9$ kg), act to facilitate fatigue-crack initiation (e.g., Fig. 5). Typically, an increasing bending stress range, $\Delta\sigma_b$ (~10 to 80 MPa) is applied until one crack begins to grow, whereupon the stress is held constant until subsequent crack arrest; the procedure is then repeated. Tests are periodically interrupted every 10^2 to 10^4 cycles to permit the use of optical microscopy or gold-coated cellulose-acetate replicas to monitor crack growth. Full details are given in ref. [25].

For the indentation crack configuration, there are generally two contributions to the stress intensity, namely that due to the applied stresses and that due to the residual stress field surrounding the indent, i.e., $K = K_{app} + K_R$. The stress-intensity factor resulting from the residual stress, K_R, can be calculated from the peak indentation load, P_{int}, and the half surface-crack length, c, using Eqs. 1,2. Linear-elastic solutions for the applied K_{app} are given in ref. [44] for three-dimensional semi-elliptical surface cracks in bending (and/or tension) in terms of the crack depth a, surface crack length $2c$, elliptical parametric angle ϕ, shape factor Q, specimen thickness t, specimen width b, and remote (outer surface) bending stress σ_b, from the applied load P_{app}:

$$K = H_c \ \sigma_b \left(\frac{\pi a}{Q} \right)^{½} F\ (a/c, a/t, c/b, \phi)\ , \tag{5}$$

where H_c is the bending multiplier and F is the boundary correction factor, the form and functional dependencies of which are detailed in ref. 44. Eq. 5 is valid for $0.2 \leq a/c \leq 1$ and $0° \leq \phi \leq 90°$ at $c/b \leq 0.5$. It is worth noting that with increasing crack length, K_R decreases monotonically, independent of the applied loads, while K_{app} typically increases dependent upon the applied load levels; correspondingly, as the crack grows out of the field of the indent, the total K may show an initial reduction followed by a steady increase.

Typical results showing the crack-growth rate behavior, derived from the extension per cycle of the surface (trace) length of small cracks (~100 to 500 μm) emanating from the indent corners in the pyrolytic-carbon coating, are plotted as a function of the externally applied ($K_{max} = K_{app}$) and total ($K_{max} = K_{app} + K_R$) stress intensities in Fig. 11a. When characterized in terms of K_{app}, small-crack growth rates display seemingly anomalous behavior, namely, a negative dependency on stress intensity, and crack growth at K levels between 0.08 and 0.36 MPa√m, well *below* the (long-crack) threshold K_{TH} of ~0.7 Mpa√m.

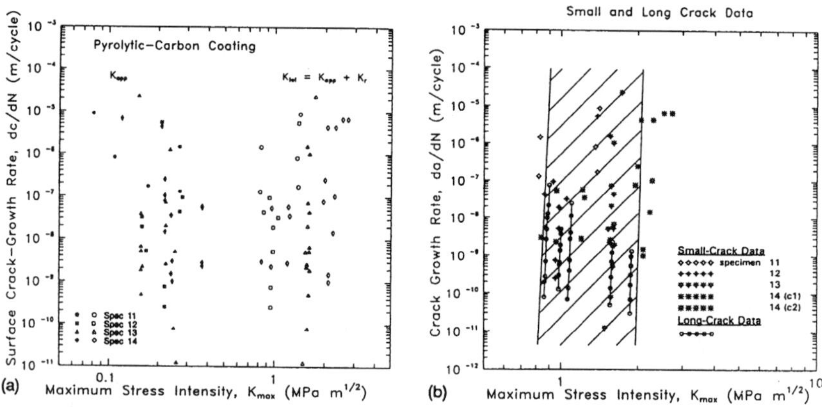

Figure 11. Experimental data showing cyclic fatigue-crack propagation rates, *da/dN*, in pyrolytic-carbon coated graphite for long, through-thickness (edge) cracks and physically-small surface cracks, showing a) small-crack data plotted as a function of K_{max}, computed from the externally applied loads ($K_{max} = K_{app}$; filled symbols) and in terms of contributions from both the applied loads and the residual stress field of the indent ($K_{max} = K_{app} + K_R$; open symbols); b) comparison of the growth rates of both long and small surface cracks as a function of $K_{max} = K_{app} + K_R$. Note correspondence between long and small crack data where K_{max} incorporates both applied and residual stresses [25].

Such behavior, which has been reported for many metallic and ceramic materials [e.g., 22,25,31,32,40-43], in general results from a diminished role of crack-tip shielding** with cracks of limited wake. Whereas the applied stress intensity increases with crack extension in the usual

** Crack-tip shielding mechanisms act to asffect crack advance by altering the local stress intensity actually experienced at the crack tip [45,46]. Such mechanisms, which act principally in the crack wake, include transformation toughening in ceramics, crack bridging in composites, and crack closure during fatigue-crack growth.

manner (as K_{app} is a function of $P_{app} a^{1/2}$), the total stress intensity, which is actually experienced at the crack tip and hence drives the crack, may have a different functional dependence on crack size as it is derived from the superposition of the applied stress intensity with that due to shielding. Here the shielding results from the residual tensile stresses around the indent, which diminish as the crack extends away (K_R is a function of $P_{int} a^{-3/2}$ from Eq. 2). These dependencies on crack size are such that K_R is largest when the crack is smallest whereas the total (near-tip) stress intensity ($K_{max} = K_{app} + K_R$) may exhibit a minimum as the crack advances. Accordingly, when small-crack data are plotted in the usual way in terms of the externally applied loads, a negative dependency on stress intensity will be seen. However, when the full driving force is calculated taking into account the contributions from both applied and shielding (residual) stresses, the true positive dependency of growth rates on K_{max} is achieved (Fig. 11a).

From the perspective of prosthesis design, of prime importance here is that a close correspondence is achieved when the appropriately characterized small surface-crack data are compared with corresponding long edge-crack results, similar to that reported recently for SiC-whisker reinforced alumina [32]. This normalization is shown for pyrolytic carbon in Fig. 11b, although both forms of data are marked by extensive scatter. At the same K_{max}, small-crack growth rates tend to be somewhat faster than corresponding long-crack rates. Although probably associated with scatter, this may in part result from an overestimation of K_R due to relaxation of the tensile residual stresses from the partial recovery of the pyrolytic carbon following removal of the indenter.

Thus, although current analyses use data on long through-thickness (edge) cracks, actual flaw growth *in vivo* is more likely to arise from small, semi-elliptical cracks which occur predominantly on the pyrolytic-carbon coating surface and propagate both along the surface and depth-wise toward the graphite core. It appears from Fig. 11, however, that a reasonable correspondence can be obtained between long and small crack behavior provided growth rates are characterized in terms of the appropriate K, incorporating both external (applied) and internal (residual) stresses. This is important as the collection of such small-crack data in pyrolytic-carbon materials is both complex and extremely labor-intensive. Nevertheless, with appropriate characterization, in principle either form of growth-rate data may be utilized for design and life-prediction analyses.

Indentation Deformation and Fracture

The observations described above of small fatigue crack growth in pyrolytic carbon are important as they are more representative of the potential in-service failures; specifically, the mechanical failure in heart-valve prostheses *in vivo* is likely to be associated with the early propagation of physically small flaws, typically with dimensions of the order of micron to tens of microns, in the pyrolytic carbon coating [16]. Accordingly, small-crack data could provide the basis for improved damage-tolerant life-prediction methodologies to define quality-control standards and more accurately predict the safe structural life of such implant devices. However, this notion has recently been challenged on the basis of claims i) that "... cracks even 1.5 mm long do not grow, and thus do not appear to effect (*sic*) the long life behavior of pyrolytic carbon..." [47-49], and ii) that the analyses used to estimate the residual stress fields surrounding the Vickers indentations (i.e., Eqs. 1,2) are incorrect for pyrolytic carbon [30], as the deformation associated with the indent is fully reversible [30].

The first question primarily results from the work of Sines and Ma [48,49] where they claim that there is clearly defined threshold for the non-propagation of surface cracks in pyrolytic carbon, and that crack growth below this threshold cannot occur because of the absence of dislocations, implying that the results presented in Fig. 11 are impossible. In particular, they report that 29 pyrolytic carbon specimens have been tested at stress-intensity ranges between 1.0 and 1.8 MPa√m for 6×10^8 cycles without any observable crack growth. There are two problems with their conclusions. First, there is now evidence of mobile dislocations in pyrolytic carbon [50,51]. Second, their results were obtained at constant alternating stress, i.e., under increasing stress-intensity conditions if crack growth occurs, whereas it is well known that thresholds in fatigue should be determined under *decreasing K* conditions. Without prior extension under decreasing K conditions, whether an initial crack will grow or not will often be compromised by how it was created and the nature of the conditions at the crack tip, e.g., from crack blunting or excessive local inelasticity. This presumably is why the small cracks in their samples did not grow at ΔK levels between 1 to 1.8 MPa√m, whereas every other investigator of fatigue-crack growth in pyrolytic carbon [20-25] was able to propagate cracks at stress intensities of 1 MPa√m or less.

The second issue results from the work of Lankford and Sines [30], who have recently reported diamond pyramid indentation experiments on glass and pyrolytic carbon; based on their observations of virtually closed indenter load-displacement hysteresis loops in the pyrocarbon, they claim that the residual material displacements and hence the residual stresses are approximately zero *for indenter load ranges up to 10 kg*. The implication here is that pyrolytic carbon is a perfectly elastic, viscoelastic material. This is certainly an interesting point as some recovery of the hardness indents is noticeable in pyrolytic carbon at small indentation loads. However, it is unlikely that there is *no* permanent deformation in the pyrocarbon following indentation at the high loads (i.e., 8-10 kg) used to nucleate the small cracks, as within the residual indentation impression (e.g., Fig. 5), there is definite evidence of microcracking, which is a clear indication of the existence of inelastic deformation.

To examine this point, high-resolution atomic force microscopy and instrumented microindentation studies were performed on a commercial pyrolytic-carbon heart-valve leaflet (manufactured by St. Jude Medical, Inc.). The material consisted of a 350-μm graphite substrate coated with a 230-μm thick layer of Si-alloyed LTI pyrolytic carbon; under scanning electron microscopy, the coating surface showed porosity with pores typically in the 2-5 μm range. Following polishing using a 1-μm diamond lapping film and subsequently Syton paste, a mean surface roughness of 4.9 nm with a maximum asperity height of 62 nm was obtained.

Three types of indentation experiments were conducted in order to verify the existence of plasticity in pyrolytic carbon and to determine its onset. First, microindentations were performed using a Berkovich-type indenter with a blue conductive diamond tip with radius of curvature of 66 nm and a nominal included angle of 35° with the vertical axis of the tip; the microindentation apparatus, developed by IBM, is described elsewhere [52]. Experiments were carried out under depth control at two different loading rates. As illustrated in Fig. 12 for a loading rate of 2.5 nm/s, at approximately 1 mN load, the onset of plastic deformation was clearly observed; on unloading, such plasticity was apparent as permanent strain in the form of a residual depth of the indent of ~25 nm.

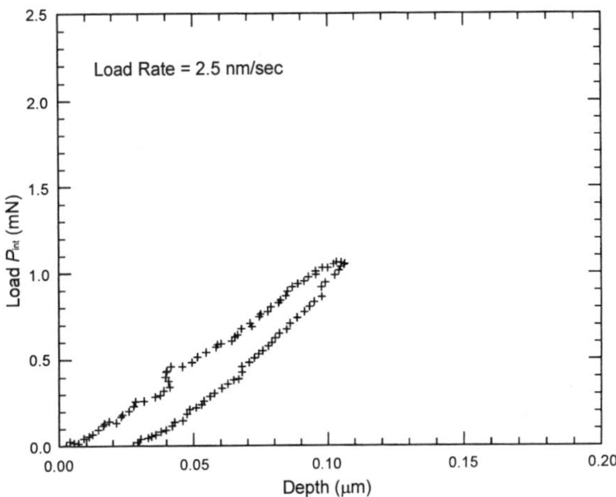

Figure 12. Indentation load P_{int} vs. depth h_c curve on pyrolytic carbon using an IBM microindenter with a Berkovich-type diamond indenter (radius of curvature 66 nm with an included angle of 35° to the vertical tip axis), used to determine the threshold load for the onset of plasticity. Using a loading rate of 2.5 nm/s, the onset of plastic deformation occurs at an approximate 1 mN load.

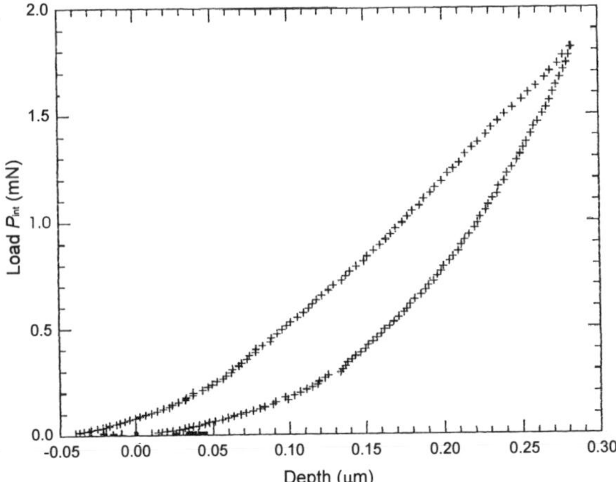

Figure 13. Indentation load P_{int} vs. depth h_c curve on pyrolytic carbon using a Digital Instruments AFM/STM (Nanoscope III) modified by Hysitron. Using the same Berkovich-type diamond indenter, the load-displacement curve was obtained for a 1.85 mN indent.

A second set of experiments was performed on a commercial (Nanoscope III) atomic force/scanning tunneling microscope (AFM/STM), manufactured by Digital Instruments, Santa Barbara, CA; the instruments was retrofitted with the capabilities to perform indentations with a STM-type tip, based on the design developed by Hysitron Inc. of Minneapolis, MN [53]. Using pre-indentation scans (set-point 1 nA, scan rate 1 Hz) to locate a sufficiently smooth area with a mean surface roughness of ~ 5 nm), indentations were made with a load of 1.85 mN (set-point 1.5 nA) using the same Berkovich-type tip, and then imaged *in situ*. The resulting load/displacement curve, shown in Fig. 13, revealed a residual depth of approximately 40 nm. This indentation was immediately imaged and was found to consist of an approximately 100 nm radius cavity, with a measured depth of 26.4 nm (Fig. 14a). Subsequent scans of the indentation area (set-point 2-2.5 nA) showed a continued decrease in the depth of the cavity (Fig. 14b). To measure this decay, the tip was withdrawn and the sample allowed to relax for 30 min; a subsequent scan of the indentation area (60 min after indentation) showed only a small amount of relaxation, with a measured depth of 22.8 nm (Fig. 14c). Periodic depth readings at intervals of 4 min on a second indentation in fact reveal a power-law decrease in cavity depth, with most of the relaxation occurring within the first 30 min (Fig. 15).

To examine further such relaxation behavior in pyrolytic carbon under an applied load, a third set of microindentation experiments was conducted using a Berkovich type indenter with a tip radius of 100 nm and an angle of 65.3° with the vertical axis. The sample was indented with a load of 1.97 mN at a rate of 5 nm/s; the load was applied for 20 min before unloading at the same rate. The resulting data (Fig. 16) show a load relaxation of approximately 4% to 1.9 mN over the initial 8 min; thereafter the load remained constant for the remaining 20 min. Such behavior can be modeled with Hertzian contact theory. Pressure is assumed to be distributed evenly over the region of contact and the contact area is assumed to be triangular and equivalent to the area of the tip. The mean pressure is given by:

$$\bar{p} = \frac{P_{\text{int}}}{A_{\text{tip}}} \quad , \tag{6}$$

where A_{tip} is the area function of the tip; typical values approach 3 GPa, i.e., of the order of $E/10$. Defining $\bar{p}$ as being equivalent to a stress causing relaxation σ_R, the strain relaxation is given by:

$$\varepsilon = \frac{1}{E} \, d\sigma_R / dt \quad . \tag{7}$$

Combining Eqs. 6,7 and using the definition of strain, the stress relaxation under no external forces is given in terms of the change in displacement $d\delta/dt$ by:

$$\frac{d\sigma_R}{dt} = -\frac{1}{E(t)} \frac{d\delta}{dt} \quad . \tag{8}$$

Using Hertzian contact theory, this yields:

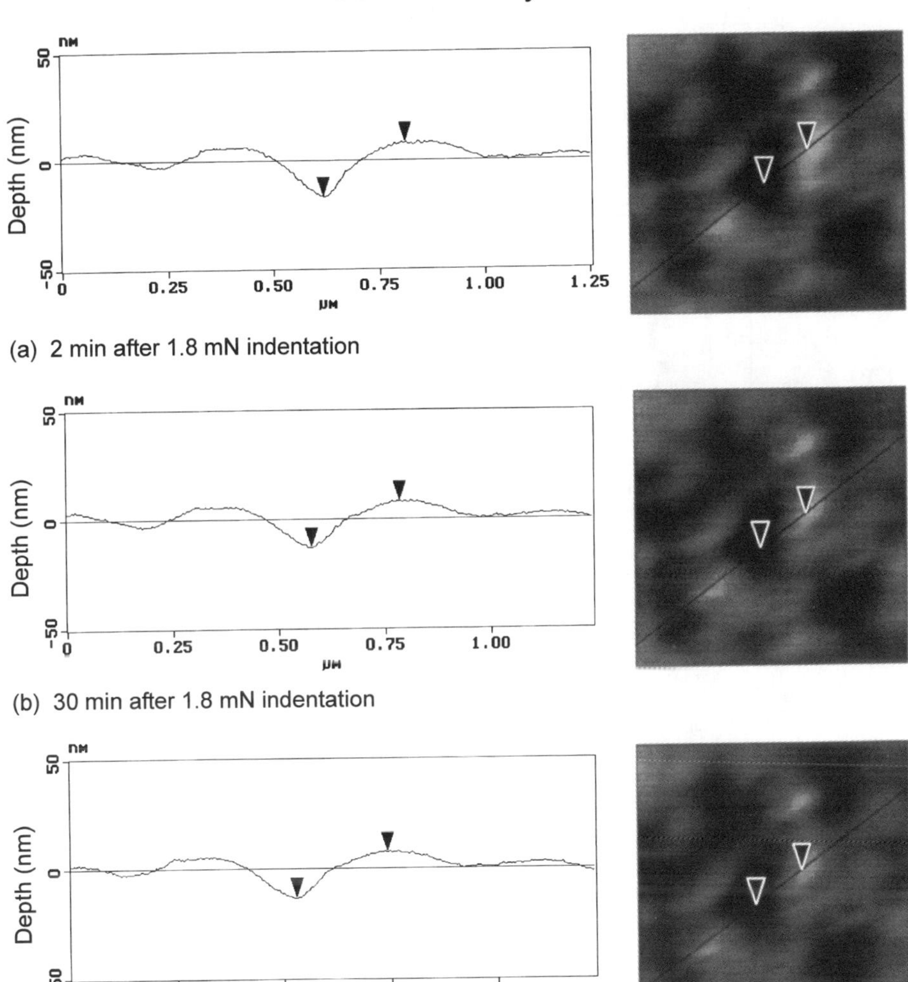

Figure 14. *In situ* scans of the 1.85 mN indent in Fig. 13 using a Digital Instruments/Hysitron AFM/STM, taken a) 2 min, b) 30 min and c) 60 min after indentation.

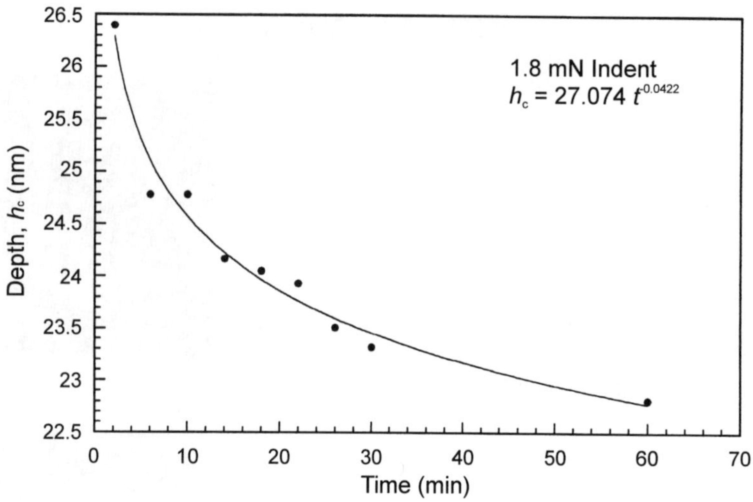

Figure 15. STM section analysis of the residual depth vs. time since indentation for a 1.8 mN indent in pyrolytic carbon, showing time-dependent relaxation of the cavity depth.

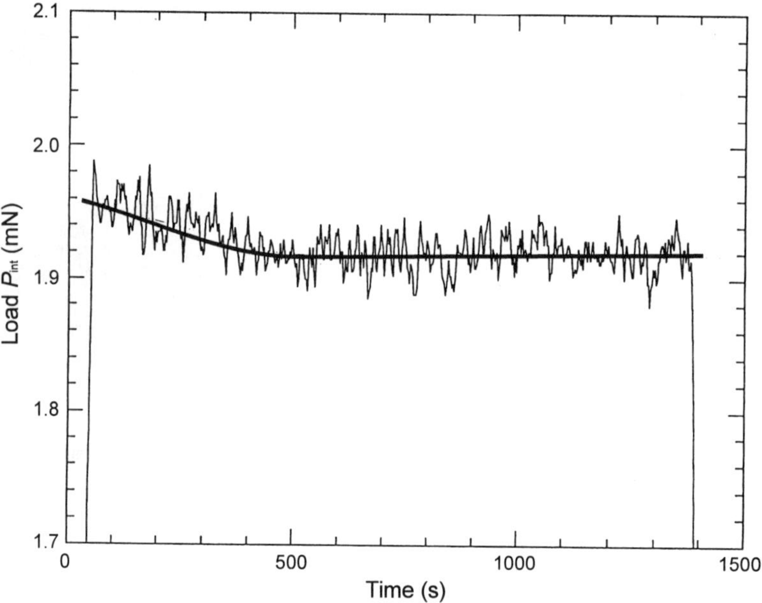

Figure 16. Load-relaxation data for pyrolytic carbon showing indentation load P_{int} as a function of time t using an IBM microindenter with a Berkovich-type diamond indenter (radius of curvature 100 nm with an angle of 65.3° to the vertical tip axis). Using a displacement rate of 5 nm/s, a 1.97 mN load was applied for 20 min, and then unloaded at the same rate.

$$\frac{dP_{\text{int}}}{dt} = \frac{dh_c}{dt}\left[\frac{P_{\text{int}}\left(49h_c + 7.25\right)}{h_c\left(24.5h_c + 7.25\right)^2} - E(t)\right] , \tag{9}$$

where P_{int} is the applied load, h_c is the contact depth and $E(t)$ is a time-dependent relaxation modulus. For the data shown in Fig. 16, load relaxation should cease at a load of 1.9 mN and a calculated contact depth of 0.086 μm. Taking into account the thermal drift, which can be as high as 25Å/°C, that is associated with such a 20-min relaxation experiment, the calculated relaxation modulus, $E(t)$ is approximately one order of magnitude less than the elastic modulus, E of the material. This implies that the material has not fully relaxed, and the material shows viscoelastic or time-dependent behavior.

This behavior can be modeled with a two-constant Maxwell model [54]. Plotting the relaxation modulus, $E(t)$, which is derived from load vs. time curves, as a function of time and finding a functional relationship for the resulting curve in terms of the relaxation moduli, E_1 and E_2, with their respective relaxation times, λ_1 and λ_2, the residual stress σ_R can be calculated at the point when the material has totally relaxed:

$$\sigma = \int_{-\infty}^{t}\left[E_1\, e^{-t/\lambda_1} + E_2\, e^{-t/\lambda_2}\right]\dot{\varepsilon}\,(t)\,dt \;\; , \tag{10}$$

where $\dot{\varepsilon}(t)$ is the strain rate that relates the change in contact depth, h_c, with respect to time.

With a no-load condition, a continued decrease in the indent depth, h_c, was imaged in the AFM within the time frame of one hour (Fig. 15); such data can be represented by power-law relationship, $h_c = 27.074\, t^{-0.0422}$, with a correlation coefficient of 0.98699. It is apparent that the load relaxed the most within the first 20 min after indentation with relaxation continuing after the load was removed.

As noted above, the load relaxation and the residual stresses associated with that relaxation are critical factors within the time frame of a fatigue test and in relation to residual stress intensity and the initiation of crack growth in small-crack experiments. A typical fatigue test involves several millions cycles; at a frequency of 25 Hz this corresponds to a couple of days. Within that time frame, the depth of the indent decreased by under 10 nm under a no load condition, with a "half life", i.e., the time it takes for the depth to return to half of its original depth, of the order of a year. Clearly, although some recovery of the indentation occurs, within the time frame of most fatigue experiments, tensile residual stresses are developed in the vicinity of hardness indentations in pyrolytic carbon and thereby contribute to the initiation and early propagation of fatigue cracks which extend from the corners.

CONCLUDING REMARKS

Pyrolytic carbon has been used successfully in hundreds of thousands of implanted heart-valve prostheses and is clearly the current material of choice for this application. However, from a traditional engineering perspective, this material is extremely brittle in the presence of cracks.

In fact, to our knowledge, the pyrolytic-carbon heart valve is one of the few, if not the only, structural application where a ceramic-like material is used in a safety-critical situation, i.e., where material failure will most likely lead to loss of human life. Furthermore, like other ceramic materials, subcritical crack-growth rates (at low homologous temperatures) are exceedingly sensitive to the stress-intensity level, i.e., the m and n exponents in Eqs. 3 and 4 are invariably in the range of 10 to 100 [e.g., 31]. This means that the life of any pyrolytic-carbon component will be highly dependent upon the pre-existing defect size and the level of applied stress (specifically in damage-tolerant life-prediction procedures, life is proportional to σ^{-m} or σ^{-n}). As a result, small changes in service stresses will lead to very large changes in projected life. Since such sensitivity is an inherent feature of low toughness materials, design requirements for heart-valve prostheses manufactured from pyrolytic carbon must be met with the highest degree of conservatism.

ACKNOWLEDGMENTS

The authors thank Dr. Avrom M. Brendzel at St. Jude Medical, Paul Schmidt and Ralph Kafesjian at Baxter for their partial support of this work, and Professor D. Kohlstedt, S. Viswanathan and J. Nelson for their assistance with the indentation studies.

REFERENCES

1. J. C. Bokros, R. J. Akins, H. S. Shim, A. D. Haubold, and N. K. Agarural, in *Petroleum Derived Carbons*, eds. M. D. Deviney and T. M. O'Grady (American Chemical Society, Washington, DC, 1976) pp. 237-65.
2. J. C. Bokros, *Carbon* **15**, 355 (1977).
3. F. J. Schoen, in *Biocompatible Polymers, Metals and Composites,* ed. M. Szycher (Lancaster, Technomic, 1983) pp. 239-61.
4. A. D. Haubold, R. A. Yapp, and J. C. Bokros, in *Encyclopedia of Materials Science and Engineering*, ed. M. B. Bever (Pergamon, Oxford/MIT Press, Cambridge, 1986) vol. 1, pp. 514-20.
5. R. H. Dauskardt and R. O. Ritchie, in *An Introduction to Bioceramics,* eds. L. L. Hench and J. Wilson (World Scientific Publ. Co., Singapore, 1993), pp. 261-79.
6. V. V. Kotlensky, *Trans. Met. Soc. AIME*, **223**, 830 (1965).
7. V. V. Kotlensky and H. E. Martens, *J. Am. Ceram. Soc.*, **48**, 135 (1965).
8. F. J. Schoen, *Carbon*, **11**, 413 (1973).
9. H. S. Shim, *Biomaterials and Medical Devices: Artificial Organs*, **2**, 55 (1974).
10. J. L. Kaae, T. D. Gulden, and S. Liang, *Carbon*, **10**, 701 (1972).
11. J. C. Bokros, L. D. LaGrange, and F. J. Schoen, in *Chemistry and Physics of Carbon*, ed., P. L. Walker (Dekker, New York, NY, 1972) pp. 103-71.
12. J. L. Kaae, *Carbon*, **13**, 51 (1975).
13. E. Pollmann, J. Pelissier, C. S. Yust, and J. L. Kaae, *Nuclear Technol.,* **35,** 301 (1977).
14. A. J. Larrieu, E. Puglia, and P. A. Allen, *Ann. Thorac. Surg.*, **34**, 192 (1982).
15. D. Lindblom, L. Rodriguez, and V. O. Björk, *J. Thorac. Cardiovasc. Surg.,* **97**, 95 (1989).

16. V. Kelpetko, A. Moritz, J. Mlczoch, H. Schurawitzki, E. Domanig, and E. Wolner, *J. Thorac. Cardiovasc. Surg.*, **97**, 90 (1989).
17. R. O. Ritchie, R. H. Dauskardt, and F. J. Pennisi, *J. Biomed. Mater. Res.*, **26**, 69 (1992).
18. R. O. Ritchie and P. Lubock, *J. Biomech. Eng., Trans. ASME*, **108**, 153 (1986).
19. D. J. Chwirut and W. F. Regnault, *Med. Prog. thr. Tech.,* **14**, 193 (1988).
20. R. O. Ritchie, *J. Heart Valve Disease*, **4** (1995) in press.
21. R. O. Ritchie, R. H. Dauskardt, W. Yu, and A. M. Brendzel, *J. Biomed. Mater. Res.*, **24**, 189 (1990).
22. R. O. Ritchie, R. H. Dauskardt, and A. M. Brendzel, in *Bioceramics. Volume 6* (*Proc. 6th Intl. Symp. Ceramics in Medicine*), eds. P. Ducheyne and D. Christiansen (Butterworths-Heinemann, 1993) pp. 229-36.
23. R. O. Ritchie and R. H. Dauskardt, "Fracture toughness and subcritical crack-growth behavior of Pyrolite in simulated physiological environments", Technical Report to CarboMedics, Inc., November (1990), *cited in ref. 34.*
24. C. B. Gilpin, A. D. Haubold, and J. L. Ely, in *Bioceramics. Volume 6* (*Proc. 6th Intl. Symp. Ceramics in Medicine*), eds. P. Ducheyne and D. Christiansen (Butterworths-Heinemann, 1993) pp. 217-23.
25. R. H. Dauskardt, R. O. Ritchie, J. K. Takemoto, and A. M. Brendzel, *J. Biomed. Mater. Res.*, **28**, 791 (1994).
26. R. B. More, A. D. Haubold, and L. A. Beavan, *Trans. Soc. Biomater.*, **15**, 180 (1989).
27. B. R. Lawn, A. G. Evans, and D. B. Marshall, *J. Am. Ceram. Soc.*, **63**, 574 (1980).
28. G. R. Anstis, P. Chantikul, B. R. Lawn, and D. B. Marshall, *J. Am. Ceram. Soc.,* **64**, 533 (1981).
29. C. B. Ponton, and R. D. Rawlings, *Mater. Sci. Tech.,* **5**, 865 (1989).
30. J. Lankford and G. Sines, *J. Biomed. Mater. Res.*, **29**, (1995) in press.
31. R. O. Ritchie and R. H. Dauskardt, *J. Ceram. Soc. Japan,* **99**, 1047 (1991).
32. R. H. Dauskardt, M. R. James, J. R. Porter, and R. O. Ritchie, *J. Am. Ceram. Soc.*, **75**, 759 (1992).
33. C. J. Gilbert, R. H. Dauskardt, and R. O. Ritchie, *J. Am. Ceram. Soc.*, **77** (1995).
34. L. A. Beavan, D. W. James, and J. L. Kepner, in *Bioceramics. Volume 6* (*Proc. 6th Intl. Symp. Ceramics in Medicine*), eds. P. Ducheyne and D. Christiansen (Butterworths-Heinemann, 1993) pp. 205-10.
35. R. O. Ritchie, *Intl. Metals Rev.*, **20**, 205 (1979).
36. W. P. Minnear, T. M. Hollenbeck, R. C. Bradt, and P. L. Walker, *J. Non-Cryst. Solids*, **21**, 107 (1976).
37. U. Soltesz and H. Ritter, in *Metals and Ceramics Biomaterials. Volume 2, Strength and Surface*, eds. P. Ducheyne and G. W. Hastings (CRC Press, Boca Raton, 1984) pp. 23-61.
38. P. H. Hodkinson and J. S. Nadeau, *J. Mater. Sci.*, **10**, 846 (1975).
39. J. L. Ely and A. D. Haubold, in *Bioceramics. Volume 6* (*Proc. 6th Intl. Symp. Ceramics in Medicine*), eds. P. Ducheyne and D. Christiansen (Butterworths-Heinemann, 1993) pp. 199-204.
40. S. Suresh and R. O. Ritchie, *Intl.. Metals Rev.*, **29**, 445 (1984).
41. R. O. Ritchie and J. Lankford (eds.), *Small Fatigue Cracks*. 665 pp. (The Metallurgical Society of AIME, Warrendale, PA, 1986).
42. K. J. Miller and E. R. de los Rios (eds.), *The Behaviour of Short Fatigue Cracks*. 560 pp. (Institut Mechanical Engineers, London, UK, 1986).

43. A. A. Steffen, R. H. Dauskardt, and R. O. Ritchie, *J. Am. Ceram. Soc.*, **74**, 1259 (1991).
44. I. S. Raju, S. N. Atluri, and J. C. Newman, Jr, in *Fracture Mechanics: Perspectives and Directions (Twentieth Symp.)*, ASTM STP 1020. eds., R. P. Wei and R. P. Gangloff (Am. Soc. Test. Matl., Philadelphia, PA, 1989) pp. 297-316.
45. R. O. Ritchie, *Mater. Sci. Eng.,* **103A**, 15 (1988).
46. A. G. Evans, *J. Am. Ceram. Soc.*, **73**, 187 (1990).
47. A. D. Haubold, *Med. Prog. thr. Tech.,* **20**, 201 (1994).
48. G. Sines and L. Ma, in *Bioceramics. Volume 6* (*Proc. 6th Intl. Symp. Ceramics in Medicine*), eds. P. Ducheyne and D. Christiansen (Butterworths-Heinemann, 1993) pp. 211-15.
49. L. Ma and G. Sines, *Materials Letters*, **17**, 49 (1993).
50. S. R. Snyder, T. Foecke, H. S. White, and W. W. Gerberich, *J. Mater. Res.*, **7**, 341 (1992).
51. S. R. Snyder, W. W. Gerberich, and H. S. White, *Phys. Rev. B*, **47**, 10 823 (1993).
52. S. Venkataraman, Ph.D. Thesis, University of Minnesota, Minneapolis (1994).
53. E. T. Lilleodden, W. Bonin, J. Nelson, J. T. Wyrobek, and W. W. Gerberich, *J. Mater. Res.*, **10**, (1992) in press.
54. C. Macosko, *Rheology*, 527 (1995).

FRACTURE TOUGHNESS OF DIAMOND MEASURED BY VICKERS INDENTER RELOADING TECHNIQUE

SERGEY N. DUB
Institute for Superhard Materials of the Ukrainian National Academy of Sciences,
2 Avtozavodskaya Str., Kiev, 254074, Ukraine

ABSTRACT

The crack initiation sequence during Vickers indenter penetration into the (100) plane of diamond was observed. It was revealed that for <110> orientation, two median cracks are initiated during the loading half of the cycle. In unloading, the median cracks grow toward the surface and turn into half-penny cracks. It was proposed to use half-penny cracks as initials and evaluate fracture toughness in second loading from a crack-starting load for these cracks. The length of initial half-penny cracks was measured after first loading cycle from opposite side of the sample. Test results obtained by the reloading technique agree well with the data on conventional fracture toughness determination by indentation from the length of cracks after unloading.

INTRODUCTION

Mode I fracture toughness (K_{IC}) is important mechanical property of brittle materials. It measure the ability to resist to initial crack propagation. For evaluation of fracture toughness precracked specimen is used. During tension or bending loading, the critical load for initial crack propagation is fixed. For hard and brittle materials, the most commonly used specimens are single-edge notched beam tested in three- or four-point bending, the double cantilever beam loaded in tension, the double torsion specimen and short-rod specimen with chevron notch [1]. For all this techniques large specimens of a certain shape are needed. Therefore for diamond fracture toughness determination, the indentation techniques are used [2-13] which applicable to diamond and cBN crystals with grain size exceeding 200 μm. These techniques are based on the phenomenon of crack formation at contact loading of brittle solids. For Vickers indentation test Evans and Charles [14] obtained

$$K_{IC}\Phi/Ha^{0.5} = 0.15k(C/a)^{-1.5} \qquad (1)$$

where Φ is Marsh' factor (≈ 3), H is Vickers load-independent hardness, a is the indentation half diagonal, $k = 3.2$, C is the indentation radial crack length (Fig.1). The value of k was determined empirically using values of K_{IC} measured by standard techniques on macroscopic specimens. Later numerous equations have been proposed to calculate the fracture toughness by the Vickers indentation technique for different materials and indentation crack types [15-20]. For half-penny radial cracks the equation derived by Anstis et. all [21] is most often used

$$K_{IC} = 0.016(E/H)^{0.5}P/C^{1.5} \qquad (2)$$

where E is the Young modulus and P is load. For materials with load-dependent hardness, it is

Mat. Res. Soc. Symp. Proc. Vol. 383 © 1995 Materials Research Society

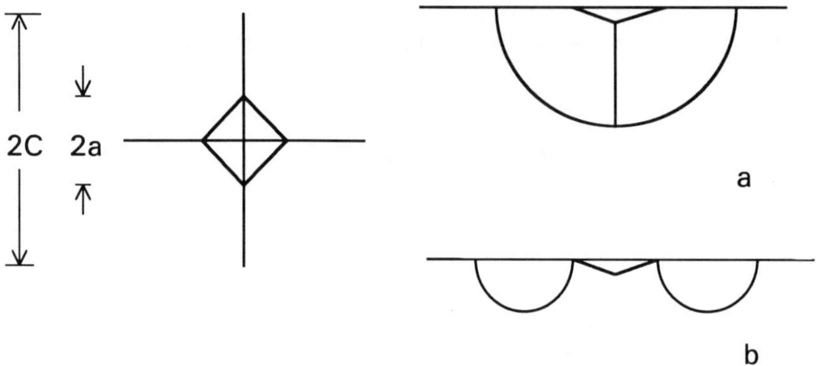

Fig.1. Vickers indentation cracks in brittle materials: (a) half-penny radial cracks; (b) Palmqvist radial cracks.

difficult to estimate E/H ratio. In this case, more convenient to use the equation which includes only load and radial cracks length [22].

$$K_{IC} = 0.0725P/C^{1.5} \tag{3}$$

In result of diamond tests by Vickers indentation method, the fracture toughness anisotropy on (100) plane was revealed. Radial cracks in <110> direction are twice as long as that in <100> at the same load [6]. Vickers indentations in diamond are characterized by the absence of lateral cracks, which are typical for indentations in other brittle materials. Around an indentation and inside it, the shallow ring cracks are observed [9]. Ring cracks are typical especially for indentations on (111) plane of diamond [23] (Fig.2). Such cracks are usually generated by the elastic

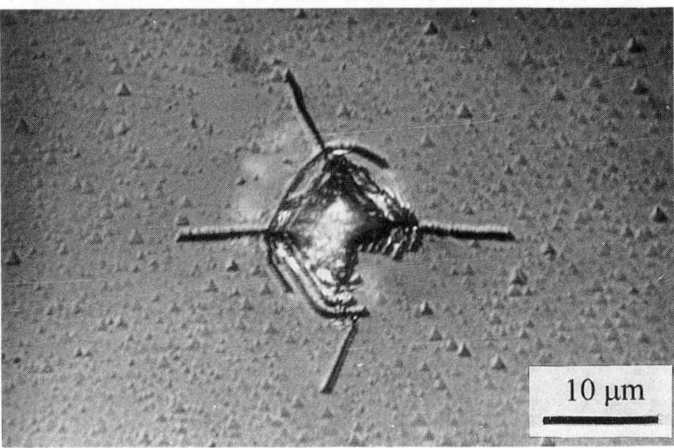

Fig.2. Optical micrographs of Vickers indentation made on (111) plane of natural diamond. and then etched. Radial cracks propagate in <110> directions.

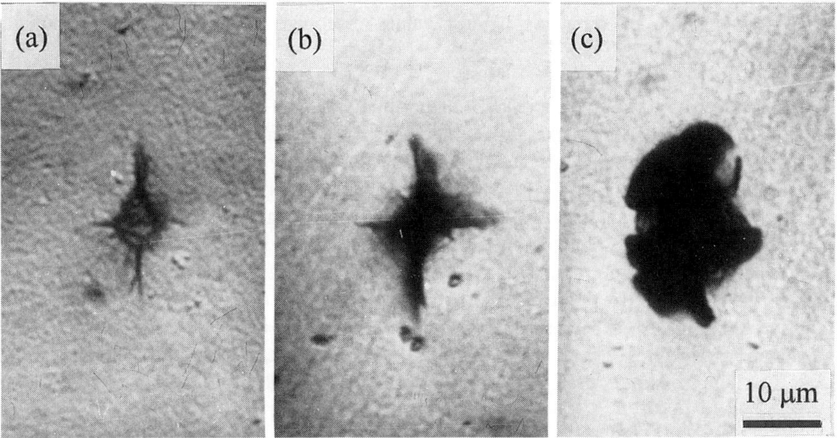

Fig.3. Optical micrographs of Vickers indentation on (110) plane of natural diamond. (a) before etching. Crack in <100> along $\{110\}_{90^o}$ plane is much longer than those in <110> along $\{100\}_{90^o}$ plane; after 15 min (b) and 30 min (c) etching.

loading with a flat-punch or a spherical indenter. During prolonged(>30 min) etching in KNO_3 melt at 650 ^oC, ring cracks around indentation grow to cone ones (Fig.3) [23]. Probably, during the etching of diamond due to a high level of resudial stresses around the indentations, a stress-corrosion cracking takes place. In crystals with an extremely high level of internal stresses, the C/a ratio is unusually high (6 and higher, sometimes radial cracks cross the whole face of a crystal) and radial cracks can deviate from the cleavage direction [9,23] (Fig.4). At room temperature, the fracture toughness of natural and synthetic diamond measured by indentation is the same and equals approximately 5 MPa $m^{1.5}$ [6]. This value is much higher than those other hard

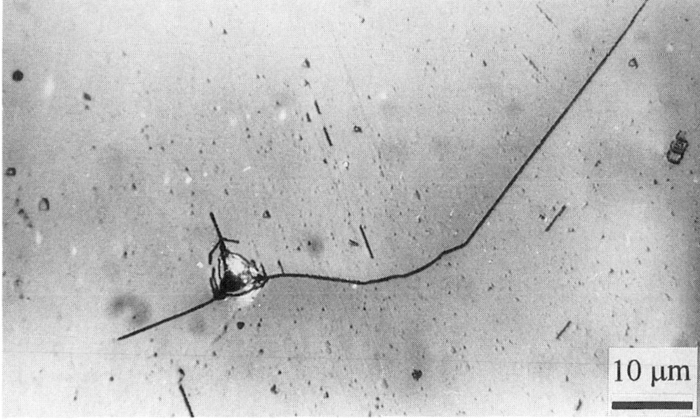

Fig.4. Optical micrograph of Berkovich indentation on a (111) plane of synthetic diamond with high level of internal stresses.

materials. For example, the fracture toughness values of sapphire [24] and α-SiC [25] single crystals don't exceed 2.5 MPa $m^{1.5}$.

At temperatures up to 1225 °C and loads higher than 2 N, half-penny radial cracks develop around indentations on (100) plane of diamond . They pass under indentation and are typical of brittle materials [8]. At temperature above 1220 °C, the mobility of dislocations in diamond become large enough and plasticity of diamond sharply increase [26,27]. The long slip lines around high-temperature indentations verify this observation (Fig.5). Increase of dislocations mobility results in brittle-to-ductile transition in diamond. Radial cracks of Palmqvist type typical for tough materials develop around Vickers indentation instead of half-penny radial cracks [8]. Later it was established that the temperature of brittle-to-ductile transition depend on internal structure of diamond [9]. It decreases for diamond with low nitrogen centres concentration and low dislocations densities. The fracture toughness of diamond changes slightly with heating up to 1225 °C. At higher temperatures, one can observe a rapid growth of fracture toughness due to crack-tip plastic flow [8].

Recently a great progress in production of thick diamond films by micro-wave plasma chemical vapor deposition using CH_4-H_2 mixture was reached. The fracture toughness of CVD diamond measured by indentation is close to that of natural diamond [10-13].

Thus, application of indentation technique allows the study of fracture toughness of diamond and diamond films both at room and high temperatures. But Vickers indentation technique in the form as it is usually used have serious shortcomings. Firstly, K_{IC} , the critical stress intensity in the crack tip for catastrophic extension, is estimated from length of arrested cracks. So, the length of radial cracks should be associated with K_A, critical stress intensity for crack arrest, rather than with K_{IC} [28]. Secondly, the shape profile of cracks can be very different for various materials [29-31]. Even for the same material, the shape profile of radial cracks depend on load,

Fig.5. Optical micrograph of Vickers indentation made at 1100 °C on a (110) plane of natural diamond. Radial cracks propagate in <112> along $\{111\}_{90^o}$ planes. Slip lines run in <110>. In this direction the $\{111\}_{35^o16'}$ planes intersect the (110) plane.

indenter orientation and test temperature [6,8,20,32,33]. Not surprisingly that indentation fracture toughness values a) are as a rule load-dependent [20,28,31], b) are in agreement with those obtained by conventional measurement techniques for some materials and aren't for the other.

In this study, the fracture toughness testing of natural diamond was carried out by Vickers indenter reloading technique. First loading and unloading of Vickers indenter produced initial half-penny cracks from indentation corners. The fracture toughness was determined at reloading from a crack-starting load for initial cracks using an appropriate expression.

THEORETICAL BACKGROUND

To evaluate the fracture toughness by Vickers indenter reloading technique, the stress intensity factor for half-penny cracks should be known. Lawn and Fuller have obtained for centre-loaded half-penny configuration (C/a >>1) [34]

$$K_I = P/[\tan \Psi (\pi C)^{1.5}] \qquad (4)$$

where Ψ is the characteristic half-angle of the indenter.
Eq.(4) is reduced for Vickers indenter (half-angle between faces is 74°) to

$$K_I = 0.052\ P/C^{1.5} \qquad (5)$$

This equation agrees completely with that obtained by Tanaka et al. [35]. They evaluated the stress intensity factor of Vickers-produced hatl-penny cracks system under the assumption of a center-loaded crack.. For a traverse crack model where the residual compressive stress field in the plastic zone is assumed to contribute to the net stress intensity factor, it was obtained

$$K_I = 0.0513\ P/C^{1.5} \qquad (6)$$

Eq.(6) successfully predicts the fracture threshold for the stable growth of a half-penny crack.

Usually C/a ratio in brittle materials changes from 2 to 5. To estimate the K_I for low C/a ratio let consider a plane circular crack of radius C in solid subjected to a $\sigma_z(r)$ tensile stress, acting normal to the crack plane over a central circular area S of radius **a**. The stress intensity factor is given by Paris and Sih [36]

$$K_I = \frac{2}{\sqrt{\pi C}} \int_0^C \frac{r\sigma_z(r)\, dr}{\sqrt{C^2 - r^2}} \qquad (7)$$

We assume that the $\sigma_z(r)$ has a maximum at the area S centre and linear reduces to zero at its edge

$$\sigma_z(r) = \frac{3P}{\pi a^2 \tan\Psi}\left(1 - \frac{r}{a}\right), \quad r \le a;$$
$$\sigma_z(r) = 0, \quad r > a. \tag{8}$$

By substitution of (8) into (7) the stress intensity factor is found to vary as

$$K_I = \frac{6P}{\pi^{3/2} C^{1/2} a^2 \tan\Psi}\left(C - \frac{1}{2}\sqrt{C^2 - a^2} - \frac{C^2}{2a}\arcsin\frac{a}{c}\right) \tag{9}$$

Figure 6 shows a plot of relative deviation of the stress intensity factor K_I from the point force solution $P/[\tan\Psi(\pi C)^{1.5}]$, with the horizontal dashed line representing the point force solution.

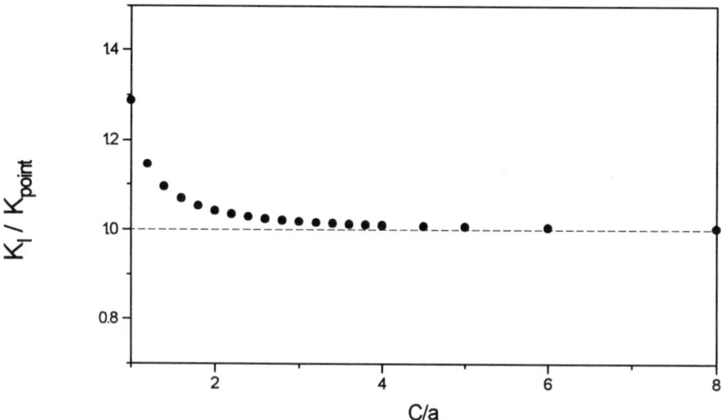

Fig. 6. Plot of relative deviation of K_I from point-force solution.

It is clear that for C/a >2 the deviation from the point-force solution is negligibly small and Eq.(5) can be used. This result agrees well with that obtained by Marshall for a half-penny crack subjected to a uniform stress [37]. In this case, the deviation from a point-force approximation is insignificant for C/a > 1.3.

EXPERIMENTAL

About 0.5 mm in thickness plate cut from (100) planes of natural Type Ia diamond with a low level of internal stresses was tested. The indentation diagonals were parallel to the <100> and <110> directions (henceforth <100> and <110> indentations). Indentation experiments were conducted using the test fixture illustrated schematically in Fig. 7. This fixture allows the simultaneous viewing of the fracture process and measurement of the indenter load. A resistance load cell with resolution of 0.05 N was used.

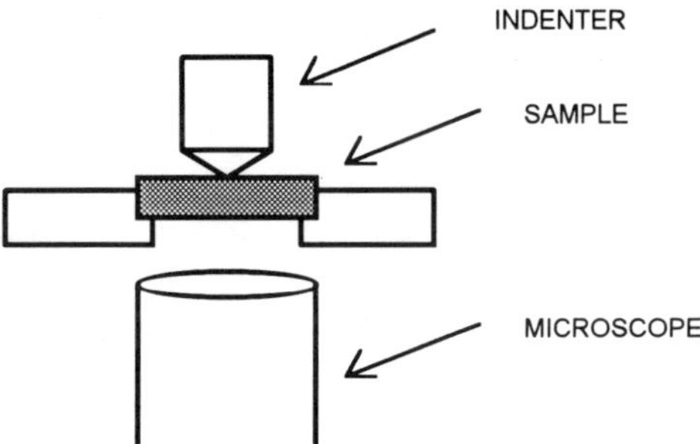

Fig. 7. A schematic representation of the attachment for an optical microscope which is used to observe the process of Vickers indenter penetration into diamond

RESULTS AND DISCUSSION

Obseravtion of Process of Vickers Indenter Penetration in Diamond

For <110> indentation under Vickers impression in diamond two orthogonal median cracks appear at load about 2 N. One of the median cracks develops along the $(110)_{90^o}$ plane, it propagates at a right angle to the surface. Second median crack extends along the $(111)_{54.7^o}$, therefore its projection can be seen (Fig. 8a). Median cracks grow with the load. On the load decreas by approximately 60%, median cracks spread toward the surface and turn into half-penny ones (Fig. 8b). On the complete unloading the half-penny cracks were wedged to open by a residual stress field (Fig. 8c). On (100) plane of diamond the radial cracks were observed at loads smaller than 2 N [6]. Probably, these cracks initiate on unloading due to a residual stress field and they are of Palmqvist type. Observed for diamond sequence of indentation crack initiation is close to a standard model described for normal (nondensifying) soda-lime glass [30]. As distinct from glass, the lateral cracks were absent in diamond. During reloading the initial half-penny cracks begin to grow on the loading half of the cycle.

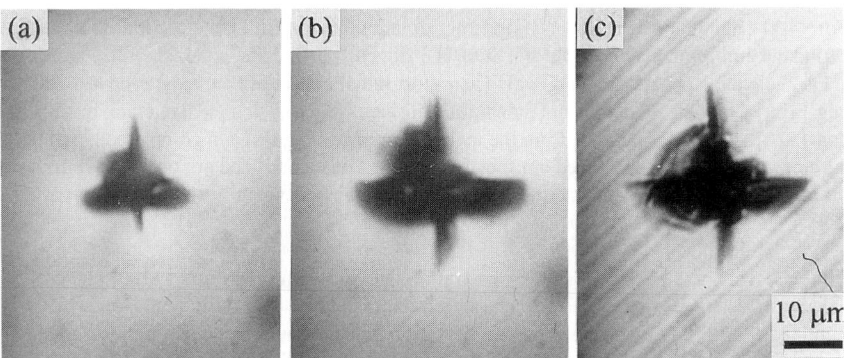

Fig. 8. Phographic sequence of an <110> indentation cycle in diamond.
(a) loading, 100% of P_{max}; (b) unloading, 40% of P_{max}; (c) complete unloading.

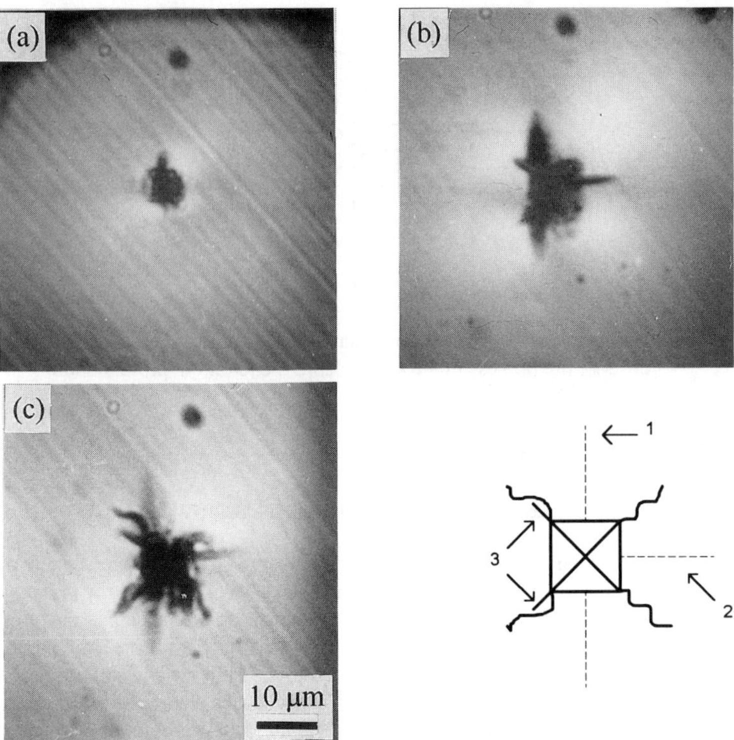

Fig. 9. Photographic sequence of an <100> indentation cycle in diamond. (a) loading, 30% of P_{max}; (b) loading, 100% of P_{max}; (c) complete unloading. 1- median crack in <110> along $\{111\}_{54.7^o}$; 2- median crack in <110> along $\{110\}_{90^o}$; 3- radial cracks in <100> along $\{100\}_{90^o}$.

For <100> indentation, diagonals of indenter do not coincide with the directions of cleavage on (100) plane and sequence of cracks initiation is changed. The median crack in <110> along $\{111\}_{54.7^o}$ plane appears first (Fig. 9a). Then short radial cracks of Palmqvist type in <100> along $\{100\}_{90^o}$ plane develop from indentation corners. The half of a median crack in <110> along $\{110\}_{90^o}$ was formed last from the indentation edge (Fig. 9b). On complete unloading the length of median cracks increased but they remain subsurface ones and are observed in transmited light only. Also, secondary noncrystallographic radial cracks develop from indentation corner (Fig.9c).

The observation of Vickers indenter penetration into soda-lime glass and a (100) plane of MgO single crystal were also performed. In glass, radial cracks grow only during unloading (for a maximum load < 15 N). During reloading of indentation in magnesium oxide either initial cracks did not increase in length and new cracks developed, or the start of initial cracks propagation was accompanied by the initiation of new large cracks. Thus, we fail to test the materials with known K_{IC} by Vickers indenter reloading technique.

Measurement of Diamond Fracture Toughness by Vickers Indenter Reloading Technique

Tests were performed using <110> orientation on (100) plane of a diamond plate. At the first loading cycle, a 4.0 N load was applied to Vickers indenter. Two median cracks 12.1 μm in length develop under indentation. On unloading to 0.6 N, median cracks turn to half-penny ones 12.7 μm in length (Table I). Thus, after the first loading cycle the initial half-penny cracks were received. At the second loading, the initial cracks start to propagate at 5.7 N load. The moment of crack start was fixed visually and the loading was stopped at once. Shortly after that the crack stopped to grow. The length of a crack increased to 13.7 μm. On unloading the length of half-penny cracks does not change. In the third loading cycle, the 13.7 μm half-penny cracks were used as initial. They began to propagate at 6.0 N. Seven loading cycles were performed on one indentation, each time with a higher load (Table I). The cracks pattern around indentation in diamond after seven loading cycles is shown in Fig. 10. Both half-penny cracks extend along $\{110\}_{90^o}$ plane. The length of a cone crack is neglectly small if compared with half-penny ones.

Table I. Reloading Indentation Measurement Results for Type Ia Diamond

Initial cracks length C_1 (μm)	Critical load P (N)	Arrested cracks length C_2 (μm)	Length of unloaded cracks C_3 (μm)	K_{IC} MPa $m^{1.5}$ Eq.(5)	K_{IC} MPa $m^{1.5}$ Eq.(3)	K_{IA} MPa $m^{1.5}$
-	4.0	12.1	12.7	-	6.4	4.9
12.7	5.7	13.7	13.7	6.5	8.2	5.9
13.7	6.0	15.5	15.5	6.1	7.1	5.2
15.5	6.4	16.1	17.4	5.4	6.4	5.1
17.4	7.1	19.4	19.4	5.0	6.0	4.3
19.4	9.6	21.8	23.4	5.8	6.2	4.9
23.5	13.1	26.2	27.8	5.9	6.5	5.1

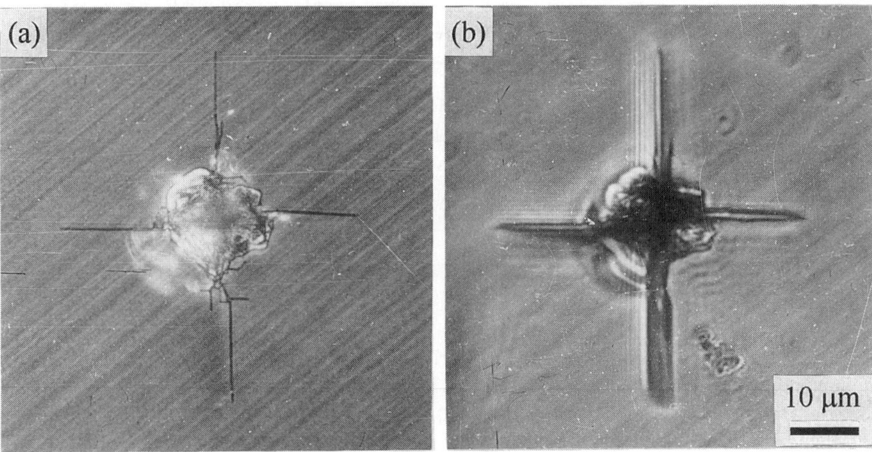

Fig.10. Optical micrograph of <110> indentation in diamond obtained after seven loading cycles. (a) reflective mode; (b) transmission mode, phase contrast.

From the length C_1 of initial half-penny cracks the fract ure toughness of diamond by reloading technique was evaluated using Eq.5. From the length C_3 of half-penny cracks after unloading the fracture toughness of diamond by conventional indentation technique was measured using Eq. 3 (Table I). Thus, at one indentation the fracture toughness of diamond were measured by two indentation techniques at seven loads. The fracture toughness of diamond measured by conventional technique is slightly higher than that of reloading technique (5.8±0.5 and 6.7±0.8 MPa $m^{1.5}$, respectively). This discrepancy is caused by the fact that Eq.3 overestimate the fracture toughness values. For example, the fracture toughness of silicon s.c. evaluated using Eq.3 is about 0.9 MPa $m^{1.5}$ (Table II). Using of conventional measurement technique on macroscopic specimens yields

Table II. Indentation Fracture Measurements Results for Silicon Single Crystal ({110} surface, one of the indenter diagonal was along <112>. Half-penny cracks for this diagonal propagate along $\{111\}_{90^o}$ plane).

P	a	C	H	K_{IC} (MPa $m^{1.5}$)	
(N)	(μm)	(μm)	(GPa)	Eq.(3)	Eq.(2)
0.10	1.9	3.7	12.6	1.00	0.85
0.20	2.7	5.8	12.5	1.02	0.86
0.29	3.5	7.7	11.3	1.00	0.85
0.39	4.1	10.0	10.8	0.90	0.76
0.49	4.7	11.8	10.3	0.88	0.75
0.59	5.1	13.3	10.5	0.88	0.75
0.69	5.6	14.5	10.2	0.90	0.77
0.78	6.1	15.9	9.8	0.90	0.76
0.98	6.7	19.3	10.1	0.84	0.71
1.18	7.6	20.7	9.5	0.91	0.77
1.47	8.7	26.2	9.0	0.79	0.68
1.96	10.0	31.0	9.1	0.82	0.70
2.94	12.1	40.1	9.3	0.84	0.71

fracture toughness values in the range from 0.64 to 0.70 MPa $m^{1.5}$ [20,21,38,39]. Test results obtained by reloading technique confirm the applicability of conventional indentation technique to measurement of diamond fracture toughness. The length C_2 of arrested half-penny cracks allow us to estimate K_{IA} value for diamond using Eq.(5). It was found that K_{IA} of diamond is about 5.0 MPa $m^{1.5}$.

CONCLUSIONS

The observation of crack initiation sequence during Vickers indenter penetration into (100) plane of diamond was performed. It was revealed that under <110> indentation two median cracks develop. On unloading the median cracks spread toward the surface and turn into surface cracks. The shape profile of these cracks is close to half-penny. Another indentation cracks were much smaller if compered with half-penny ones. During reloading the half-penny cracks start to propagate on loading half of cycle. It is proposed to use these cracks as initial and evaluate fracture toughness of diamond from a crack-starting load. The length of initial cracks was measured from opposite side of the specimen. The length of cracks after unloading were used to measure the fracture toughness of diamond by conventional indentation technique. Test results obtained by the reloading technique agree well with the data on conventional fracture toughness determination by indentation from the length of radial cracks after unloading.

ACKNOWLEDGEMENT

The author would like to thank Prof. A.L. Maistrenko and Dr. V.I. Kushch for heplful discussions and Dr. V. I. Mal'nev for his help in making of indentation device.

REFERENCES

1. D.J. Rowcliffe and S.M. Johnson, SPIE Vol.**681** Laser and Nonlinear Optical Materials, 143 (1986).
2. A.N. Nalyetov, Yu.A. Klyuev, O.N. Grigoryev, Yu.V. Mil'man and V.I.Trefilov, Rep. USSR Acad. Sci. **246,** 83 (1978).
3. S.N. Dub, A.L. Maistrenko and V.I. Mal'nev, in Problems of Fracture in Metals (MDNTP, Moskow, 1980) p. 83.
4. J.E. Field and C.J. Freeman, Philos. Mag A **43**, 595 (1981).
5. S.N. Dub and V.I. Mal'nev, in The Methods of Study of Superhard Materials Properties (ISM AN USSR, Kiev, 1981) p. 21.
6. N.V. Novikov and S.N. Dub, J. Hard Mater. **2**, 3 (1991).
7. Yu.A. Klyuev, A.M. Naletov and V.I. Nepsha, in New Diamond Science and Technology, edited by R. Messier, J.T. Glass, J.E. Butler and R. Roy (Mater. Res. Soc., Pittsburg, 1991) p. 149.
8. N.V. Novikov, S.N. Dub and V.I. Mal'nev, J. Hard Mater. **4**, 19 (1993).
9. N.V. Novikov, S.N. Dub, V.I. Mal'nev and V.V. Beskrovanov, Diamond and Relat. Mater. **3**, 198 (1994).
10. M. Drory, Cl.F. Gardinier and J.S. Speck, J. Am. Ceram. Soc. **74**, 3148 (1991).
11. M. Drory and Cl.F. Gardinier, J. Mater. Res. **7**, 781 (1992).
12. J.J. Mecholsky, Y.L. Tsai and W.R. Drawl, J. Appl. Phys. **71**, 4875 (1992).
13. R.S. Sussman, J.R. Brandon, G.A. Scarsbrook, C.G. Sweeney, T.J. Valentine, A.J. Whitehead and C.J.H. Wort, Diamond and Relat. Mater. **3**, 303 (1994).
14. A.G. Evans and E.A. Charles, J. Am. Ceram. Soc. **59**, 371 (1976).
15. B.R. Lawn, A.G. Evans and D.B. Marshall, J. Am. Ceram. Soc. **63**, 574 (1980).
16. K. Niihara, R. Morena and D.P.H. Hasselman, J. Mater. Sci. Lett. **1**, 13 (1982).
17. D.K. Shetty, I.G. Wright, P.H.Mincer and A.H. Claner, J. Mater. Sci. **20**, 1873 (1985).
18. M.T. Laugier, J. Mater. Sci. Lett. **6**, 355 (1987).
19. K.M. Liang, G. Orange and G. Fantozzi, J. Mater. Sci. **25**, 207 (1990).
20. S.N. Dub and A.L. Maistrenko, in Fracture Mechanics of Ceramics Vol.10 edited by R.C. Bradt, D.P.H. Hasselman, D. Munz, M. Sakai and V. Ya. Shevchenko (Plenum Press, New York and London, 1992), p.109.
21. G.R. Anstis, P. Chantikul, B.R. Lawn and D.B. Marshall, J. Am. Ceram. Soc. **64**, 533 (1981).
22. K. Tanaka, J. Mater. Sci. **22**, 1501 (1987).
23. S.N. Dub and V.G. Malogolovets, presented at the 4th Europ. Conf. "Diamond Films 94",

Il Ciocco, Italy (unpublished).
24. R. Nowak, K. Ueno and M. Kinoshita, in Fracture Mechanics of Ceramics Vol.10 edited by R.C. Bradt, D.P.H. Hasselman, D. Munz, M. Sakai and V. Ya. Shevchenko (Plenum Press, New York and London, 1992), p.155.
25. J.L. Henshall, D.B. Rowcliffe and J.W. Edington, J. Am. Ceram Soc. **60**, 373 (1977).
26. A.A. Aptekman, O.V. Bakun, O.N. Grigoryev, and V.I. Trefilov, Rep. USSR Acad. Sci. **290**, 845 (1986).
27. C.A. Brookes, Diamond and Relat. Mater. **1**, 13 (1991).
28. Z. Li, A. Ghosh, A.S. Kobayashi and R.C. Bradt, J. Am. Ceram. Soc. **72**, 904 (1989).
29. J.D. Sullivan and P.H. Lauzon, J. Mater. Sci. Lett. **5**, 247 (1986).
30. R.F. Cook and G. M. Pharr, J. Am. Ceram. Soc. **73**, 787 (1990).
31. M.-O. Guillou, J.L. Henshall, R.M. Hooper and G.M. Carter, J. Hard Mater. **3**, 421 (1992).
32. H.P. Kirchner and T.J. Larchuk, J. Am. Ceram. Soc. **65**, 506 (1982).
33. S.L. Jones, C.J. Norman and R. Shahani, J. Mater. Sci. Lett. **6**, 721 (1987).
34. B.R. Lawn and E.R. Fuller, J. Mater. Sci. **10**, 2016 (1975).
35. K. Tanaka, Y. Kitahara, Y. Ichinose and T. Iimura, Acta Metall. **32**, 1719 (1984).
36. G.P. Cherepanov, Mechanics of Brittle Fracture (Nauka, Moscow, 1974), p. 399.
37. D.B. Marshall, J. Amer. Ceram. Soc. **66**, 127 (1983).
38. S. Danyluk, Wear **103**, 149 (1985).
39.M. Brede, K.J. Hsia and A.S. Argon, J. Appl. Phys. **70**, 758 (1991).

SYNTHETIC DIAMOND STRENGTH ENHANCEMENT THROUGH HIGH PRESSURE/HIGH TEMPERATURE ANNEALING

W.E. JACKSON AND STEVEN W. WEBB
GE Superabrasives, 6325 Huntley Rd., Worthington, OH, 43085

ABSTRACT

Annealing of low inclusion, single crystals of synthetic Type I diamond at 1200-1700°C and 50-60kbar, was observed to increase average crystal compressive fracture strength by as much as 20%. The increase in strength is associated with the healing of lattice dislocations, coalescence of small (< 400 nm) metal inclusions, and dissociation of aggregate-nitrogen. Photoluminescence and FTIR spectroscopic data indicate H3 and "A"-type defect centers are unstable relative to NV_x and "C"-type defect centers under the conditions of these experiments. Dissociation rate is greater in the (111) growth sector than in the (100). Changes in residual lattice strain of the samples could not be detected using Raman spectroscopy and optical microscopy. A model reaction of the type, $2V + N_2V = NV_2 + NV$, in which N = nitrogen and V = vacancy (i.e., N_2V = H3 center) is proposed which provides a qualitative thermodynamic understanding of the observations reported in this study consistent with previous results of other researchers. In addition, photoluminescence examination of smaller crystals produced from a quench of early stage non-equilibrium growth reveal a lower concentration of aggregate nitrogen than those quenched from late stage growth suggesting that nitrogen may aggregate via *in situ* annealing processes. Ultimately, understanding mechanisms to improve synthetic single crystal diamond strength will lead to improved tools which utilize such crystals (e.g., drill bits and saws for cutting stone and concrete) which constitute a market of ≈ $2b worldwide.

INTRODUCTION

Crystal strength is a prime factor in determining the performance of industrial diamond tools such as saws, drills, and grinding wheels. While high temperature annealing is commonly used to improve the mechanical toughness of ceramics, metals and diamond thin films, little knowledge exists on the benefits of annealing saw-grade, low-inclusion, single crystals. Annealing helps relieve local strain by vacancy diffusion and defect rearrangements to lower energy states while large-scale strain is relieved by plastic flow within the lattice. Overall, the effect is a reduction in the number density of high stress planes which would otherwise present potential fracture surfaces. Annealing of diamond, however, requires both high temperatures to achieve plastic flow and high pressures to stabilize the lattice and avoid backconversion of diamond to graphite especially at internal metal interfaces commonly present in synthetic crystals. DeVries[1] found that the minimum annealing temperature required for lattice mobility (i.e., irreversible deformation) was 1000-1200°C at 50kbar.

Brozel et al. (1978) observed that when Type Ia diamonds initially containing high levels of aggregate nitrogen were annealed at 1960°C and 85kbar, the aggregates partially dissociated to form single substitutional centers[2]. Chrenko et al. (1977) observed that annealing at 1900°C and 55kbar significantly changed the optical properties of Type Ib diamond crystals[3]. They found that single substitutional nitrogen atoms can aggregate to form

Mat. Res. Soc. Symp. Proc. Vol. 383

A-centers if initially present in the diamond in a supersaturated condition. Evans and Wild (1965)[4] studied the plastic flow of diamond single crystals (1800°C) and hypothesized that aggregate platelet nitrogen impeded motion of dislocations, thus slowing deformation of the crystal under stress. Brookes et al.[5] observed microslipping in diamond due to limited dislocation movement at 100kbar and 1000°C.

EXPERIMENTAL

Annealing experiments were done in a belt-type apparatus at ~ 55kbar and 1200-1800°C for 15-60 minutes using a pressure medium of pre-densified graphite powder. Diamond samples were all synthetic, single crystals, with cubo-octahedral morphologies approximately 300-425 microns in diameter and from FTIR spectra contained approximately 10% aggregate nitrogen. Compressive fracture strength (CFS) measurements of 300 crystals/sample were made using a roll crusher[6].

Photoluminescence (PL) and Raman experiments directed argon ion laser light at either 514.5nm (2.41eV) or 457.9nm (2.71eV) through an Olympus microscope. Laser power at the sample was maintained at approximately 100mW. Luminescence was recollected by the lens and analyzed with a U-1000 ISA scanning monochromator and PMT. Photoluminescence data were collected at 10 cm^{-1} steps using a 3 second counting time[7].

RESULTS

Figure 1 shows PL spectra collected from the (111) surfaces of crystals annealed at 1100°C and 1 atm (HT conditions) for 2hrs under H_2 versus the original as-grown crystals. Aggregation of singlet nitrogen centers to form H3 centers (2.46eV) has occurred uniformly on both faces of the crystals due to annealing. The (100) surface showed similar results. Figure 2 shows PL data from crystals annealed at 1200°C and 50kbar for 1 hour. A decrease in the intensity of the zero phonon line at 2.46eV and associated sideband (~2.35eV) is observed. Conversely, a strong increase in the intensity of the 1.945eV ZPL (NV centers[8]) and 2.15eV ZPL (nitrogen-2 vacancy centers[8]) is observed.

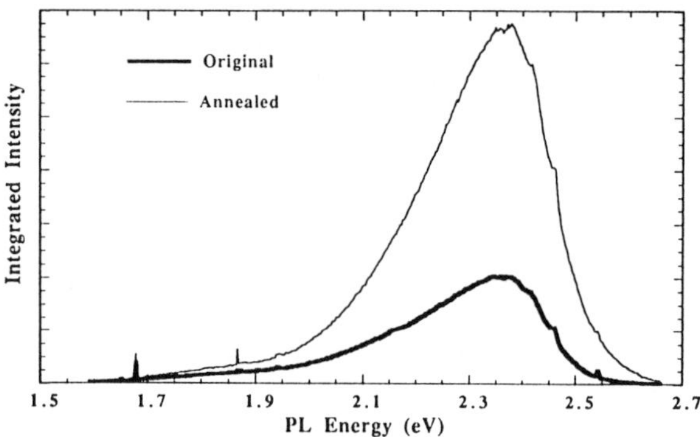

Figure 1. (111) PL spectra of HT Annealed Sample

Raman spectra collected from the HPHT annealed sample showed no shift in the 1332cm^{-1} band and no bands associated with graphite. The former observation suggests the relief of plastic strain is less than 0.2cm^{-1} in magnitude. In addition, optical strain birefringence observations (crossed polarized light) could detect no difference between as-grown and annealed crystals.

The average CFS of samples with lowest inclusion levels (~800 ppm metal) was increased by 10-20% after HPHT annealing. Samples with higher levels of metal inclusions did not develop any increase in strength from HPHT annealing. In addition, samples annealed under HT conditions did not exhibit an increase in average CFS.

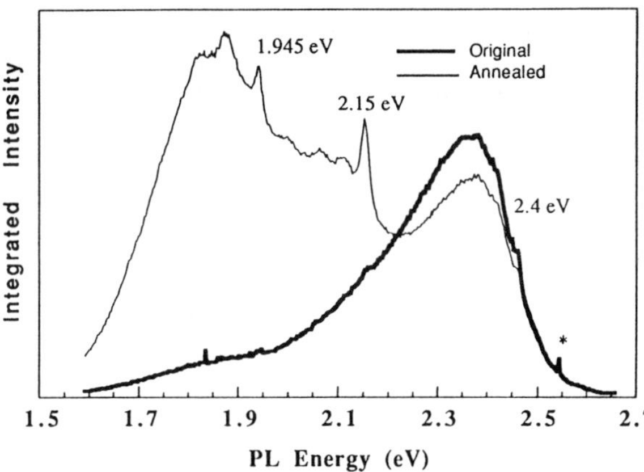

Figure 2. (111) PL Spectra of HPHT Annealed Sample (blue line)

SEM/EDS examination of the HPHT annealed sample revealed the presence of new, subsurface metal-rich inclusions (~50 μm) in some crystals increasing sample bulk magnetic susceptibility. Such changes were not observed for the HT annealed sample suggesting the formation of aggregate metal is favored at high-pressure. It is believed that the strongest crystals of the population are further strengthened by HPHT annealing. The weaker crystals, destroyed in the process, are those containing the most metal which was aggregated by annealing.

By etching/oxidizing diamond in molten KNO_3 it is possible to evaluate the density of dislocations and defects on the surface of diamond. High-energy defects will be preferentially attacked in the molten salt. Resulting surface pits provide a measure of the number density of internal dislocations terminating at the surface. Both SEM and optical microscopy suggest HPHT annealing reduces high-energy surface sites and possibly, by extension, the number of internal dislocations.

DISCUSSION - ANNEALING

PL data confirm the short-range nature of nitrogen conversion in diamond at 1200-1700°C. Conversion is slower in the (100) surface than the (111) surface, for both HT and HPHT annealing, which agrees with results of Satoh et al.[9]. The intensity change of the 2.46eV peak is greater on the (111) surface versus the (100) for the same annealing conditions. Furthermore, it is observed that aggregate nitrogen dissociates under HPHT annealing conditions while singlet nitrogen aggregates under HT annealing conditions. These data qualitatively agree with that of Brozel et al.[2] and suggest that under HPHT annealing conditions mobile vacancies and aggregate nitrogen defect centers (H3) combine and dissociate to form single substitutional nitrogen-vacancy plus nitrogen-defect complexes in the form of: $2V + N_2V = NV + NV_2$, in which 2V represents two mobile vacancies or holes in the lattice (perhaps associated with dislocations), N_2V the H3 center, NV the single nitrogen-vacancy defect, and NV_2 the multiple-vacancy-bearing single nitrogen-defect. At low pressure (1 atm) and 1200 °C the equilibrium described in eq. 1 is driven to the left describing nitrogen aggregation and vacancy formation (i.e., entropy favorable). At high pressure (> 50kbar) and

1200 ^{o}C, and high initial activity of H3 centers in the crystal, the reaction proceeds to the right, dissociating nitrogen and consuming vacancies (i.e., volume favorable). This hypothesis is supported by the observation that the HPHT annealed crystals appear to have fewer internal dislocations and higher compressive strength. Vacancies obviously play a role in activating what is observed to be a very slow rate process.

This model reaction may be generally applicable to describe prior studies of nitrogen behavior under annealing conditions through: $\Delta G = V\Delta P - S\Delta T + RT\Sigma_i \Delta \ln a_i$ (a_i = activity of nitrogen-defect center *i*) . At equilibrium, deviations in chemical potential, P, T, and a_i are zero. Positive deviations ($\Delta G>0$) indicate movement to the left ($K_{eq}<0$), favoring formation of aggregate nitrogen centers. Negative deviations imply movement to the right favoring dissociation of nitrogen ($K_{eq}>0$). For the equation as written both V and S are negative; volume and entropy are decreased moving to the right. Vacancies are consumed and configurational mobility is lost by converting nitrogen to a singlet state.

This model can be tested with literature observations. A decrease in pressure ($\Delta P<0$) at nominal temperature and activity will cause a positive deviation, favoring aggregate nitrogen formation. This was observed in low-pressure annealing at 1200 ^{o}C (this study). Increasing temperature to 2000 ^{o}C[4] and/or increasing the concentration of singlet nitrogen centers will also produce a positive deviation, favoring aggregate nitrogen[10]. Starting with mainly aggregate nitrogen (Type Ia), at higher temperature, will produce negative deviation in free energy, favoring dissociation[3]. In this case the entropic effect is outweighed by chemical potential differences.

Clearly V and S are small such that large alterations in P and T are necessary to move the equilibrium. Volume and entropy can be overwhelmed by the nitrogen activity of the starting crystals. The linear equilibrium model can explain the seemingly disparate literature reports of nitrogen aggregation or dissociation at HPHT.

The observation of metal aggregation upon annealing is consistent with earlier studies. Naletov et al.[11] reported an increase in magnetization and ferromagnetism after annealing at 850-1000 ^{o}C. Vishnevskii et al.[12] reported a large increase in susceptibility and resistance at 700 ^{o}C. It is proposed that HPHT metal aggregation in synthetic diamond crystals is derived from lattice compression, causing local coalescence of metal. This process would be diffusion controlled, facilitated by vacancies in the fashion that nitrogen aggregation is accelerated by defects introduced by electron irradiation[13]. A reaction of the type: $nM = M_n$, facilitated by temperature and vacancy content, is postulated, where nM is the number of metal atoms in isolated clusters and M_n the aggregate metal content. This reaction will tend to shift right at high pressure and temperature and for crystals of high metal content. This implies that formation of aggregate metal produces a volume reduction and may be linked to the nitrogen-vacancy rearrangements in these crystals. Lawson and Kanda[14] suggest an optical center associated with Ni forms and grows in intensity as substitutional nitrogen aggregates under HPHT annealing conditions of 60kbar and 1700 ^{o}C. According to our model such a reaction would be expected to free up vacancies which would assist in metal coalescence and might help explain these results.

The increase in diamond average fracture strength effected by HPHT annealing is correlated to changes in local strain, surface defects and dislocation/defect densities. Nitrogen may increase fracture strength by pinning dislocations thereby reducing the activation energy required for crack propagation[2]. Chrenko et al.[4] supports this theory by observing high nitrogen aggregation rates around dislocations.

Surface dislocation density observations indicate that annealing may heal local defect structures in the crystal. These dislocations may provide the latent crack sites for initiation of crystal failure. Eliminating dislocations would produce the observed increase in fracture strength. Noyikov et al.[15] recently proposed a correlation between dislocations in diamond, fracture toughness and lattice nitrogen type, at 1200 ^{o}C where diamond is slightly plastic. They propose that dislocations lead to local tensile stresses which reduce fracture toughness, consistent with the observations of this work.

DISCUSSION - GROWTH

Evans[16] and Klyuev et al.[17] suggest that nitrogen incorporated in natural diamond in an atomically dispersed state would slowly migrate and aggregate during diamond growth at the high pressures and temperatures of Earth's mantle. Kanda and Yamaoka[18] also hypothesize that in-situ annealing occurs during growth of synthetic diamond which progressively aggregates nitrogen in the crystal with the concentration being highest at the crystal center. However, they do not preclude the possibility that aggregated nitrogen centers formed in the metal solvent-catalyst and were simply incorporated into the crystal during growth. To further study these issues Raman and PL data from two sets of crystals were collected: (1) crystals that achieved nominal size, referred to as 'mature' and, (2) crystals that were collected just after growth commenced referred to as 'nascent'.

Figure 3 shows PL data for these crystals. The nascent crystals contain a higher ratio of NV centers to H3 centers relative to the mature crystals. This observation agrees with the first hypothesis of Kanda and Yamaoka[18] and suggests that nitrogen is grown into the crystal in singlet form and then slowly aggregates during subsequent growth. Consequently it would be expected that the relative H3 concentration should decrease radially from the center of the diamond providing the melt from which it is being grown contains excess nitrogen and growth conditions are not altered substantially.

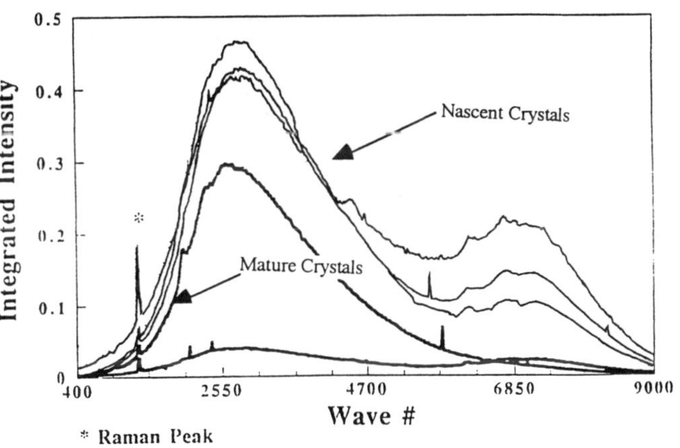

Figure 3. (111) PL Spectra for Mature Crystals

From the Raman spectra for the same pair of crystals, it is observed that the nascent crystals are under significant surface tensile strain. It is suggested that during initial nucleation and growth the rapid volume change leads to compressive strain frozen into the crystal center. Strain is relieved as the crystal grows out and becomes tensile, eventually relaxing to an unstrained surface state observed for the mature crystals.

Acknowledgments

Thanks are extended to Terri McCormick and Bob Nemanich (North Carolina State University) for performing the PL and Raman measurements.

References

1. R.C. DeVries, Mater. Res. Bull., 10, 1193-1200 (1975).
2. M.R. Brozel, T. Evans, and R.F. Stephenson, Proc. Roy. Soc. London, A361, 109-127 (1978).
3. R.M. Chrenko, R.E. Tuft and H.M. Strong, Nature, 270, 141-144 (1977).
4. T. Evans and R.K. Wild, Phil. Mag. **12**, 479 (1965).
5. C.A. Brookes, V.R. Howes and A.R. Parry, Nature, **332**(10), 139-141 (1988).
6. W.E. Jackson and S.C. Hayden, Finer Points, **5**, 26 (1993).
7. T.L. McCormick, M.S. Thesis, North Carolina State University, (1994).
8. J.E. Field, **The Properties of Natural and Synthetic Diamond**, Academic Press (1992).
9. S. Satoh, H. Sumiya, K. Tsuji and S. Yazu, in Science and Technology of New Diamond, edited by Saito et al., KTK, Tokyo, 351 (1990).
10. T. Evans and Z. Qi, Proc. Roy. Soc. London, A**381**, 159-178 (1982).
11. A.M. Natetov, V.I. Nepsha and S.D. Zvonkov, Tr.-Vses. Nauchno-Issled. Konstr.-Tekhnol. Inst. Prir. almazov Instrum., **3**, 45-50 (1974).
12. A.S. Vishnevskii et al., Fiz.-Khim. Probl. Sint. Sverkhtverd. Mater. **77-1**, 138-41 (1978).
13. A.T. Collins, J. Phys. Chem., **13**, 2641 (1980).
14. S.C. Lawson and H. Kanda, Diamond and Related Materials, **2**, 130-135 (1993).
15. N.V. Noyikov, S.N. Dub, V.I. Mal'nev and V.V. Beskrovanov, Diamond and Related Materials, **3**(3), 198-204 (1994).
16. T. Evans, Contemp. Phys. **17**, 45-70 (1976).
17. Y.A. Klyuev, V.I. Nepsha and A.M. Naletov, Sov. Phys. Solid State, **16**, 2118-2126 (1974).
18. H. Kanda and S. Yamaoka, Diamond and Related Materials, **2**, 1420-1423 (1993).

THRESHOLD CRACK SIZE IN A PYROLYTIC CARBON FOR NO GROWTH UNDER CYCLIC STRESS

LING MA, GEORGE H. SINES AND C. BARCLAY GILPIN*
Department of Materials Science and Engineering
University of California, Los Angeles, CA 90095, USA
*Department of Mechanical Engineering
California State University, Long Beach, CA 90840, USA

ABSTRACT

In this work, the fracture toughness of monolithic pyrolytic carbon and sandwich composites of this carbon on a graphite substrate were measured by a disk-shaped compact specimen and by using a tensile specimen. On this material, a large number of specimens that had small cracks induced by a Vickers diamond indenter were tested under cyclic stress. The existence of a true threshold stress intensity factor range for non-propagation of the cracks was demonstrated. Statistical confidence on pre-cracked specimens was obtained through testing a large number of specimens to a very long lifetime of the order of 10^9 cycles.

INTRODUCTION

A chemical vapor deposited isotropic pyrolytic carbon (PyC), alloyed with silicon, has been widely used as components in mechanical heart valve prostheses with success for decades. However, very little data has been published on its strength and its resistance to cyclic stressing. The design of the valve and the choice of material must provide for survival for lifetimes of the order of 10^9 cycles. More extensive test experience is clearly needed at the service stress level and to the requisite lifetime. New methods have had to be developed to reach the extreme cyclic lifetime and to obtain the extensive replication of tests needed for statistical confidence.

Three separate studies have demonstrated that this form of carbon resists the initiation of fatigue cracks at stresses that approach the static fracture strength; the fatigue strength is within the scatter band of the fracture strength [1-3]. Two of the studies terminated the tests at 10^7 cycles, but the third study carried the tests to 10^8 cycles [3]. Because of the necessity for extreme reliability at very long life, we have carried out tests at stresses, which are several times the service stress, at cycles greater than or equal to 6×10^8 cycles and with large scale replication to give a probability of survival greater than 92.2% with 95% confidence [4]. (The service strain is only 2.13% of the fracture strain). There have been no fatigue crack initiations. What remained to be investigated is the resistance of this carbon to the growth of extrinsic flaws under cyclic stress.

In this study we sought to find if a true threshold stress intensity factor existed rather than determining the crack growth rate on compact tensile specimens, which have an induced crack length of the order of 10 mm, and extrapolating it to some chosen initial crack length. In the following study small cracks were induced with a Vickers diamond indenter and very

Mat. Res. Soc. Symp. Proc. Vol. 383

long life tests were conducted to see at what range of stress intensity factor no growth would occur.

EXPERIMENTAL

Material and Specimen

The material and specimens were made by Carbon Implants, Inc. (now known as Medtronic Carbon Implants, Inc.) of Austin, Texas. The material is a chemical vapor deposited carbon in a fluidized bed on graphite substrate to give an isotropic pyrolytic coating. The carbon was alloyed with 5 to 8 wt% silicon. The Young's modulus of PyC is 31.2 GPa with a standard deviation of 1.9 [5]. The modulus of rupture of sandwich specimens by four point bending test is 400 MPa with a standard deviation of 40.2. The Poisson's ratio of PyC ranges from 0.22 to 0.30.

Fatigue Testing Method

Conventional Krouse plate bending fatigue machines which run at 1725 rpm were modified to accommodate as many as twenty specimens at a time on a 6.35 mm thick triangular plate. The plate is shaped so that it experiences the same stress throughout when it is bent as a cantilever beam. The specimens were fastened by adhesive to the plate in a way that the center portion (6 mm x 6 mm) of the specimen was free from the plate to ensure that the strain was predominantly uniaxial [4].

To obtain increased testing speed, cantilever rotating beam fatigue machines, which run up to 10,000 rpm, were modified to test up to four specimens at a time. The 8 mm square-sectioned rotating loading beam was made so that a specimen could be affixed to each flat surface. The specimens were also installed by adhesives in the same way as was used for the plate machine to give uniaxial loading [4].

Measurement of K_{IC} by Disk-Shaped Compact Specimen Test

The specimens were in the form of a modified ASTM E399 disk-shaped compact specimen DC(T). The specimens were slightly different in that there was no flat portion, i.e. the "c" dimension was not used. The machined crack was 4.8 mm long, 0.2 mm wide with a tip radius of 0.1 mm. The machined crack was sharpened under the displacement control mode. Five monolithic PyC samples with a thickness of 0.69 mm were tested. Four sandwich specimens with a total thickness of 0.87 mm having the graphite 0.27 mm and each PyC layer 0.30 mm thick were also tested. All testing was done with a MTS Model 810 closed loop hydraulic test machine. Crack length was measured on the specimen surface with a microscope having a resolution of 0.02 mm [6].

Measurement of K_{1C} of Monolithic PyC by Tensile Test

Monolithic PyC specimens were prepared by coating graphite substrates and then they were split along the midplane of graphite, and the graphite removed leaving monolithic PyC samples 0.47 mm thick.

Three monolithic pyrolytic carbon strips had cracks introduced by the Vickers diamond indenter and then doublers were attached to the ends for tensile testing. The specimens measured 12 x 24 x 0.47 mm. The test was performed on an Instron at a crosshead speed of 0.254 mm/min.

Pre-cracked Sandwich Specimens under Cyclic Stress

The sandwich specimens had a coating thickness of 0.25 mm on a 0.30 mm graphite substrate. The specimens measured 6 x 30 x 0.8 mm.

The first group, thirty pre-cracked sandwich specimens, were tested in uniaxial stress up to and beyond 6 x 10^8 cycles under a strain level of 500 $\mu\varepsilon$ ($\mu\varepsilon \equiv 10^{-6}$ strain) with R=-1 (R is the ratio of minimum stress and maximum stress) or more severe conditions. The cracks were induced by a Vickers diamond indenter with loads from 89 N to 133 N (20 lb to 30 lb). The longest crack was 1.84 mm and the shortest one was 0.60 mm. The mean value of the crack length was 1.11 mm with the population standard deviation of 0.35 mm. The crack length was measured with a B&L metallograph to an accuracy of 10 microns.

The second group, three sandwich specimens and a monolithic pyrolytic carbon specimen with slightly longer cracks (2.04 to 2.44 mm), were tested in uniaxial loading at the strain levels of 1,000 $\mu\varepsilon$ with R=0.05 and 500 $\mu\varepsilon$ with R=-1.

The third group, seven pre-cracked sandwich specimens, were tested in uniaxial loading at 1,000 $\mu\varepsilon$ with R=-1. The cracks ranged from 0.7 to 1.88 mm.

RESULTS

Measurement of K_{IC} by Disk-Shaped Compact Specimen Test

The mean value of K_{IC} of monolithic PyC was 1.17 MPa$\sqrt{m}$ with the standard deviation of 0.17. The mean value of K_{IC} of sandwich specimens was 1.84 MPa$\sqrt{m}$ with the standard deviation of 0.22. The procedures of ASTM E399 were followed.

Measurement of K_{IC} of Monolithic PyC by Tensile Test

The measured K_{IC} of the monolithic PyC specimens is listed in Table I. The crack was modeled as a central through crack in the calculation.

Table I K_{IC} of monolithic PyC by tensile test

Specimen number	490-4A	490-15A	490-17A
K_{IC} (MPa$\sqrt{m}$)	1.19	1.39	1.25
Average K_{IC} (MPa$\sqrt{m}$)		1.28	

Pre-cracked Sandwich Specimens under Cyclic Stress

The first group, thirty pre-cracked sandwich specimens were tested under cyclic uniaxial stress up to and beyond 6 x 10^8 cycles. The strain levels were 500 $\mu\varepsilon$ with R=-1 and 1,000 $\mu\varepsilon$ with R=0.05. Thirty pre-cracked specimens survived 6 x 10^8 cycles and beyond, with no crack propagation or failure, giving 90.5% probability of survival with 95% confidence if all the specimens had the least crack length of 0.6 mm [7]. This exceeds the FDA requirement that a fatigue strength of 90% survival with 95% confidence be achieved on uncracked specimens at 6 x 10^{8} cycles with a safety factor of two. It is to be noted that the FDA requirement was met here with pre-cracked specimens. In the spot test to simulate the physiological conditions of the valve, specimen 181-11 with an induced crack of 1.58 mm survived 8.1 x 10^{8} cycles at 1,000 $\mu\varepsilon$ with R=0.05 in Ringer's solution.

The second and third group were tested under more severe conditions, i.e. longer crack length or higher strain level. All specimens experienced rapid crack growth to failure.

The results of the fatigue test of pre-cracked specimens are listed in Table II. The ΔK_1 was calculated by the following equation [5]:

$$\Delta K_1 = E \, \nu_O \, (\pi \, a)^{1/2} \, / \, h \tag{1}$$

where E is the Young's modulus (31.2 GPa), ν_O is the normal displacement, a is the half crack length and h is the effective half length of the specimen. The ΔK_1 value in Table II has been modified for the sample width [8, 9].

The dynamic stress analysis of the bileaflet heart valve by the finite element analysis indicated that the peak service strain of the valve is less than 250 $\mu\varepsilon$ with R=0. The strain levels of the fatigue tests used in this work were at 500 $\mu\varepsilon$ and 1,000 $\mu\varepsilon$; thus making the testing strain level higher than the estimated peak service strain by a factor of two and four, respectively. The loading level is given in terms of elastic strain instead of stress, because elastic strain was used to set the testing machines and monitor the tests. Results tabulated in Table II were collected by conducting the accelerated test at speeds of 1,725 rpm and 10,000 rpm. We found no difference in results from the testing speeds of 1,725 rpm and 10,000 rpm.

Table II Fatigue test of pre-cracked specimens

No. of specimens		strain ($\mu\varepsilon$)[1]	R	Crack 2a (mm)	K_1 max (MPa$\sqrt{m}$)	ΔK_1 (MPa$\sqrt{m}$)	No failure[2] (10^6 cycles)	Failure cycles
28	s[3]	500	-1	0.60-1.84	0.50-0.90	1.00-1.80	≥600[5]	-
1	s	1,000	0.05	1.63	1.69	1.61	600	-
1	s	1,000	0.05	1.58	1.66	1.58	810	-
1	s	1,000	0.05	2.44	2.14	2.03	-	<100
1	m[4]	500	-1	2.32	1.04	2.08	-	$3x10^6$
1	s	500	-1	2.29	1.03	2.06	-	$5x10^6$
1	s	500	-1	2.04	0.95	1.90	-	$4x10^6$
7	s	1,000	-1	0.70-1.88	1.08-1.83	2.16-3.66	-	1

1) strain given is the maximum of the cycle.
2) tests terminated with no failure.
3) s - "sandwich" pyrolytic carbon coated graphite.
4) m - monolithic pyrolytic carbon.
5) one specimen exceeded 2.4 x 10^{9} cycles.

DISCUSSION

Vickers Diamond Indenter Induced Cracks

By controlling the load, cracks ranging from 0.5 mm to several millimeters can be induced by the Vickers indenter. In most cases, the corner cracks emanated from the impression longitudinally and transversely to the axes of the specimen and were well suited for this study because they all had a very sharp tip. Under the impression secondary cracks also form; a pyramidal crack and within the pyramid are conical cracks (see Figure 1). It was found that the cracks penetrated the pyrolytic carbon coating but were arrested at the carbon-graphite interface [5]. In this study, the pre-cracked specimens were modelled as a through, central crack under constant displacement. The tip to tip length of the corner crack was regarded as the crack length 2a in the fracture calculations.

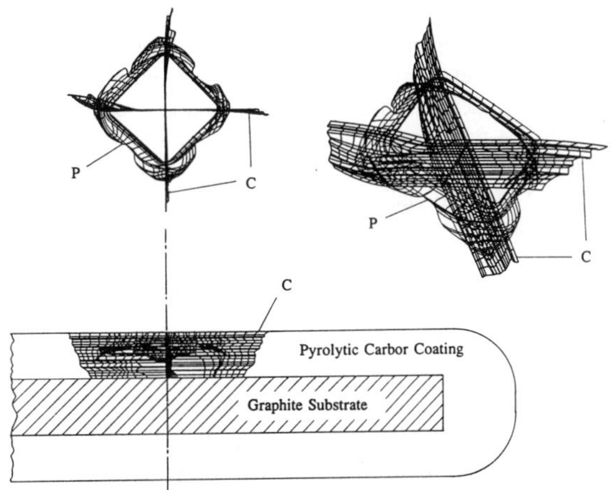

Figure 1 Cracks in the PyC made by the Vickers diamond indenter. C represents corner cracks; P represents the pyramid crack. Some secondary cracks are omitted.

For the monolithic PyC specimens, the cracks were induced after the graphite substrates had been removed and we do not know the morphology of the secondary cracks beneath the impression. Since the crack did not penetrate to the back of the specimen, the K_{IC} value listed in Table I is higher than the true value because in the calculation of the stress intensity factor it was treated as a through crack.

Fracture Calculation

When calculating the crack growth rate under cyclic stress, it is a common practice to use only the positive portion of the cycle for reversed strain conditions. The justification of doing so is that the crack closes during the negative portions of cycles; however we feel strongly that this may not be the case for the pyrolytic carbon. During the fatigue test, it was observed that carbon powder was pushed out to the specimen surface from the crack. This would prevent the crack from closing during the negative portion of the cycle, therefore, the full range of strain should be used. Table II shows the results when the full range of the strain is considered in ΔK_1.

It is demonstrated in Table II that all specimens that survived with no crack growth had ΔK_1 values less than 1.80 MPa$\sqrt{}$m, while all specimens that failed had ΔK_1 values greater than 1.90 MPa$\sqrt{}$m. The measured K_{IC} value for the sandwich specimens by disk-shaped compact specimen was 1.84 MPa$\sqrt{}$m with the standard deviation of 0.22. Ritchie *et al* reported K_{IC} of 1.90 MPa$\sqrt{}$m on a similar material [10]. Some pre-cracked specimens did not fail immediately after the load was applied even though the calculated ΔK_1 exceeded K_{IC}. This can be attributed to the complex shaped cracks introduced by the Vickers diamond indenter.

It was also observed during the compact disk test that the long transverse cracks (6 to

8 mm) in the PyC of sandwich disks had grown under ΔK_I of 0.89 to 1.29 MPa$\sqrt{m}$. It was not clear whether the crack in the PyC layer was leading the one in graphite or vice versa. However direct comparison between the stress intensity factors for the graphite coated carbon sandwich materials obtained by the compact tensile specimen and the Vickers indented strip may not be feasible because of the overly simplified models used in their calculation. It is to be noted that calculations in this work were based on the assumption that the specimens were isotropic and homogeneous; despite the fact that the modulus of elasticity of the two materials is quite different. Better models to analyze cracks in layered composite materials are needed.

Threshold Crack Size in a Pyrolytic Carbon

In spite of the difficulties encountered in analyzing the cracks in layered composite materials, we would like to bring attention to our experimental observations. As it is demonstrated in Table II, cracks longer than 2 mm grew and those less than 1.80 mm did not grow at a strain level of 500 $\mu\varepsilon$ with R=-1. This indicates that there is a threshold size for cyclic fatigue crack non-propagation. We did not detect any crack growth for all the pre-cracked specimens in the first group up to 6 x 10^8 cycles and beyond at the strain levels that are two and four times higher than the estimated peak service strain. For the cracks that survived 6 x 10^8 cycles without growth, the growth rate was less than 10^{-14} m/cycle for the accuracy of the crack measurement of 10 microns. This is to *assume* that in the worst case the crack propagated up to our limit of detection of 10 microns by the end of 6 x 10^8 cycles. This is a factor four orders of magnitude lower than the threshold value defined by ASTM [11].

The existence of the threshold size of fatigue crack non-propagation of pyrolytic carbon is of importance in terms of service reliability within the designed lifetime of the heart valve and the quality control during the manufacturing process. When the flaw size is less than the threshold size, the flaw would not propagate during the service lifetime provided that the valve is operated under the designed stress. The threshold size of fatigue crack non-propagation makes the proof test of the components and the whole assembly for the valve very effective to screen out the flaws because proof tests can be performed at a level one order of magnitude higher than the service stress.

REFERENCE

1. F. J. Schoen, *Carbon*, **11**, 413 (1973).
2. H. S. Shim, *Biomaterials, Medical Devices and Artificial Organs*, **2** (1), 55-65 (1974).
3. J. C. Bokros, A. D. Haubold, R. J. Akins, L. A. Campbell, C. D. Griffin and E. Lane, in *Replacement Cardiac Valves*, edited by E. Bodnar and R. W. M. Frater (Pergamon Press, New York, 1991), pp. 21-48.
4. L. Ma and G. Sines, "Fatigue of Isotropic Pyrolytic Carbon Used in Implanted Mechanical Heart Valves", submitted to *The Journal of Heart Valve Disease*.
5. L. Ma and G. Sines, *Materials Letters*, **17**, 49, (1993).
6. C. B. Gilpin, A. D. Haubold and J. L. Ely in *Bioceramics* **6**, edited by P. Ducheyne and D. Christiansen (Proceedings of the 6th International Symposium on Ceramics in Medicine, Philadelphia, PA, 1993) pp. 217-223.
7. P. F. Packman, S. J. Klima, R. L. Davies, J. Malpani, J. Moyzis, W. Walker, B. G. W. Yee and D. P. Johnson, *Metals Handbook: Nondestructive Inspection and Quality Control*, Vol. **11**, (American Society for Metals, Metals Park, OH, 1976),

pp. 414-424.
8. D. P. Rooke & D. J. Cartwright, *Compendium of Stress Intensity Factors*, (Her Majesty's Stationary Office, London, 1976), p. 14.
9. M. Isida, *International Journal of Fracture Mechanics*, **7** (3), 301 (1971).
10. R. O. Ritchie, R. H. Dauskardt and Weikang Yu, *Journal of Biomedical Materials Research*, **24**, 189 (1990).
11. "Standard Test Method for Measurement of Fatigue Crack Growth Rates, E 647-93," *Annual Book of ASTM Standards: Metals Test Methods and Analytical Procedures*, **03.01**, (ASTM, Philadelphia, 1993), p. 679.

THE UNIQUE STRENGTH OF DIAMOND

JOHN J. GILMAN
Dept. of Materials Science and Engineering
Boelter Hall, UCLA, Los Angeles, CA 90024

ABSTRACT

The features of diamond that make it unique are: 1] its efficient tetrahedral structure which is stabilized by resonance of the bonding electrons among adjacent bonds; 2] the exceptionallly high ratios of its shear moduli to its bulk modulus; 3] the large gap between the top of its valence band and the bottom of its conduction band which makes dislocation mobility very low in it; 4] its high surface energy. Although diamond is nearly elastically Isotropic, the ratio of its C_{44} shear constant to its bulk modulus is 1.3 which is the highest measured value for this ratio. It is quite unusual for it to be greater than unity. This high shear stiffness appears to be associated with the bimodal distribution of electrons along the lengths of the C-C bonds. Also, for centrally directed interatomic forces, the ratio C_{44}/C_{12} is expected to be unity. But In diamond it is 4.6, indicating that the behavior of diamond cannot be described even approximately by pair-potentials. Because of its unique strength diamond can perform exceptionally in mechanical devices such as: pressure vessels, ultracentrifuges, springs, flywheels, and cutting (or drawing) tools.

NATURE OF STRENGTH

"Strength" in the context of this paper means stability in the face of arbitrary mechanical disturbances. Therefore, a "strong" material must be able to withstand various modes of strain including: tension, compression, torsion, and bending. The theme of the discussion is that diamond, of all known materials, is uniquely capable in this regard. The reason being that the ratio of its shear stiffness to its bulk stiffness is exceptionally high; and this ratio is a measure of "solidity". Since other opinions have been expressed in the literature, the evidence is the subject of this paper.

Structural materials can become unstable in a variety of ways. They can buckle elastically, flow plastically, twin, or fracture. Therefore, no single property determines their stability (strength). The most improtant properties (setting chemical stability aside) are: micro-geometry, bulk stiffness, shear stiffness, dislocation mobility, and surface energy.

At the molecular level symmetry is important. For general strength it needs to be as nearly isoaxial as possible (i.e., cubic or isotropic). Fibers and membranes can be strong for special kinds of loading, but they are not generally strong.

Elastic stiffness is needed to minimize the deflections caused by loads; particularly the deflections that lead to buckling. Shear stiffness is especially important for resisting torsion. In fact, it can be said that the ratio of the shear to the bulk modulus is a measure of "solidity". If this ratio is zero, one has a liquid, and if it is very large, a

Mat. Res. Soc. Symp. Proc. Vol. 383

material is not elastic but rigid. By allowing large deflections to occur, plastic flow can lead to instability and failure, and the most effective way to avoid plastic flow is have low dislocation mobility in a material. Finally fracture requires cracks, and cracks represent the formation of surfaces, so they are more difficult to form if the surface energy is high.

UNIQUE MECHANICAL PROPERTIES OF DIAMOND

The very structure of diamond is unique. The open tetrahedral arrangement of the atoms in it forms the most efficient of all possible frameworks because each node (atom) is surrounded by just four struts (bonds). This is the minimum number of struts needed to fix the position of a point in three dimensions.

The bonds of the diamond framework are also unique in at least two ways: the distribution of bonding electrons in them; and in their connectedness. They can be considered to form puckered rings [1]. Each of the puckered rings can be decomposed into six identical "elbows" (Figure 1). One of these is indicated by an arrow in the figure. The elbows are shared by two bonds each, and six "arms" constitute each bond. There are four bonds per carbon atom, so twelve rings are associated with each carbon. Thus there is ample opportunity for the electrons to resonate among the bonds, thereby adding stability to the structure.

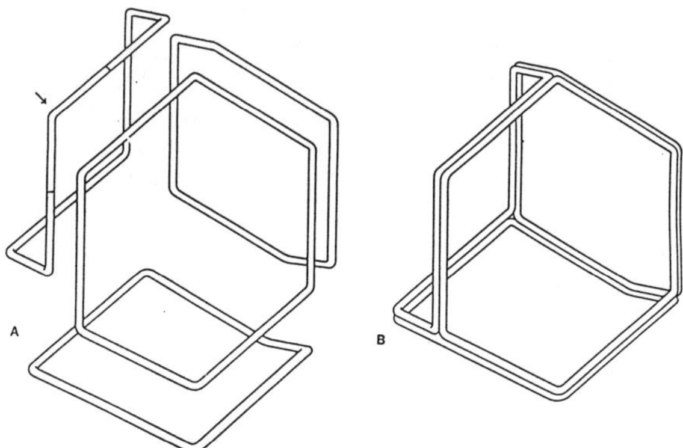

Figure 1 - Formation of the diamond framework from six-sided puckered rings (chairs). The carbon atoms reside at the bends of the rings (bend angles equal 109.5° = tetrahedral angle).

The tetrahedral framework of diamond results from the electronic structures of carbon atoms, and the Pauli principle. With atomic number six, carbon is the first atom that has enough electrons to form a tetrahedral framework from electron pair bonds. Also, their principal quantum numbers are small (1 and 2), so the size of the atoms is small making the bonding electron density high. The low-lying 1s orbitals which contain two of the six electrons are too isolated to participate in the bonding, leaving 2s-orbitals and 2p-orbitals. These are close enough in energy to combine into four sp-hybrid orbitals lowering the net bonding energy. The electrostatic energy of this set of four bonds can be minimized if they are directed toward the corners of a tetrahedron where they are equidistant from one enother. In diamond the hybrid orbitals of the carbon atoms combine in pairs to form the tetrahedral framework.

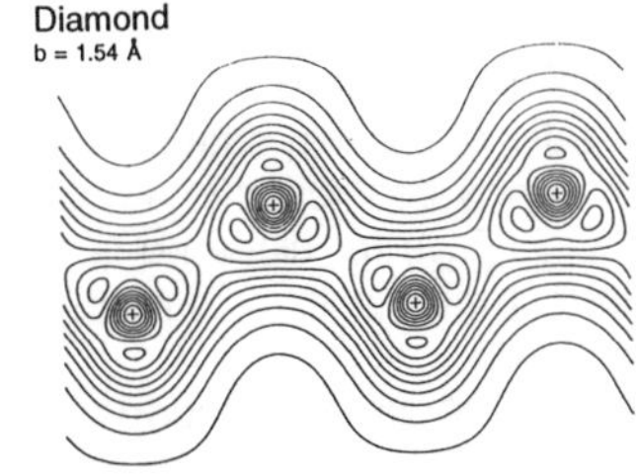

Figure 2 - Electron densities along the bonds in diamond and silicon. The scales of the two drawings are in proportion to the bond lengths, b.

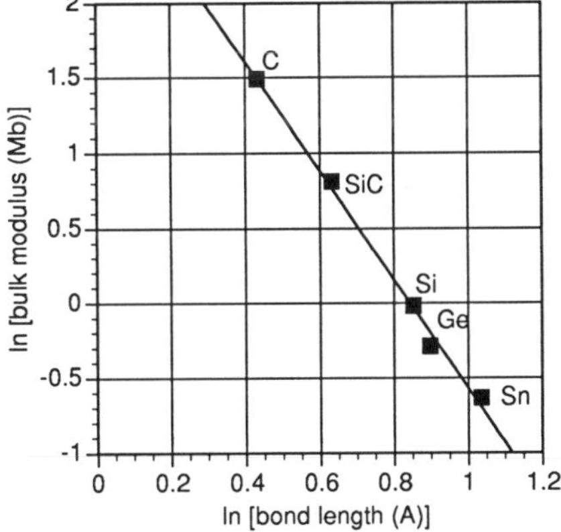

Figure 3 - Bulk modulus vs. bond length.

Experimental X-ray scattering [2] has confirmed that these hybrid bonds consist of directed concentrations of the bonding electrons (Figure 2).

The distribution of electrons in the bonds of carbon is different than in other tetrahedrally bonded crystals such as silicon. For carbon the distribution is strongly bimodal along the length of the C-C bond whereas it is unimodal in the case of Si-Si; and of Ge-Ge. This is not just a curiousity, but an important influence in making the properties of diamond unique. Theoretical calculations also yield these distributions [3].

Elasticity

The most important factor in determining elastic stiffness is the interatomic distance. For covalent crystals, this was emphasized long ago by Keyes [4]. The reason is that the bonding forces are electrostatic (once the electron distribution is determined) and according to Coulomb's law are proportional to $1/d^2$. Then, since stresses are forces per unit area, and elastic stiffness is stress per unit strain, it is proportional to $1/d^4$. This is approximately what is observed, but not exactly because the charge distribution may vary with d.

F igure 3 shows how the bulk moduli of the Group IV crystals depend on their atomic spacings. The scales of the graph are logarithmic showing that indeed the dependence is a power function, but the exponent is 3.6 rather than 4.0. Thus the electron distribution changes from one element to another in addition to the change in atomic spacing.

Figure 4 shows how $C_{44,}$ the modulus for simple shear varies with d. The dependence is again a power function with the exponent = 4.6. The modulus

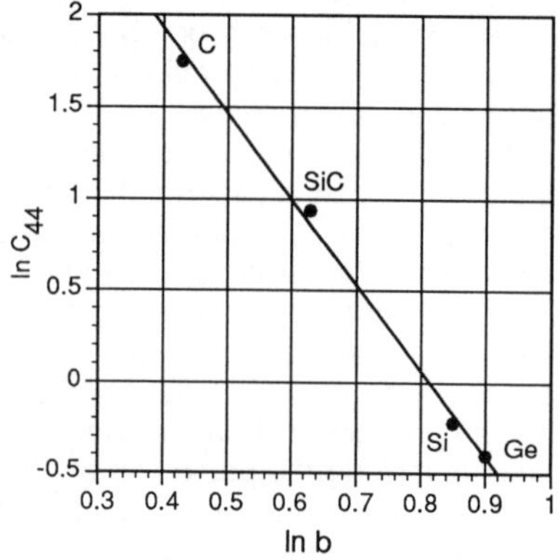

Figure 4 - Shear modulus vs. bond length.

for compound shear, $C^* = (C_{11} - C_{12})/2$ varies similarly with d.

The ratio of the two shear moduli (i.e., the anistropy coefficient = $A = C_{44}/C^* = 2C_{44}/(C_{11} - C_{12})$) for diamond is 1.21 indicating that the elasticity is nearly isotropic.

The uniqueness of diamond appears when the ratios of the shear moduli to the bulk modulus are considered. These are listed in Table 1 for all crystals that are isoelectronic with diamond (except for AlP for which there are no data). Note that, for the metallic elements, these ratios are invariably less than one. In strong contrast, C_{44}/B (the "solidity" coefficient) is 1.30 for diamond, and this is larger than for the other electronically similar crystals. Also, the data of Table 1 show that the deviation from the Cauchy condition ($C_{44}/C_{12} = 1$) is very large. This means that central forces between atoms cannot be used to describe the bonding in diamond. In other words the forces that resist bond bending are unusually large in diamond; and the bonds are highly directed.

Table 1 - Comparison of Important Elastic Ratios (units = GPa)

Crystal	C_{44}/B	C^*/B	C_{44}/C_{12}	C_{11}	C_{12}	C_{44}	B
C	1.30	1.06	4.5	1080	130	580	447
BN	1.20	0.79	2.5	820	190	480	400
SiC	1.10	0.50	1.7	352	140	233	211
Si	0.82	0.52	1.3	166	64	80	98

Overall, the reason that diamond has unique strength is because of its large shear stiffness, not because its bulk modulus is exceptionally large. Its bulk modulus is only 12% larger than that of BN, while its shear modulus is 21% larger.

Plasticity

In simple bending, dislocation motion in diamond has not been observed below about 1600 °C [5]. Observations near indentations made in compression must be interpreted carefully because of the possibility that a phase transition has occurred [6]. The temperature dependence of the plastic flow that is associated with dislocation motion is shown in Figure 5. The graph indicates that little dislocation motion occurs below the Debye temperature, θ. This is similar to the behavior of Si and Ge. The

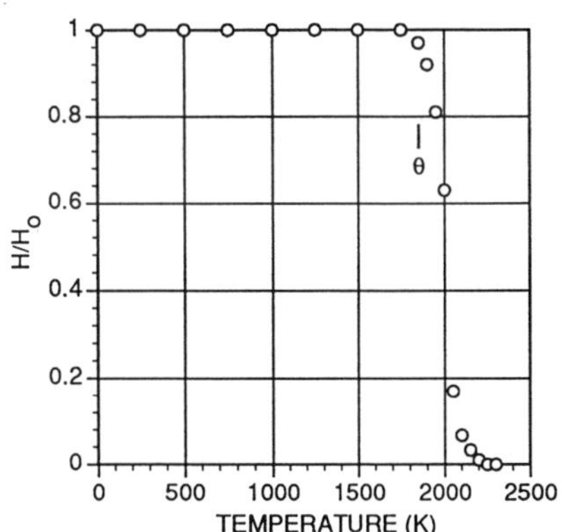

Figure 5 - Dependence of relative hardness on temperature for diamond. Data of Evans and Sykes [5].

activation energy for dislocation motion at high temperatures is 10.7 eV. (twice the minimum energy gap). This correlates with the behavior of other Group IV crystals (Figure 6). The behavior shown in Figures 5 and 6 indicates that dislocation mobility in diamond is determined by electronic processes; that is, by quantum rather than classical mechanics. The barrier to the motion is the LUMO-HOMO energy gap for the C-C bonds as described elsewhere for silicon [6]. The temperature dependence is determined by Fermi quantum statistics because the kinks on dislocation lines that allow motion have unpaired electrons associated with them, so they are Fermi quasi-particles. This has also been discussed elsewhere for silicon [7].

Low dislocation mobility makes diamond very resistant to plastic yielding and therefore contributes to its mechanical stability (strength).

According to the quantum theory, the parameters that determine the low mobility are the large LUMO-HOMO energy gap and the small bond length. This combination leads to large energy gradients at dislocation kinks and therefore to large resistance to motion.

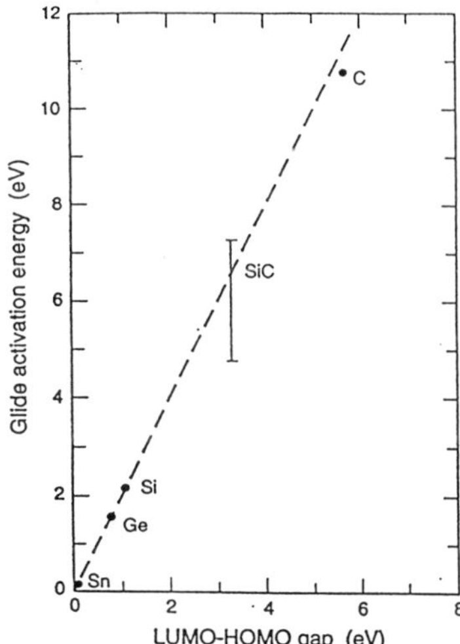

Figure 6 - Showing that the glide activation energy equals twice the LUMO-HOMO gap for Group IV crystals.

Fracture

Diamond does not cleave easily because of its high surface energy. For its cleavage plane, {111} this is higher than for the cleavage plane of any other known material. It is 5.5 ± 0.15 J/m^2 [8]. Since the tip region of a sharp crack is in a state of approximately uniaxial strain, the cleavage surface energy is related to the elastic stiffness constant, $C_{11}(111) = C_{cl}$ where (111) is the cleavage plane. In terms of this constant the surface energy is given the given approximately by: $C_{cl}d/\pi^2$

APPLICATIONS OF THE UNIQUE STRENGTH OF DIAMOND

There are several mechanical devices whose performance depends on the absolute strength of the material from which they are constructed. Others depend on the weight specific strength; and a few on the volumetric specific strength. Diamond pertains to the first category. Examples are [9]: ultracentrifuges; containers for magnetic, mechanical and electrostic fields; and devices for storing power such as flywheels, or for controlling power such as rocket engines. They have been used for very demanding cutting (ultramicrotomes) and shaping (wire-drawing) operations for a long time. Their use for containing mechanical fields (diamond anvil pressure vessels) is relatively recent.

Diamond based ultracentrifuges have interesting potential properties. They should be able to produce acceleration fields in the ten billion g range which is about 10^4 times larger than conventional ultracentifuges. The resulting gravitational potential might be as large as 9×10^{10} $(cm/sec)^2$ at the edge of a 2 mm. rotating disk. This could cause frequency shifts of atomic processes as large as $\Delta\nu/\nu = 10^{-10}$.

SUMMARY

The crystallographic, and chemical bond structures of diamond are unique because carbon is the smallest atom capable of forming a uniform set of tetrahedral bonds. The result is a quite unusual set of fundamental strength parameters, including:

1. the largest known ratio of shear-to-bulk elastic stiffness; 1.3.
2. the largest known glide activation energy; 10.7 eV.
3. a very large Cauchy ratio; 4.5.
4. largest known cleavage surface energy; 5.5 J/m^2.
5. very small Poisson's ratio; 0.07.

This makes it unlikely that compound crystals can be found that have superior strength (hardness).

REFERENCES

1. J. J. Gilman, "Novel Method for Constructing Tetrahedral Frames", in MRS Symposium Proceedings, Vol. 120, High Temperature/ High Performance Composites, Edited by Lemkey, Fishman, Evans, and Strife, p. 369, Materials Research Society, Pittsburgh, PA (1988).

2. M. A. Spackman, "The Electron Distribution in Diamond: a Comparison Between Experiment and Theory", Acta Cryst., A47, 420 (1991).

3. M. L. Cohen and J. R. Chelikowsky, Electronic Structure and Optical Properties of Semiconductors, Chap. 7, Springer-Verlag, New York (1988).

4. R. W. Keyes, J. Appl. Phys., 33, 3371 (1962).

5. T. Evans and J. Sykes, Phil. Mag., 29, 135 (1974).

6. J. J. Gilman, "Why Silicon is Hard", Science, 261, 1436 (1993).

7. J. J. Gilman, "Physical Chemistry of Intrinsic Hardness", Proc. Fifth Inter. Conf. Sci. Hard Materials, to be published in Mat. Sci. and Eng. (1995).

8. The Properties of Natural and Synthetic Diamond, Ed. by J. E. Field, p.679, Academic Press, London (1992).

9. J. J. Gilman, "Ultrashigh Strength Materials of the Future", Mech. Eng., 83, 55 (1961).

FRACTURE TOUGHNESS AND SUBCRITICAL CRACK GROWTH IN CVD DIAMOND

A. KANT,[1] M. D. DRORY,[2] AND R. O. RITCHIE[1]

[1]Materials Sciences Division, Lawrence Berkeley Laboratory, and Department of Materials Science and Mineral Engineering, University of California, Berkeley, CA 94720-1760
[2]Crystallume Corporation, Santa Clara, CA 95054

ABSTRACT

The fracture toughness, stress corrosion and cyclic fatigue properties of polycrystalline chemical vapor deposited (CVD) diamond have been investigated on thick (~100 to 300 μm) free-standing films. Specifically, the fracture toughness, K_c, of diamond was determined using indentation methods and for the first time by the tensile testing of pre-notched fracture-mechanics type compact-tension samples. Measured K_c values were found to be between 5 and 7 MPa-m$^{1/2}$ by either method and to be apparently independent of grain size and shape. Studies on subcritical crack growth (i.e., at stress intensities less than K_c) indicated that CVD diamond is essentially immune to stress-corrosion cracking under sustained loads in room air, water and acid environments. Corresponding experiments to examine susceptibility to cyclic fatigue are currently being performed using indentation-precracked cantilever beams cycled in three-point bending.

INTRODUCTION

The extreme properties of synthetic, polycrystalline diamond, such as the highest values of hardness, stiffness (elastic modulus) and room-temperature thermal conductivity, together with a low coefficient of friction, make it a prospective material for many structural applications, including ultra-hard coatings on hard disks, cutting tools and bearings, and for semiconductor packaging.[1,2] Although most of these projected uses involve both sustained and repetitive mechanical or thermally-induced loading, surprisingly little is known about the deformation and fracture properties of diamond, and it is unclear at this time whether the polycrystalline material is susceptible to either stress-corrosion cracking or cyclic fatigue. The problem is particularly acute for diamond processed by chemical vapor deposition (CVD) as the material has a columnar microstructure with large differences in grain size, shape, orientation, and grain-boundary reaction layers, all features which could lead to wide variations in mechanical behavior. Accordingly, this ongoing study was instigated to examine the fracture toughness, stress-corrosion and cyclic fatigue properties of CVD diamond in order to understand the salient microstructural damage and crack-growth mechanisms which can result in premature failures.

EXPERIMENTAL PROCEDURES

Material

CVD diamond samples (~100 to 300 μm thick) were grown in a microwave plasma reactor on [100] polished silicon wafers using an excitation of 2.45 GHz in a mixture of 1% methane in

Mat. Res. Soc. Symp. Proc. Vol. 383

hydrogen at a total flow rate of 200 std. cc/min. The chamber pressure was 30-70 torr and applied microwave power was 500-1500 W. After growth, the substrate was removed using acid etchants[3] and the free-standing films were laser cut to the required geometry.

Fracture Toughness

Approximate fracture toughness, K_c, values were obtained using indentation toughness measurements on both the nucleation and growth surfaces of the CVD films using a Vickers pyramid indenter. The value of K_c was calculated from the indentation load P, the impression size $2a$, and the mean radial length (surface tip-to-tip) $2c$ of the cracks emanating from the corners of the indent (Fig. 1a):

$$H = P/(2a^2) \quad , \tag{1}$$

$$K_c = \xi (E/H)^{1/2} (P/c^{3/2}) \ , \tag{2}$$

where E is the Young's modulus, and H is the hardness.[4] The empirical constant ξ is somewhat uncertain for diamond; in general, a value of 0.016 ± 0.004 is chosen, derived from comparisons of indentation and conventional fracture-mechanics K_c measurements on several ceramics.[5] Experiments were performed at three different load levels, with the impression sizes and crack lengths measured using scanning electron microscopy (SEM) in the backscatter mode (Fig. 1b).

More accurate measurements of the fracture toughness were conducted using pre-notched disk-shaped compact-tension DC(T) specimens (Fig. 2). Procedures essentially conformed to ASTM E-399, standard,[6] only with crack lengths monitored *in situ* using electrical-potential measurements across a thin NiCr gauge sputtered on the specimen surface; full details are given in ref. [7]. Lower-bound K_c values are determined at point of crack initiation from an atomistically sharp flaw; however, as attempts to pre-crack the present diamond samples by fatigue were unsuccessful, corrections were applied to the results to allow for both the effect of the notch[8] and for subsequent crack deflection,[9] as described elsewhere.[7]

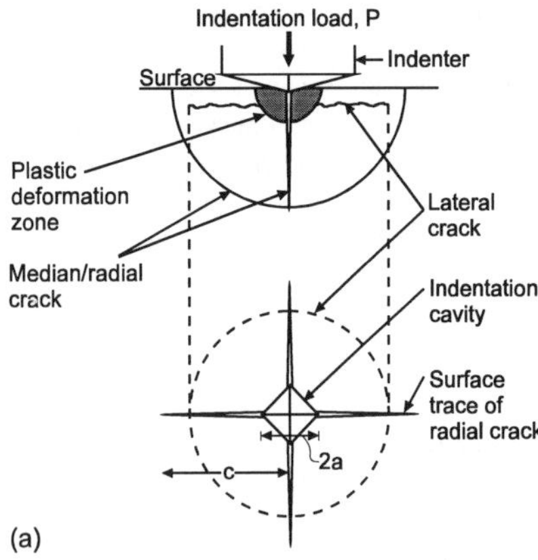

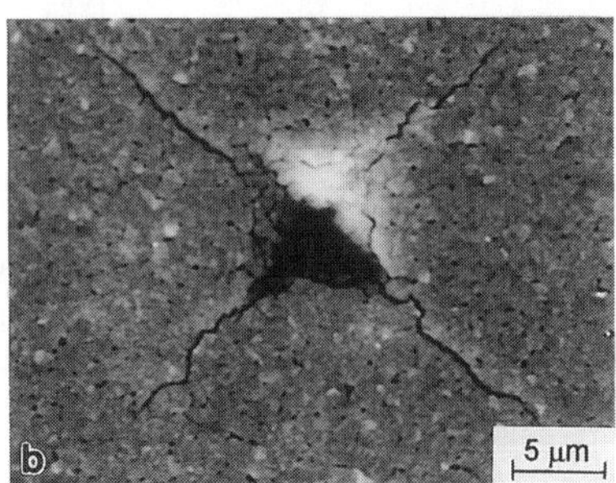

Fig. 1: a) Schematic illustration of radial and lateral cracks generated in a brittle material due to the Vickers hardness indentation, b) backscatter SEM image of surface indentation on a CVD diamond film.

Stress Corrosion

Subcritical crack growth (i.e., at stress intensities less than K_c) can be investigated under sustained loads using the indentation toughness

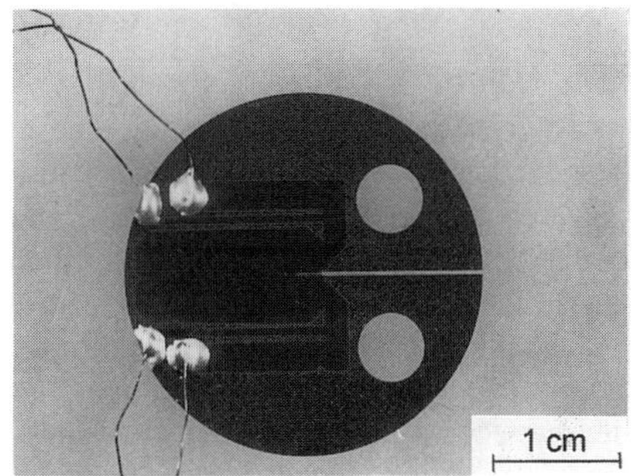

Fig. 2: Disk-shaped compact-tension DC(T) specimen used for measuring fracture toughness of CVD diamond by conventional ASTM-style fracture mechanics tests.[7]

method; since the radial-median cracks form due to the residual tensile stress field surrounding the indent, any subsequent extension over time after their initial arrest (at K_c) can be interpreted as evidence of stress-corrosion cracking.[10] The present experiments on CVD diamond were performed in oil to suppress any environmental-assisted cracking immediately after removal of the Vickers indenter. After measurement and cleaning, the indents were subsequently exposed to environments of room air (22°C, 45% relative humidity), distilled water, and acids (concentrated HCl and H_2SO_4), and monitored in a metallograph for evidence of any time-dependent extension of the surface cracks.

Cyclic Fatigue

Attempts to study fatigue-crack growth using DC(T) samples were unsuccessful as instability always preceded any stable crack initiation or extension. In addition, samples were too thin for cyclic far-field compression tests. To minimize these problems, cantilever-beams of ~100-µm thick CVD diamond, loaded in three-point bending, are now being used; here crack initiation is achieved by the presence of Vickers indentations spaced at regular intervals along the tensile surface of the beam (Fig. 3). This technique, which has been used successfully to monitor fatigue-crack growth in other brittle systems such as pyrolytic carbon,[11] has the advantage that crack extension is driven by both the far-field bending stresses and the residual tensile stresses surrounding the indent; as the latter field dominates at small crack sizes, initial crack growth proceeds into a decreasing stress field and is thus less likely to go unstable.

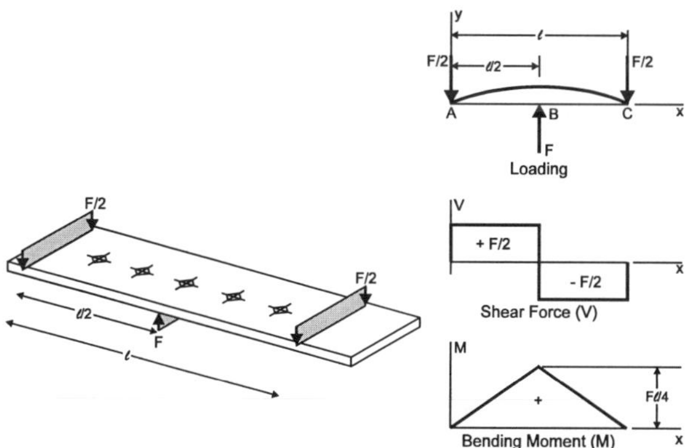

Fig. 3: Schematic illustration of the set-up for the fatigue experiment using three-point bend cantilever beam specimens. Vickers indentations are placed on the tensile (top) surface of the sample, at increasing distance from the point of maximum bending moment at the center. F is the applied load and ℓ the loading span.

RESULTS AND DISCUSSION

Material

The microstructure of the CVD diamond films, as characterized using through-thickness SEM, consisted of long columnar grains, with a cross-sectional grain size of less than 1 μm on the nucleation side of the film to over 20 μm on the growth side. The extent of non-diamond carbon was minimal, as evidenced by the absence of the ~1500 cm^{-1} peak in Raman spectroscopy spectra. More precise Raman studies also revealed a small shift in the characteristic ~1332 cm^{-1} sp^3-diamond peak (Fig. 4), indicative[12] of compressive stresses in the film of the order of 0.3 GPa.[13] As such measurements yielded comparable values on both the nucleation and growth sides, such residual stresses can be attributed to the presence of defects and grain boundaries within the film.

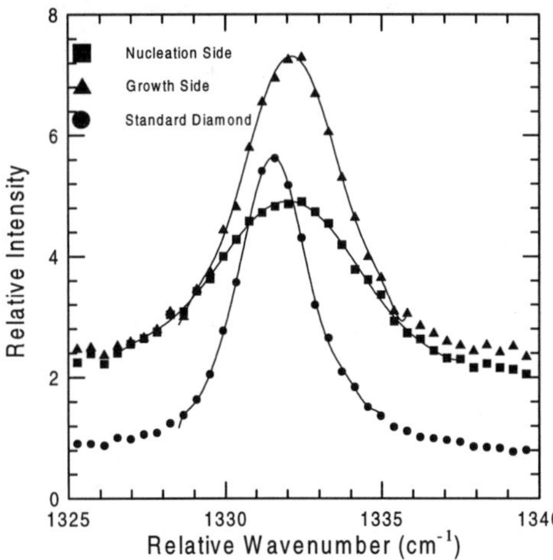

Fig. 4: Comparison of shift in Raman peaks for sp^3-diamond from the nucleation and growth sides of the sample with a standard (~1332 cm^{-1}) diamond peak.

Fracture Toughness

The fracture toughness provides the most realistic assessment of the resistance to failure of a brittle material in the presence of a pre-existing crack. Previous measurements of K_c for synthetic diamond have exclusively involved approximate methods, such as indentation[3,14] and fracture mirror[15] techniques. In the present work, attempts were made to verify such techniques, specifically the indentation method, by conducting for the first time full ASTM standard[6] toughness tests.

Results from the two fracture-mechanics type DC(T) specimens yielded K_c values of 5.3 and 7.3 MPa-$m^{1/2}$, after allowance for the effect of the radius of the notch and the extent of crack deflection.[7] Associated fractography in the SEM revealed that the fracture mechanism was primarily intergranular. Since with this geometry, the crack extends through the film thickness, these toughness values represent crack growth perpendicular to the length of the columnar grains with the full range of grain sizes sampled.

Corresponding indentation toughness measurements yielded K_c values of 4.8 to 5.5 MPa-$m^{1/2}$ on the nucleation (fine-grained) side of the film, compared to previous measurements[3] of 5.3 MPa-$m^{1/2}$ on the growth (coarse-grained) side. Crack length and impression size data for the current measurements are shown in Fig. 5 for three different indentation loads. Clearly the mean values of indentation toughness of 5.2 to 5.3 MPa-$m^{1/2}$ compare closely with the compact-tension results of 5.3 to 7.3 MPa-$m^{1/2}$.

With respect to microstructure, a variation in K_c with grain size might be expected if appreciable non-diamond carbon (which has significantly lower Young's modulus[16]) is present in the grain-boundary region, as the relative volume of this phase increases inversely with grain

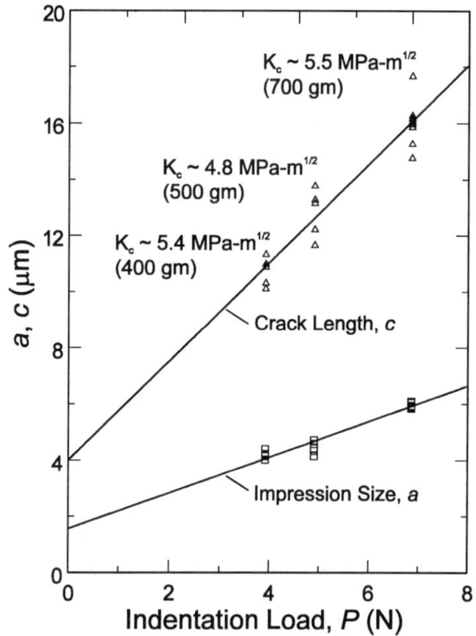

Fig. 5: Plot of radial crack length and hardness impression size due to various indentations at three different loads. The mean indentation fracture toughness for each load has been indicated.

size. However, for the present free-standing films, the toughness appears (within experimental scatter) to be independent of the grain size and orientation, consistent with the absence of the ~1500 cm^{-1} peak in the Raman spectra.

Stress Corrosion

Despite one report in the literature to the contrary,[17] the present experiments failed to detect any subcritical crack growth in the free-standing CVD diamond films under sustained (non-cyclic) loads in a variety of oxidizing environments. However, due to the inherent difficulties in conducting stress-corrosion tests using indentation techniques (e.g., it is extremely difficult to image crack growth *immediately* after removal of the indenter), further studies should be performed using, for example, precracked DC(T) samples under dead load or load relaxation conditions.

Cyclic Fatigue

Cantilever-beam experiments (Fig. 3) to assess the susceptibility of synthetic diamond to cyclic fatigue failure are currently in progress. Since in three-point bending, the stress intensity for the indent crack is decreased for indents further away from the centerline, the current approach is to examine for possible crack extension at each indent following catastrophic failure at the centerline indent (which experiences maximum bending moment). To date, stable crack extension has been seen under monotonic loading, but the generation of definitive results under cyclic loading has been frustrated by the non-uniform thickness of the films, which leads to local point loading and premature failure.

CONCLUSIONS

From a study of critical and subcritical cracking in thick (~100 to 300 μm), free-standing, CVD diamond films, the following conclusions can be made:

- Using Raman spectroscopy, residual stresses in the free-standing films were found to be compressive with a magnitude on the order of 0.3 GPa.

- Fracture toughness, K_c, values were found to lie between 5 and 7 MPa-$m^{1/2}$, based on both indentation measurements and using ASTM-style compact-tension tests. Within experimental scatter, K_c values were independent of grain size and orientation.
- No evidence of subcritical crack growth under sustained loading (stress-corrosion cracking) could be detected in a variety of oxidizing environments. No definitive result has been obtained to date on the corresponding susceptibility of polycrystalline diamond to subcritical crack growth by cyclic fatigue.

ACKNOWLEDGMENTS

This work was supported by the Director, Office of Energy Research, Office of Basic Energy Sciences, Materials Sciences Division of the U.S. Department of Energy under Contract No. DE-AC03-76SF00098. We thank Dr. Joel Ager for help with the Raman spectroscopy, and Prof. Reiner Dauskardt and Dr. Jim McNaney for assistance with the mechanical testing.

REFERENCES

1. J. E. Field, ed., *The Properties of Diamond* (Academic Press, 1979).
2. J. E. Field, ed., *The Properties of Natural and Synthetic Diamond* (Academic Press, 1992).
3. M. D. Drory, C. F. Gardinier and J. S. Speck, *J. Am. Ceram. Soc.* **74**, 3148 (1991).
4. G. R. Anstis, P. Chantikul, B. R. Lawn and D. B. Marshall, *J. Am. Ceram. Soc.* **64**, 533 (1981).
5. C. B. Ponton and R. D. Rawlings, *Mater. Sci. Tech.* **5**, 865 (1989).
6. ASTM Standard E-399 "Standard Test Method for Plane-Strain Fracture Toughness," *Annual Book of ASTM Standards* (ASTM, Philadelphia, PA, 1993), vol. 3.01, Section 3.
7. M. D. Drory, R. H. Dauskardt, A. Kant and R. O. Ritchie, *J. Appl. Phys.* (Sept. 1995), in press.
8. M. Creager and P. C. Paris, *Int. J. Fract. Mech.* **3**, 247 (1967).
9. B. Cotterell and J. R. Rice, *Int. J. Fract.* **16**, 155 (1980).
10. B. R. Lawn, *Fracture of Brittle Solids* 2nd ed., (Cambridge University Press, Camabridge, U.K., 1993).
11. R. H. Dauskardt, R. O. Ritchie, J. K. Takemoto and A. M. Brendzel, *J. Biomed. Mater. Res.* **28**, 791 (1994).
12. J. W. Ager and M. D. Drory, *Phys. Rev. B* **48**, 2601 (1993).
13. H. Boppart, J. van Straaten and I. F. Silvera, *Phys. Rev. B* **32**, 1423 (1985).
14. N. V. Novikov and S. N. Dub, *J. Hard Mater.* **2**, 3 (1991).
15. J. J. Mecholsky, *Proc. 2nd Intl. Conf. Diamond and Related Materials*, M. Yoshikawa *et al.*, eds. (1992), p. 199.
16. M. E. O'Hern, C. J. McHargue, R. E. Clausing, W. C. Oliver and R. H. Parrish, in *Materials Research Society Extended Abstracts* (Materials Research Society, Pittsburgh, PA, 1989), EA-19, p. 131.
17. M. D. Drory and C. F. Gardinier, *J. Mater. Res.* **7**, 781 (1992).

Part V

Friction and Wear

DIAMOND FILMS FOR TRIBOMECHANICAL SYSTEMS

H. J. BOVING, W. HÄNNI, M. MAILLAT, J. P. DAN AND P. NIEDERMANN
CSEM Swiss Center for Electronics and Microtechnology, Inc.
Rue Jaquet-Droz 1, 2007 Neuchâtel, Switzerland

ABSTRACT

Progress made in the field of CVD diamond has enabled to strengthen and diversify mechanical applications of these coatings. Although CVD diamond is presently mainly in the laboratory stage, it is possible to recognize the main applications in the field of tribology as follows: cutting tools, metal working tools, bearings and mechanical components. In these 4 areas diamond films can be beneficial, but there are still several problems related to nucleation, adhesion and surface roughness of the coating, to be solved.

INTRODUCTION

Diamond has considerable industrial interest because of its outstanding mechanical, thermal, optical and electrical properties:

- highest known hardness (10 times harder than steel),
- highest thermal conductivity (5 times more than copper),
- excellent optical transmittance from ultraviolet to infrared (0.25 to 6 μm), and from far infrared to radiofrequency waves,
- good electrical insulator; semiconductor properties if doped (large band gap: 5.5 eV)

This paper deals with mechanical aspects of CVD diamond. Specific characteristics of CVD diamond products are as follows:

- polycrystalline (PCD) coating (< 100 nm to several mm thick) with fine grained microstructure (< 0.1 to 10 μm or more),
- single crystals of dimensions up to several mm
- large (4 to 8 inch wafer size) parts can be coated,
- thin self sustaining membranes (thickness down to 0.5 μm; Ø up to several cm)

K. E. Spear, in his book "Synthetic Diamond", presents a comprehensive review on the chemistry, properties and applications of CVD diamond, where it is possible to find details on the several topics presented below [1] .

Pertinent properties of diamond and several CVD processes capable of producing diamond are commented. To increase the use of CVD diamond efforts are needed in: controlled nucleation density, improved adhesion and polishing of diamond films.

PERTINENT MECHANICAL AND THERMAL PROPERTIES OF DIAMOND

Differences observed in the published diamond data are due to limited sample sizes, and to impurities. Based on earlier experience with CVD and PVD coatings, it is assumed that most mechanical and thermal properties of natural, high-pressure high-temperature synthetic (HPHT), and CVD diamond are similar [2] .

With its **hardness** of 100 GPa, diamond is the hardest known material. This property provides the outstanding wear resistance which constitutes the basis for most mechanical applications.

Mat. Res. Soc. Symp. Proc. Vol. 383

The **compressive strength** (90 GPa) of diamond is much higher than its **tensile strength** (4 GPa). For ideal behavior, the diamond coating must have compressive residual stresses. Therefore, the thermal expansion coefficient of diamond ($1.10^{-6}\ K^{-1}$) must be below that of the substrate materials. When this is the case the residual stresses in the coating, after the high temperature CVD treatment, are of the compressive type. This is the situation with the substrates considered for mechanical applications: cemented carbide ($5.10^{-6}\ K^{-1}$), Si_3N_4 ($4.10^{-6}\ K^{-1}$) and tool steels ($12.10^{-6}\ K^{-1}$).

Diamond has the highest known **thermal conductivity** (1500 W/m.K). This property is extremely beneficial when diamond films are used for tool applications, where the heat generated by the tool must quickly be drained away and transferred to the substrate.

CVD DIAMOND PROCESSES (low-pressure vapor-phase synthesis)

Several CVD diamond processes have been developed, and a detailed description can be found in the literature [3] . Table I brings specific aspects of 5 CVD diamond processes.

Table I: Review of the main CVD diamond processes.

Process	Substrate Temperature (° C)	Pressure (mbar)	Gas Mixture	Growth Rate (µm/h)
High frequency microwave plasma	500-1000	30-100	H_2/CH_4	0.1-10
DC or RF plasma arc	650-1050	150-300	$H_2/CH_4(Ar)$	20-1000
Thermal filament*	700-1000	10-100	H_2/CH_4	0.1-10
RF Plasma torch	700-1000	100-1000	$Ar/H_2/CH_4$	1000
Flame	700-1100	1000	C_2H_2/O_2	200

* filament temperature 2000° C

Essentially all processes are based on the interaction between activated carbon (in the gas phase as well as on the substrate) and atomic hydrogen. Hereafter are simplified reactions which are representative of the different CVD diamond processes:

H_2 fi $H^* + H^*$
$CH_4 + H^*$ fi $CH_3^* + H_2$
$CH_3^* + H^*$ fi $CH_2^* + H_2$
$CH_2^* + H^*$ fi $CH^* + H_2$

The activated CH_x* radicals in the gas phase react with activated C* radicals of the substrate, to form diamond.

For mechanical applications the CVD diamond is essentially of the polycrystalline (PCD) type, which has the required properties. Figure 1 presents a view in a thermal filament reactor where a diamond film is deposited on a 5 inch diameter.

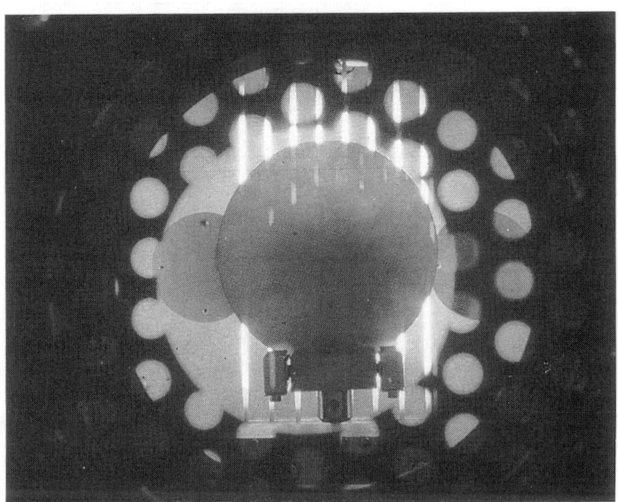

Figure 1: View of CVD diamond deposition on a Si wafer, by the thermal filament method.

The process mostly applied to produce CVD diamond on tools is DC (or RF) plasma arc, because the growth rate is high, and because many parts can be coated in one batch.

MECHANICAL APPLICATIONS OF CVD DIAMOND

The use of CVD diamond in mechanical applications is in its early stages. The market is presently extremely competitive, with the consequence that very little data are published. Potential and market mature mechanical applications of CVD diamond fall essentially into the following categories: tools and mechanical components.

1. Tools

The functional requirements for cutting, or metal working tools are basically different, therefore they are considered separately.

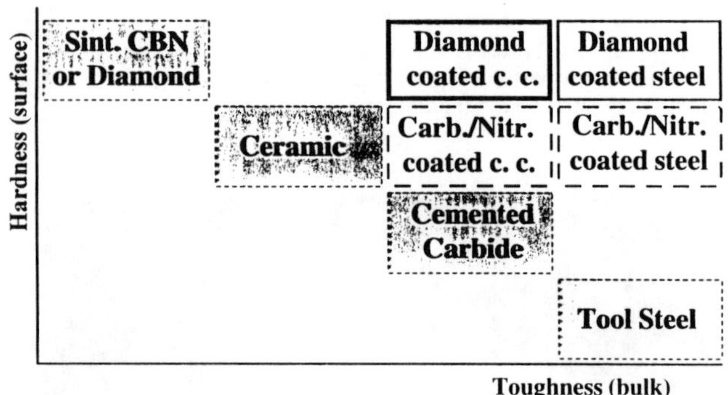

Figure 2: Hardness-toughness relationship for cutting tools.

1.1. Cutting tools

The general hardness-toughness relationship for cutting tools is shown in figure 2. Going from steel to sintered CBN or diamond, tools become more wear resistant, but less tough, and much more expensive. Benefits are obtained by combining tough substrates with hard wear resistant coatings. The first step in that direction was made when CSEM developed the CVD of TiC on cemented carbide cutting inserts; the tool lifetime increased by an order of magnitude [4] .

In cutting technology there is a rule of thumb claiming that the wear resistance of PCD is 10 times greater than that of polycrystalline cubic boron nitride (PCBN), and 100 times greater than that of cemented carbide (WC with Co binder). This means that there is a tremendous potential for diamond coatings in cutting tool applications. The coatings are essentially used in the "as deposited" condition. Two coating types are used: thin films, which are below 10 μm, and thick films which are above 10 μm.

CVD diamond for cutting tools use essentially cemented carbide substrates for technical (adhesion) and economic reasons. Diamond coating of ceramic (Si_3N_4) and tool steels is investigated as well [5,6], but faces difficulties such as nucleation and adhesion. To improve the adhesion issue metallic interlayers, with good bonding to both steel and diamond, are proposed [7].

Presently, R/D concentrates on controlled nucleation, adhesion and surface roughness. Hereafter these 3 aspects are considered separately.

Nucleation of diamond on normally ground cemented carbide is not good as can be seen from figure 3 [8]. In less than one half hour CVD TiC covers 100 % of the cemented carbide surface, whereas CVD diamond needs approximately 6 hours for the same coverage. By careful preparation of the substrate surface, however, a suitable nucleation density can be obtained.

Adhesion of CVD diamond on cemented carbide is hindered by the presence of Co on the substrate surface. Attemps to remove Co from the surface and to a depth of a few microns [9], and to decarburise the cemented carbide [10], prior to the CVD, have been made. Probably the optimum solution has not been found yet.

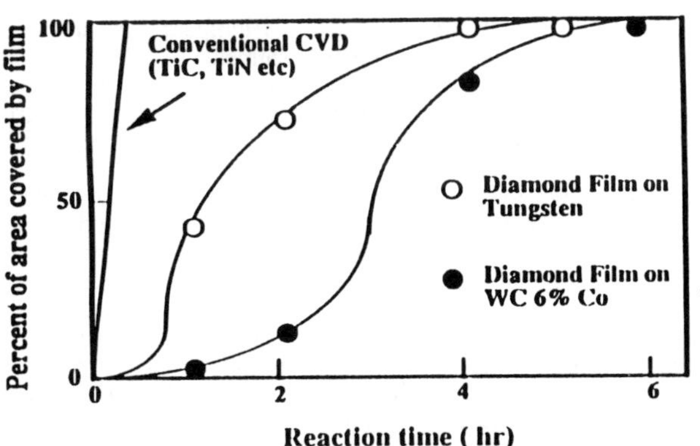

Figure 3: Area covered by CVD coatings as a function of time.

The **surface roughness** of as deposited diamond can be 10 times or more larger than that of the cemented carbide substrate. The roughness increase depends on the substrate surface structure, the coating thickness and the nucleation density. For economical reasons the coated tools must operate in the as coated condition, therefore the substrates must be as smooth as possible (10 to 50 nm) and the coating thickness should be limited to a few microns.

Diamond can be used for machining materials of very different natures. However, it is not suited for machining carbide forming metals such as Ti, V or Cr, nor for machining metals such as Mn, Ni, Fe or Co, in which carbon forms a solid solution. These chemical interactions with the workpiece accelerate tool wear. Hereafter are examples of materials that are currently machined with diamond:

- concrete
- abrasive wood/plastic composites
- green ceramics
- ceramic and glass products
- Al and Al alloys
- natural stone
- cemented carbide
- carbon fiber composites
- abrasive plastics
- Cu and Cu alloys

CVD diamond coated cemented carbide cutting tools are presently produced on an industrial basis by several companies; the US and world market for diamond and diamond-like films and coated products is presently evaluated at approximately 50 million US $ [11].

1.2. Metal forming tools

Metal working tools operate without chip removal, but most often with intensive friction between workpiece and tool. If cutting tools can operate with "as coated" surfaces, metal working tools operating under strong friction, on the contrary, must have smooth "polished" surfaces. Because of its extreme hardness and low friction, diamond is a perfect material for coating metal working tools; both cemented carbide and steel substrates can be used. The following difficulties have to be resolved:

- the "as coated" surfaces are too rough and there is need for polishing,

- there are adhesion and nucleation problems as with the cutting tools.

When steel substrates are used, the elevated CVD temperature constitutes a risk for tool distortion, and in addition there is need for heat post-treatment.

To the knowledge of the authors there are presently no widespread industrial applications of CVD diamond coated metal forming or working tools.

There are isolated cases where such tools are produced for special dedicated purposes. CSEM is presently engaged in development work on the CVD diamond coating of cemented carbide wire bonding tools used in the manufacture of integrated circuits. This tool operates under small friction conditions. For a good wear resistance these tools are normally made of cemented carbide. Interaction between the Al wires and the Co binder in the cemented carbide tool leads to adhesive wear. A CVD diamond thin film (approximately 1 μm) coating on the cemented carbide tool turns out to be extremely beneficial; the first results show that the tool lifetime increases approximately by 1 order of magnitude. This application is particularly advantageous since the coating can be used in the "as deposited" condition, if the substrate surface is of good quality. Figure 4 shows a SEM view of cemented carbide wire bonding tool head.

Figure 4: SEM view of wire bonding tool.

2. Mechanical Components

CVD diamond coated parts, as with most innovative processes, will need 5 to 10 years to create a market. At this time, 2 applications are emerging, e.g. advanced bearings and micromechanical components.

2.1. Bearings

The hardness, wear resistance and low friction of diamond make this material ideal for slide, roller or ball bearings. For these applications the required characteristics of diamond coatings are: low friction, good adhesion and extremely low surface roughness. In spite of the potential tribological qualities of the diamond coatings, there is very little published on their use in bearings, except for the work by M. Drory [12].

It is difficult to make direct comparisons between the **tribological data** of diamond found in the literature, because they are obtained, for scientific reasons, under a wide variety of conditions of sample morphology, speed, load, geometry and environment (temperature, gas, pressure). The interested reader is referred to the complete and comprehensive review by M. Gardos on the tribology of different types of diamond in [13]. CSEM performed pin on disc tribotests on CVD diamond, obtained by thermal filament method. These measurements were performed under normal laboratory environment; the air humidity was kept constant. The results of the different evaluations are presented in table II.

The following test conditions were used:

- disc: 1 μm "as coated" CVD diamond on Si wafer (roughness 50 to 100 nm)
- pins: 6 mm balls of resp. 52100 steel, Al_2O_3 and Si_3N_4
- applied load: 0.5 N
- sliding speed: 10 cm/s
- lubricant: none (dry friction)
- environment: air (relative humidity 36%), 24 °C

Table II: tribological data obtained on as deposited CVD diamond coatings.

PIN	DISC	COEFFICIENT OF FRICTION		PIN WEAR $10^{-15} m^2/N$
52100 Steel	CVD Diamond on Si	start	0.64	38
		end	0.62	
Polycryst Al_2O_3	CVD Diamond on Si	start	0.60	4.9
		end	0.19	
Polycryst Si_3N_4	CVD Diamond on Si	start	0.51	10
		end	0.32	

The initial coefficient of friction is 0.5 to 0.6, regardless the partner material. Except for the steel partner, the friction coefficient goes down to 0.1 to 0.2 after the break-in phase. These results are confirmed by data from the literature [14]. During mechanical break-in the roughness is reduced by mechanical wear until the area of contact is large enough to lower the contact stress below the fracture limit. The more elevated coefficient between steel and diamond is assumed to be due to the "chemical" reaction between the partners.

The tribological behavior of perfectly smooth diamond surfaces (5 to 10 nm) has been reported to be different [15].

The **surface morphology** of CVD diamond is of critical importance for the tribological performance of diamond coated bearings. Advanced bearing technology took advantage since several years, of CVD TiC coatings on bearing balls [16,17]. Since diamond is harder than TiC, and has a lower friction than TiC, thin diamond films on bearing balls are expected, under certain ciconstances, to perform even better. Research and development work in this direction is going on. The limitations here are the need for an extremely smooth polished surface (better than 5 nm), and the possibility of chemical reaction between diamond coated balls and steel raceways. There is presently an impressive amount of work dealing with chemical or mechanical polishing of diamond. Hereafter is a list of technologies which are being investigated, beside the normal abrasive type lapping or polishing.

- polishing by reaction with oxygen or fluorine ions or gas [18]
- laser ablation [19]
- ion beam milling [20]
- hot metal plate lapping [21]
- electrical discharge [22]
- diffusional reactions of metals with diamond [23]
- molten rare-earth metals [24]

Most of these processes apply to plane or revolution type surfaces. For bearing races and in particular for balls, the polishing of diamond coatings is a complicated and difficult task. It will take more time and development until CVD diamond coated balls will reach production level.

3. Micromechanical parts

Atomic force microscopy (AFM) is an increasingly popular, versatile and simple method to obtain topographic images of a surface up to high resolution (atomic scale). The sensing element of such a microscope is a microfabricated cantilever with a sharp tip at its end. This tip has a height of 3 - 15 μm and an apex radius of typically less than 100 nm. These tips, usually made of Si or Si_3N_4, are subject to mechanical wear. The application of a CVD diamond coating on Si tips provides an important protection against mechanical wear, resulting in improved tips for AFM, and enlarging the applicability of AFM in materials testing and related measuring techniques. CSEM has demonstrated that such tips can be coated conformally with fine geometry control. The tips are presently undergoing testing, and first results have already demonstrated high resilience of the tips under severe conditions. Figure 5 shows a thin (100 nm), conducting CVD diamond coating on a silicon tip of an AFM cantilever [25].

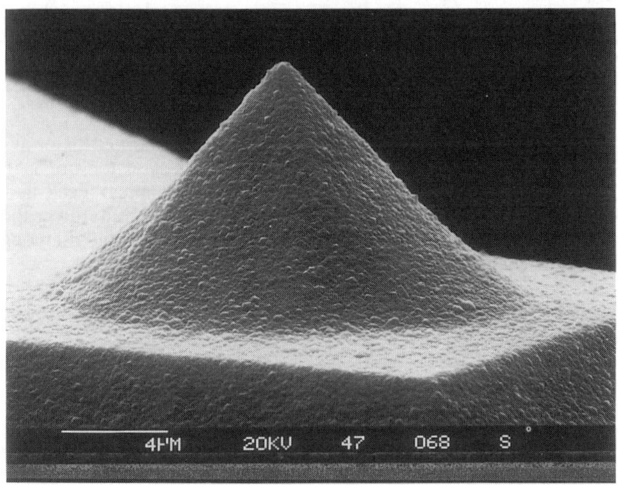

Figure 5: CVD diamond-coated Si AFM tip.

CONCLUSIONS

Despite the outstanding and promising qualities of CVD-diamond, the market penetration of coated mechanical components is slow. This is mainly due to technical difficulties which are related to: adherence - nucleation and morphology.

CSEM is active in potential thin film diamond application in precision mechanics and microsystem markets.

For further commercial development of diamond films, there is an urgent need for new mechanical applications.

REFERENCES

1. K. E. Spear and J. P. Dismukes, *Synthetic Diamond*, The Electrochemical Society, John Wiley & Sons, Inc. (1994).

2. J. E. Field, *The Properties of Diamond*, Academic Press (1979).

3. H. E. Hintermann and A. K. Chattopadhyay, Low Pressure Synthesis of Diamond Coatings, CIRP Annals (1993) 42.

4. H. E. Hintermann, H. Gass, Obenfläche - Surface, **12** (1971) 177 - 180.

5. L. Schafer, M. Sattler and C. P. Klages, *Application of Diamond Films and Related Materials*, Elsevier, New York (1991) 453-460.

6. M. Nesladek, Proc. 3rd Int. Conf. on New Diamond Science and Technology, Heidelberg, 1992, Diamond and Related Materials, **2** (1993) 357.

7. H. C. Shih, C. P. Sung, C. K. Lee, W. L. Fan and J. G. Chen, Diamond and Related Materials, **1**(1992) 605-611.

8. N. Kikuchi and H. Yoshimura, (1988), New Diamond 42.

9. M.Yagi, Proc. 1st Int. Conf. on New Diamond Science and Technology, Tokyo, 1988, p. 399.

10. K. Saijo, M. Yag, K. Shibuki and S. Takatsu, Surf. and Coatings Technology, **47** (1991) 646-653.

11. Advanced Ceramics Report, Elsevier Science Ltd., Sept. 1994, 12 - 13.

12. M. Drory, McClelland and A. Ryder, 2nd International Conference, *Applications of Diamond Films and Related Materials*, Ed. Yoshikawa, from Y. Tzeng and W. A. Yarbough, MYU, Tokyo 1993, Printed in Japan, 207-213.

13. M. N. Gardos, Tribology and Wear Behavior of Diamond, *Synthetic Diamond*, K. E. Spear and J. P. Dismukes, The Electrochemical Society, Inc., John Wiley & Sons, Inc. (1994) 419-504.

14. M. L. Languell, M. A. George, J. J. Wert and J. L. Davidson, Journal of Metals, **46**, 7 (1994) 66-70.

15. M. Kozaki, K. Higuchi, S. Noda and K. Uchida, Diamond and Related Materials, **2** (1993) 612-616.

16. H. J. Boving and H. E. Hintermann, Thin Solid Films, **153** (1987) 253-266.

17. R. M. Walker, H. J. Boving, H. E. Hintermann, R. Price and E. P. Kingsbury, Ceramic Coatings as Wear Inhibitors in Slow-Rolling Contact, Proc. Int. Conf. on Metallurgical Coatings and Thin Films, San Diego, USA, 19-23 April 1993.

18. A. B. Harker, from Y. Tzeng and al., (eds) *Applications of Diamond Films and Related Materials*, Elsevier, New York, 1991, p. 223 - 225.

19. A. Boudina, E. Fitzer, G. Wahl, H. Esrom, Diamond and Related Materials, **2** (1993) 678-682.

20. A. Hirata, H. Tokura and M. Yoshikawa, from Y. Tzeng and al., (eds) *Applications of Diamond Films and Related Materials*, Elsevier, New York, 1991, p. 227 - 232.

21. H. Tokura and M. Yoshikawa, from Y. Tzeng and al., (eds) *Applications of Diamond Films and Related Materials*, Elsevier, New York, 1991, p. 241- 248.

22. P. Silveri, Ind. Diam. Rev., **46** (1986) 108.

23. S. Jin, J. E. Graebner, T. H. Tiefel and G. W. Kammlott, Diamond and Related Materials, **2** (1993) 1038 - 1042.

24. M. McCormack, S. Jin , J. E. Graebner, T. H. Tiefel and G. W. Kammlott, Diamond and Related Materials, **3** (1994) 254-258.

25. N. Blanc, University of Neuchâtel, Switzerland. Private Communication.

MORPHOLOGICAL CONTROL OF DIAMOND THIN FILMS: ITS INFLUENCE ON FRICTION AND WEAR

ANDREW L. YEE, HOCKCHUN ONG AND R.P.H. CHANG
Department of Materials Science & Engineering
Northwestern University, Evanston, IL 60208

ABSTRACT

Microwave plasma enhanced chemical vapor deposition was used to grow diamond films with different morphologies and surface roughnesses. With the proper choice of deposition parameters (111) faceted, octahedral, flat (100) and microcrystalline diamond films were obtained. Scanning electron microscopy, atomic force microscopy and stylus profilometry were used to assess the surface topography for each type of film. Raman spectroscopy and x-ray diffraction were also used to determine the purity of the diamond phase and growth orientation of the films, respectively. Single pass friction and wear tests were conducted on each film in order to determine the effect of surface morphology on the coefficient of friction and wear of the counterface materials and/or diamond films. Counterface materials included alumina, tungsten carbide, zirconia, and the (100) face of a synthetic diamond single crystal. Results showed a decrease in the coefficient of friction as the film roughness decreased. Specific wear of the non-diamond counterface materials showed a marked decrease for the flatter and smoother diamond surfaces. For diamond on diamond, the coefficient of friction also decreased as film topography became smoother. Wear of the diamond films occurred by fracture or shearing of asperity tips which was most severe for the rougher films. Control of diamond morphology is shown to be of paramount importance in tribological applications in order to reduce abrasive wear, material transfer, and diamond film fracturing.

INTRODUCTION

Natural diamond possesses many unique material properties, some of which are ideal for tribological applications. Diamond possesses unparalleled hardness, an extremely low specific wear when sliding upon itself[1], as well as a low intrinsic coefficient of friction[2]. However, the use of diamond films grown using chemical vapor deposition (CVD) methods has been limited in the coatings industry. This is due in part to the polycrystalline nature of CVD diamond which poses serious problems especially for tribology. There is a definite need to control the surface roughness, surface topography and purity of diamond films in order for them to be viable tribological coatings. As a result, unique growth techniques have been used in order to tailor diamond morphologies to enhance their friction and wear performance.

Previous authors have studied the friction and wear of diamond thin films as a function of environment[3-4], surface roughness[5-8] and counterface material[9-10]. However, in almost all cases long term wear testing was conducted resulting in extensive wear of both the diamond film and counterface. Final friction values showed no dependence on the initial diamond roughness values[5-10]. Therefore, in this report we used single pass friction and wear testing in order to focus on the initial stages of tribological behavior of our films. During the initial stages the "true" coefficient of friction can be measured, the initial wear of the counterface or diamond film can be observed, and mechanisms of friction and wear can be determined.

EXPERIMENTAL PROCEDURE

The diamond films used in this study were synthesized using a quartz tube microwave plasma chamber described previously[11]. The substrates were (001) n–type silicon wafers (~1

Mat. Res. Soc. Symp. Proc. Vol. 383

cm^2) which were ultrasonically treated in a DIH_20: <0.25 μm diamond powder solution for 30 minutes, rinsed in acetone and methanol and blown dry with dry nitrogen. Conditions used to grow four unique types of diamond film morphologies are listed in Table 1. No post-deposition mechanical or chemical polishing were performed on any of our films.

Table I. Diamond deposition parameters.

Condition	I	II	III	IV
Methane %	0.5	0.7	0.7=>0.5	0
Freon %	0	0	0	3
Oxygen %	0.5	0	0	1
Hydrogen %	Balance	Balance	Balance	Balance
Pressure (Torr)	30	30	30	30
Power (Watts)	360	320	320=>360	360
Temp. (°C)	850	690	690=>850	850

Each of the films produced by conditions I-IV were examined using scanning electron microscopy (SEM), atomic force microscopy (AFM) and stylus profilometry. SEM provided qualitative assessment of the surface roughness and prominent topological features. AFM images were taken on a Nanoscope II with a silicon nitride cantilever tip with a tip radius of 20–50 nm. Stylus profilometry using a Tencor P–10 with a 1–2.5 μm diamond tip measured film roughness using 50 μm scan lengths . Raman spectra were taken of each film using the 488^+ nm line of an argon laser. The laser power at the sample surface was ~100 mW and scans were taken from 1000 to 1800 cm^{-1}. Finally, x–ray diffraction data was taken with a step size of 0.05° and a dwell time of 1 second.

Friction and wear tests were carried out on a dead–weight loaded friction tester. Each pass was made in single direction at a speed of 0.095 mm/s for a total of 30 seconds. A load cell transducer measured the force due to friction. Dividing the measured frictional force by the applied normal force gave the coefficient of friction. A load of 355 g or 3.47 N were used in all cases. All tests were performed in air with a relative humidity of about 50–60%. Counterface materials included zirconia (MgO stabilized), alumina, and tungsten carbide each in the shape of 0.5" diameter spheres with surface finishes <50 nm. Hertzian contact pressures were calculated to range from 0.8-1.4 GPa. For diamond on diamond testing a synthetic yellow diamond single crystal was used. The (100) face was aligned parallel to the sliding surface and slid in the <110> or so–called "hard direction"[2]. The contact pressure for this case was estimated to be ~0.3 GPa.

Wear volumes were calculated by measuring the width of the wear flat worn on the counterface material using optical microscopy. The wear volume (W_v) is given by[12]

$$W_v = \frac{\pi}{3} h^2 (3R - h) \tag{1}$$

where h is defined as the linear wear given as

$$h = R - \left(R^2 - \frac{d^2}{4}\right)^{\frac{1}{2}} \tag{2}$$

where d is the diameter of the wear flat and R is the radius of the spherical counterface.

RESULTS AND DISCUSSION

Figure 1 shows SEM micrographs of each of the four morphologies of diamond grown using conditions from Table 1. The film morphology resulting from condition I (Figure 1a) was quite rough with large-grained jagged (111) and (110) faceting dominating the topography. Figure 1b shows the octahedral morphology resulting from using condition II. This film was characterized by four sided pyramid shaped facets most of which were pointing normal to the substrate surface. Condition III, the two-step process, yielded a flat (100) tile-like morphology as shown in Figure 1c. AFM roughness measurements of the individual facets indicated atomic smoothness, but facets were inclined with respect to the horizontal (~5-10°) thereby increasing the overall roughness of the film. Finally, the freon grown film exhibited a ball-like but smooth fine-grained surface. Stylus profilometry measured roughnesses for each film to be ~200, ~100, ~35 and ~46 nm respectively. Raman spectroscopy revealed the characteristic peak for diamond centered at ~1332 cm^{-1}. The full width half maximum was on the order of 6 cm^{-1} for the rough film, 8-10 cm^{-1} for the octahedral and flat diamond and 12-14 cm^{-1} for the freon grown film. Broad peaks around 1550 cm^{-1} due to amorphous or graphitic carbon were also observed and were most prominent for the microcrystalline films.

X-ray diffraction determined whether each film had any preferred growth orientation. By comparing the intensity ratios of the low index planes of the deposited films versus that of the JCPDS 6-675 card file for a randomly oriented diamond standard, a qualitative assessment on film fiber texture can be made. The data revealed that the rough films are strongly (111) growth oriented, the flat and octahedral films showed a pronounced (100) texture, while the microcrystalline films showed no preferential growth orientation.

Figure 1. Diamond film morphologies resulting from conditions listed in Table I. a) rough (condition I), b) octahedral (condition II), c) flat (condition III) and d) microcrystalline (condition IV).

The coefficient of friction of the non-diamond counterface materials versus film morphology is shown in Figure 2. Friction on the (100) face of natural IIb diamond is given for reference.

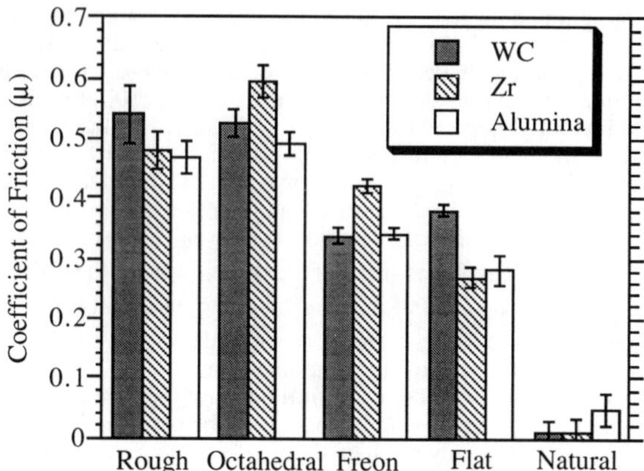

Figure 2. Friction of spherical counterfaces versus diamond morphology.

There were two coefficient of friction regimes evident. There was a high friction regime for the rough and octahedral faceted films ranging from 0.48-0.6 and low friction regime for the flat and microcrystalline films ranging from 0.28-0.44. It is interesting to note how the coefficient of friction was so high for the octahedral films even though its measured roughness value was half that of the rough films. This is due to the specific geometry of the facets (i.e. asperities) present on the octahedral films. In a model proposed by Tabor[2], the observed friction is a function of the true friction and the slope of the asperities (θ) given as:

$$\mu_{avg} = \frac{\mu_t (1 + \tan^2 \theta)}{1 - \mu_t^2 \tan^2 \theta} \tag{3}$$

The slope of the asperities for the octahedral faceted film consistently ranged between 50-60° from the horizontal as determined by AFM. Therefore, the unique surface topography of this film results in a large contribution to the overall observed coefficient of friction as predicted by (3).

Examination of the wear flats worn on the spherical counterfaces yielded information on the specific wear volume (W_V/Nm) for each ball material versus diamond film morphology. The results are shown in Figure 3. Each of the counterface materials followed the same trend in that there was a decrease in specific wear as the morphology of the film went from rough to octahedral to flat to microcrystalline. Examination of the wear scars left on the diamond films after single pass testing revealed the mostly likely mechanisms of wear. Abrasion due to the sharp asperity tips of the rough and octahedral films was responsible for wear of the spherical counterfaces. Large amounts of material transfer occurred for the rough and octahedral films. This material transfer will eventually contribute to the mechanism of adhesive friction and wear for multiple pass testing. Wear scars on the flat diamond showed that abrasive wear occurred at the edges of the tilted (100) facets as the pile up of wear debris along the leading edges of the facets suggested. Even the microcrystalline film exhibited abrasive wear mechanisms as indicated by the material transfer located at the high points on the film surface.

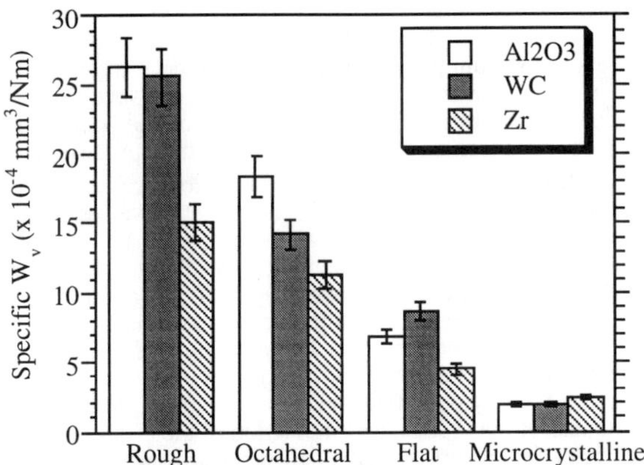

Figure 3. Specific wear volume versus counterface material versus diamond film morphology.

Results from the diamond stylus on diamond film tests are shown in Figure 4. As was seen previously the coefficient of friction had a high (0.3-0.4) regime and a low (~0.25) regime corresponding to the rough/octahedral films and the flat/freon films, respectively. Also noteworthy was the fact that the noise in the data decreased as the films became smoother. Examination of the diamond films after testing showed that noise in the friction measurements was possibly related to fracturing events occurring on the film. The rough and octahedral films showed significant amounts of shearing or fracturing of the tips of sharp protruding facets. Even though the overall contact pressure was ~0.3 GPa, local contact pressures at asperity tips exceeded the fracture strength of diamond resulting in facet shearing. There were some isolated cases of fracturing of the corners of some of the tilted (100) facets in the flat films. No direct evidence of fracturing was observed in the freon films due to limitations in the SEM resolution.

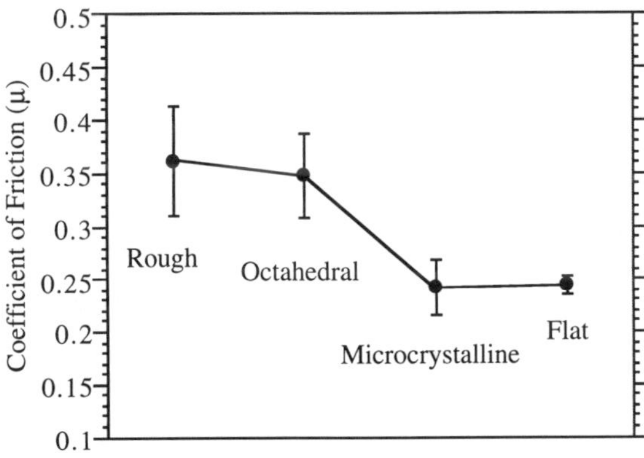

Figure 4. Coefficient of friction of diamond on diamond versus film morphology.

CONCLUSIONS

Diamond films each with unique topographies, surface roughnesses and diamond purities have been grown on silicon substrates by controllng the specific carbon precursor percentage and deposition temperature. Flat tile-like morphologies can be achieved using condition III, a two step deposition process. Details of this process will be reported elsewehere[13]. Single pass friction and wear testing is shown to be effective in evaluating the tribological performance of each type of film. Coefficient of friction values decreased from high values for the rough and octahedral films to lower values for the flat and microcrystalline films using both diamond and non-diamond counterface materials. Wear volumes were drastically decreased for tests on flat and microcrystalline morphologies. Testing with harder non-diamond counterface materials such as silicon nitride and silicon carbide would be of interest. Abrasion at asperity tips is responsible for wear of the non-diamond counterfaces while asperity or tip fracturing is the dominant wear mechanism for diamond on diamond testing. With proper control of diamond film morphology potentially viable tribological coatings can be achieved. Due to the atomic smoothness of the (100) facets, reduction of the (100) facet tilting will be of special interest due to its potential for superior tribological performance.

ACKNOWLEDGMENTS

We gratefully acknowledge Dr. Z. Yang and Prof. Y. W. Chung for use of the friction and wear tester. We acknowledge A. A. Setlur for his assistance in the diamond depositions and Dr. K. Grannen of Seagate for providing the freon grown films. This work was supported by the Basic Energy Sciences Division of the Department of Energy. This research utilized MRL Central Facilities supported by the National Science Foundation at the Materials Research Center of Northwestern University.

REFERENCES

1. M. Seal, Ind. Diam. Rev. **25**, 111 (1965).
2. D. Tabor, in The Properties of Diamond, edited by J. E. Field (Academic Press, London, 1979) pp. 325-350.
3. M.N. Gardos and B.L. Soriano, J. Mater. Res. **5**, 2599 (1990).
4. S.S. Perry, J.W. Ager, III and G.A. Somorjai, J. Mater. Res. **8**, 2577 (1993).
5. M. Kohzaki, K. Higuchi, S. Noda and K. Uchida, J. Mater. Res. **7**, 1769 (1992).
6. K. Miyoshi, R.L.C. Wu, A. Barscadden, P.N. Barnes, and H.E. Jackson, J. Appl. Phys. **74**, 4446 (1993).
7. B.K. Gupta, A. Malshe, B. Bhushan and V.V. Subramaniam, J. of Tribology **116**, 445 (1994).
8. C. Lai, Y.C. Wang, P. Lu, J.B. Wachtman Jr. and G.H. Siegel Jr, Mater. Sci. Eng. **A183**, 257 (1994).
9. S. Jahanmir, D. E. Deckman, L.K. Ives, A. Feldman and E. Farabaugh, Wear **133** 73 (1989).
10. A.K. Gangopadhyay and M.A. Tamor, Wear **169**, 221 (1993).
11. R. Meilunas, M.S. Wong, T.P. Ong, and R.P.H. Chang, in Laser and Particle-Beam Modification of Chemical Processes on Surfaces, edited by A.W. Johnson, G.L. Loper, and T.W. Sigmon (Mater. Res. Soc. Symp. Proc. **129**, Pittsburgh, PA, 1989), p. 533.
12. S.J. Bull, P.R. Chalker, C. Johnston, V. Moore, Surf. Coat. Technol. **68/69**, 603 (1994).
13. A.L. Yee, H.C. Ong, R.P.H. Chang to be published, in Fourth International Symposium on Diamond Materials at the 187th Meeting of the Electrochemical Society, Reno, NV, May 21-26, 1995.

FRICTIONAL BEHAVIOR OF C_{60} MICROPARTICLE-COATED STEEL

WEI ZHAO, JINKE TANG, ASHOK PURI, ALEXANDER U. FALSTER* AND WILLIAM B. SIMMONS, JR.*
Department of Physics, and *Department of Geology and Geophysics, University of New Orleans, New Orleans, LA 70148, USA

ABSTRACT

The frictional behaviors of 304 stainless steel disks coated with C_{60} microparticles, both containing benzene and free of benzene, have been studied under different loads and sliding speeds with a pin-on-disk configuration in ambient air atmosphere at room temperature. The results indicated that the coating containing benzene, benzene-solvated C_{60} microparticles ($C_{60}{\cdot}4C_6H_6$), reduced friction as well as wear. The coated samples showed a 50-70% reduction in friction coefficient in comparison to uncoated samples. Neither the coated nor the uncoated sample showed significant change in friction coefficient for different sliding speeds. Under different loads, the uncoated sample had almost the same friction coefficient. However, with the increase of load, the friction coefficient of $C_{60}{\cdot}4C_6H_6$-coated disk showed a minimum value of 0.25 at 25 g load and then reached the uncoated values beyond 50 g load. The coefficient of friction of the disk coated with benzene-free C_{60} showed a slight increase with load, reaching the value of uncoated 304 stainless steel disk at about 40 g. The reduced friction of the solvated-C_{60} coated 304 stainless steel is probably due to the lowered shear strength of the hcp structure of $C_{60}{\cdot}4C_6H_6$ molecular crystal in which the benzene molecules are intercalated. The results of this study suggest the importance of the presence of second component, in addition to C_{60}, in the coating materials in order for them to form a preferred crystal structure with low shear strength as far as using C_{60} as a solid lubricant is concerned.

INTRODUCTION

It is most interesting that soccer ball-like fullerene C_{60} molecule was speculated to be a solid lubricant because of its unique spherical shape, weak intermolecular bonding, high chemical stability, low surface energy and high load bearing capacity. Pristine C_{60} solid is a molecular crystal which is of well-known face-centered-cubic (fcc) structure at room temperature [1]. It can readily sublimate at temperatures about 450°C in a vacuum and dissolves in a wide range of nonpolar solvents such as toluene and benzene, which allows C_{60} molecules to easily attach to the surface of a substrate in the form of films. To date, the frictional behaviors of C_{60} were only studied by a few groups whose results have showed some contradictions. Blau and Haberlin [2] reported a high friction coefficient (~0.6) of C_{60} powder with 90% purity (10% C_{70}). Bhattacharya et al.[3] obtained similar results for sublimed C_{60} film on Si_3N_4 substrate. In contrast, Bhushan and Gupta et al.[4,5] found a low friction coefficient (~0.12) of sublimed C_{60} film on Si. We noticed that in their experiments, the measurement conditions such as load and sliding speed were not the same, which may influence their results. In addition, in Blau and Haberlin's work [2], the coating of C_{60} particles on aluminum through evaporation of a drop of C_{60}-toluene solution showed a low friction coefficient characteristic of the substrates which was explained to be due to the discontinuous deposition of microcrystals on the aluminum. To our knowledge, the evaporation of C_{60} solution results in solvated C_{60} microcrystals which consist

Mat. Res. Soc. Symp. Proc. Vol. 383 © 1995 Materials Research Society

of C_{60} and solvent molecules. The solvated C_{60} microcrystals have structures different from that of pristine C_{60} crystal [6-9]. Therefore, the solvated C_{60} crystals may show different frictional behaviors which are worthwhile to be explored further. In this paper, we have studied the frictional behavior of 304 stainless steel (304SS) disks coated with C_{60} microparticles, both containing benzene and free of benzene, under different loads and linear speeds with a pin-on-disk configuration in ambient air atmosphere at room temperature. A reduced friction coefficient by 70% to 0.25 at 25 g load was observed for $C_{60}{\cdot}4C_6H_6$ coated 304SS. The mechanism of low friction coefficient of $C_{60}{\cdot}4C_6H_6$ coating and the role of benzene in C_{60} are discussed.

EXPERIMENTAL

The 99.9% pure C_{60} powder was from the Bucky USA Co. (Houston, TX). The coating of benzene containing C_{60} microparticles on 304SS disks was by dropping saturated solution of C_{60} in benzene on disks and subsequent evaporation of benzene. The composition of the evaporated microparticles has been reported to be $C_{60}{\cdot}4C_6H_6$ [6]. The C_{60} distribution on disks was about 95 $\mu g/cm^2$. At room temperature $C_{60}{\cdot}4C_6H_6$ is stable, and benzene can not be driven off even in vacuum. DSC measurements showed an endotherm beyond 200°C. Some of the disks coated with $C_{60}{\cdot}4C_6H_6$ microparticles were later heat-treated at 200 to 300°C in vacuum about 5 mins in order to drive off the residual solvent benzene.

The IR spectra of C_{60} were measured with a Perkin-Elmer 1760 Infrared Fourier Transform Spectrometer. The data were recorded by a Perkin-Elmer 7700 Professional Computer. A NaCl crystal was used as a substrate.

A scanning electron microscope (SEM) and an optical microscope were used to obtain topographic micrographs of C_{60} microparticles and the wear tracks, respectively.

The tribological measurements were carried out by using an ISC-200PC pin-on-disk tribometer (Implant Sciences Co.) which allows the tests with different loads and sliding speeds. The disk specimens of 304SS, 1.5x1.5 cm large by 0.5 cm thick, were mechanically polished to a surface roughness of less than 0.3 μm. The 3.2 mm diameter spheres of Al_2O_3 were used as pins. Pins were not coated and were highly polished. Applied loads ranged from 5 to 100 g. Friction coefficients were measured continuously, but only the steady state values reached near the end of 1000 cycles are plotted in the figures for comparison under different loads and sliding speeds. All data were collected by a PC computer.

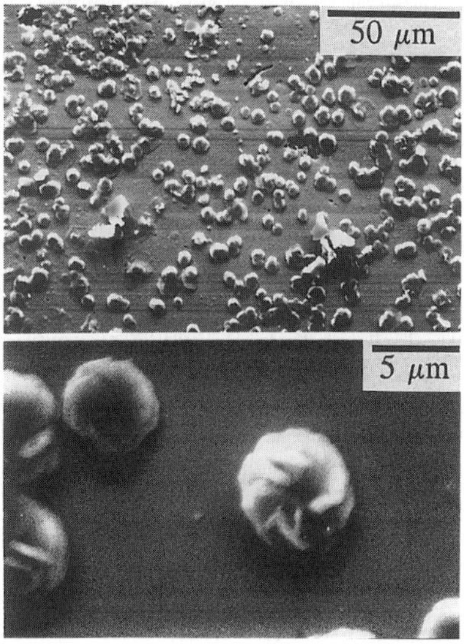

Figure 1. SEM micrographs of $C_{60}{\cdot}4C_6H_6$ microparticles on 304 stainless steel under different magnifications.

RESULTS AND DISCUSSION

SEM micrographs of $C_{60}{\cdot}4C_6H_6$ microparticles on 304SS are shown in Fig. 1. Most of the particles were in the shape of a ball composed of petal-like microtwins. Their average size was about 5 μm. A few particles were flake-like crystals. Fig. 2 shows the IR spectra of the microparticles before and after heat treating. There are IR lines of benzene at 671, 1034, 1460, 1477, 2847 and 2915 cm^{-1} in the as-prepared microparticles besides the four IR lines of pristine solid C_{60} at 526, 576, 1182, 1428 cm^{-1} [10], in agreement with the idea that as-prepared C_{60} microparticles were in the form of $C_{60} \cdot 4C_6H_6$ [6] which is of well-known hexagonal close-packed (hcp) structure [10]. After heating at 200°C in vacuum for 5 mins., the sample only showed the four IR lines of pristine C_{60} (the IR lines of benzene had disappeared) which indicated that the residual solvent benzene was driven off and the $C_{60} \cdot 4C_6H_6$ microparticles turned into the pristine C_{60} microparticles whose structure is fcc [1].

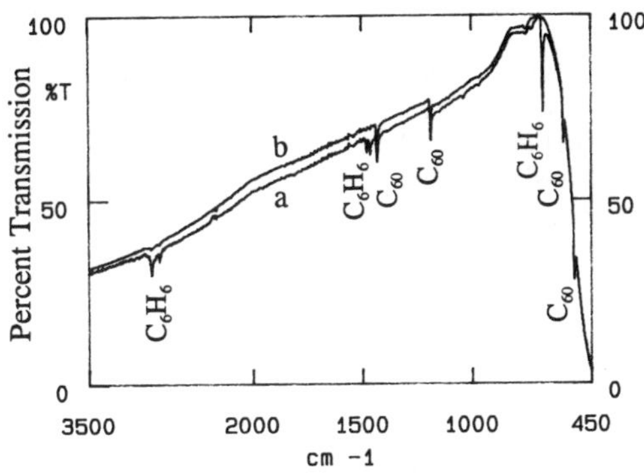

Figure 2. Infrared spectra of as-prepared microparticles before (a) and after heat treating at 200°C in vacuum (b).

Figure 3 shows the friction coefficients of the 304SS disks uncoated and coated with $C_{60} \cdot 4C_6H_6$ versus sliding distances (or wear cycles) as well as corresponding wear tracks.

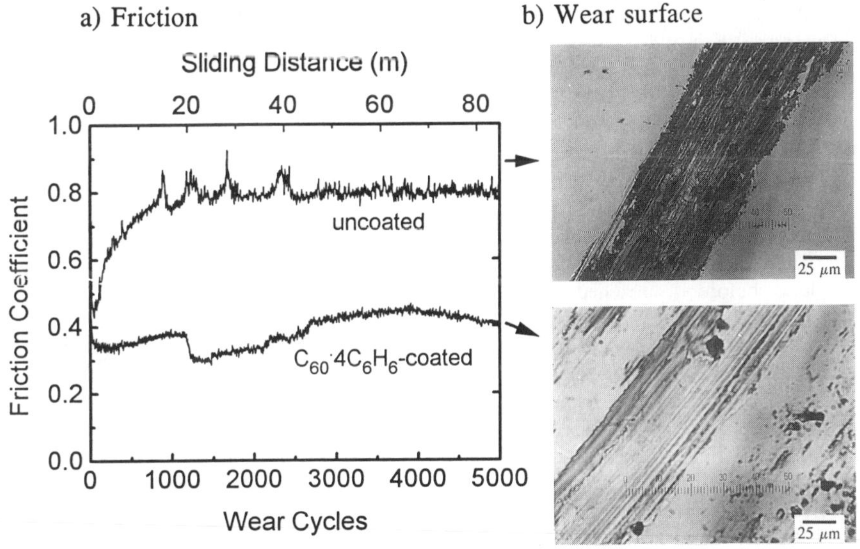

Figure 3. a) Friction traces from 304 stainless steel disks uncoated and coated with $C_{60} \cdot 4C_6H_6$ microparticles under load 15 g and sliding speed 2 cm/s. b) Optical micrographs of the wear tracks.

The loads used were 15 g. For the first several hundred cycles, the friction coefficient of uncoated 304SS increased, and it reached the maximum (~0.8) near 1000 cycles. The friction coefficient of $C_{60}\cdot 4C_6H_6$-coated 304SS kept almost the same value (~0.4) before 1160 cycles. After 1160 cycles its friction coefficient was reduced to about 0.3 and then gradually went back to about 0.4 during the following cycles. The optical micrographs of the wear tracks showed that much less wear occurred on the $C_{60}\cdot 4C_6H_6$-coated 304SS than the uncoated 304SS and these reductions were accompanied by a change from an adhesive wear mode, in which material was torn out of the wear track, to an abrasive wear mode for which only sliding grooves were seen. The above results indicated that the coating of $C_{60}\cdot 4C_6H_6$ on 304SS reduced friction as well as wear. The friction coefficients (~0.4) were reduced by ~50% in comparison to uncoated values (~0.8).

The load and sliding speed dependences of the friction coefficients of 304SS disks uncoated, coated with $C_{60}\cdot 4C_6H_6$ and heat treated are shown in Fig. 4. All samples showed little change with linear speed. Under different loads, the friction coefficient of uncoated 304SS disk almost the same (~0.75). That of the heat treated sample, coated with solvent-free C_{60}, increased with increasing load and reached the value of uncoated 304SS disk at about 40 g. At 5 g load, the friction coefficient was about 0.56, consistent with the previously reported results [2,3] and the value of our sublimed C_{60} film on Si [11]. For the $C_{60}\cdot 4C_6H_6$-coated 304SS disk, its friction coefficient was about 0.44 at 5 g load. Upon increasing loads, the friction coefficients of $C_{60}\cdot 4C_6H_6$-coated disks showed a minimum value of 0.25 at 25 g load before it reached the uncoated values beyond 50 g load. The low coefficient of friction observed at lighter loads represents a reduction of 50-70% in comparison to the uncoated value due to the presence of $C_{60}\cdot 4C_6H_6$ coating. Similar results were also observed on 304SS coated with solvated C_{70} microparticles [11].

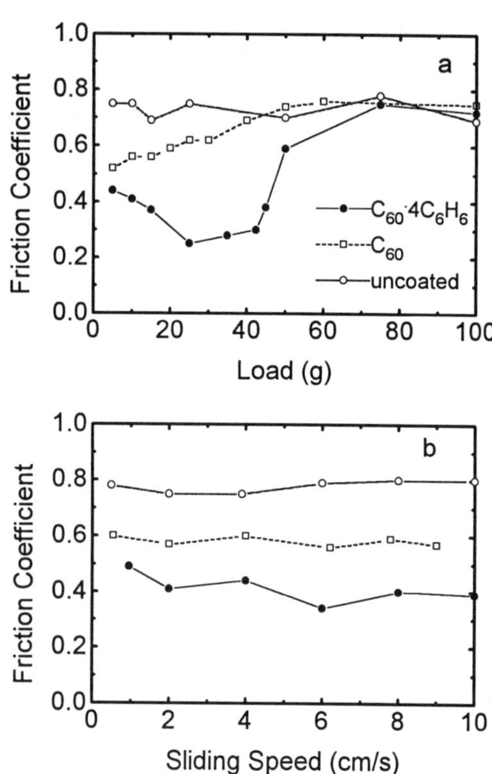

Figure 4. Friction coefficient at the end of 1000 cycle pin-on-disk test of 304 stainless steel disks uncoated, coated with $C_{60}\cdot 4C_6H_6$ and heat treated (C_{60}), plotted as functions of load (a) (2 cm/s sliding speed) and sliding speed (b) (10 g load).

The relatively high friction coefficient of benzene-free C_{60} particles layer was thought to be due to the tendency of the C_{60} particles to clump and compress into a high shear strength layer which is hard to deform [2]. In our case, the high shear strength layer was formed on the pin surface which separated the pin and disk, as observed by SEM [11]. Instead, in the case of solvated-C_{60} microparticles layer, its reduced friction may be related to its preferred structure which prevents the formation of the high shear strength layer.

The $C_{60}{\cdot}4C_6H_6$ crystal can be visualized in terms of a hexagonal close-packed arrangement of C_{60} molecules with the hexagonal axis along *a*. The C_{60} spheres within a close-packed layer are then moved well apart and the benzene molecules are inserted, and the close-packed layers moved towards each other along the *a* axis until adjacent molecules (along the *a* axis) touch. Three of the four benzene rings lie parallel to the C_{60} molecular surface, and the other appears to fill an interstice between the other molecules [6]. Therefore the intercalation of the benzene molecules into the close-packed layers may produce a low shear strength for $C_{60}{\cdot}4C_6H_6$, which makes C_{60} molecules act like tiny ball bearings working at the interface. Further investigation is under way.

CONCLUSIONS

The above results indicated that the coating of $C_{60}{\cdot}4C_6H_6$ on 304SS reduced both friction and wear. The friction coefficients ($\sim$0.4 at 15 g load and 0.25 at 25 g load) were reduced by $\sim$50-70% as compared with uncoated values ($\sim$0.8). The coated and uncoated samples showed less change upon different linear speeds. For the load dependence, the uncoated 304SS disk showed almost the same friction coefficient with different loads. As a fuction of applied load, the friction coefficient of $C_{60}{\cdot}4C_6H_6$-coated disk decreased, reaching a minimum of 0.25 at 25 g load and then showed an increase in value, reaching the uncoated values beyond 50 g load. The disk coated with benzene-free C_{60} showed a coefficient of friction of about 0.56 at 5 g load and then increased with increasing load, reaching the value of uncoated 304SS disk at about 40 g. The high friction coefficient of benzene-free C_{60} particles layer was due to a layer of high shear strength C_{60} formed on the pin surface which had a direct contact with the disk. In the case of solvated-C_{60} microparticle layer, the benzene molecules were intercalated into the C_{60} lattice, which may play the role of a molecular lubricant among C_{60} ball molecules, and the C_{60} molecules acted like tiny ball bearings at the interface.

Our present study on the tribological properties of C_{60} suggests that, as far as using C_{60} as a solid lubricant is concerned, it may be necessary to have a second component, besides C_{60}, present in the lubricant in order for them to form a preferred structure with low shear strength. Because C_{60} can dissolve in a wide range of solvents and form various solvated molecular crystals, further experiments are needed in order to find an excellent solid lubricant involving fullerenes.

ACKNOWLEDGEMENT

The authors wish to thank the grant supports from DOE, Louisiana Education Quality Support Fund and Research Corporation.

REFERENCES

1. P.A. Heiney, J.E. Fischer, A.R. McGhie, W.J. Romanow, A.M. Denenstein, J.P. McCauley, and A.B. Smith III, Phys. Rev. Lett. **66**, 2911 (1991).
2. Peter J. Blau and Christian E. Haberlin, Thin Solid Films **219**, 129 (1992).

3. R.S. Bhattacharya, A.K. Rai, J.S. Zabinski, N.T. McDevitt, J. Mater. Res. **9**, 1615 (1994).
4. B. Bhushan, B.K. Gupta, G.W. Van Cleef, C. Capp, and J.V. Coe, Appl. Phys. Lett. **62**, 3253 (1993).
5. B.K. Gupta, B. Bhushan, C. Capp, and J.V. Coe, J. Mater. Res. **9**, 2823 (1994).
6. M.F. Meidine, P.B. Hitchcock, H.W. Kroto, R. Taylor and D.R.M. Walton, J. Chem. Soc., CHEM. COMMUN. 1992, 1534.
7. R.M. Fleming, A.R. Kortan, B. Hessen, T. Siegrist, F.A. Thiel, P. Marsh, R.C. Haddon, R. Tycko, G. Dabbagh, M.L. Kaplan, and A.M. Mujsce, Phys. Rev. B, **44**, 888 (1991).
8. S.M. Gorun, K.M. Creegan, R.D. Sherwood, D.M. Cox, V.W. Day, C.S. Day, R.M. Upton and C.E. Briant, J. Chem. Soc., CHEM. COMMUN. 1991, 1556.
9. X.D. Shi, A.R. Kortan, J.M. Williams, A.M. Kini, B.M. Savall, and P.M. Chaikin, Phys. Rev. Lett. **68**, 827 (1992).
10. W. Krätschmer, L.D. Lamb, K. Fostiropoulos, and D.R. Huffman, Nature (London) **347**, 354 (1990).
11. W. Zhao, J. Tang, A. Puri, Y.X. Li and L.Q. Chen, presented at the 187th ECS Spring Meeting, Reno, Nevada, May 21-26, 1995.

FRICTION AND WEAR OF AMORPHOUS HYDROGENATED CARBON

S.L. Heidger*
National Aeronautics and Space Administration, Lewis Research Center, Cleveland Ohio 44135

ABSTRACT

Uniform amorphous hydrogenated carbon (a-C:H) films with surface roughnesses ranging between 1 nm and 4 nm were produced by radio frequency self biased plasma enhanced chemical vapor deposition (rf PECVD) on <111> Silicon substrates using 100% methane precursor gas mixture, rf power densities ranging between 0.11 W/cm^2 and 1.07 W/cm^2, and pressures ranging between 0.67 Pa and 40 Pa. Reciprocating sliding friction experiments were conducted on the a-C:H films with hemispherical, silicon nitride pins in dry nitrogen and in 60% relative humidity. The coefficients of friction and the wear rates of the a-C:H were very low in dry nitrogen, ranging from 0.03 to 0.05, and from 1.1 x 10^{-8} mm^3/Nm to 2.3 x 10^{-6} mm^3/Nm, respectively. In 60% relative humidity, the initial coefficients of friction were approximately 0.30. However, the steady state coefficients of friction of the a-C:H films ranged from 0.10 and 0.30, depending on the deposition conditions. The wear rates ranged from 2.0 x 10^{-9} mm^3/Nm to 8.9 x 10^{-8} mm^3/Nm in 60% relative humidity. Raman microprobe spectroscopy and Auger electron spectroscopy (AES) revealed that sliding friction was transforming the a-C:H films into a material primarily composed of sp^2 bonded carbon with increasing short range order. Qualitatively, the amount of wear which occurred corresponded to the extent that the structural changes progressed. The a-C:H films were further characterized by scanning electron microscopy (SEM) and surface profilometry.

INTRODUCTION

Hard, amorphous hydrogenated carbon (a-C:H) can be produced with properties such as very low static and dynamic friction in a variety of environments, abrasion resistance, and chemical inertness.[1-3,8,23] However, the properties of a-C:H are strongly dependent on the deposition conditions. By manipulation of the deposition parameters, a-C:H films can be produced with a broad range of properties ranging from soft and polymer-like[1,4] to very hard and "diamondlike"[1]. Because they can engineered to possess excellent tribological properties, a-C:H films are attractive materials for applications such as overcoats for magnetic recording media[5,6], and as dry lubricants for engine parts and mechanical components like bearings, shafts and sleeves[1-6,8,9,16,23].

Raman and Auger electron spectroscopy (AES) are commonly used to characterize the structure of carbon materials, including a-C:H[7-11]. The first order Raman spectra of diamond and graphite each display a sharp peak representing their Brillouin zone center modes, T_{2g} at 1332 cm^{-1} and E_{2g} at 1580 cm^{-1}, respectively. An additional peak, denoted the D (disorder) band, arises at 1350 cm^{-1} in the first order Raman spectrum of polycrystalline graphite and disordered sp^2 bonded carbon materials, resulting from the loss of translational symmetry and relaxation of the $k = 0$ selection rules.[12] In disordered carbon materials, the frequencies of the D(disorder) and G(graphitic) bands, as well as the ratios of their intensities and widths have been

*National Research Council—NASA Research Associate at Lewis Research Center.

Mat. Res. Soc. Symp. Proc. Vol. 383 © 1995 Materials Research Society

qualitatively related to the degree of structural order, and to a lesser extent, the amount of sp^3 and sp^2 bonded carbon present[1,7,9,16] because the intrinsic intensity of the sp^2 bonded carbon is at least 50 times that of the sp^3 bonded carbon[9].

Differences in the fine structure of the carbon KLL Auger peaks of diamond and graphite have been interpreted as resulting from the difference of their valance density of states, and hence, are related to the amount of sp^3 and sp^2 bonded carbon present.[19,20] In figure 1, the fine structure of the carbon KLL Auger peak of diamond, graphite and amorphous carbon as reported by Lurie and Wilson[17] is shown. The fine structure of the carbon KLL peak consists of two peaks. The peak closest to the primary carbon KLL peak (P1) is attributed to a KV_1V_1 Auger transition while the peak occurring at a slightly lower energy (P2) is attributed to a KV_2V_2 Auger transition.[17,18] Qualitatively, the amount of sp^3 and sp^2 bonded carbon present on the surfaces of the a-C:H have been inferred from the fine structure of the carbon KLL Auger peak[21].

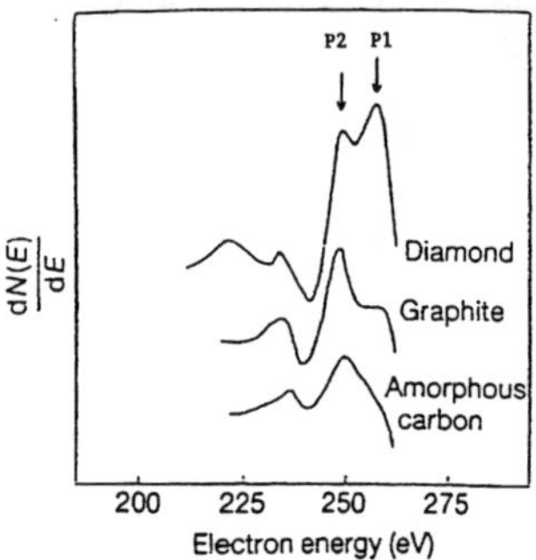

Figure 1.— Carbon KLL Auger fine structure of diamond, graphite and amorphous carbon from Lurie and Wilson[17].

In this investigation, a-C:H films were produced on silicon <111> substrates by rf PECVD using 100% methane as the precursor gas and varying the rf power density and deposition pressure. These films were examined to determine the range of parameters in which our deposition system can produce adherent a-C:H films with optimal tribological properties in dry nitrogen and humid air environments. In addition, the chemical and structural changes to the surfaces of the a-C:H films which occur during sliding contact in dry nitrogen and humid air were studied using Raman microprobe spectroscopy and scanning AES.

EXPERIMENT

Nine a-C:H films were deposited on <111> silicon by rf PECVD using a rf generator operating at a frequency of 13.56 MHz inductively coupled to a 15.24 cm diameter (powered) electrode which also served as the substrate mount. The deposition conditions for the a-C:H films are summarized in Table 1.

Several analytical techniques were used to characterize the a-C:H films including: SEM to determine the surface morphology; surface profilometry to determine the surface roughness; Raman microprobe spectroscopy and AES to characterize the structure and surface chemistry of the as-deposited, wear tracks and wear debris of the a-C:H films.

Reciprocating sliding friction experiments were conducted on the a-C:H films in dry nitrogen and in humid air environments. For a detailed description of the reciprocating friction testing apparatus see Miyoshi[2]. Friction experiments were conducted with the a-C:H films in contact

Table 1 Deposition conditions for a-C:H films.

Technique: Rf PECVD — Substrate: Silicon (111)

Precursor Gas: 100% Methane — Flow rate: 5 to 40 sccm

SAMPLE	RF POWER DENSITY W/sq.cm	PRESSURE Pascals	SELF BIAS VOLTAGE Volts	DEPOSITION RATE nm/min	THICKNESS nm
A	0.27	0.67	-700	2.06	248
B	0.27	3.33	-780	4.23	254
C	0.27	6.67	-800	10.65	426
D	0.19	10.00	-500	9.99	400
E	0.27	13.33	-750	12.90	516
F	0.11	13.33	-460	6.56	263
G	0.55	13.33	-1100	21.8	872
H	0.92	13.33	-1200	16.22	746
I	1.07	40.00	-1250	22.45	449

with a hemispherical Si_3N_4 pin with a radius of curvature of 1.6 mm. The experimental parameters for reciprocating sliding friction experiments were: 0.98 N load (approx. initial mean Hertzian contact pressure of 910 MPa); sliding velocity ranging from 1.4 mm/s to 5.2 mm/s; room temperature (23°C); in either dry nitrogen (< 1% relative humidity) or humid air (60% relative humidity). The friction force was monitored continuously during the friction experiments.

RESULTS AND DISCUSSION

Surface Morphology, Structure and Chemistry

From SEM, the a-C:H films were observed to be very smooth, uniform and free of pin holes. The surface roughness of the a-C:H film ranged from 1 nm rms to 4 nm rms as measured by surface profilometry.

AES of the surfaces of the a-C:H films reveal the presence of a small amount of oxygen. No other contaminants are observed. Hydrogen concentrations of the films were not measured yet. However, a-C:H films deposited using similar deposition techniques reportedly contain greater than 20% hydrogen.[1,16] The fine structure peaks, P1 and P2, of the carbon KLL Auger spectra for the a-C:H films are all very similar. Both P1 and P2 are very broad and overlap. The relative intensity of P2, the lower energy peak, is greater than P1. The fine structure of the carbon KLL spectra from the a-C:H films resembles that of amorphous carbon, as shown in figure 1.

The Raman spectra provide more detailed information on the structure of a-C:H film produced with different deposition conditions. Figure 2 shows an example of a typical Raman spectrum of the a-C:H films. The two, very broad overlapping bands are denoted the D (disorder) and the G (graphitic) bands, respectively, for their similarity to the first order Raman spectrum of polycrystalline graphite and disordered sp^2 bonded carbon materials. However, the frequency of the G band has decreased by more than 40 cm^{-1} from that of graphitic carbons.[7,9] The ratio of the integrals of the D band and G band, which has been observed to be qualitatively related to the bonding in a-C:H films[1,8-12,16], correlates with the deposition conditions. As the

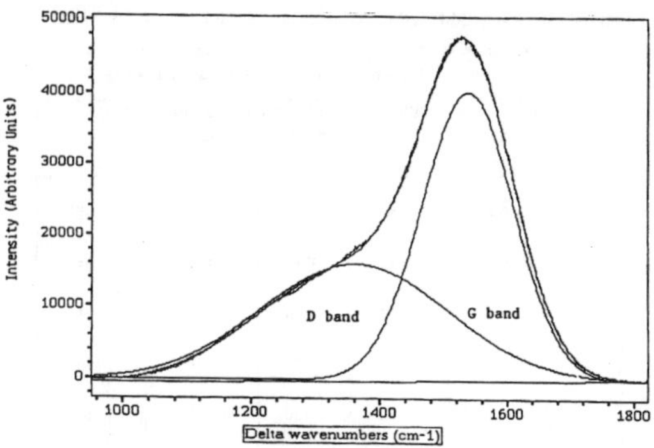

Figure 2.— Deconvolution of the Raman spectrum of a typical a-C:H film.

rf power density is increased, the D/G intensity ratio increases, the frequency of the G band has increased and the G bandwidth is reduced which implies the short range order of sp^2 bonded carbon clusters in the amorphous carbon network increases[16] as a function of increasing rf power density. As the deposition pressure is increased, the converse is observed.

Friction Behavior

Figure 3 shows examples of the coefficients of friction of the a-C:H films sliding against the Si_3N_4 pin in dry nitrogen. The steady state coefficient of friction for the a-C:H films sliding against a Si_3N_4 pin was very low. It ranged between 0.03 and 0.05 in dry nitrogen.

Figure 4 shows examples of the coefficients of friction of the a-C:H films sliding against the Si_3N_4 pin in 60% relative humidity. Initially, the median coefficients of friction of the a-C:H films sliding against a Si_3N_4 pin in 60% relative humidity were quite high, ranging from 0.26 to 0.30. However after only 7,000 passes, the median coefficient of friction of the sample which was deposited at 0.27 W/cm^2 and 0.67 Pa was 0.09 in 60% relative humidity. This value remained steady for approximately 37,000 passes after which it gradually increased to 0.15 at 80,000 passes. In addition, the median coefficients of friction of four other a-C:H films gradually decreased to a steady state value. The steady state coefficients of friction of a-C:H films deposited at 0.27 W/cm^2 and 3.33 Pa, 6.67 Pa, and 13.33 Pa were 0.24, 0.16 and 0.15,

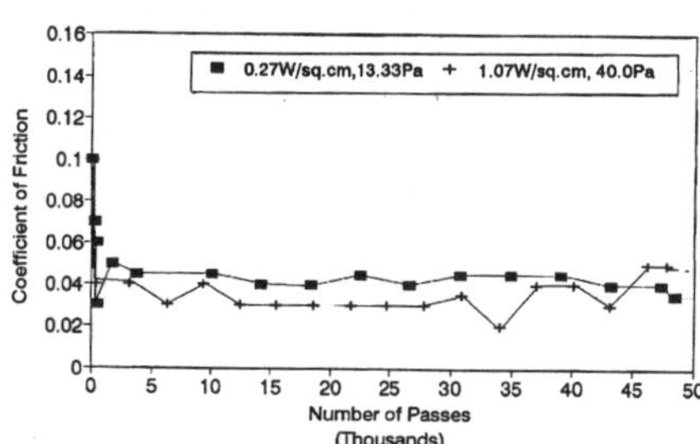

Figure 3.— Examples of the coefficients of friction of the a-C:H films sliding against the Si_3N_4 pin in dry nitrogen as a function of the number of reciprocating cycles. (a) a-C:H deposited with 100% CH_4, 0.27 W/cm^2 at 13.33 Pa and (b) a-C:H deposited with 100% CH_4, 1.07 W/cm^2 at 40 Pa.

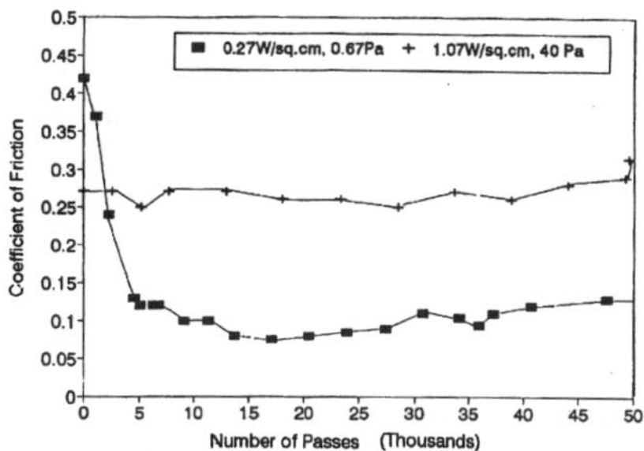

Figure 4.— Examples of the coefficients of friction of the a-C:H films sliding against the Si_3N_4 pin in 60% relative humidity as a function of the number of reciprocating cycles. (a) a-C:H deposited with 100% CH_4, 0.27 W/cm^2 at 0.67 Pa and (b) a-C:H deposited with 100% CH_4, 1.07 W/cm^2 at 40 Pa.

respectively. The steady state coefficient of friction of the a-C:H sample which was deposited at 0.55 W/cm^2 and 13.33 Pa, was 0.19.

Wear Behavior

Figures 5a and 5b are examples of the wear tracks and wear debris which formed on the a-C:H films as a result of sliding contact with the Si_3N_4 pin in dry nitrogen, respectively, while figures 6a and 6b are examples of the wear tracks and wear debris produced in humid air. The wear tracks were generally smooth with most of the loose debris accumulated outside of the wear tracks. Smeared patches of wear debris were present throughout the wear track produced in dry nitrogen. However, the wear track produced in humid air contained very little wear debris. In general, the wear scars on the Si_3N_4 pins were smooth, regardless of the sample or the test environment (dry N_2 or humid air). Figure 5c presents an example of the wear scar produced on a Si_3N_4 pin in dry nitrogen. Most of the loose and smeared wear debris accumulated outside the wear scar smeared out in streaks parallel to the sliding direction.

Table 2 is a summary of the results of the friction and wear experiments performed on the a-C:H films. The wear behavior of the a-C:H films is dependent on the deposition conditions. Wear rates of films produced both with low power and low pressure, and with high power and high pressure are greater in dry nitrogen than in humid air. However, the wear rates of a-C:H films produced at low power and high pressure are greater in humid air than in dry nitrogen. Regardless of the deposition conditions, a-C:H films have very low wear in dry nitrogen and in humid air when sliding against Si_3N_4.

Figure 5.—Micrograph of a-C:H deposited at 0.55 W/cm^2 and 13.33 Pascals; (a) wear track, (b) wear debris, and (c) wear scar on Si_3N_4 pin produced in dry nitrogen.

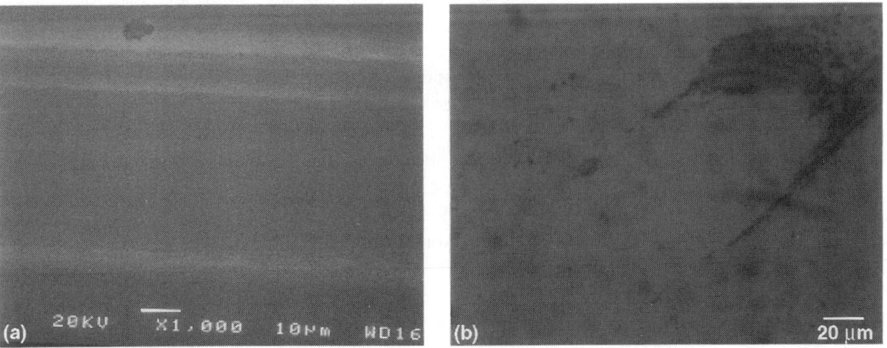

Figure 6.—Micrograph of a-C:H deposited at 0.55 W/cm^2 and 13.33 Pascals; (a) wear track, (b) wear debris produced in 60% relative humidity.

Table 2 Results of reciprocating sliding friction experiments on a-C:H films.

SAMPLE	RF POWER DENSITY W/sq.cm	PRESSURE Pascals	Steady State Friction Coefficient		Film Wear Rate (x 10^-6 cu.mm/Nm)	
			Dry Nitrogen	60% Rel.Humidity	Dry Nitrogen	60% Rel.Humidity
A	0.27	0.67	0.05	0.10	0.084	0.015
B	0.27	3.33	0.05	0.30	2.3	0.0084
C	0.27	6.67	0.04	0.18	0.011	0.089
E	0.27	13.33	0.04	0.15	0.014	0.020
F	0.11	13.33	0.03	0.29	0.018	0.018
G	0.55	13.33	0.04	0.19	0.22	0.0060
I	1.07	40.00	0.03	0.26	0.373	0.0020

Raman and AES of Wear tracks

Chemical and structural examinations of the wear tracks and debris particles produced on the a-C:H films in both dry nitrogen and humid air were conducted by Raman microprobe spectroscopy and AES. The changes observed in the Raman and Auger spectra which were obtained on the as-deposited samples, on the smooth areas inside the wear tracks and on the debris particles located near the wear tracks were similar for all the a-C:H films. Therefore, the spectra obtained from these regions on the sample which was produced at 0.55 W/cm^2 and 13.33 Pa, are presented in figure 7.

The figure 7a(*i*) is the Raman spectra of the as-deposited a-C:H film. The frequencies of the D(disorder) and G bands are 1370 cm^{-1} and 1541 cm^{-1}, respectively, and the halfwidths are 209 cm^{-1} and 103 cm^{-1}, respectively. The D/G intensity ratio is 0.88. No other bands are distinguishable in the Raman spectra of the as-deposited film. In comparison with the spectra of the as-deposited films, the most prominent difference in the Raman spectra of the wear tracks and debris particles is the emergence of a third band which appears between 1590 cm^{-1} and 1610 cm^{-1}. Since this band is located where the E_{2g} phonon of disordered graphite appears,[22] henceforth, it will be referred to as the G_d band and the Raman band located between 1500 cm^{-1} and 1560 cm^{-1} will be referred to as the G_a band.

Narrowing of the D bandwidth is observed on the tracks formed both in dry nitrogen and in humid air. However, on the dry nitrogen wear track, figure 7a(*iv*), the G_d band is not observed and the D/G_a intensity ratio has decreased slightly, while on the humid air wear track, figure 7a(*ii*), the D/G_a intensity ratio has increased by 58% and the G_d band is present. Narrowing of the Raman bands indicate the short range order of carbon bonds has increased in the wear tracks. The presence of the G_d band in the spectrum from the humid air wear track and the increase in the D/G_a intensity ratio indicate that a graphitic structure was forming as a result of sliding friction. The Raman spectra of the debris particles which accumulated outside the wear tracks, figures 7a(*iii*) and 7a(*v*), have similar structure. The G_d band is more prominent than the G_a band in their Raman spectra. Narrowing of the D and G_a bands and the presence of the distinct G_d band, as well as the increases in the D/G_a band intensity ratios imply that the wear debris formed by sliding friction is composed mostly of sp^2 bonded carbon with increased short range order.

In figure 7b(*i*), the carbon KLL Auger peak on the as-deposited a-C:H film is presented. The fine structure peaks, P1 and P2, of the carbon KLL Auger spectra are very broad. The relative intensity of P2 is greater than P1, and the position of the peak closest to the main carbon Auger line, P1, cannot be distinguished. This line shape resembles that of amorphous carbon (see figure 1).

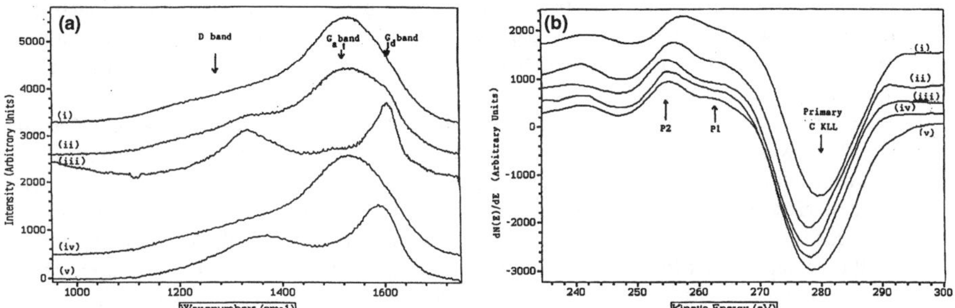

Figure 7.—(a) Raman spectra of a-C:H film; (b) Carbon KLL Auger fine structure of a-C:H film. Deposited at 0.55 W/cm^2 and 13.33 Pascals; (i) as-prepared, (ii) smooth area of humid air wear track, (iii) wear debris located near humid air wear track, (iv) smooth area of dry nitrogen wear track, and (v) wear debris located near dry nitrogen wear track.

The primary carbon KLL peak is located at a lower energy in spectra from the wear tracks produced both in dry nitrogen and in humid air implying that there was less charging during the Auger survey of the wear tracks than there was during the survey of the as-deposited film which is consistent with an increase in the sp^2 bonded carbon. However, in comparison with the Auger spectra of the as-deposited film, no difference the relative intensity of P2 to P1 is observed in the fine structure of carbon KLL peak obtained the dry nitrogen wear track, figure 7b(*iv*). In the humid air wear track, the carbon KLL fine structure peaks, figure 7b(*ii*), are also quite broad, but P1 appears as a distinct shoulder on the primary carbon KLL peak and the relative intensity of P2 to P1 is greater indicating a relative increase in the number of sp^2 bonded carbon. The fine structure peaks of the carbon KLL Auger spectra for the debris particles formed in both environments, figures 7b(*iii*) and 7b(*v*), are similar to those of the humid air wear track, although the width of P2 is less and the relative intensity of P2 to P1 is greater indicating a further increase in sp^2 bonded carbon. However, the increase in the relative intensity of P2 to P1 is not as great in the spectra obtained from the wear debris formed in dry nitrogen which implies that the degree of graphitization of the debris particles formed in dry nitrogen is less than the debris particles formed in humid air.

Both the Raman and Auger spectra from locations on the a-C:H film where humid air and dry nitrogen friction tests were performed provide evidence that sliding friction gradually transformed the a-C:H into a disordered material composed primarily of sp^2 bonds with greater short range order than the as-deposited film. This material was soft and wore off quickly accumulating outside the wear tracks. The structure of the material which remained on the wear tracks was similar to the as-deposited a-C:H film although there was evidence of increasing sp^2 ordering.

SUMMARY

Over the range of deposition parameters examined in this paper, a-C:H films produced by rf PECVD using 100% methane precursor gas mixture, 0.27 W/cm^2 rf power density and 0.67 Pa deposition pressure had the best overall tribological performance. The steady state coefficients of friction in dry nitrogen and in 60% relative humidity were 0.05 and 0.10, respectively, while the wear rates were 8.4 x 10^{-8} mm^3/Nm and 1.5 x 10^{-8} mm^3/Nm, respectively. In general, the friction and wear of all the a-C:H films tested against Si_3N_4 pins were very low in dry nitrogen, ranging from 0.03 to 0.05, and from 1.1 x 10^{-8} mm^3/Nm to 2.3 x 10^{-6} mm^3/Nm, respectively.

The steady state, the coefficient of friction of a-C:H films in 60% relative humidity ranged from 0.10 to 0.30 and was dependent on the deposition conditions. Although friction was always higher in humid air than in dry nitrogen, the wear was still very low, ranging from 2.0 x 10^{-9} mm^3/Nm to 8.9 x 10^{-8} mm^3/Nm.

The Raman spectra and the fine structure of the carbon KLL Auger spectra revealed that sliding friction was transforming the a-C:H films into a material composed primarily of sp^2 bonded carbon with increasing short range order. Qualitatively, the amount of wear that an a-C:H film experienced corresponded to the extent that the structural changes progressed. Spectra from the wear tracks and debris of films experiencing the least amount of wear most resembled those of the as-deposited film. Raman spectra from wear tracks of a-C:H films experiencing greater wear contain a third band located between 1590 cm^{-1} and 1610 cm^{-1} which becomes more distinct, while the fine structure peaks of the carbon KLL Auger spectra have a stronger resemblance to those of graphite with greater wear. The debris particles located near the wear tracks consisted of disordered, mostly sp^2 bonded carbon.

ACKNOWLEDGEMENTS

The author would like to thank Dr. J. Lauer of the University of California-San Diego and Dr. K. Miyoshi of NASA-Lewis Research Center for helpful discussions, and the members of the Surface Science Branch of NASA-Lewis Research Center for their assistance and continued encouragement.

This work has been supported by the National Research Council, the Vehicle Propulsion Directorate of the U.S. Army Research Laboratory and the Surface Science Branch of NASA-Lewis Research Center.

REFERENCES

(1) J.C. Angus, P. Koidl and S. Domitz, in Plasma Deposited Thin Films, J. Mort and F. Jansen eds., Ch.4 (CRC Press, 1986) 89.
(2) K. Miyoshi, Adv. Info. Storage Syst., Vol. **3**, (Amer. Soc. of Mech. Engrs. 1991) 147.
(3) K. Miyoshi, R.L.C. Wu and A. Garscadden, Surf. & Coat. Technol., **54/55** (1992) 428.
(4) R. Memming, H.J. Tolle and P.E. Wierenga, Thin Solid Films, **143** (1986) 31.
(5) S. Agarwal, E. Li and N. Heiman, IEEE Trans. on Magnetics, **29(1)** (1993) 264.
(6) J.K. Lee, M. Smallen, J. Enguero, H.J. Lee and A. Chao, IEEE Trans. on Magnetics, **29(1)** (1993) 276.
(7) K. Angoni, Carbon, **31 (4)** (1993) 537.
(8) R.J. Nemanich, J.T. Glass, G. Lucovsky and R.E. Shroder, J. Vac. Sci. Technol., **A 6 (3)** (1988) 1783.
(9) D.S. Knight and W.B. White, J. Mater. Res., **4(2)** (1989) 385.
(10) F. Rossi, B. Andre', A. vanVeen, P.E. Mijnarends, H. Schut, M.P. Delplancke, W. Gissler, J. Haupt, G. Lucazeau and L. Abello, J. Appl. Phys., **75 (6)** (1994) 3121.
(11) P.V. Huong, Diam. & Rel. Mat., **1** (1991) 33.
(12) R.J. Nemanich and S.A. Solin, Phys. Rev. B, **20 (2)** (1979) 392.
(13) K. Enke, H. Dimigen and H. Hubsch, Appl. Phys. Lett. **36** (1980) 291.
(14) D.R. McKenzie, B.C. McPhedran, N. Savvides and D.J.H Cockayne, Thin Solid Films, **108** (1983) 247.
(15) B. Dischler, A. Bubenzer and P.Koidl, Solid State Commun., **48** (1983) 105.
(16) J. Robertson, Advances in Physics, **35 (4)** (1986) 317.
(17) P.G. Lurie and J.M. Wilson, Surf. Sci., **65** (1977) 476.
(18) S.V. Pepper, Appl. Phys. Lett., **38** (1981) 344.
(19) A.M. Bonnot, Surf. Coat. Technol., **45** (1991) 343.
(20) D.E. Ramaker, Phys. Scr., **T41** (1992) 77.
(21) W. Zhu, J.E. deVries, M.A. Tamor and K.Y. Simon Ng, Surf. & Coat. Technol., **71** (1995) 37.
(22) P.V. Huong, B. Marcus, M. Mermoux, D.K. Veirs and G.M. Rosenblatt, Diam. & Rel. Mater., 1 (1992) 869.
(23) F. Richter, K. Bewilogua, H. Kupfer, I. Mhling, B. Rau, B. Rother and D. Schumacher, Thin Solid Films, **212** (1992) 245.
(24) B.E. Williams and J.T. Glass, J. Mater. Res., **4(2)** (1989) 373.
(25) H.S. Hu, A. Joshi and R. Nimmagadda, Mater. Res. Soc. Ext. Abstract, **EA-19** (1989) 103.
(26) T.J. Moravec and T.W. Orent, J. Vac. Sci. Technol., **18** (1981) 226.

THE EROSIVE BEHAVIOUR OF DIAMOND

C S JAMES PICKLES, E J COAD, G H JILBERT AND J E FIELD
PCS Group, Cavendish Laboratory, Cambridge University, CB3 OHE, UK

ABSTRACT

Diamond exists in many forms including natural diamond, synthetic diamond produced at high temperatures and pressures, polycrystalline diamond composites (PCD's) and most recently chemical vapour deposited (CVD) diamond. The liquid impact behaviour of these materials has been characterised in terms of a single and multiple impact damage threshold velocity which was obtained using a jet technique capable of reaching velocities of 600 m s^{-1}. The multiple impact damage threshold velocity is a highly reproducible parameter which is invaluable in comparing the rain erosion resistances of different materials.

The sand erosion resistances of the various forms of diamond have also been evaluated using a specially designed sand blasting apparatus at velocities up to 270 m s^{-1}. The extent of damage is quantified by mass and transmission loss as a function of impact velocity.

The results of these experiments reveal that the exceptional mechanical properties of diamond cause it to have an equally exceptional response to erosion.

INTRODUCTION

The specific application being considered in this paper is the aerospace erosion due to high velocity flight such as would be experienced by a missile or jet aircraft. Diamond's excellent transmission in the 8-12 μm waveband makes it a suitable candidate to replace the materials currently in operation. These materials, such as zinc sulphide and germanium, have extremely low resistances to sand particle and raindrop impact and this can be the limiting factor in the performance of the missile. Diamond is therefore being considered as a coating material for the existing substrates or, ideally, as a bulk window.

The fame that diamond enjoys is due not only to its high price and aesthetic qualities but also to its exceptional mechanical and thermal properties and these properties have a profound influence on its response to high velocity impact. Table I compares some of the mechanical and thermal properties of natural and CVD diamond, with those of zinc sulphide and germanium. The following sections describe how each of these properties influence the response of diamond to the aerospace erosion environment. The lessons learned by considering these specific erosion conditions can be applied to many other erosion phenomena.

Mat. Res. Soc. Symp. Proc. Vol. 383

Table I Mechanical and thermal properties of diamond [1,2].

	natural diamond	CVD diamond	zinc sulphide	germanium
Knoop hardness ($kg\ mm^{-2}$)	9000 [(111) face]	8265	250	850
Fracture toughness ($MPa\ m^{1/2}$)	3.4 - 5.0 [(111) face]	5.3-8.0	1.0	0.7
Fracture stress (MPa)	2500	200 - 1000	100	90
Young's modulus (GPa)	1050	*	74	103
Density ($kg\ m^{-3}$)	3520	*	4090	5320
Poisson's ratio	0.07	*	0.30	0.28
Thermal conductivity ($W\ m^{-1}\ K^{-1}$)	2000 + IIa at 293K	1700-2600 (293 K)	19	59
Thermal expansion coefficient ($10^{-6}\ K^{-1}$)	0.8	*	7.0	6.1
Heat capacity ($J\ g^{-1}\ K^{-1}$)	0.52	*	0.47	0.31

* = same value as natural diamond

+ (3,500 $W\ m^{-1}\ K^{-1}$ for isotopically pure diamond, 15,000 $W\ m^{-1}\ K^{-1}$ for Type IIa at 80 K)

LIQUID AND SOLID PARTICLE IMPACT THEORY

There are a large number of different damage mechanisms observed in the field of aerospace erosion but they can be divided into two broad categories; ductile failure and brittle failure. Diamond and most of the materials used in the aerospace applications under consideration fail in a predominantly brittle fashion. This type of failure is initiated by the stress waves that propagate through the target material at impact (see figure 1). The shaded widths of the shear and compressional waves (the two bulk waves) in figure 1 represent the relative amplitudes of particle motion. The Rayleigh surface wave contains the bulk of the impact energy[4] and is seen in the diagram as a ripple behind the shear wave (it is typically ~10% slower). The vertical and horizontal arrows in the diagram represent the two components of displacement associated with the Rayleigh wave. The powers of **r** indicate the geometrical attenuation of the waves with radial distance. The Rayleigh wave has the least attenuation, the greatest energy and is also concentrated at the specimen surface where it can exploit the largest flaws in a typical brittle target. It is thus the major cause of damage in brittle materials under particle impact conditions[5] although it will later become clear that in the case of diamond the bulk waves can also play a large part. The depth to which the Rayleigh wave penetrates depends on its wavelength which in turn depends on the duration of the impact.

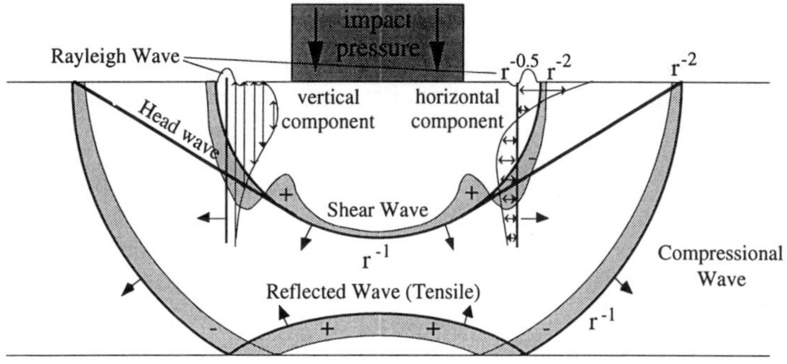

Figure 1 Stress waves caused by an impact[3].

The major damage mechanisms at the front surface of brittle and ductile targets are shown in figure 2. An intermediate damage mechanism (elastic-plastic) between purely ductile and brittle failure can be seen in figure 2b which has a deformed central zone surrounded by radial and lateral cracks caused by hoop stresses around the crater. Plastic and elastic-plastic behaviour occurs for soft targets and sharp erodents. Brittle (elastic) failure, especially for hard targets and blunt erodents, often results in a complete ring crack around the contact region. The crack propagates into the bulk of the material resulting in a "Hertzian" cone[6]. If the impact is short (for small or weak particles) there is insufficient time for the crack to grow into a complete ring and short circumferential cracks are observed. This pattern of damage is seen with liquid impact, in which the water drop is initially, and briefly, compressed causing high "water hammer" pressures[7] which reduce to Bernoulli stagnation pressures when release waves propagate in from the edge of the contact zone. The impact severity is determined by the intensity and duration (usually less than 1 μs) of these water hammer pressures and these in turn depend on the profile of the front of the drop.

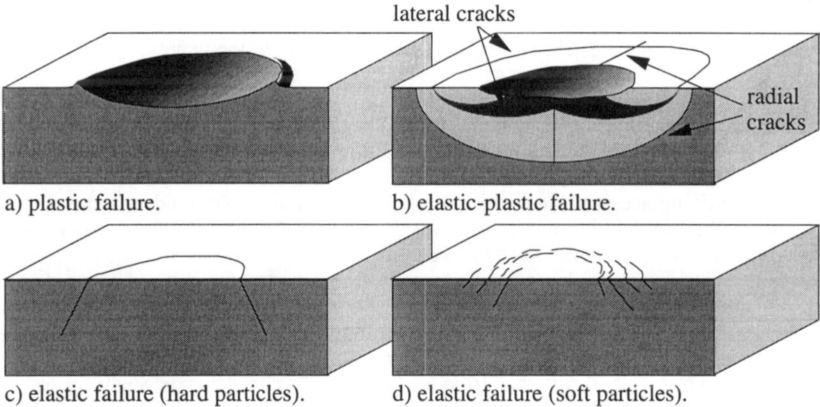

Figure 2 Damage mechanisms observed after liquid and solid particle impact.

There is an additional damage mechanism observed during impact with a liquid drop when water jets laterally across the target (often at many times the impact velocity) and attacks any surface flaws and imperfections. It is this mechanism which dictates that for a sample to have a high rain erosion resistance it must have no surface relief.

EXPERIMENTAL

The apparatus for simulating sand and rain erosion both rely on projecting an impacting particle at a stationary target thereby reversing the experimentally complex scenario of a rapidly moving object travelling through a sand or rain field.

Sand Erosion

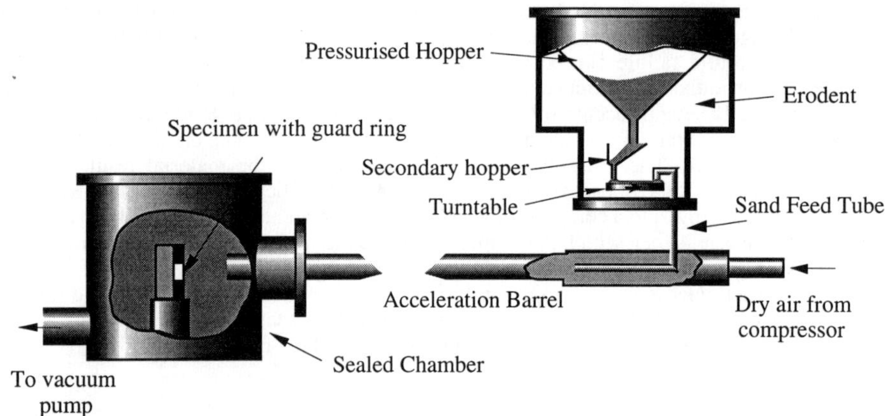

Figure 3 The Cavendish laboratory sand erosion facility.

The sand erosion rig is shown in figure 3 [8, 9]. The flux is controlled over the range 0.05 - 25 kg m^{-2} s^{-1} by selecting which groove on the rotating turntable the sand is fed into and by adjusting the speed of rotation. The sand is collected off the turntable and delivered to the end of the 4 m long acceleration barrel. The long barrel allows the sand particles time to accelerate to the gas velocity before they reach the sample chamber which gives a high impact velocity (up to ~250 m s^{-1}) and low velocity spread.

Several techniques have been used to measure the erodent velocity. Andrews[10] and Scullion[11] cross-correlated the photomultiplier signals resulting from sand particles passing through two curtains of laser light. More recently series of streaks have been obtained in a high shutter speed video camera, each streak corresponding to an illuminated sand particle. The velocity can be simply determined from the length of the streak and the shutter speed.

The erodent used in these experiments was a sieved quartz sand (C25/52) with a mean mass of 190 μg and linear dimension in the range 300 to 600 μm (mean 500 μm). The erodent was discarded after a single experiment.

Analysis of the erosion damage is by Scanning Electron Microscopy (SEM), mass loss and (because of the particular application) infrared transmission loss.

Rain Erosion

The simulation of rain erosion is not so simply achieved since spherical drops disintegrate when projected at the velocities of interest. The solution is to duplicate the crucial compressible water-hammer phase, which dictates the threshold velocity for damage of a material, by using a jet of water which becomes rounded by its passage through the air so that when it impacts the target its front profile is that of a raindrop. The apparatus used to produce these high velocity jets is the Multiple Impact Jet Apparatus[12-14] (MIJA) shown in figure 4. The jets created by this device are highly reproducible, with a small spread in velocities (<1.5%) and can be delivered onto a site specified by the computer at a rate of one every 3 s. The velocity of each jet fired is determined by the time taken to break the light beams from a series of optical fibres. The velocity range achievable by MIJA is 80 - 600 m s^{-1}, which is sufficient to cause damage in all of the materials of interest. In all of the experiments described in this paper a 0.8 mm nozzle was used. The diameter of spherical raindrop which is equivalent to this diameter jet varies with impact velocity and the graph of this variation is plotted in Hand et al.[15].

The parameter that is investigated in MIJA studies is the Absolute Damage Threshold Velocity (ADTV, the velocity below which there is no circumferential damage regardless of the number of times the site is impacted). This is assessed by selecting a series of sites on a sample (typically a 25 mm disc) and assigning an impact velocity to each. Every site is impacted once and examined under a microscope. The velocity at which damage first becomes visible is noted on a graph. The sample is then returned to MIJA and each site impacted a second time. This process is repeated with the Damage Threshold Velocity for a particular number of impacts (DTV (X impacts)) decreasing with impact number. A threshold curve for a 2 mm thick CVD diamond sample is given in figure 5 and shows how the curve flattens out at ~300 impacts suggesting that the ADTV ~ DTV (300 impacts).

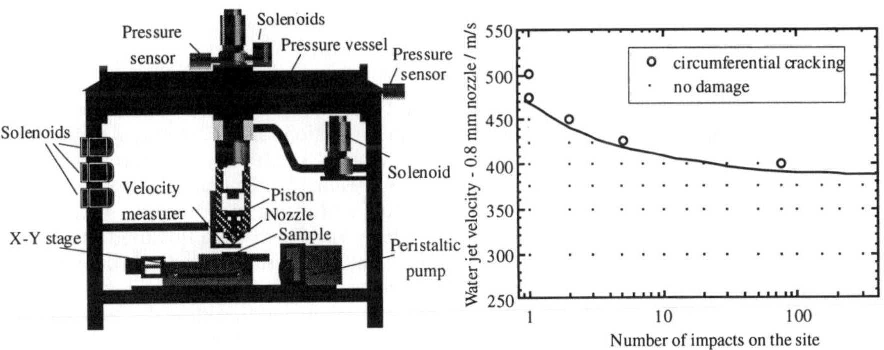

Figure 4 The Multiple Impact Jet Apparatus Mk IV (MIJA).

Figure 5 The threshold velocity curve for a 2 mm thick sample of CVD diamond (poor quality, growth side).

THE MECHANICAL PROPERTIES OF DIAMOND

Hardness

Perhaps surprisingly the exceptionally high hardness of diamond (the highest known to man) does not, in itself, dictate that the material will have an exceptionally high sand or rain erosion resistance. It is true that for good sand erosion performance a material must have a hardness higher than that of the erodent so that the incoming projectile is defeated. If the target hardness is sufficiently high then there will be little or no ductile deformation in the target and the material fails through a purely elastic, brittle mechanism (figure2c, d). In this case the threshold velocity for damage is determined by the material's fracture toughness rather than its hardness[6].

Diamond's high hardness means, therefore, that it will tend to fail under erosive conditions through the elastic Hertzian cone damage mechanism (figure 2c). However the shape of this cone is distorted by the cleavage planes in a natural single crystal (predominantly {111}[16]) and by the grain boundaries in a CVD diamond see figure 6.

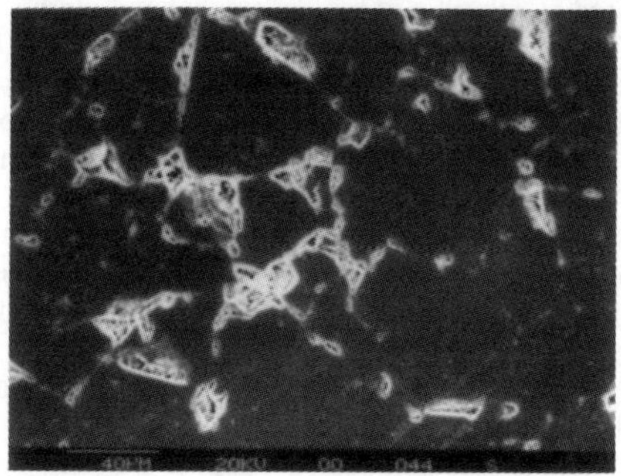

Figure 6 Sand erosion damage on CVD diamond (growth side, 260 s at 35 m s^{-1}, flux=10.5 kg m^{-2} s^{-1}).

Fracture Toughness

As mentioned in the previous section it is primarily the fracture toughness that determines the extent of damage in brittle materials exposed to the conditions of sand erosion. Diamond's high fracture toughness therefore gives it a high resistance to sand erosion. At low velocities only partial ring cracks form and it takes prolonged exposures for the cracks to intersect and material to be removed. A graph of exposure time required for ring crack formation on Type I diamond flats can be seen in figure 7 from data in Feng and Field[16] (the flux used was 10.5 kg m^{-2} s^{-1}). The same authors also assessed a commercial polycrystalline diamond composite material (Syndite, supplied by De Beers Industrial Diamond Division) and found that at 140 m s^{-1} it had a similar erosion resistance to the natural material, although the erosion mechanism involves the removal of the metal binder and subsequent fracture of the diamond skeleton. Kaye and Field[17] found that the ratio of gravimetric erosion rates (the mass lost from the target per mass of erodent) between Syndite, tungsten carbide and stainless steel at 250 m s^{-1} was 1:160:19500 and this highlights diamond's impressive sand erosion resistance.

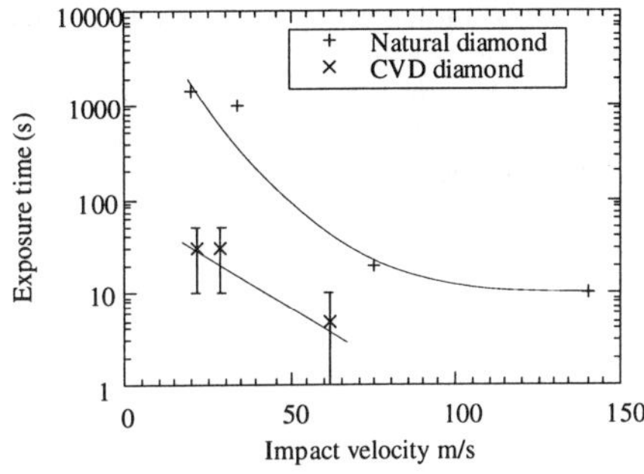

Figure 7 The exposure time required for ring crack formation in natural and (intermediate quality) CVD diamond.

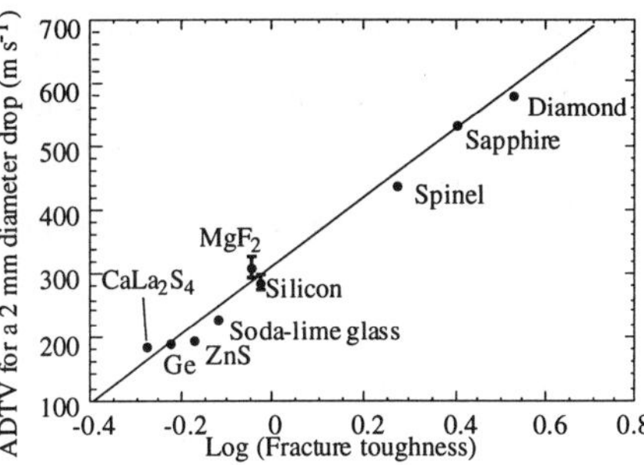

Figure 8 the dependence of liquid impact threshold velocities on fracture toughness.

Rain erosion resistance is also extremely dependent on fracture toughness. This is demonstrated by plotting threshold velocity for circumferential fracture (obtained using a 0.8 mm jet on MIJA and then converted to a standard 2 mm diameter spherical drop, equations for this conversion can be found in Hand et al.[15]) against log fracture toughness[14] (see figure 8). The data point for diamond is for natural Type IIa material (~515±15 m s^{-1} for a 0.8 mm jet ≡ 570 m s^{-1} for a 2 mm drop). It was obtained using a 6 mm diameter specimens which, for reasons which will be discussed below, will give an ADTV lower than that of a larger specimen. A full threshold curve for a 2 mm thick specimen of CVD diamond can be seen in figure 5 and suggests a 0.8 mm jet threshold velocity for this sample of 388±12 m s^{-1}. Figure 8 and the quoted fracture toughness for CVD diamond in table I (it is common for polycrystalline materials to have fracture toughnesses up to approximately twice that of single crystals, because of the increased crack surface area[18]) suggest that it should have an ADTV higher than that of the natural single crystal. The reason that it does not is because of the large flaws in the polycrystalline material.

Flaw size

When a stress is applied to a material there are two parameters which will determine whether it will fail or not; the fracture toughness and the flaw size. For good quality natural diamond Field[2] estimates the size of critical sharp ended flaws to be as small as 0.5 μm. In contrast a fractography study by Hehn et al.[19] detected flaws in CVD diamond of 170±90 μm. This study suggested that the flaw size increases with sample thickness (although it was found to be independent of the size of the grains, which are larger on the growth surface than on the nucleation surface) and revealed that processing dependent bulk cracks could significantly reduce the material's fracture toughness.

The results of sand erosion experiments on CVD diamond, confirm those of the MIJA liquid impact study, i.e. that the material is significantly weaker than natural diamond, again due to the large flaws. The considerable difference between the exposure times required to cause ring crack formation for CVD diamond and natural diamond can be seen in figure 7.

Infrared transmission curves from these experiments indicated that the finer grained nucleation side is more resistant to sand erosion than the growth side at these velocities.

In both the rain and sand erosion studies evidence of both intergranular and transgranular fracture was observed, suggesting that the grain boundaries are quite strong. It is clear however that the grain size is a crucial parameter when considering the erosion resistance of CVD diamond and indeed many of its other properties.

Fracture Stress

The fracture stress of natural diamond was reported by Field (chapter 12)[2] to be ~3.75 GPa. The strengths reported for CVD diamond have varied from ~746-1138 MPa[20] to 200-400 MPa[21] and this reflects the wide diversity in quality of CVD diamond available. However it should be emphasised that different strength measurement techniques can give widely varying results. For instance the indentation technique used to determine the fracture stress of natural diamond stresses a very small surface area and tends to give high values. Also the samples tested by Sussman et al. were all <300 μm in thickness (c.f. the results reported by Harris which were from samples of thicknesses up to 1.2 mm) which limits the critical flaw dimensions. The current low strength of CVD diamond is attributed to large flaws, residual stresses and occasional low fracture toughnesses[22].

One of the consequences of the high fracture stress of natural diamond is that, in applications where the strength is a factor, thinner diamond samples can be used than would be necessary for other materials[23]. However the thin specimens, while potentially capable of surviving the static loads imposed on it are susceptible to another erosion damage mechanism. This is the tensile failure at the rear of the sample caused by reflection of the dilatational wave, which can be seen as a crack pattern within the ring crack in figure 9. This failure pattern was observed in sand erosion experiments on 0.57 mm thick plates at velocities of 21 m s^{-1} and above (see figure 10) and in the MIJA experiments at threshold velocities of 265±75 m s^{-1} (in a study of 8 x 20-25 mm diameter discs, 0.5-1 mm thick). The spread in MIJA threshold velocities is between different samples (different sites on the same sample give good agreement) and this again reflects the variation in material quality. This variation obscures the expected correlation between thickness and threshold velocity for the damage mechanism. Rear surface failure significantly reduces the damage threshold velocity for a thin diamond sample. It was not observed in the 2 mm thick sample in figure 5.

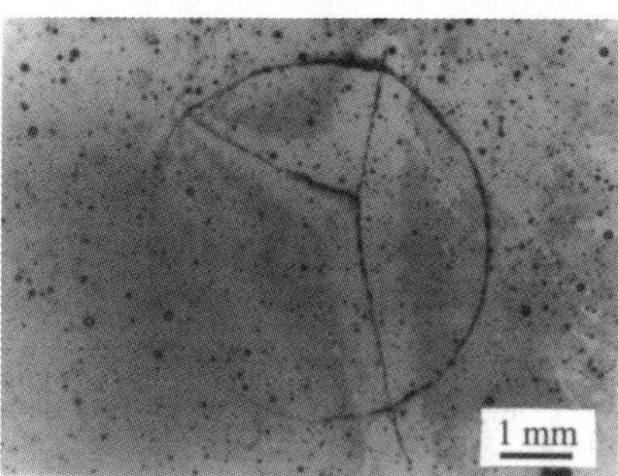

Figure 9 Liquid impact damage in 0.75 mm thick CVD diamond (150 impacts at 400 m s^{-1}).

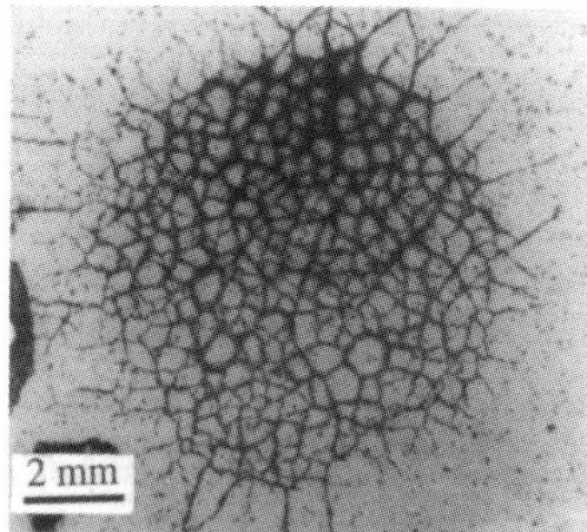

Figure 10 Rear surface tensile failure in CVD diamond after sand erosion (30 s at 61 m s^{-1}, nucleation side, 10.5 kg m^{-2} s^{-1}).

Young's Modulus

From an impact standpoint the most important ramification of diamond having a high value of Young's modulus is that it gives the material a very high stress wave velocity. For diamond the dilatational wave speed is ~18 km s^{-1} and therefore information about the impact can return rapidly from any nearby boundaries and there is a high possibility of stress waves reinforcing in small samples. This effect is enhanced by the low stress wave attenuation in diamond. Both of these factors will increase the rear surface tensile failure described above. Examples of some impact experiments dominated by stress wave reflection effects are given below:-

Reflection from the edges

Figure 11 shows a fracture in CVD diamond (A) which occurred when the compressive bulk wave reflected as a tensile wave from the disc edge (B) and met the slower Rayleigh wave. The two waves interfere to give fracture which can initiate at an impact velocity which is below that of the semi-infinite material. The edges of a diamond disc can be considerably weaker, in an impact situation, than the remainder of the sample because of this effect.

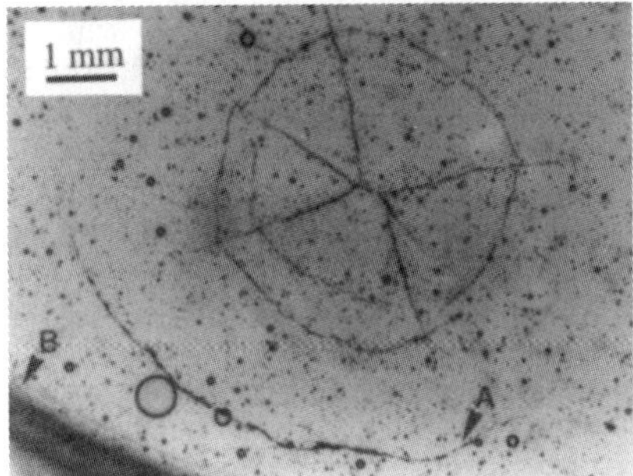

Figure 11 Fracture (A) caused by stress wave interaction (0.75 mm thick 10 impacts at 400 m s^{-1}).

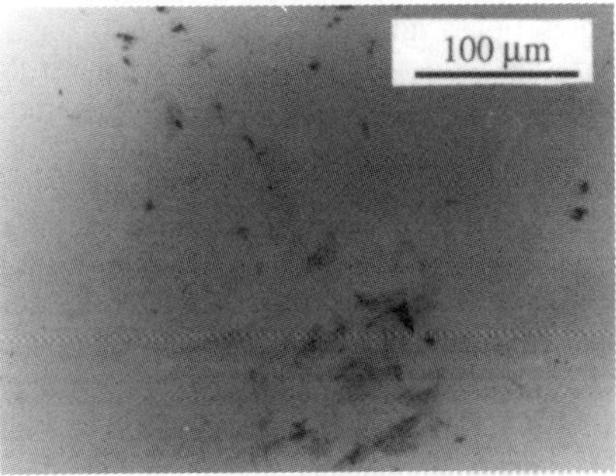

Figure 12 central damage on natural diamond caused by overlapping stress waves (300 impacts at 475 m s^{-1}).

Reflection from the rear surface

The bulk of the samples tested on MIJA have been less than 1 mm in thickness and consequently exhibit an unusual circumferential damage pattern. A typical single liquid impact consists of a number of short, circumferential fractures at the edge of the water hammer zone (<0.8 mm for the nozzle used in these experiments). The length of these cracks increases with the number of impacts so that for multiple impacts full ring cracks can be seen. The damage pattern observed in the thin CVD diamond layers is shown in figure 9 and the circumferential crack can be seen to be much larger than expected. Calculations by Seward et al.[24] have shown that this radius corresponds closely to the predicted radius at which a dilatational bulk wave has changed mode at the rear surface and reflected as a distortional wave which interferes with the Rayleigh wave propagating along the front surface. This phenomenon has previously been recorded by Bowden and Field[5] in the form of secondary bands of fracture outside a typical fracture pattern, but only on diamond have they been observed in isolation without the inner ring of damage.

The threshold velocity for the outer ring of circumferential damage for 8 x 20-25 mm diameter discs, 0.5-1 mm thick, was 420 ± 60 m s^{-1}. Again the variation in quality obscures the expected dependence on specimen thickness. The 2 mm thick sample from which a full threshold curve was obtained (figure 5) showed only the inner damage ring but at a relatively low threshold velocity (388±12 m s^{-1}) because of the low quality (due in part to its thickness).

Reflection from the edges and rear surfaces

The sample of natural Type IIa diamond which provided the data point in figure 8 was 6 mm in diameter and 1 mm thick. A second sample of the same dimensions was tested subsequently and failed catastrophically at 512±12 m s^{-1} (in good agreement with the first) but some very fine central fractures were observed at 462 ± 12 m s^{-1} formed by stress wave overlap (figure 12). Several sapphire and 2 CVD diamond samples of the same dimensions were also obtained for comparison. Stress wave interference is important in all of these small samples tested, but the effect is clearest in the case of the CVD diamonds. Both samples were impacted centrally, one on the growth face (a) and one on the nucleation face (b), but both samples failed initially on the growth surface (theoretically the weaker surface) with (a) having the lowest 300 impact, 0.8 mm jet threshold velocity (325±25 m s^{-1} for (a) versus 375±25 m s^{-1} for (b)). If the failure (a central crack) was simply caused by the dilatational wave becoming tensile at the rear surface then it would be expected that the sample where the rear surface is weakest (b) would have the lowest threshold velocity. In fact the front surface has the greatest concentration of stresses. This is because the dilatational wave reflects off the disc edge and returns to focus in the disc centre at the same time as the reflection off the rear surface returns to the front surface for the third time (0.33 μs). It is thus the sample with the weakest front surface (a) which has the lowest threshold velocity. The damage pattern at the centre of the Type IIa natural diamond caused by the overlapping stress waves can be seen in figure 12.

The data from these experiments give us a direct comparison between the impact resistances of CVD and natural diamond. The CVD thresholds (325±25 and 375±25 m s^{-1}) are considerably lower than the 462-515 m s^{-1} of the natural diamond. A sample of artificially grown high temperature high pressure Type Ia diamond (3.45 mm thick and 9.51 x 8.42 mm across) was also tested and gave a threshold velocity of 525±25 m s^{-1}. This emphasises the poor strength of current CVD diamond samples.

A second consequence of the high stress wave velocity is that diamond has a very high acoustic impedance (product of density and wave speed). This parameter determines the amount and nature of the stress waves reflecting at a boundary and since any material that diamond is mounted on (for instance in the form of the protective coatings discussed below) is likely to have a lower acoustic impedance, the compressive bulk waves will reflect as tensile waves and cause stress wave interaction effects such as those discussed above. The samples tested under MIJA were mounted on sapphire discs to partially acoustically back them but sufficient tensile stresses were still set up to cause the rear surface tensile crazing failure.

THE THERMAL PROPERTIES OF DIAMOND

Of the thermal properties given in table I both the thermal conductivity and thermal expansion coefficient are important for high velocity aerospace applications, since these properties (along with fracture stress, Young's modulus and Poisson's ratio) determine the thermal shock resistance of the material (exceptionally high in the case of diamond). However the low thermal expansion coefficient, while aiding thermal shock resistance, can in one case in particular have the opposite effect on erosion resistance. This is the case of diamond coatings on existing aerospace substrates. Such a system would have superior optical properties to solid diamond (especially in the 3-5 μm waveband) and the high cost and slow growth of CVD diamond films would not be so crucial. The major problem is with the adhesion of the coating to the substrate. The low thermal expansion coefficient of diamond means that as the sample cools from the high deposition temperatures the substrate contracts far more than the diamond and its upper surface is put into tension (while the coating is compressed). This thermal stress has two consequences. The first is that the fracture stress of the substrate is reduced (in extreme cases the sample fractures as it cools). The second is that the coating will more easily delaminate.

A range of CVD coated samples have been tested on MIJA. The problem of adhesion has been addressed by introducing a patterned interlayer to reduce the tensile stresses in the substrate and to allow the coating to 'lock' onto the sample. Samples tested included coated silicon (25-85% improvement in threshold velocity over the uncoated substrate(7 samples)), germanium (92-130% improvement (5 samples)), zinc sulphide (62% improvement (1 sample)) and sapphire (25% improvement (1 sample)). Total coating thicknesses were in the range 5 - 12 μm. Note however that the first two substrates are opaque and the coating may conceal damage at lower velocities. In most cases coating stripping due to the thermal expansion mismatch and lateral jetting was the first form of damage observed.

CONCLUSIONS

Diamond's high hardness dictates that impact damage will be of a brittle nature. The fracture toughness is therefore an important parameter, although the large surface flaws and possible regions of twinning in CVD diamond severely reduce its erosion resistance. There is a large variation in the quality of CVD diamond samples available and this is reflected in the wide spreads in strength and damage threshold velocity.

The primary damage mechanisms observed in impact conditions are:-

1. Complete or partial Hertzian (circumferential) fracture at the edge of the loaded region.
2. Secondary outer ring of failure (from the interaction between bulk and Rayleigh waves).
3. Rear surface tensile crazing.

It is tempting to use thin (and thus cheap) CVD diamond samples if at all possible, but the last two damage mechanisms are thickness dependent and for thin samples will greatly reduce their threshold velocities. This is a particular problem in the case of diamond because of its high modulus and consequent high stress wave velocity. Thicker samples are obviously desirable, unfortunately however, with current growth technology the quality decreases as the thickness increases.

ACKNOWLEDGEMENTS

This work has been sponsored by the Defence Research Agency. The authors would like to thank Dr J A Savage of DRA Malvern for overseeing the work and R. Marrah for technical assistance. Thanks are also due to GEC Marconi, DRA and De Beers who provided the samples for the study

REFERENCES

1. D.C. Harris, Infrared Window and Dome Materials, (SPIE Optical Engineering Press, 1992)
2. J.E. Field (editor), Properties of Natural and Synthetic Diamond, (Academic Press, 1992)
3. R.D. Woods, J. Soil Mech. Founds. Div. Am. Soc. Civ. Engnrs. **94**, 951 (1968).
4. G.F. Miller and H. Pursey, Proc. Roy. Soc. **A233**, 55 (1955)
5. F.P. Bowden and J.E. Field, Proc. Roy. Soc. **A282**, 331 (1964)
6. A.W. Ruff and S.M. Wiederhorn, in Treatise on Materials Science and Technology, Vol. 16: Erosion, edited by C.M. Preece (Academic Press, New York, 1979), p. 69
7. S.S. Cook, Proc. Roy. Soc. **A119**, 481 (1928)
8. J.E. Field, Q. Sun and H. Gao, SPIE **2286**, 301 (1994)
9. D.R. Andrews, S.M. Walley and J.E. Field, (Proc. Erosion by Liquid and Solid Impact **6**, Cambridge UK, 1983) paper 36
10. D.R. Andrews, PhD thesis, Cambridge University, 1980
11. I.M. Scullion, PhD thesis, Cambridge University, 1987
12. C.R. Seward, C.S.J. Pickles and J.E. Field, SPIE **1326**, 280 (1990)
13. C.R. Seward, C.S.J. Pickles, R. Marrah and J.E. Field, SPIE **1760**, 280 (1992)
14. C.R. Seward, E.J. Coad, C.S.J. Pickles and J.E. Field, SPIE **2286**, 285 (1994)
15. R.J. Hand, J.E. Field and D. Townsend, J. Appl. Phys. **70** (11), 7111 (1991)
16. Z. Feng and J.E. Field, J. Hard Mater. **1** (4), 273 (1991)
17. P. L. Kaye. and J.E. Field, J. Hard Mater. **4**, 167 (1993)
18. R.W. Davidge, Mechanical Behaviour of Ceramics, (Cambridge University Press, 1979)
19. L.P. Hehn, Z. Chen, J.J. Mecholsky, Jr., P. Klocek, J.T. Hoggins and J.M. Trombetta, J. Mater. Res., 9 (6), 1540 (1994)
20. R.S. Sussman, J.R. Brandon, G.A. Scarsbrook, C.G. Sweeney, T.J. Valentine, A.J. Whitehead and C.J.H. Wort, Diamond and Related Materials **3**, 303 (1994)
21. D.C. Harris, SPIE **2286**, 218 (1994)
22. A.B. Harker, D.G. Howitt, S.J. Chen, J.F. Flintoff and M.R. James, SPIE **2286**, 254 (1994)
23. C.A. Klein, Diamond and Related Materials, **2**, 1024 (1993)
24. C.R. Seward, J.E. Field and E.J. Coad, J. Hard Mater. **5**, 49 (1994)

VIBRATIONAL SIGNATURES OF DIAMOND SURFACES

Th. Köhler, Th. Frauenheim, * D. Porezag, M. Pederson, **
* Theoretische Physik III, Technische Universität, 09107 Chemnitz, Germany
** NRL, Complex Systems Theory Branch, NRL, Washington D.C.20375

ABSTRACT

Stable clean and hydrogenated diamond (100) and (111) surface reconstructions found by density-functional molecular-dynamics (DF-MD) are characterized by their vibrational properties. For sufficiently large surface slab supercells we have calculated the surface top-three-layer projected vibrational spectra and reproduced the main features obtained from experiments. The various reconstructions of the diamond surface yield rich spectra of surface modes involving excitations of dimers, trimers and chains. We classify most characteristic surface modes and discuss correlations to peaks observed in surface sensitive experiments. The vibrational spectra represent signatures of the considered surfaces that might be used to understand and to classify as grown diamond surfaces.

INTRODUCTION

One of the key problems in surface analysis and the understanding of diamond growth related properties is an unequivocal relation of experimental surface sensitive features to well defined theoretical signatures obtained from *ab initio* atomic scale surface modelling. Whereas in earlier studies the diamond surfaces were aimed only at gathering more insight into silicon surfaces, the successful high-rate synthesis of diamond films by CVD is now redirecting growing attention to the diamond surfaces themselves.

Surface sensitive experimental techniques like high resolution electron-energy-loss spectroscopy (HREELS) [1, 2], and scanning tunneling microscopy (STM) [3] allied to theoretical simulations [4, 5, 6, 7] evermore reveal the secret of diamond surfaces. However, there has been limited theoretical work on diamond surface vibrations [8, 9].

In the following paper, theoretical predictions of vibrational signatures of diamond surfaces will rely on *ab initio* calculations of forces based on density-functional (DF) theory using an *ab-initio* nonorthogonal tight-binding (TB) Hamiltonian within a minimal basis representation of localized valence electron orbitals (LCAO). For a detailed description of the theoretical background we refer the reader to Porezag et al.[6, 10]. All diamond surface reconstructions discussed throughout this paper have been previously obtained by applying the same DF-concepts to molecular-dynamics (MD) annealing investigations of two-dimensional periodic surface slab models. The stable surface reconstructions of diamond (100) and (111) are summarized in detail by Frauenheim, et al.[6].

STABLE DIAMOND SURFACE RECONSTRUCTIONS

All MD annealing simulations for determining stable surface reconstructions have been performed by using periodic boundary conditions for surface slabs of finite thickness. In the case of the clean and hydrogenated $C(100)-(2\times1)$ we used 8 C atoms per layer whereas 12 C atoms per layer were used in all cases of C(111) reconstructed surfaces. The bottom double layer of the slab has been fixed to simulate the infinite crystalline substrate by saturating the dangling bonds with hydrogen whilst the stability and dynamic restructuring are studied on the remaining top layers.

Mat. Res. Soc. Symp. Proc. Vol. 383 © 1995 Materials Research Society

(100) surface
Considering first the (100) situation, it is well established that the clean surface is highly unstable with respect to a (2×1) reconstruction, spontaneously favouring strongly π-bonded dimer chains with bond order 1.75 and a dimer bond length of 1.41 Å. The energy gain per surface atom relative to the bulk structure is about 1.5 eV. Studying the effect of atomic hydrogen approaching the surface under thermal conditions, we find that all π-bonds are reactive to hydrogen bonding. As a result, a highly stable monohydrogenated surface is formed that maintains the (2×1) reconstruction, but now with σ-bonded dimers with slightly elongated bonds of 1.61 Å. A conversion of this surface into a bulk (1×1) structure with even higher hydrogen coverage upon heating to 1200 K and further hydrogen supply could not be obtained.

(111) surface
Turning to the (111) surface experiments [11], the observation of rather different features indicates a complicated and rich reconstruction behaviour. As confirmed by STM[3], the 1×1 bulk structure is strongly stabilized by hydrogen termination of the single dangling bond sites. Up to temperatures of 2000 K no hydrogen dissociation could be observed in a short simulation period of 10 ps. In the absence of hydrogen, the π-bonded Pandey-chain has been established as the most stable configuration[4, 6] giving an energy gain of 0.7 eV/surface atom relative to the unreconstructed bulk situation. Though there is experimental evidence for these Pandey-chains on as grown facets [11], no dynamical reconstruction path upon heating the clean (111) surface could be confirmed. In recent MD annealing studies we found that the surface, with an energy gain of 0.3-0.4 eV/surface atom, first tends to graphitize by weakening the bonds to the substrate layer rather than forming the Pandey-chain [15]. Hydrogenating the chains, the $C(111) - (2 \times 1)$ Pandey chain reconstruction is maintained, now forming a metastable surface state that is higher in energy than the hydrogenated bulk C(111) by 0.6 eV/surface atom. Another stable surface reconstruction spontaneously appears if a 1 to 1 adsorption of CH-radicals instead of the H-coverage is given. Due to the presence of two radical electrons, each adsorbed CH tends to bond with two other adsorbed CH, forming single chains at a separation of 4.4 Å within a (2×1) reconstruction of the surface. This situation completely changes if the CH-radical are adsorbed more sparsely on the surface, e.g. by successive nucleation of groups of 3 CH-radicals at three neighboring radical surface sites. In this case symmetric trimers, C_3H_3, are formed as a metastable adsorbate surface pattern. If the coverage with adsorbed trimers is completed a nice $(\sqrt{3} \times \sqrt{3})$ R30° (111) hexagonal surface cell evolves and confirmes recent STM experiments [3, 11]. The (2×1) reconstructed single chains are more stable than the related $(\sqrt{3} \times \sqrt{3})$ R30° configuration by about 0.6 eV/surface atom.

To study the surface projected phonon spectra of all reconstructed surfaces we applied a conjugate gradient method to the stable structures found in order to obtain the minimum energy configurations.

SURFACE PROJECTED VIBRATIONAL DENSITY OF STATES (VDOS)

The vibrational properties have been calculated within the harmonic approximation by constructing the dynamical matrix. By displacing each atom $i, (i = 1, N)$ by δr (in our case $\delta r = 0.02\, a_B$) from its equilibrium position into the directions of the three basis vectors $\vec{e}_\alpha$ of the cartesian coordinate sytem and into the corresponding opposite directions, one can calculate the elements $H_{ij}^{\alpha\beta}$ of the dynamical matrix using the forces $F_{j,\|\vec{\mathbf{e}}_\beta}$ acting on each atom j into the directions $\vec{\mathbf{e}}_\beta$:

$$H_{ij}^{\alpha\beta} = \frac{F^-_{j,\|\vec{\mathbf{e}}_\beta} - F^+_{j,\|\vec{\mathbf{e}}_\beta}}{2\delta r}.$$

The signs refer to the two possible displacements of atom i in the direction $\pm\vec{\mathbf{e}}_\alpha$. After symmetrizing the dynamical matrix and projecting out translational and rotational modes, we solve the general eigenvalue problem

$$\mathbf{H}\vec{\mathbf{Y}} = \omega^2 \mathbf{M}\vec{\mathbf{Y}},$$

where $\mathbf{M}$ denotes a matrix with the atomic masses on the main diagonal while ω and $\vec{\mathbf{Y}}$ are the eigenvalues and their corresponding eigenvectors.

To simulate the transition from surface-like to bulk-like behavior and to eliminate the influence of the H-termination at the bottom C layer in the vibrational density of states, we have fixed these atoms by giving them infinite masses. To separate further the surface phonon modes from coupled excitations of carbon atoms near the surface from bulk-like modes, we used a projection technique. If $\mathcal{E}$ denotes the set $\mathcal{E} = \{\nu_i\}_{i=1}^{M \le 3N}$ of all indices of coordinates of atoms defined as surface-atoms we can calculate the surface density of states $g_S(\omega)$ as

$$g_S(\omega) = \frac{1}{3N} \sum_{s\in\mathcal{E}} \sum_{i=1}^{3N} \delta(\omega - \omega_i) |\langle \vec{\mathbf{p}}_s | \vec{\mathbf{Y}}_i \rangle|^2,$$

where we have to sum over all elements of the index set and $\langle \vec{\mathbf{p}}_s | \vec{\mathbf{Y}}_i \rangle$ is the scalar product of a vector $\vec{\mathbf{p}}_s = (\underbrace{0,\ldots,0}_{s-1}, 1, \underbrace{0,\ldots,0}_{3N-s})$ with the i-th eigenvector $\vec{\mathbf{Y}}_i$.

We found that it is reasonable to project out the topmost three layers for calculation of the surface phonon density of states. In the case of H-termination *of the upper surface* the H layer is also included. Due to the experimental resolution of the HREELS spectrometer [2], we have obtained our spectra by using a Lorentzian of width 125.66 cm^{-1} for the broadening of our eigenvalues.

C(100) – (2 × 1) reconstruction

The surface phonon power spectra are shown in Figure 1, with the most characteristic modes is summarized in the corresponding Table.

Both spectra of the clean and hydrogenated C(100) – (2×1) surfaces show characteristic broad bands below ≈1500 cm^{-1}. In the case of the clean surface we find an additional weak band centered at about 1630 cm^{-1} and in the hydrogenated case a distinct peak at 3078 cm^{-1}. Taking into account the difference in the carbon and hydrogen masses, it is reasonable that the spectrum of the hydrogenated surface appears to be more intense.

In analyzing the vibrational eigenstates we have identified all important surface modes as various excitations of dimers coupled with vibrations of the subsurface layers. For the clean C(100) – (2 × 1), the highest frequency determined at 1667 cm^{-1} belongs to the C–C stretching motion of atoms within the dimers. This value corresponds to the π-like character of the dimer bond which is found in between the values of singly bonded ethane (1116 cm^{-1}) and doubly bonded ethylene (1928 cm^{-1}), obtained by the same DF-TB method.

The C–C stretching vibrations in coupling to subsurface excitations produce the weak band features in the C(100) – (2 × 1) spectrum as shown in Figure 1. Except for this stretching band we can split the spectrum of the clean C(100) – (2 × 1) into three regions with broad bands at 440–820, 820–1120 and 1120–1500 cm^{-1}, respectively. Phonon modes belonging to these bands represent all kinds of spatially hindered C–C dimer vibrations which are coupled with subsurface vibrations. We may characterize bouncing and swinging modes as hindered translations and rocking and twisting modes as hindered rotations of the dimers. Due to coupling between adjacent dimers and the phase relationship of displacements of neighboring dimers we observe these modes twice, once in-phase and once out-of-phase. The same behaviour is observed at the H-terminated surface for which the most important modes are listed in the Table, too.

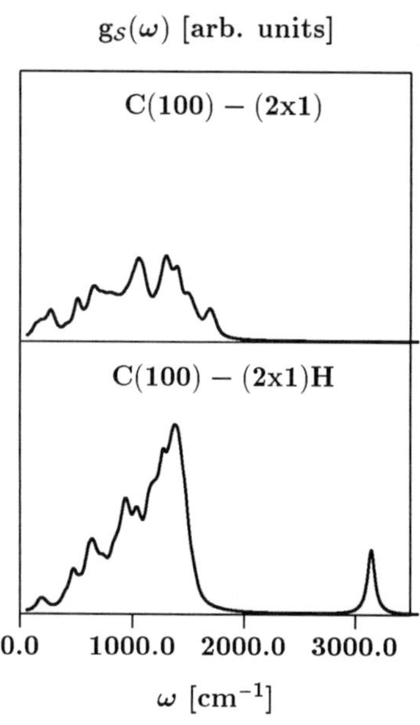

Feature	ω [cm^{-1}]
C(100) – (2 × 1)	
dimer stretching	1667
dimer twisting (in-phase)	1463
dimer twisting (out-of-phase)	1336
dimer swinging (in-phase)	1345
dimer swinging (out-of-phase)	964
dimer bouncing (in-phase)	728
dimer bouncing (out-of-phase)	559
dimer rocking (in-phase)	1322
dimer rocking (out-of-phase)	662
C(100) – (2 × 1)H	
C–H stretching (symmetric)	3078
C–H stretching (asymmetric)	3078
C–H bending (in-phase)	1408
C–H bending (out-of-phase)	1367
C–H bending (in-phase)	1346
C–H bending (out-of-phase)	1342
C–H bending (out-of-phase)	1162
C–H rocking (in-phase)	1309
C–H rocking (out-of-phase)	1307
C–H scissoring (out-of-phase)	1182
C–H scissoring (in-phase)	1052
CH–CH stretching (out-of-phase)	847
CH–CH bouncing (out-of-phase)	580

Figure 1: Surface projected vibrational density of states of (2 × 1)- reconstructed (100) diamond surfaces; the most characteristic surface modes are listed in the related Table.

The region of 440–820 cm^{-1} is clearly dominated by bouncing modes at 559, 728 cm^{-1} and out-of-phase rocking modes at 662 cm^{-1} including various kind of mixing between them. The broad feature around 820–1120 cm^{-1} can be assigned to an intense out-of-phase swinging mode, 964 cm^{-1}. Furthermore, the third band region 1120–1500 cm^{-1} contains all remaining surface modes, such as the dimer twisting modes at 1336, 1463 cm^{-1}, the in-phase rocking mode at 1322 cm^{-1} and the dimer in-phase swinging mode at 1345 cm^{-1}. As the result of a stronger coupling of in-phase motions to the subsurface layer, all in-phase modes are found to lie at higher frequencies compared to the out of phase ones. In comparison with detailed HREELS studies of the clean C(100) – (2 × 1) surface performed by Lee, Apai et al. [2], who found 3 distinct peaks at 700, 1015 and 1225 cm^{-1} and to rescaled Si(100) – (2 × 1) phonon frequencies at 540, 910 and 1260 cm^{-1} obtained by Mele et al. using a tight binding scheme [12], these results are clearly assignable to the broad band regions obtained in our calculation. Only the high frequency modes seem to be slightly shifted to higher frequencies relative to the experiment and to recent theoretical data of Alfonso et al. [9].

In the case of the hydrogenated C(100) – (2 × 1) spectrum, the dominating contributions to the surface VDOS are due to excitations of the H-terminated surface dimers, mostly described as hindered rotations and translations of the dimers and their related hydrogens. The C-H vibrations coupled to the subsurface modes represent the main features on the surface. The most characteristic difference to the spectrum of the clean surface is the existence of a distinct peak centered at 3078 cm^{-1}. This peak is related to the C–H bond stretching vibration. In analyzing the eigenvectors, we find the existence of symmetric and asymmetric C–H bond stretching. Due to a very strong coupling both vibrations mix together within an interval of only 1 cm^{-1}. The features in between 1120–1500 cm^{-1} have been assigned to the various C-H bending and C-H rocking modes. Additionally, we can identify almost pure in-phase and out-of-phase C-H scissoring modes at 1052, 1182 cm^{-1} and there are a lot of intermediate-type coupled vibrations. For the lower frequencies, we further identify an out-of-phase CH-CH stretching at 847 cm^{-1} and one corresponding C-H bouncing mode at 580 cm^{-1}. The mixing of various types of vibrations on the H-terminated dimers and their coupling to the subsurface, makes it more and more difficult to classify the modes within simple terms.

C(111) – (2 × 1) reconstructions

In a second part of our discussion we will concentrate on surface phonon modes of the clean π-bonded C(111) – (2 × 1) (PC) *Pandey chain*, the hydrogenated *Pandey chain*, C(111) – (2 × 1)H (PC) and the hydrogenated *single chain*, C(111) – (2 × 1)H (SC). The formation of parallel chains in all (2 × 1) reconstructions dominates the most characteristic surface phonon modes as various excitations of the chains in coupling with subsurface vibrations. Depending on the phase relation of atoms belonging to two adjacent chains, we again distinguish in- and out-of-phase behaviour and find a great variety of planar and spatial chain excitations. The corresponding modes represent bond-angle preserving and -changing motions of atoms as well as vibrations showing oscillating bond lengths between neighboring atoms in the chains. There are also vibrations in which the entire chains or parts of them perform hindered translations or rotations with respect to the subsurface layers. All modes show a strong interchain coupling generating different periodic behavior. The surface projected phonon spectra of the C(111) – (2 × 1) reconstructions are shown in Figure 2 and a collection of the most significiant modes is given in the related Table.

In analyzing the vibrational properties of the clean π-bonded $C(111)-(2\times1)$ PC, we distinguish between two kinds of chains at the surface, the topmost π-bonded sp^2- and the second layer sp^3-bonded chains. The vibrational spectrum of this surface shows broad bands below 1500 cm^{-1}, at 300–900, 900–1250 and 1250–1480 cm^{-1}. A distinct single peak centered at 1550 cm^{-1} is caused by a symmetric planar sp^2-chain compression mode and a second one, breaking the symmetry of the chain, at 1549 cm^{-1}. Both excitations are in-phase between neighboring sp^2-chains and the bond-angles are not preserved. Due to the strong π- bonding character of the chains, these vibrations occur at the high frequency edge of the spectrum. The sp^2-chains are also responsible for the dominantly planar modes between 1250–1480 cm^{-1}. As the most intense modes in this range in-plane chain angular vibrations between neighboring sp^2-chains at 1407 cm^{-1} and 1405 cm^{-1} appear. Towards lower frequencies other dominating planar shearing modes became visible, in which each row of atoms forming one chain perform a translation against the other. The in-phase mode is located at 1351 cm^{-1} and the out-of-phase one at 1277 cm^{-1}. The lower frequency range, 900–1250 cm^{-1}, is occupied by common excitations of neighboring sp^3- bonded chains and coupled sp^2 and sp^3 chains. Due to the weakening in the σ-chain bonding relative to the π-bonded chains all sp^3 chain vibrations are shifted to lower frequencies. As a very interesting feature, we identify the in- and out-of-phase coupling of neighbouring sp^3 chains via the sp^2 chains on the top layer, which in this particular case is almost stationary. At lower frequencies within this band combined vibrations between neighboring sp^2- and sp^3 - bonded chains are detected. Considering lastly the band at 300–900 cm^{-1}, the modes are clearly dominated by spatial excitations of the sp^2 chains and by translational modes of these chains. We find very intense surface modes from an asymmetric out-of-plane bouncing vibration of the sp^2 chains at 690 cm^{-1} and an in-phase rocking mode at 656 cm^{-1} representing an example of a hindered translation of the chains against the surface. At the low frequency band we always find translational modes of the sp^2 chains. We can assign an out-of-phase symmetric bouncing of the chains at 408 cm^{-1} as an example for a translation perpendicular to the surface and a translational shearing between two sp^2 chains as a planar vibration parallel to the surface at 382 cm^{-1}. The selected modes in the related Table can give only a few illustrations of the rich variety of chain vibrations on such type of surface reconstruction. In HREELS experiment Lee, Apai et al. have identified a very intense feature at 1235 cm^{-1}, which by use of a rescaling technique may be related to a highly dipole-active mode at 58 meV on the $Si(111)-(2\times1)$ surface.

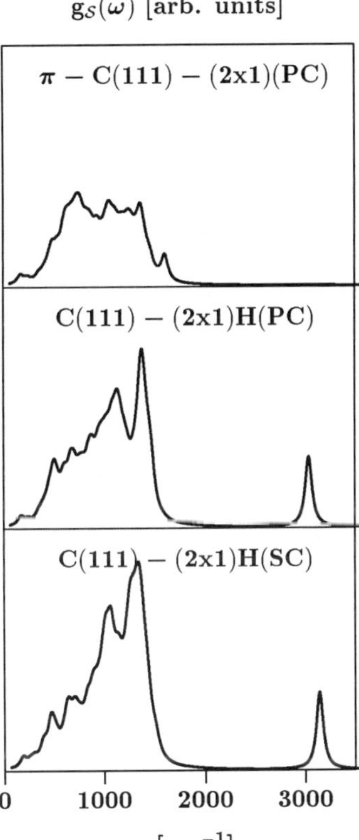

Feature	ω [cm^{-1}]
C(111) − (2 × 1)(PC)	
sym.sp^2 - chain compr. (i)	1550
asym. sp^2 - chain compr. (i)	1549
sp^2 - chain-angular vibr.planar (i)	1407
sp^2 - chain-angular vibr.planar (o)	1405
sp^2 - chain shearing (i)	1351
sp^2 - chain shearing (o)	1277
sp^3 - chain-angular vibr.planar (o)	1210
sp^3 - chain shearing (o)	1201
sym.sp^2-sp^3-chain str. + compr.	1126
asym.sp^2 - chain bouncing (o)	690
sp^2 - chain rocking (o)	656
sp^2 - chain transl.planar (o)	572
sym.sp^2 - chain bouncing (o)	408
sp^2 - chain shearing transl.planar (o)	388
C(111) − (2 × 1)H (PC)	
C–H stretching, compr.	2967-2975
C–H rocking (o)	1416
asymmetric C-H bending (i)	1340
asymmetric C-H bending (o)	1320
symmetric C-H bending (i)	1298
(CH) - chain-angular vibr.planar (i)	1082
(CH) - chain shearing (o)	1078
(CH) - chain bouncing (o)	396
(CH) - chain shearing (betw.chains)	345
(CH) - chain rocking (o)	320
C(111) − (2 × 1)H (SC)	
symmetric C-H str. + compr.	3079
symmetric C-H stretching (i)	3078
asymmetric C-H stretching (o)	3071
asymmetric C-H bending (i)	1391
asymmetric C-H bending (o)	1385
C–H rocking (i)	1332
C–H rocking (o)	1301
C–H scissoring (i)	1211
symmetric C–H bending (i)	1197
symmetric C–H bending (o)	1192
(CH) - chain-angular vibr.planar (i)	1087
(CH) - chain rocking (i)	429

Figure 2: Surface projected vibrational density of states of (2 × 1)- reconstructed (111) diamond surfaces, (i) in-phase, (o) out-of-phase; the most characteristic surface modes are listed in the related Table.

In performing angular dependent electron-energy-loss spectroscopy (EELS) measurements, DiNardo and co-workers identified this dipole-active mode as a longitudinal optical surface phonon polarized in-plane along the π-bonded chain[14]. Converted to the $C(111)-(2\times1)$ *Pandey chain* model, the planar in- and out-of-phase π-chain shearing modes at 1277 and 1337 cm^{-1} represent such possible dipole-active modes in good agreement with the experiment.

Finally, we turn to a discussion of the hydrogen terminated $C(111)-(2\times1)$, on which the σ-bonded C–H chains will be responsible for most of the observable phonon modes. Analysis of the spectra of the hydrogen terminated (2×1) reconstructions shows that both spectra are of similar shape,with broad bands below 1500 cm^{-1}. The single distinct peaks at high frequency in each case are due to symmetric and asymmetric C–H stretching and compression modes. In the case of $C(111)-(2\times1)H$ (PC), there is a continous transition from mostly asymmetric C–H stretching and compression to mostly symmetric vibrations in a range of 2975–2967 cm^{-1}. The situation on the $C(111)-(2\times1)H$ (SC) however, seems to be more clear. The highest frequency at 3079 cm^{-1} appears in a simultaneous appearence of a C–H compression mode on one and a C–H stretching on the neighboring chain. Furthermore, an in-phase C–H stretching mode between neighboring chains has been found at 3078 cm^{-1}. Moving to lower frequencies within this band we obtain a gradual transition from symmetric to asymmetric stretching and compression behavior.

The main part of the spectrum of the hydrogenated Pandey chain can be split into three characteristic bands at 300–800, 800–1200 and 1200–1500 cm^{-1}. The highest frequencies in the third band refer to pure rocking motion of the C–H components perpendicular to the chain. The entire high frequency band is dominated by corresponding hindered rotational vibrations of the (C–H)-complexes. We find a transition from rocking- to scissoring-like vibrational behavior within the chains including all kinds of coupling between them. The most intense modes, see eg. the peak maximum at 1318 cm^{-1}, which dominate the band, are due to (C–H)-complex bending vibrations. For illustration we denote a symmetric (C–H) bending mode at 1298 cm^{-1}, which is an in-phase shearing translation of the chain atoms. The most significant modes belonging to the region 800–1200 cm^{-1} can be assigned to chain excitations in which the (C–H)-complexes move as a unit. As representative examples we identify are chain bond-angle changing and chain shearing (C–H)-complex vibrations at 1082 cm^{-1} and 1078 cm^{-1}, respectively. Moving towards lower frequencies, the "planar" vibrations are replaced to a large extent by spatial vibrations. The (C–H)-complex chains also participate in most of the soft planar or spatial vibrations in the frequency range 300–800 cm^{-1}. As examples for hindered translational motions of the chains we allocate an almost pure symmetric bouncing mode at 396 cm^{-1} and a planar shearing translation of neighboring chains at 345 cm^{-1}. Finally, we will mention a very soft rotational mode of the (C–H)-complex chains moving against the other in an out-of-phase rocking mode at 320 cm^{-1}.

Because of the structural similarity of the chain structures, the $C(111)-(2\times1)H$ (SC) surface spectrum is similar to the *Pandey chain* case. In a related table we briefly summarize the main vibrational properties of this model, too. Compared to the HREELS experimental studies of Aizawa et al. [13], which also gives evidence for the high frequency modes in terms of C–H stretching vibrations our theoretically predicted frequencies are slightly shifted to higher values. Comparing further the exp. data in the range of 1000 – 1500 cm^{-1} the (C–H)-complex bending vibrations are very well confirmed by our studies. All C–H bonding deformation vibrations like symmetric and asymmetric bending, rocking, scissoring modes are found within the experimental determined frequency intervall.

CONCLUSIONS

The understanding of the static and dynamic properties of surfaces has been one of the main aims of surface science over the past decade. Powerful methods like inelastic neutral atom scattering, high resolution electron-energy-loss spectroscopy as well as atomic scale imaging by STM provide a lot of experimental data about the physical behavior of surfaces. In order to explain the results of these experiments for given diamond surfaces, we have calculated the surface projected vibrational spectra within a density-functional molecular-dynamics scheme. We have given a detailed description of specific phonon modes which may be discussed as being particularly responsible for adsorption of various growth species.

We gratefully acknowledge support from the Deutsche Forschungsgemeinschaft within the trinational German-Austrian-Swiss D-A-CH project and we want to thank D. Alfonso and D. A. Drabold at OU-Athens for many interesting discussions and for drawing our attention to the surface projection technique.

REFERENCES

1. B. J. Waclawski, D. T. Pierce, N. Swanson, R. J. Celotta, J. Vac. Sci. Technol. **21** (1982) 368.
2. S.-T. Lee, G. Apai, Phys. Rev. B **48** (1993) 2684.
3. H.-G. Busmann, W. Zimmermann-Edling, S. Lauer, H. Hertel, Th. Frauenheim, P. Blaudeck, D. Porezag, Surf. Sci. **295** (1993) 340.
4. S. Iarlori, G. Galli, F. Gygi, M. Parinello, E. Tosatti, Phys. Rev. Lett. **69** (1992) 2947.
5. S. H. Yang, D. A. Drabold, J. B. Adams, Phys. Rev. B **48** (1993) 5261.
6. Th. Frauenheim, U. Stephan, P. Blaudeck, D. Porezag, et al., Phys. Rev. B **48** (1993) 18189.
7. J. Furthmüller, J. Hafner, Europhys. Lett. in print.
8. B. N. Davidson, W. E. Pickett, Phys. Rev. B **49** (1994) 11253.
9. D. R. Alfonso, D. A. Drabold, S. E. Ulloa, Phys. Rev. B, in print.
10. D. Porezag, Th. Frauenheim, Th. Köhler, G. Seifert, R. Kaschner, submitted to Phys. Rev. B, Oct. 1994.
11. W. Zimmermann-Edling, Thesis, Universität Freiburg/Breisgau, 1994.
12. E. J. Mele, D. C. Allan, D. L. Alerhand, D. P. DiVincenzo, J.Vac.Sci.Technol. **34** (1985) 1068.
13. T. Aizawa et al., Phys. Rev. B **48** (1993) 18348.
14. N. J. DiNardo, W. A. Thompson, A. J. Schell-sorokin, J. E. Dermuth, Phys. Rev B **34** (1986) 3007.
15. G. Jungnickel, D. Porezag, Th. Frauenheim, W. R. L. Lambrecht, B. Segall and J. C. Angus, MRS proceedings 1995 , same Volume .

DIAMOND {111} SURFACE: GRAPHITIZATION OR RECONSTRUCTION?

G. Jungnickel, D. Porezag, Th. Frauenheim * W. R. L. Lambrecht, B. Segall** and J. C. Angus†
*Institut für Physik, Theoretische Physik III, Technische Universität Chemnitz, D-09009 Chemnitz, Germany
**Department of Physics, Case Western Reserve University, Cleveland OH 44106-7079, U.S.A.
†Department of Chemical Engineering, Case Western Reserve University, Cleveland OH 44106-7217, U.S.A.

ABSTRACT

The reconstruction of the diamond {111} surface is re-examined by means of density functional theory based tight-binding molecular dynamics. Evidence is found for competition between a graphitizing tendency leading to an unreconstructed but relaxed 1×1 surface and a π-bonded chain-like 2×1 reconstruction. The implications of the possible co-existence of these two distinct surface phases for diamond growth are discussed.

INTRODUCTION

The diamond {111} surface is well known to undergo a 2×1 reconstruction upon annealing at 1100–1300K [1, 2, 3, 4]. This reconstruction is related to the desorption of hydrogen and the original 1×1 surface before annealing is thus widely believed to be completely saturated with hydrogen (terminating each of the dangling bonds). The basic model for this reconstruction obtained by total energy minimization [5] is the Pandey π-bonded chain model [6]. Although recently there has been some discussion of refinements of this model, addressing the question of whether the π-bonded chain is symmetric [5, 7, 8], buckled as in Si [9], or dimerized [10, 11], the basic model appears to be well established. Less well established is the question of how the reconstruction actually forms. While Northrup and Cohen [12] and Ancilotto *et al.* [9] for Si {111} proposed a smooth bond-switching path from 1×1 to 1×2, which was claimed [9] to be related to a soft-phonon mode of a saddle-point in the total energy surface of the 1×1 model, Frauenheim *et al.* [7] proposed a two step model in which successive dimer rows of additonal C are absorbed on the surface as a possible alternative.

One motivation for revisiting this problem is the recent paper by Davidson and Pickett [13] in which they found stepped {111} surfaces to graphitize instead of undergoing the 2×1 reconstruction. Graphitization of diamond {111} had been suggested earlier by Phillips [14] but this model was abandoned when the 2×1 reconstruction was discovered. The new findings of Davidson and Picket raise a number of questions: 1) is this graphitic model a metastable state or a true groundstate for the stepped surface? 2) is such a graphitic layer model also a possible metastable state for the step-free surface, or are steps essential in stabilizing this model, and, if so, how large can the terraces be? 3) if a graphitic model is possible as transition metastable state, then what is its role in the dynamic formation of 2×1?

An additional reason for our interest in these questions is that graphite was proposed by Lambrecht *et al.* [15] to be a likely precursor to de-novo nucleation of diamond by epitaxial nucleation on the graphite prism planes (at the edges of graphite particles). If the diamond

Mat. Res. Soc. Symp. Proc. Vol. 383

surface under growth conditions has graphitic overlayers, then these may well be involved in the growth mechanism as a conversion layer in a similar manner as described earlier for the nucleation problem. A graphitic conversion layer growth model has already been proposed previously by Tamor and Hass [16]. These are sufficient reasons to re-examine the {111} surface reconstructions.

In the present paper, we present some initial results from molecular dynamics simulations at constant high temperature, suggesting that at the {111} surface in fact there is a competion between a tendency to graphitize and to undergo a 2×1 reconstruction. We find that these two surface phases may co-exist side by side on the surface. Although we suggest that these may actually be non-equilibrium states, they may well play a role in the growth because we find that adsorbates of carbon act very much like steps in promoting the graphitization.

COMPUTATIONAL METHOD

Molecular dynamics simulations were carried using interatomic forces calculated using a tight-binding type expression for the total energy

$$E_{tot} = \sum_{i}^{occ} \epsilon_i + \frac{1}{2} \sum_{\mathbf{R},\mathbf{R}'} \Phi(|\mathbf{R} - \mathbf{R}'|), \quad (1)$$

with Φ a semi-empirically adjusted pair-potential and $\sum_i^{occ} \epsilon_i$ the sum of occupied eigenvalues of a Hamiltonian based on local density functional theory. The Hamiltonian makes the two-center approximation and uses matrix elements calculated from "compacted" self-consistently calculated free atoms using a Gaussian minimal basis set. The compaction of the atomic charge densities improves the transferability of the Hamiltonian matrix elements to the solid state environment. The pair-potential is adjusted so as to obtain the correct bond length, bond energy and vibrational frequency of the diatomic molecule as well as lattice constants, bulk moduli, and cohesive energies of diamond and graphite. The method was described in more detail elsewhere [17, 18] and tested on clusters, hydrocarbon molecules[17], C_{60} [19] and was found to give excellent results for modelling of amorphous carbon [20] and previous diamond surface studies.[7].

The models used for the present simulations consist of 4-6 double layers with periodic boundary conditions. Each layer contains $N \times N$ primitive {111} surface unit cells with $N = 4, 6$. The bottom layer is capped with hydrogen and kept fixed. The Verlet algorithm is used with timesteps of ~1 femtosecond and typical runs are for about 200-600 timesteps. Calculations were carried out at various constant temperatures from 300 K to 4500 K in steps of 600 K with initial random velocities and starting from the ideal 1×1 hydrogen free surface after relaxation with steepest descent. Some runs were restarted from specific initial configurations appearing as snapshots in the above runs in an attempt to stabilize structures of possible interest. Some results were also obtained on 5 layer models with some isolated carbon or hydrocarbon rings adsorbed on them in the middle of the 2D-unit cell.

RESULTS

The structure of a typical simulation is shown in Fig. 1. This picture shows an average over snapshots at 10 different times. One may see that the surface layer has relaxed slightly outwards and has a reduced buckling. Bonds in the range $1.85\text{Å} < d < 2.30\text{Å}$ are indicated

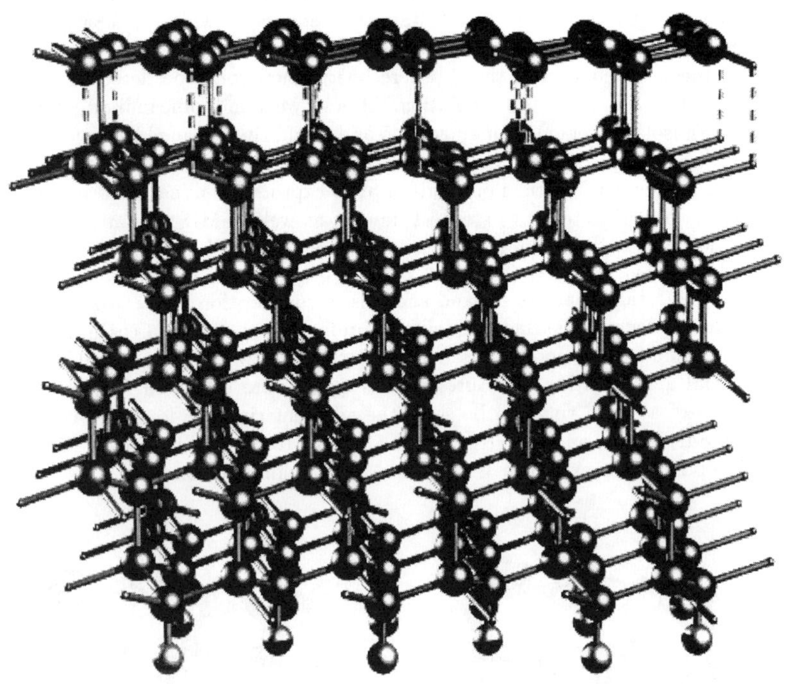

Figure 1: Structure of a diamond {111} surface model averaged over several snapshots after 0.6 ps at 1200K. Note broken bonds and flattening of top surface.

by dashed lines. The breaking of bonds as well as the tendency to flatten out the hexagonal network of the top double layer indicates a clear tendency towards graphitization. We find that with increasing temperature this tendency increases. The average interplanar distance between the centers of the double layers for the surface to subsurface layers (d^s_{111}) increases relative to that in the bulk (d^b_{111}) by 7 to 22 % and varies almost linearly with temperature. The buckling of the surface layer measured as the average width of the double layer initially is reduced but at higher temperatures increases because of the thermal fluctuations. The number of broken bonds increases with increasing temperature. By 2100 K the number of in tact bonds between surface and subsurface layer was reduced to ~16 % and the surface layer became totally detached (in the sense of having only weak bonds) by 2700 K. Bonds in the lower layers only started breaking after the first layer was completely "graphitized", i.e. when all its bonds to subsurface were converted into "weak" bonds. This differs from the recent results of de Vita *et al.*[21]. The surface energy of these models increases almost linearly with temperature by about 20±5 meV/Å^2 per 100K. The ideal bulk terminated surface energy was calculated to be 2.32 eV/(surface atom) higher than that for the bulk, while the relaxed model (obtained by steepest descent) from which we start had an energy lower by 0.17 eV /(surface atom). All our models have thus higher energies than the minimum energy π-bonded chain 2×1 state, the energy of which is 0.85 eV/(surface atom) lower than that of the ideal surface.

No spontaneous formation of the 2×1 reconstruction could be detected within the admittedly short time frame of the simulation. At best, we found some individual snapshots that indicate an isolated region with adjacent 5 and 7 fold rings formed by bond switching. These are characteristic of the Pandey π-bonded chain model. We took this model as starting point for further MD runs in an attempt to capture the dynamic formation of the reconstruction. We subjected it to steepest descent as well as MD runs at low T (300K). However, it never locked into the Pandey π-bonded chain model.

We concluded from this that the size of the 5-7 ring structure may have been too small a fraction of the total surface to expand and was in competition with the 1×1 structure. We then examined smaller surface models. A similar snapshot with a chain-like structure reminiscent of the Pandey model was found. In this case we were indeed able to quench it. The initial and resulting structures of this quench are shown in Fig. 2. Although the formation of a 5-fold ring can be seen, the adjacent ring is 6-fold, not 7-fold, one of its bonds is quite weak, and the structure is adjacent to a graphitic region. This picture suggests the possible co-existence of 1×1 and 2×1 chain-like domains.

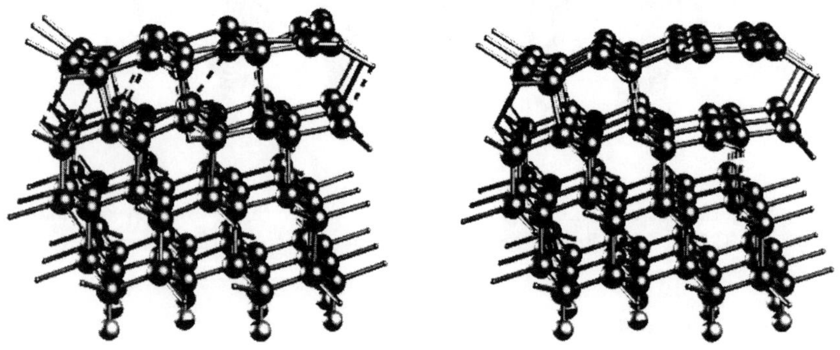

Figure 2: Initial (left hand side) and quenched model of {111} surface showing the co-existence of structures reminescent of the Pandey π-bonded chain and a graphitic 1×1 domain.

In an attempt to simulate some aspects of growth conditions, we also investigated models with one or three fused C_6 rings with or without hydrogenation of their edges on top of the surface. The initial coordinates of these models were obtained simply by eliminating part of the surface layer of the previous models. We thus start out from a configuration in which the ring is already incorporated in the diamond structure, but the top surface layer has not yet formed a complete mono-layer. The model with hydrogenation corresponds to C_6H_6 (i.e. a benzene like fragment, although again, we started from a buckled configuration of the ring). Although these models require further study, a few initial interesting observations could be made. Already, after 200 time steps and at a temperature as low as 300 K, substantial bond breaking took place in the surface layer, except directly underneath the adsorbed atoms. At higher temperatures, or, after longer runs, the rings usually broke up into fragments and the surface and subsurface layer had many broken bonds. The region directly underneath the adsorbed species continued to serve as "anchoring" points for these graphitic structures that were forming on the surface.

The example shown in Fig.3 was chosen because it illustrates two things at the same time. It has a C_6H_6 ring adsorbed onto the top layer. The regions away from the ring are clearly seen to graphitize. One may also notice on the left of the fragment a bond flipping event which might lead to the formation of a 5-membered ring. Typically, in such structures, we found the graphitic surfaces to be slightly curved, very similar to the structure of the top layer of the stepped surfaces studied by Davidson and Pickett.[13]

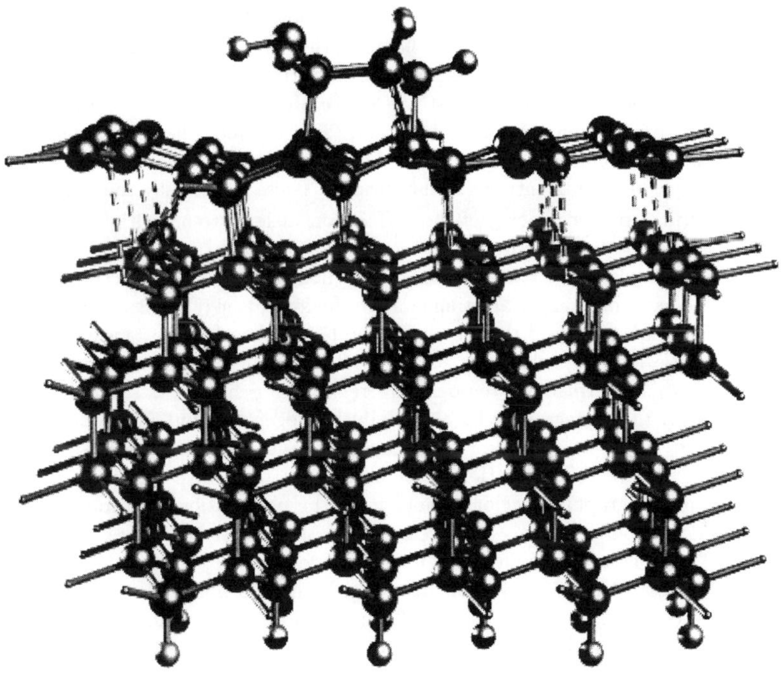

Figure 3: Model of diamond {111} with adsorbed C_6H_6 fragment after 0.2 ps at 300K. Note broken bonds and bond flip event on the left.

DISCUSSION

Our results are at first sight rather different from the prevalent point of view of the formation of the 2×1 reconstruction. We note, however, that, to our knowledge, this is the first attempt to model the behavior of the diamond {111} surface at constant elevated temperatures. Previous work, including the MD work of Ancilotto *et al.* [9] on Si {111} was primarily directed towards finding the minimum energy state of the surface. To this end, Ancilotto *et al.* [9] used an inital heating up of the surface (so as to overcome barriers), after which the system was slowly cooled to the ground state. Under these circumstances, the 2×1 reconstruction was found to occur spontaneously in about 1 ps, following a transition path similar to the one proposed earlier by Northrup and Cohen [12] and according to Ancilotto *et al.* [9] related to a soft-phonon instability of the saddle-point like 1×1 relaxed state. Our present calculations indicate that for C at constant elevated temperature, this transition clearly does not happen as easily and spontaneously. Instead, we find basically a trend towards graphitization.

The pseudo-graphitized 1×1 state of our surfaces is somewhat reminiscent of Vanderbilt and Louie's [5] relaxed but unreconstructed surface. We find that at higher temperatures, this state progressively becomes more graphitic in the sense that more bonds to the subsurface are broken and that the surface layer floats farther and farther above the surface and loses its diamond-like buckling. It is clearly not a static structure at these temperatures and shows rather floppy behavior with strong fluctuations of the atomic positions normal to the surface and with bonds flopping back and forth between the 5-7 ring and 6-6 ring configurations. This floppiness suggests that the higher energy we obtain for these models may be offset by vibrational entropy of the surface. These states may thus correspond to a local minimum in the free energy of the surface, or, at least to an initial transition state after hydrogen has been removed.

Although our high-T 1×1 models are clearly not ground state structures, we believe they have some legitimacy as a transition state in the formation of the 2×1 reconsstruction. In fact, Hamza *et al.*[4] showed that the hydrogen leaves the {111} surface already at 1100 K while the half-order spots characteristic of the 2×1 reconstruction only start appearing at 1270 K. In this intermediate temperature regime, which is of particular interest for diamond growth, the surface may be essentially hydrogen free but unreconstructed. We believe our models at 1200 K to be of particular relevance to this state.

A second experimental indication that the surface may have exhibit a co-existence of 2×1 and pseudographitic regions comes from the observation that much less than a monolayer of hydrogen, as low as 5 % according to Mitsuda *et al.*[22], is required to force reversion to 1×1. This follows naturally if the surface is already partially pseudographitic 1×1 and only the 2×1 parts need to be undone by hydrogenation. However, an alternative theory for how coverages of only about $\theta = 0.3$ can undo the 2×1 reconstruction was given for Si by Ancilotto and Selloni [23]. One may think that when the reconstruction is undone locally, stress must build up at the boundaries between (hydrogenated) unreconstructed and (hydrogen-free) reconstructed domains. This may force the dereconstruction before the rehydrogenation is complete. While the prevailing point of view is that the equilibrium state of the hydrogen-free diamond surface is 2×1 reconstructed, and the equilibrium state of the 1×1 surface is completely saturated with hydrogen, there are clearly intermediate stages in which parts of the surface could be basically hydrogen free and 1×1.

We interpret our difficulties to stabilize the π-bonded chain fragments as an indication that a minimal critical size of such flipped bond regions is required before the system locks into the lower energy 2×1 reconstruction. This assertion is based on the fact that we could only stabilize a 2×1 region in the small cell model where it was a substantial fraction of the surface. This suggests that a nucleation and growth type of dynamics is involved as is often the case in the competition between two phases. Further investigations of the process of locking into the minimum energy Pandey reconstruction are necessary. In view of Ancilotto *et al.* 's work [9] a slow cooling scheme will be needed. In the future, we also plan to see whether C and Si show significantly different behavior in this respect.

Finally, our calculations with adsorbed diamond fragments, suggest that the latter facilitate the graphitization. The structure near such fragments is similar to that near steps. This conclusion is thus in accord with Davidson and Pickett's [13] that steps lead to graphitization of the top layer. Apparently, the region directly underneath the adsorbed atoms is more rigid and hence forms an anchoring point for the graphitic regions which otherwise would become detached from the surface. This anchoring is in some sense a generalization of the structures found at the prism plane interfaces between graphite and diamond [15]. The buckled rings of diamond can apparently rather smoothly join up with the flat ring network of graphite, eventually curving the graphitic sheet between the anchoring points. The fact that this anchoring is observed with various adsorbate structures on the top surface suggests that it is not restricted to a very specific local electronic or bonding structure but rather to a mechanical effect. One may think of the region directly underneath the adsorbed diamond fragment as a stiff object (in which tetrahedral bonding is conserved because it is locally a subsurface layer instead of a surface layer) embedded in a softer medium which is at the point of breaking bonds. Vibrational waves in this softer region between the top layer which has no adsorbates on top of itself may then tend to form standing waves between the stiff objects and hence may more easily break, i.e. graphitize.

The possible occurence of these "graphitic" structures on the surface at the growth temperature suggests that these may be involved in the growth proces. If some breaks occur in this graphitic surface layer and they expose edges, the hydrogenation conversion mechanism suggested in [15] would apply and may convert the growing graphitic layer into diamond.

Indirect experimental evidence for the possibility of a graphitic conversion layer comes from the observation by Fallon and Brown [24] of sp^2 carbon at the grain boundaries in polycrystalline diamond films by scanning electron microscopy using the electron energy loss signal. Although other explanations may be offered for this observation, it is at least suggestive that these may be a remnant of such a conversion layer. The fact that CVD diamond films immediately after growth typically have a higher conductivity than natural IIa diamond also suggests that the diamond surface may have a conductive (and hence possibly graphitic) surface layer during growth. However, it was also found that hydrogen passivation of traps can increase the conductivity of natural and CVD diamond by several orders of magnitude.[25, 26, 27] Other references to graphitic carbon being present during growth of diamond may be found in Ref. [28] and [29].

CONCLUSIONS

Molecular dynamics simulations at constant elevated temperatures show a tendency for the diamond surface to graphitize instead of undergoing a 2×1 reconstruction. This trend becomes progressively stronger with higher temperatures. Although this is clearly not a minimal energy state, it is suggested to be a dynamic transition state in the formation of the reconstruction. This suggests that the dynamics of the 2×1 reconstruction may be slower than previously thought and may involve nucleation and growth of competing (surface) phases. In fact, we find evidence that such structures may exist side by side on the surface. Clearly, this will have important impact on our understanding of the growth of diamond on {111}. We also found that adsorbates facilitate graphitization. All of this suggests that diamond growth on {111} may take place through a graphitic conversion layer as was previously suggested by Tamor and Hass.[16] Some indirect experimental evidence for the existence of a pseudographitic 1×1 surface phase was mentioned.

This work was supported by the Deutsche Forschungs Gemeinshaft and the National Science Foundation. W. R. L. wishes to thank B. Pate for a useful discussion on the rehydrogenation issue and A. de Vita for a discussion of mechanical aspects of the surface layer diamond/graphite interface and his studies of graphitization.

REFERENCES

1. J. J. Lander and J. Morrison, *Surf. Sci.* **4**, 241 (1966)
2. B. B. Pate, *Surf. Sci.* **165**, 83 (1986).
3. S. V. Pepper, *Surf. Sci.* **123**, 47 (1982).
4. A. V. Hamza, G. D. Kubiak, and R. H. Stulen, *Surf. Sci.* **206**, L833 (1988).
5. D. Vanderbilt and S. G. Louie, *Phys. Rev. B* **30**, 6118 (1984).
6. K. C. Pandey, *Phys. Rev. Lett.* **47**, 1913 (1981); **49**, 233, (1982).
7. Th. Frauenheim, U. Stephan, P. Blaudeck, D. Porezag, H.-G. Busmann, W. Zimmermann-Edling, and S. Lauer, *Phys. Rev. B* **48** 18189 (1993).
8. W. G. Schmidt and F. Bechstedt, *Phys. Rev. Lett.*, to be published.
9. F. Ancilotto, W. Andreoni, A. Selloni, R. Car, and M. Parinello, *Phys. Rev. Lett.* **65**, 3148 (1990).
10. S. Iarlori, G. Galli, F. Gygi, M. Parrinello, and E. Tosatti, *Phys. Rev. Lett.* **69**, 2947 (1992).
11. C. Kress, M. Fiedler, and F. Bechstedt, *Europhys. Lett.* **28**, 433 (1994)
12. J. E. Northrup and M. L. Cohen, *Phys. Rev. Lett.* **49**, 1349 (1982).
13. B.N. Davidson and W. Pickett, *Phys. Rev. B* **49**, 14770 (1994).
14. J. C. Phillips, *Surf. Sci.* **40**, 459 (1973).
15. W. R. L. Lambrecht, C. H. Lee, B. Segall, J. C. Angus, Z. Li, and M. Sunkara, *Nature* **364**, 607 (1993).
16. M. Tamor and K. C. Hass, *J. Mater. Res.* **5**, 2273 (1990).
17. P. Blaudeck, Th. Frauenheim, D. Porezag, G. Seifert, and E. Fromm, J. Phys. Condens. Matter **4**, 6389 (1992).
18. D. Porezag, Th. Frauenheim, Th. Köhler, G. Seifert and R. Kashner, *Phys. Rev. B*, to be published.

19. P. Blaudeck, Th. Frauenheim, H.-G. Busmann, and T. Lill, *Phys. Rev. B* **49**, 11409 (1994).

20. Th. Frauenheim, P. Blaudeck, U. Stephan, and G. Jungnickel, *Phys. Rev. B* **48**, 4823 (1993); Th. Frauenheim, G. Jungnickel, Th. Köhler, and U. Stephan, *J. Non-Cryst. Solids* **182**, 186 (1995), and refs. therein.

21. A. De Vita, G. Galli, R. Car, and A. Canning, *Bull. Am. Phys. Soc.* **40**, 597 (1995); and private communication.

22. Y. Mitsuda, T. Yamada, T.J. Cuang, H. Seki, R. P. Ching, J. Y. Huang, and Y. R. Shen, *Surf. Sci* **257** L633 (1991).

23. F. Ancilotto and A. Selloni, *Phys. Rev. Lett.* **68**, 2640 (1992).

24. D. J. Fallon and L. M. Brown, *Diamond & Rel. Mater.* **2**, 1004 (1993).

25. B. R. Stoner, J. T. Glass, L. Bergman, R. J. Nemanich, L. D. Zoltal, and J. W. Vandersande, *J. Electron. Materials* **21**, 629 (1992).

26. M. I. Landstrass and K. V. Ravi, *Appl. Phys. Lett.* **55**, 975 (1989); *ibid.* **55**, 1391, (1989).

27. S. Albin, and L. Watkins, *IEEE Electron. Device Lett.* **11**, 159 (1990).

28. J. C. Angus and E. A. Evans, in *Novel Forms of Carbon*, ed. C. L. Renschler, D. Cox, J. Pouch, and Y. Achiba, *Mater. Res. Soc. Symp. Proc.* Vol. 349 (1994) p. 385, and refs. therein.

29. A. R. Badzian, T. Badzian, R. Roy, R. Messier,and K. E. Spear, *Mat. Res. Bull.* **23**, 531 (1988).

Part VI

Applications

PROCESSING-MICROSTRUCTURE-TENSILE PROPERTY OF VAPOR GROWN CARBON FIBER REINFORCED CARBON COMPOSITE

JYH-MING TING
Applied Sciences, Inc., 141 West Xenia Avenue, Cedarville, OH 45314

ABSTRACT

In contrast to the form in which other carbon fibers are produced, vapor grown carbon fiber (VGCF) is produced from gas phase precursors in the form of individual fibers of discrete lengths. VGCF can be harvested as a mat of semi-aligned, semi-continuous fibers, with occasional fiber branching and curling. The use of VGCF mats as reinforcement result in composites which exhibit unique microstructure and physical properties that are not observed in other types of carbon composites. This paper describes the processing of VGCF mats reinforced carbon composites, and its unique microstructure and properties. Utilization of fiber tensile properties, as well as thermal conductivity, in the composites is discussed. Comparison of experimental results from various VGCF composites to theory indicates that mechanical properties are more strongly affected by characteristics of VGCF mat than are thermal conductivity. The implications of this relationship favors applications for thermal management where structural demands are less stringent.

INTRODUCTION

High-power, high-density electronic devices are required in advanced electronic systems for improved performance, endurance, and survivability. Various integrated technology developments are therefore being pursued in order to meet the requirements. One of the important technologies being sought is to improve thermal management technique while achieving weight and size reduction. This involves the use of high performance materials in electronic packaging. A high performance material is depicted by its high thermal conductivity, light weight, matching coefficient of thermal expansion (CTE), and desired mechanical properties. These characteristics can only be met by composite materials, which have recently become attractive for applications in thermal management. [1 2 3] This paper presents a high performance vapor grown carbon fiber reinforced carbon composite exhibiting thermal conductivity only second to that of diamond. Processing, microstructure, and tensile properties of the composite are discussed.

VAPOR GROWN CARBON FIBER (VGCF)

Vapor grown carbon fiber (VGCF) is produced through the pyrolysis of hydrocarbon gas in the presence of a metal catalyst at temperatures near 1000 °C. The purity of the carbon source and the mechanism of fiber growth result in a highly

Mat. Res. Soc. Symp. Proc. Vol. 383

graphitizable fiber with a unique lamellar morphology, as shown in Fig. 1, in which the graphitic basal planes have a high degree of preferred orientation and are nearly parallel to the fiber axis. VGCF therefore possesses excellent physical properties as given in Table I. [4] It is noted in Table I that VGCF exhibits an outstanding thermal conductivity which is approaching that of natural diamond and the highest among all carbon fibers as shown in Fig. 2. [5] Composites based on VGCF thus have thermal conductivities that cannot not be achieved by using any other types of reinforcement, except for diamond fiber that is currently under development. [6,7]

Fig. 1. Cross sectional view of VGCF.

Although growth of carbon fibers from vapor phase hydrocarbon was observed more than 100 years ago, [8] methods for producing research quantities had not been developed until the early 1970's. [9,10,11] More recent work has been directed to methods of production of VGCF with different morphologies and in larger quantities. [12,13,14] For example, VGCF can be produced on catalyst-seeded substrates, where the catalyst is a transition metal particle. Such fibers can be grown in a

Table I. Physical properties of single vapor grown carbon fiber.

Property		Value		
Length		10^{-3} - 30 cm		
Diameter		10^{-3} - 0.3 cm		
		As Grown	Annealed @ 2800 °C	
Density		1.8	2.0 g/cc	
Tensile Modulus*	lower bound	230(33)	360(52)	GPa (Mpsi)
	upper bound	400(58)	600(87)	GPa (Mpsi)
Tensile Strength*	lower bound	2.2(0.31)	3(0.43)	GPa(Mpsi)
	upper bound	2.7(0.39)	7.0(1.0)	GPa(Mpsi)
Ultimate Strain		1.5	0.5	%
Electrical Resistivity		10^{-3}	6.10^{-5}	Ω-cm
Thermal Conductivity		20	1950	(W/m-K)

(* Data from testing VGCF with a diameter of 7μm.)

three-stage process so that lengthening of the fiber can be independent of the thickening of the fiber. [15] In contrast to the form in which other carbon fibers are made, VGCF thus produced can be in the form of interwoven mat supported on a substrate. [16] A VGCF mat consists of semi-aligned, semi-continuous, multi-layer fibers. Fibers often branch or terminate within the mat. Although the majority of the fibers are straight, some have the appearance of helices or are curled. As discussed later, these features of VGCF mat have different effects on composite thermal and mechanical properties, respectively.

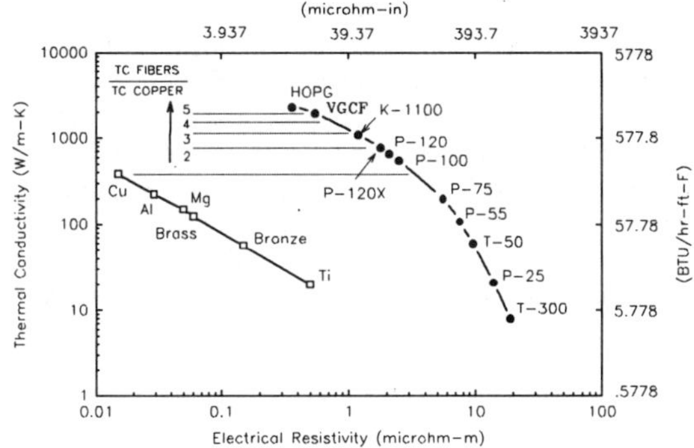

Fig. 2. Thermal conductivity of various carbon fibers and selected metals.

VGCF REINFORCED CARBON (VGCF/C) COMPOSITE

Owing to its unique mat form, weaving of fibers to produce a fiber cloth is not required which is in contrast to other carbon fibers. As-grown VGCF mats can be used to prepare fiber preforms by hand lay-up of the mats with desired dimensions and orientations into a mold. Furfuryl alcohol is usually used as binder in preforms. Molding of preform is normally performed at a temperature near 120 °C and a pressure between 35 MPa to 70 MPa.

After the molding process, rigidized preforms are subjected to two heat treatment cycles. The first heat treatment is to carbonize the binder at a temperature near 900 °C. The second heat treatment is to graphitize the preforms at a temperature greater than 2800 °C.

Preform thus made can be densified by chemical vapor infiltration (CVI) and/or pitch infiltration (PI). Prolonged densification time or multi-cycle densification is required to achieve a high composite density. Fig. 3

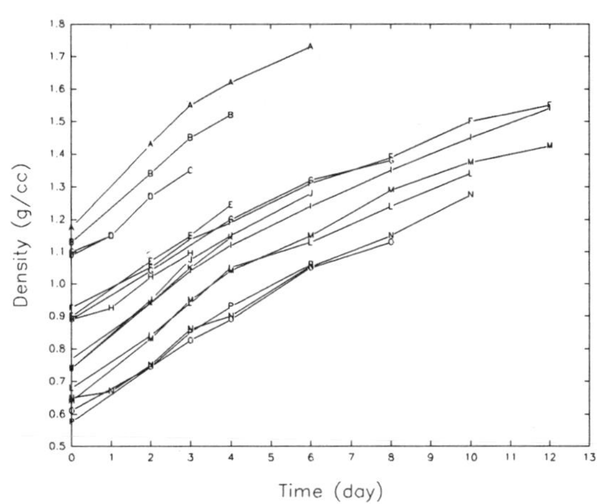

Fig. 3. Composite density as a function of CVI densification time.

shows such an example. Shown in Fig. 3 is VGCF/C composite density as a function of densification time for a number of specimens. Preforms for these specimens have different fiber volume fractions and densities. Density was calculated from weight gain recorded by a micro-balance attached to a CVI furnace. As expected, densification of preforms depends on the fiber volume fraction and density of a preform. For composite specimens prepared by pitch infiltration, heat treatment, at a temperature near 1800 °C, between densification cycles is carried out to generate open pores for further pitch infiltration. After final densification, all composite specimens are subjected to a final high temperature heat treatment in argon atmosphere for graphitization of the matrix carbon.

MICROSTRUCTURE OF VGCF/C COMPOSITE

Microstructure of VGCF/C composite was examined using an optical microscope (OM) under polarized light. Specimens for OM analysis were polished using a standard metallurgical process. In general, composite densification by either CVI or PI densification is homogeneous.

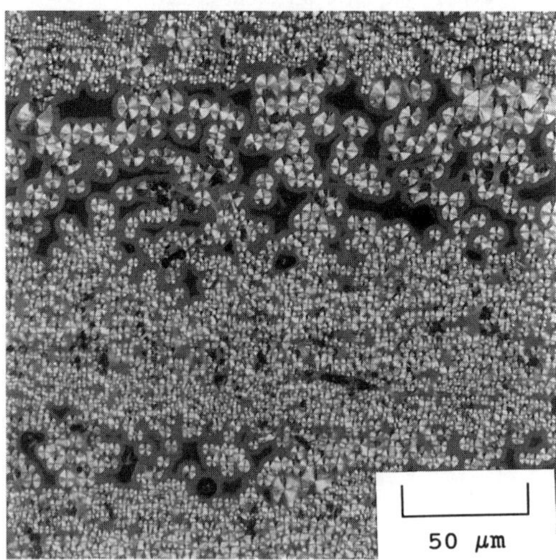

Fig. 4. VGCF/C composite showing pockets are associated with large-diameter fibers.

However, large pockets are seen and associated with large-diameter VGCF, as shown in Fig. 4 for a composite specimen densified by CVI. [17] It is believed that the occurrence of these large pockets is a result of molding or compacting of fibers with various diameters. Although compact efficiency is independent of fiber diameter, individual void by larger diameter fiber is larger than that by smaller diameter fiber. Larger, un-filled space therefore exists around large diameter fibers after the densification. It is noted in Fig. 4 that each cross features a fiber end of VGCF. This unique appearance is seen under polarized light due to the highly anisotropic nature of VGCF. Surrounding VGCF is matrix carbon obtained from CVI.

Depending on processing parameters, composite matrix exhibits different degrees of isotropicness and graphiticness. Fig. 5 compares two CVI composite specimens obtained at different CVI temperatures. Matrix carbon obtained at a CVI temperature lower than 1000°C is isotropic (Fig. 5A). When the CVI temperature is raised to above 1100 C°, matrix carbon becomes more anisotropic, as sown in Fig. 5B. However, CVI matrix carbon is not as graphitizable as VGCF. On the other hand, VGCF/C composite

(A) (B)

Fig. 5. CVI composite specimens obtained at temperatures (A) below 1000 °C and (B) above 1100 C°, respectively.

specimens obtained using pitch infiltration generally have matrix carbon exhibiting anisotropic microstructure, as shown in Fig. 6. Similar to CVI matrix carbon, pitch matrix carbon is not as graphitizable as VGCF, and therefore exhibits a lower thermal conductivity than VGCF.

THERMAL CONDUCTIVITY OF VGCF/C COMPOSITE

The most attractive property of VGCF/C composite is its thermal conductivity, which has just been explored recently. [18 19 20 21 22 23] Room temperature thermal conductivities of various uni-directional (1D) VGCF/C composite specimens, obtained from CVI processing, are given in Table II along with the fiber volume fractions (V_f) and densities. Most of the thermal conductivity measurements were performed parallel to the fiber direction (defined as the "X" direction), while some measurements were in the in-plane orthogonal direction ("Y" direction) and the through-the-thickness direction ("Z" direction). It is apparent that VGCF/C composite having a higher fiber volume fraction or a higher density exhibits a higher thermal conductivity. It was also found that increasing fiber loading was much more effective in enhancing composite thermal conductivity than increasing composite density by CVI densification. Thus, for these composites, the specific thermal conductivity decreases with increasing density which is an evidence that the matrix carbon's contribution to the composite thermal conductivity is

Table II. Thermal conductivities of VGCF/C composite specimens with different fiber volume fraction (V_f) and densities (ρ).

ID	V_f (%)	ρ (g/cc)	Conductivity (W/m-K)
L1	25%	1.26	326 (X) 36 (Y), 12 (Z)
L2	25%	1.32	344 (X)
L3	25%	1.51	372 (X) 38 (Y), 16 (Z)
M1	29%	1.15	362 (X) 49 (Y), 12 (Z)
M2	29%	1.35	374 (X) 52 (Y), 14 (Z)
M3	29%	1.49	431 (X)
H1	36%	1.32	502 (X)
H2	36%	1.48	528 (X) 72 (Y), 18 (Z)
H3	36%	1.59	564 (X) 75 (Y), 19 (Z)

Fig. 6. VGCF/C composite prepared by pitch infiltration

very limited. This can be further demonstrated by using different densificatioin methods at processing conditions. Table III compares the thermal conductivities of several 1D VGCF/C composite specimens prepared using four different CVI conditions and three different pitch infiltration conditions. Preforms for these composite specimens were cut from a master preform with a volume fraction of 39%. It is apparent that composite thermal conductivity is dominated by the reinforcing VGCF. A higher composite thermal conductivity can be achieved using a higher fiber volume fraction of VGCF, as shown in Table IV. A composite thermal conductivity of 910 W/m-K represents the current record, but not a limit.

TENSILE PROPERTIES OF VGCF/C COMPOSITE

Tensile testing was performed on various composite specimens with different fiber volume fractions, fiber architecture, and densities. Table V summaries data obtained from the testing. Each datum shown in Table V represents an average value of at least three data points. The experimental data are significantly lower than predicted by theories based on the data given in Table I. This is explained by considering effective length of individual fiber and/or effective modulus of VGCF mat. Only 1D composite is considered.

It is known that a carbon fiber exhibits highly orthotropic properties. Therefore, when modeling composite mechanical properties, one has to consider the different

Table III. Thermal conductivities (κ) of 1D VGCF/C composite specimens prepared by different densification methods at different processing conditions. All the specimens have the same fiber volume fraction of 39%. Also shown are the densities (ρ) and specific thermal conductivities (κ/ρ).

Densification Condition	None (Preform)	Chemical Vapor Infiltration A	B	C	D	Pitch Infiltration a	b	c
κ (W/m-K)	481	559	460	590	568	463	647	736
ρ (g/cc)	1.13	1.55	1.55	1.62	1.60	1.56	1.70	1.79
κ/ρ	428	361	297	364	355	297	381	411

Table IV. Thermal conductivities of high fiber volume VGCF/C composites.

ID	V_f (%) X	Y	Thermal Conductivity (W/m-K) X	Y	Z	ρ (g/cc)
14	55	0	824	89	24	1.70
01	65	0	910	84	33	1.88
03	45	15	635	373	21	1.80

Table V. Tensile modulus, tensile strength, and strain of failure for VGCF/C composite specimens.

ID	ρ (g/cc)	V_f (5)	Architecture	Modulus (GPa)	Strength (MPa)	Strain of Failure (%)
A	1.58	15	1D	26.2	36.5	0.14
B	1.72	41	1D	51.0	69.6	0.16
C	1.79	52	1D	53.1	66.8	0.15
D	1.78	61	1D	86.1	57.2	0.07
E	1.71	42	2D	30.0	40.7	0.16
F	1.83	52	2D	48.2	37.9	0.10
G	1.72	62	2D	26.9	37.2	0.16

behaviors of fiber along the fiber axis and transverse to the fiber axis. This is particularly important when carbon fibers are used to reinforced a ductile matrix, e.g. aluminum, exhibiting a very different mechanical behavior as compared to carbon fiber. However, in the case of a carbon matrix composite, shear modulus and bulk modulus of matrix carbon may be comparable to the transverse shear modulus and bulk modulus of the reinforcing carbon fiber, respectively. In addition, Poisson's ratio of matrix carbon is thought to be comparable to the transverse Poison's ratio of carbon fiber. Therefore, for carbon fiber reinforced carbon composites, the prediction assuming isotropic fiber properties and the prediction considering orthotropic fiber properties give very similar results. [24 25 26]

As mentioned above, a VGCF mat consists of semi-aligned, semi-continuous, multi-layer fibers. Fibers in the mat have been observed to have lengths ranging from 2 cm to 5 cm. Properties of short fiber reinforced composites have been theoretically modeled by Cox. [27] The theory has been recently applied to predict mechanical behavior of VGCF/epoxy composite. [28] According to Cox's theory, composite modulus, E, can be expressed as

$$E=E_m(1-V_f)+f(\theta)\,g(a_f)\,E_f V_f \qquad (1)$$

where E_m, E_f, and V_f are the matrix modulus, fiber modulus, and fiber volume fraction, and a_f is the fiber aspect ratio (length to diameter). The function $f(\theta)$ is a fiber distribution function and

$f(\theta)$ = 1 for perfectly aligned fibers,
$f(\theta)$ = 1/3 for 2D randomly oriented fibers, and
$f(\theta)$ = 1/6 for 3D randomly oriented fibers.

The function $g(a_f)$ is defined as

$$g(a_f)=1-\frac{\tanh B}{B} \qquad (2)$$

where

$$B=2a_f\sqrt{\frac{G_m/E_f}{\ln(\pi/2V_f)}}$$

Another popular model for predicting short fiber reinforced composite is the modified Halpin-Tsai's equation. [29]

$$E=\frac{E_m[E_f+\beta E_m+\beta(E_f-E_m)V_f]}{[E_f+\beta E_m-(E_f-E_m)V_f]} \qquad (4)$$

where $\beta = 2a_f$.

Based on Eqs. (1) and (4), and data given below, the tensile modulus of VGCF/C composite may be calculated.

E_m = 27.5 GPa [30], a_f = 2 to 5 cm/7 μm, G_m = 2 GPa [30], and E_f = 360 and 600 GPa, representing high and low bounds as shown in Table I.

Shown in Fig. 7 is composite modulus as a function of fiber volume fraction. Modulus was calculated using Eq. (1). It is noted that both Eqs. (1) and (4) give almost identical results. However, none of the model prediction compares well with the experimental data. We explain the discrepancy using the concepts of effective length and effective modulus of VGCF. In the following analysis, only Eq. (1) is used.

In a VGCF mat, the following features can be seen: (1) fibers are semi-aligned, (2) fibers are discontinuous, (3) fibers can be straight or curved, (4) fiber can have an appearance of helices, (5) fibers are over-grown, (6) fibers branch or terminate within the mat, and (7) fiber crenulation occurs. As a result, we define an effective fiber length so that within the effective length, a fiber is free from the above features and behaves as a defect-free, straight short fiber. Also, due to these features, one can assign an effective modulus for a VGCF mat in the fiber axial direction. This effective modulus would be lower than that of single VGCF given in Table I. Let us first consider the use of effective fiber length in the modeling only. Taking the average modulus of VGCF from Table I, i.e. 480 GPa, to be that of a VGCF mat, Eq. (1) predicts an effective length to be in the range of 50 μm to 100 μm. This is shown in Fig. 8. On the other hand, if only effective modulus is considered and the fiber length is taken to be 2 cm to 5 cm, as observed using SEM, in the modeling, Eq. (1) predicts that VGCF mat would have an effective modulus in the range of 50 to 150 MPa. This is shown in Fig. 9. Also shown in Figs. 8 and 9 are the experimental data. It appears

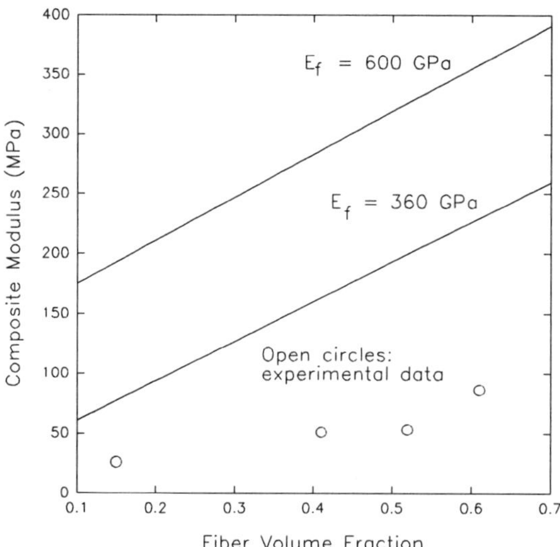

Fig. 7. Composite modulus estimated using Eq. (1) based on data given in Table I.

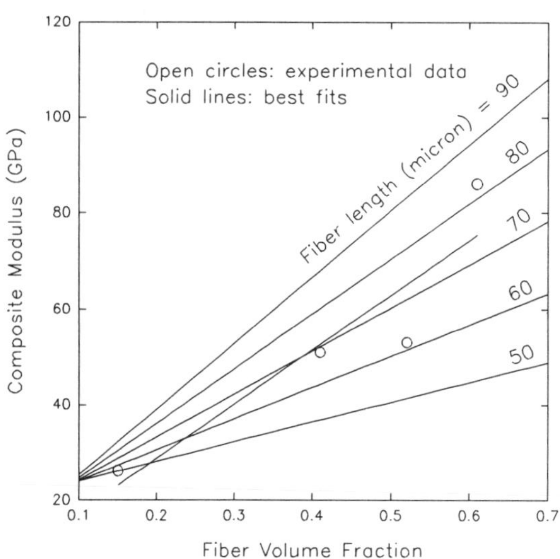

Fig. 8. Composite modulus calculated using Eq. (1) and assuming effective fiber length.

that Cox's theory predicts the composite modulus well when either effective fiber length or effective mat modulus is introduced.

TRANSLATION OF VGCF PROPERTIES IN COMPOSITE

It appears that the tensile modulus of individual VGCF cannot be translated well into a composite. As stated above, to explain the tensile behavior of the composite, two effective properties are used separately. In one case, a single reinforcing fiber is treated as a fiber having a length between 50 to 100 μm and independent of other fibers. In the other case, VGCF mat is thought to have an effective modulus in the range of 50 to 150 MPa. The question now is that which approach is more appropriate. We answer this question by using both approaches to examine the composite thermal conductivity.

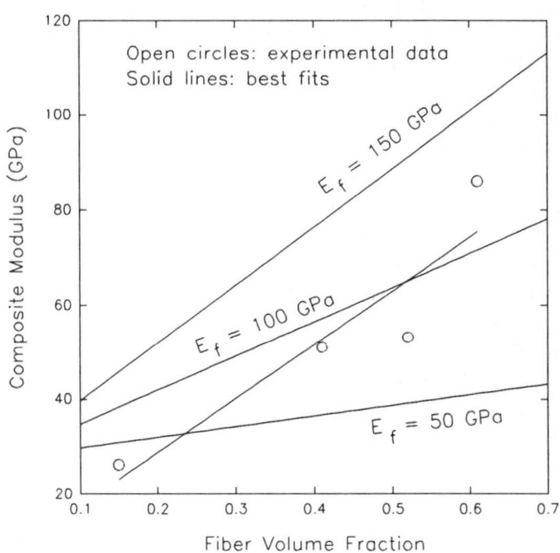

Fig. 9. Composite modulus calculated using Eq. (1) and assuming effective modulus.

Our previous studies have indicated that the thermal conductivity of single VGCF cannot be fully translate into a composite either. When the concept of effective length is used, it can be shown that composite thermal conductivity depends on fiber

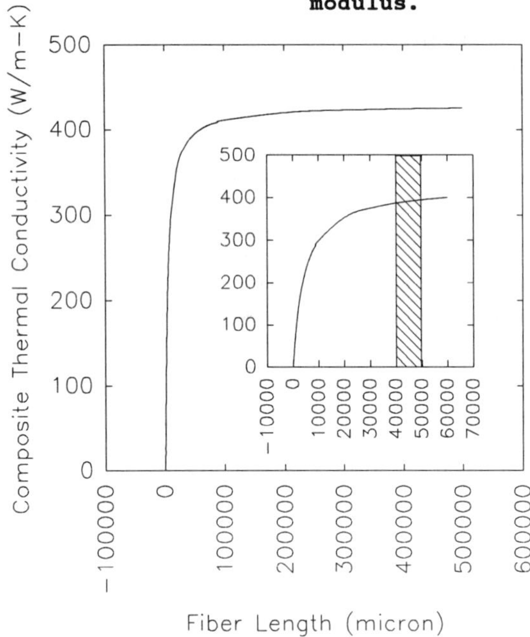

Fig. 10. Composite thermal conductivity as a function of reinforcing fiber length.

length in a way as described in Fig. 10 . [31] As seen in Fig. 10, to fully exploit the thermal conductivity of VGCF in a composite, the length of VGCF has to be in the range of tens of cm. However, analysis of data obtained from thermal conductivity measurements of numerous VGCF reinforced composite specimens has clearly indicated that individual fiber has an "effective" length of 4 or 5 cm, according to Fig. 10. This "effective" length is, in fact, very compatible to the average individual fiber length observed using SEM, i.e. 2 cm to 5 cm. On the other hand, by assigning an effective thermal conductivity in the range of 1460 to 1600 W/m-K, one can predict very well the composite thermal conductivity. [23] This seems to suggest that the use of effective modulus is more favorable. There are also additional reasons to support the use of effective modulus. It can be shown that the modulus of matrix carbon is almost linearly proportional to porosity. [32] Since the VGCF/C composite specimens were not fully densified, matrix carbon is porous and exhibit a lower modulus than that used in the analysis above, i.e. 27.5 GPa. Also, the fiber distribution function, $f(\theta)$, has been taken to be one, i.e. assuming perfect fiber orientation. However, as stated previously, fibers in the composite are semi-aligned. This indicates that the fiber distribution function has a value smaller than one. As a result, using a lower matrix modulus and a smaller number of $f(\theta)$ would lead to predicted data that are closer to the experimental data, provided effective fiber modulus is used.

Finally, it appears that composite strength cannot be described as is composite modulus. This is thought due to the high sensitivity of VGCF/C composite strength to surface or volume flaws, e.g. porosity. Fractured surfaces would have to be examined to fully understand the mechanisms that govern the strength of VGCF/C composite.

CONCLUSION

Owing to its unique growth mechanism and the resulting microstructure, vapor grown carbon fiber exhibits physical properties that are equal to or exceed those of other carbon fibers. Vapor grown carbon fibers can be produced in a form of mat, in which fibers are semi-aligned, semi-continuous, and over-grown. Carbon composites based on vapor grown carbon fiber can exhibit an excellent thermal conductivity approaching 1000 W/m-K. Tensile properties of such composites are lower than expected. However, the composite tensile modulus is comparable with that of aluminum. Although the tensile modulus of single VGCF is in the range of 360 GPa to 600 GPa, a VGCF mat exhibits an effective modulus near 150 GPa due to the uncommon features found in the mat. It appears that composite mechanical properties are more strongly affected by these features than are thermal properties. The implications favored applications for thermal management where structural demands are less stringent.

ACKNOWLEDGEMENT

This work was supported by the Department of Energy under Grant Number DE-FG02-90ER80886.A001 and by the U.S. Air Force under Contract Number F29601-93-C-0165.

REFERENCES

1. C. Zweben, JOM, pp. 15-23, July, 1992.
2. C. Zweben and K.A. Schmidt, Electronic Materials Handbook, Vol. 1, Packaging, ASM, Materials Park, OH (1989).
3. B. Nysten and J.-P. Issi, Composite **21**, 339 (1990).
4. G.G. Tibbetts, M. Endo, and C.P. Beetz, SAMPE Journal, Sep/Oct, 30-5 (1986).
5. J.-M. Ting, M.L. Lake, and D.C. Ingram, Diamond & Related Materials, **2** [5-7] 1069 (1993).
6. M. L. Lake, J.-M. Ting, and J.F. Phillips, Jr., Surf. & Coat. Tech., **62**, 367 (1993).
7. J.-M. Ting and M.L. Lake, J. Mat. Res., **9** [3] 636 (1994).
8. T.V. Hughes and C.R. Chamber, U.S. Patent No. 405480, June 18 (1889).
9. T. Koyama, Carbon, **10** 757 (1972).
10. T. Koyama, M. Endo, and Y Onuma, Japanese Journal of Applied Physics, **11,** no.4, April, 1972.
11. T. Koyama and M. Endo, Ohyo Butsuri, **42** 690 (1973).
12. G.G. Tibbetts, Carbon **30** [3] 399 (1992).
13. M. Endo and M. Shikata, Ohyo Butsuri, **54** 507 (1985).
14. G.G. Tibbetts and D.W. Gorkiewicz, Carbon, **31** [7] 1039 (1993).
15. G.G. Tibbetts, Carbon Fibers, Filaments, and Composites, 73-94, Kluwer Academic Publishers, The Netherlands, 1990.
16. J.-M. Ting and M.L. Lake, in Processing, Fabrication, and Applications of Advanced Composites, edited by K. Upadhya (ASM International, Materials Park, OH 1993), p. 117.
17. J.-M. Ting and M.L. Lake, Procd. 16th Ann. Conf. Comp., Mat. & Structures, Cocoa Beach, FL, January, 1992.
18. J.-M. Ting and M.L. Lake, Procd. 16th Ann. Conf. Comp., Mat. & Structures, Cocoa Beach, FL, January, 1992.
19. J.-M. Ting and M.L. Lake, p. 355, Procd. 17th Ann. Conf. Comp., Mat. & Structures, Cocoa Beach, FL, January, 1993.
20. J.-M. Ting and M.L. Lake, J. Nuclear Mat., 212-215 (1994) 1141-1145.
21. J.-M. Ting and M.L. Lake, J. Mat. Res., **10** [2] 247 (1995).
22. J.-M. Ting and M.L. Lake, Carbon **33** [5] (1995), ***in press.***
23. J.-M. Ting, M.L. Lake, and D.R. Duffy, J. Mat. Res., **10** [6] (1995), ***in press.***
24. Z. Hashin, NASA CR-1974, NASA, 1972.
25. Z. Hashin and B.W. Rosen, J. Appl. Mech., Vol. 31, 1964, pp.223-232.
26. Z. Hashin,, J. Appl. Mech., Vol. 46, 1979, pp.543-550.
27. H.L. Cox, British J. Appl. Phys., Vol. 3, March 1952, pp. 72-79.
28. W.J. Baxter, Rpt No. PH-1717, GM Research Lab, Warren, MI, January, 1992.
29. J.C. Halpin, Primer on Composite Materials: Analysis, Technomic Publishing Co., Lancaster, PA, 1992.
30. J.-M. Ting and M.L. Lake, Extended Abs., 21st Bien. Conf. Carbon, Buffalo, NY, June, 1993.
31. J.-M. Ting and M.L. Lake, Procd. DOE Plasma Facing Materials and Components Task Group Meeting, West Dennis, MA, September, 1992.
32. J. Whitney, EMTEC Qrt. Rpt. CT-46, January, 1995.

DEPOSITION OF DIAMOND-LIKE CARBON FILMS BY PECVD

Kyu Chang Park, Soo Chul Chun, Kyo Jun Song, Min Park, Myung Hwan Oh*, Seong Soo Choi** , Jung Hae Park***, In Sang Yang***, and Jin Jang

Department of Physics, Kyung Hee University, Dongdaemoon-ku, Seoul 130-701, Korea
* Korea Institute of Science and Technology, Seoul 136-792, Korea
**Department of Physics, Sun Moon University, Chung Nam 337-840, Korea
*** Department of Physics, Ewha Womans University, Seoul 120-750, Korea

ABSTRACT

We have studied the structural properties of hydrogenated carbon films deposited by plasma enhanced chemical vapor deposition (PECVD). The substrate holder in reaction chamber could be biased and be heated. The Raman peak intensity at 1350 cm^{-1} was increased by reducing CH_4 flow rate. The film structure changed from soft a-C:H to hard carbon with decreasing CH_4 flow rate, resulted from increased self-bias. The 1520 cm^{-1} peak shifts to higher frequency by reducing the CH_4 flow rate, probably resulted from the increased internal stress.

INTRODUCTION

Diamond-Like Carbon (DLC) film has become considerable interests for optical or protective coating due to its favorable optical and mechanical properties[1]. One widely used method to deposit a DLC film is a plasma enhanced chemical vapor deposition (PECVD)[2]. The structural and mechanical properties of a DLC film can be controlled in a wide range by changing the deposition conditions such as self-bias voltage, substrate temperature and gas pressure. Dischler et al., observed that for their system, at a substrate bias voltage(V_b) of 400-1800 V, they would have hard carbon film, whereas polymerlike films were obtained at lower substrate bias voltage[3-4]. Jiang et al., found that in the range of $0V < |-V_b| < 100V$ polymerlike films, in the range of $100V < |-V_b| < 600V$ diamond-like hard carbon films with high internal stress and in the range $600V < |-V_b| < 1400V$ graphite-like soft films with low stress were deposited. Energetic ion bombardment has a similar effect as thermal annealing on diamond-like a-C:H films[5].

Recently, amorphic diamond is applied to fabricate diode type Field Emission Displays (FED) device[6-7]. A typical material for field emission array (FEA) is Spint cathode, which can be made with either a metal or Si tip. However, some serious problems appeared in this case, such as difficulties of fabrication technique for sub-micro sized micro-tip, high voltage driving, low hardness, high vacuum sealing and tip oxidation. DLC films have some advantages compared to Si due to its high hardness, low work function and non oxidation characteristics.

In this study, DLC films were fabricated by PECVD using a $CH_4/H_2/He$ mixture. The effects of total pressure and hydrogen dilution on the structural properties of DLC films have been investigated. The hydrogen content and internal stress of the deposited films depend on the gas pressure. With increasing bias voltage, the DLC peak increases and hydrogen content decreases.

Mat. Res. Soc. Symp. Proc. Vol. 383

EXPERIMENTAL DETAILS

We used a conventional PECVD system, in which rf power was applied to the substrate holder. CH_4 and He were introduced separately into vacuum chamber. The base pressure of reaction chamber was $3x10^{-7}$ mbar, pumped by turbomolecular pump(Balzers TPH 330).

Table I shows deposition conditions for DLC film. CH_4 flow rate was changed from 0.3 to 10 sccm and the bias voltage of the substrate holder was measured. The hydrogen flow rate was changed from 5 to 30 sccm in order to study the hydrogen dilution effect. Raman spectroscopy was performed by backscattering from the sample using an Ar-ion laser operating at 514.5 nm. These spectra were analyzed quantitatively to reproduce the data as a sum of two (diamond and graphite) different lines. And the calculated spectrum was fitted to measured data and calculate the volume fraction. The stretching mode absorptions due to CH_n (n=1–3) were measured by FT-IR spectrophotometer (Bomem MB-Series) and were deconvoluted into sp^3 and sp^2 modes assuming Gaussian forms. Inter-band optical absorption coefficients were measured using Hewlet-Packard UV-VIS spectrophotometer.

Table I. Deposition conditions for DLC film

RF power (W)	**20 ~ 100**
Pressure (mbar)	**0.4**
Flow rate (sccm)	
He	**50**
CH_4	**0.3 ~ 10**
H_2	**5 ~ 30**

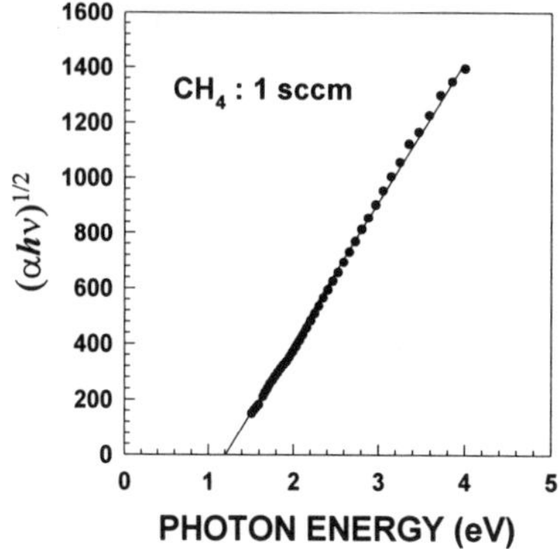

Fig.1 A typical Tauc plot of DLC film.

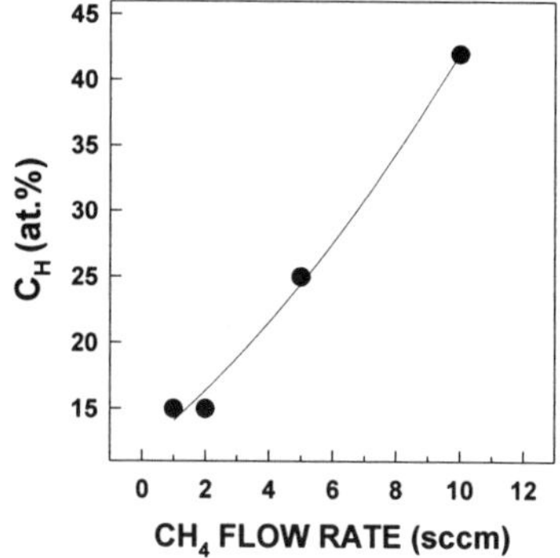

Fig. 2 Hydrogen content of DLC films versus CH_4 flow rate.

RESULTS AND DISCUSSION

Figure 1 shows the Tauc' plot of DLC film deposited at 1 sccm of CH_4. The measured optical band gap is nearly independent of CH_4 flow rate and is 1.2 eV.

Figure 2 shows the hydrogen content of DLC films plotted against CH_4 flow rate, obtained from the integration of C-H_n vibrational absorptions. The hydrogen content decreases with decreasing CH_4 flow rate, caused by increased ion bombardment effect.

Figure 3 shows the self-bias voltage plotted against CH_4 flow rate. The negative self bias was increased by decreasing CH_4 flow rate. The increase of bias voltage will change a structure of DLC films. The kinetic energy of accelerated positive ions (dominantly, CH_3^+ and $C_2H_5^+$) were increased by the increased bias voltage, resulted in bombardment on the substrate surface. This etches weakly bonded C-H_n and C-C bonds[1], leads to enhanced hardness at higher bias.

Figure 4 shows the Raman spectra of DLC films at various CH_4 flow rates. Typical Raman peaks of DLC appear at 1350 cm^{-1} and 1540 cm^{-1} for D-line and G-line of graphite phases, respectively. The peak at ~ 1350 cm^{-1} appears when the CH_4 flow rate is less then 5 sccm and tends to increase with decreasing CH_4 flow rate.

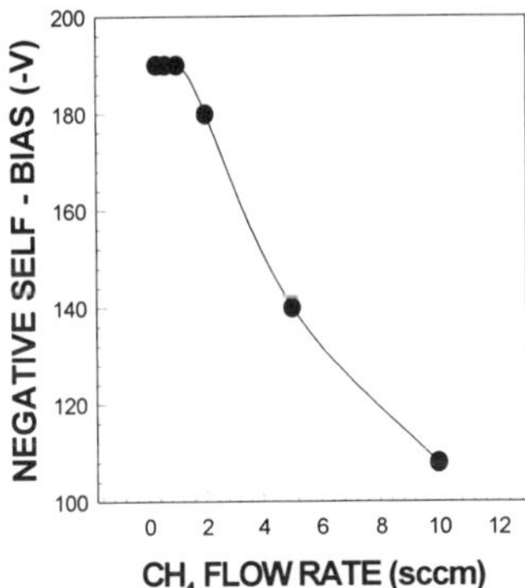

Fig. 3 Self bias voltage versus CH_4 flow rate.

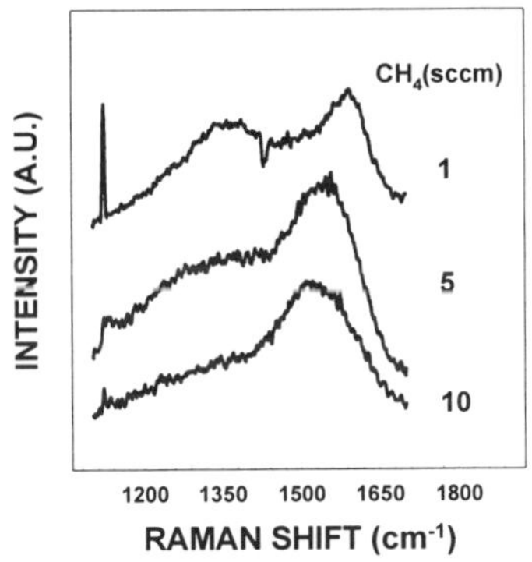

Fig.4 Raman spectra of DLC films deposited with various CH_4 flow rates.

The microcrystalline diamond could be traditional or a polytype in the wurtzite structure. The frequencies of wurzite diamond have not been reported, but by comparison with the frequency of SiC we could deduce a vibrational frequency of the strongest mode to be ~ 1175 cm^{-1}[9]. The DLC film deposited with 1 or 5 sccm CH_4 shows Raman peak at 1140 cm^{-1}. The origin of this peak is not clear, but we guess that it is a peak due to microcrystalline diamond.

Figure 5 shows the Raman spectrum for DLC film deposited at 1 sccm of CH_4, deconvoluted into two phases. The curve fitting was done by searching the peak positions and FWHMs of two phases and minimizing the difference between measured and fitted values. The peak position appears at 1360 and 1580 cm^{-1}, respectively, for disorder induced D and ordered G lines of sp^2 phases.

Table II. Possible Raman peaks of TO modes in DLC films

Frequency (cm^{-1})	Assignment	Reference
1100~1600(broad)	soft amorphous carbon	[8]
1140	microcrystalline diamond	[9] [10]
1270	amorphous diamond	[9]
1350 and 1540	typical DLC film	[11] [12]
1332	crystalline diamond	[9] [10]
1582	crystalline graphite	[9] [10]
1350	microcrystalline graphite	[9]
1460	amorphous graphite	[13]

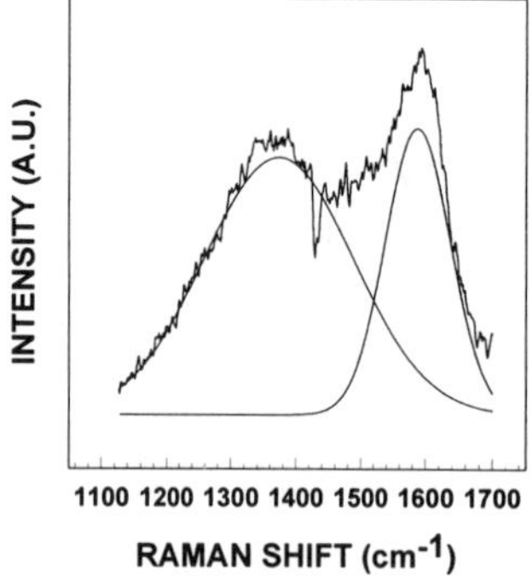

Fig.5 Raman spectrum deconvoluted into D and G lines for the DLC film deposited with 1 sccm of CH_4 flow rate.

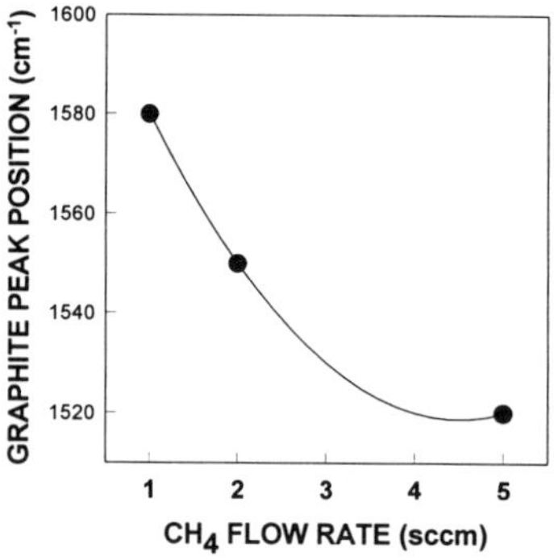

Fig. 6 Deconvoluted Raman peak corresponding to G line.

Figure 6 shows the deconvoluted graphite phase peak position plotted against CH_4 flow rate. The G line peak position decreases with increasing CH_4 flow rate.

Figure 7 shows the Raman spectra of DLC films deposited with hydrogen or no hydrogen. The hydrogen flow rate was 5 sccm. The intensity at 1350 cm^{-1} was decreased by hydrogen dilution. The frequency of G line shifted to lower value by addition of hydrogen. Abelo at. al., reported that the peak frequency shift is related to internal stress[11]. The intensity of the microcrystalline diamond mode, appeared at 1140 cm^{-1}, is increased by 5 sccm hydrogen dilution. The Raman peak assignment of various carbon films are appeared in Table II.

Figure 8 shows the Raman spectra of DLC films deposited with CH_4/H_2 mixture versus rf power. The flow rates of H_2 and CH_4 were 5 sccm and 1 sccm, respectively. The DLC peak intensity appeared at 1350 cm^{-1} increases with rf power. But internal stress appears to increase with rf power, resulting in shift of graphite peak position. We did not measure the internal stress of DLC film, however, when the film thickness was above 700nm, the film peeled from the substrate at the rf power of 100 W. On the other hand, we could deposited DLC film of 1000nm with rf power of 50 W. This means that the internal stress is strongly related with rf power used to deposition. The microcrystalline diamond peak intensity at 1140 cm^{-1} was found to decrease with increasing rf power.

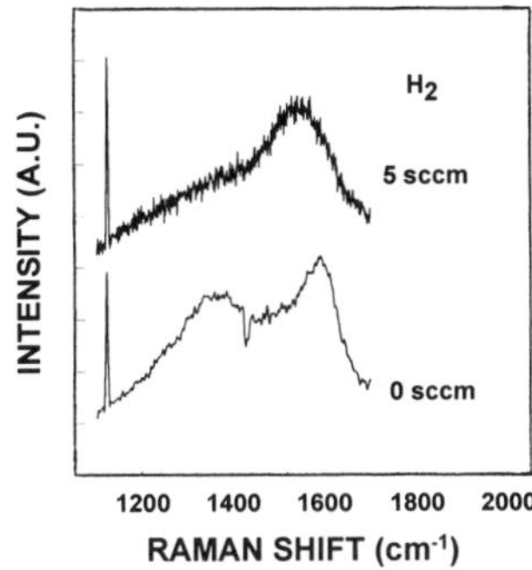

Fig.7 Effect of hydrogen dilution on the Raman shift of DLC film.

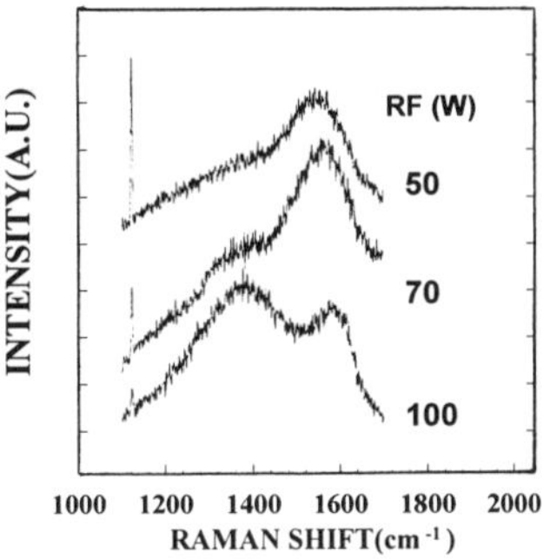

Fig. 8 The Raman spectra of DLC films deposited with various rf powers and 5 sccm hydrogen.

CONCLUSION

We studied the growth of the DLC films by PECVD. The film characteristics strongly depend on CH_4 flow rate, hydrogen dilution, and self-bias voltage increases with decreasing CH_4 flow rate. The optical band gap turns out to be independent of the deposition condition and to be ~ 1.2 eV. With decreasing CH_4 flow rate, the incorporated hydrogen content decreases and the structure changes more to hard carbon rather than soft carbon. A new Raman peak at 1140 cm^{-1} was found, and the DLC films deposited with 5 sccm hydrogen dilution and with smaller CH_4 flow rate showed strong peaks.

ACKNOWLEDGEMENT

This work was supported by the Korea Science and Engineering Foundation through the Semiconductor Physics Research Center.

REFERENCES

[1] N. Mutsukura, S.I. Inoue and Y. Machi, J. Appl. Phys.**72**, 43(1992).

[2] N.H. Chou, J. Appl. Phys. **72**, 2027(1992).
[3] J.W. Zou, K. Reichelt, K.Schmidt and B. Dischler, J. Appl. Phys. **65**, 3914(1989).
[4] J.W.Zou, K. Schmidt, K. Reichelt and B. Dischler, J. Appl. Phys. **67**, 487(1990).
[5] X. Jiang, J.W. Zou, K. Reichelt and P. Grunberg, J. Appl. Phys. **66**, 4729(1989).
[6] N. Kumar, C. Cie, N. Potter, A. Krishman, C. Hilbert, D. Eichman, Proc. SID Dig., 1009(1993).
[7] N. Kumar, H.K. Schmidt, M.H. Clark, A. Ross, B. Lin, L. Fredin and B. Baker, Proc. SID Dig., 43(1994).
[8] B.S. Elman, M.S. Dresselhaus, G. Dresselhaus, E.W. Maby, H. Mazurek, Phys. Rev. **B24**, 1027(1981).
[9] R.J. Nemanich, J.T. Glass, G. Lucovsky and R.E. Shroder, J. Vac. Sci. Technol. **A6**, 1783(1988).
[10] R.O.Dillon and J.A. Woollan, Phys. Rev. **B29**, 3482(1984).
[11] L. Abello and G. Lucazeau, Proc. 2nd European Conf. on Diamond, Diamond Like and Related Coating, 512(1991).
[12] L.C. Nistor, J. Van Landuyt, V.G. Ralchenko, E.D. Obraztova, V.E. Strelnisky, Appl. Phys. **A58**, 137(1994).
[13] S.A. Solin and R.J. Kobliska Amorphous and Liquid Semiconductors, edited by J. Stuke (Taylor and Francis, London, 1974) p.1251.

MECHANICAL PROPERTIES OF DIAMOND FIBRES

N.M. EVERITT, R.A. SHATWELL*, E. KALAUGHER AND E. NICHOLSON**

Department of Aerospace Engineering, University of Bristol, Queens Building, University Walk, Bristol, BS8 1TR, UK
*DRA Sigma, Sunbury-on-Thames, Middx TW16 7LN, UK
**Department of Chemistry, University of Bristol, Cantock's Close, Bristol, BS8 1TS, UK

ABSTRACT

Tungsten and silicon carbide fibres have been coated with diamond using the HFCVD technique. The diamond volume fraction varied between 26% and 73%. Resonance in bending tests gave a Young's modulus of 880 GPa for the diamond coating. Tensile testing indicated that the diamond fracture strength was between 600 MPa and 2000 MPa, depending on the coating thickness, and thus the grain size, of the diamond. The strain to failure of the diamond coating in bending was approximately 0.15% for 25 μm thick films.

INTRODUCTION

Polycrystalline diamond film deposited on thin cores makes an attractive material for high modulus fibres if the stiffness of the single crystal can be reproduced. The specific stiffness of diamond compares well with that of high modulus carbon fibre. In addition, the fibre is electrically insulating if deposited on a non-conducting core, which could be an advantage for stealth applications. The commercial potential of the stiff fibre will depend on its fracture strength and strain to failure compared to other existing fibre systems such as carbon fibres and silicon carbide or other ceramic fibres.

Diamond fibres have been grown on silicon carbide (SiC) and tungsten (W) cores using hot filament CVD. The following paper reports the modulus, tensile fracture strength, and strain to failure (from bend testing) of diamond coatings of such fibres.

EXPERIMENTAL DETAILS AND THEORY

Fibre growth

The diamond coating for this work was carried out in specially designed growth reactor, constructed in collaboration with Thomas Swan[1]. It is based on conventional hot filament CVD reactors, but instead of planar substrates long metallic or ceramic wires are arranged around a vertical central filament so that they are parallel to the long axis of the filament.

The source gases used were methane and hydrogen in the ratio 0.75:100. Total gas flow rate was 200 std cm^3 min^{-1}, and the chamber pressure was 20 Torr. Diamond growth rates were 0.5-1 $\mu m\ h^{-1}$.

[1]Thomas Swan, Scientific Equipment Division, Unit 1c, Button End, Harston, Cambridge, CB2 5NX, UK.

Mat. Res. Soc. Symp. Proc. Vol. 383 © 1995 Materials Research Society

The filament was made from tantalum wire and the filament to substrate distance was 5-6 mm. As a quality control tool, the temperature of the filament was read using a two colour pyrometer, imaging 2-3 coils (of 0.5 mm wire diameter) in the reading area, and the current adjusted to keep the temperature readings within the range 2050°C-2150°C. (These should not be taken as an absolute values of temperature since the emissivity of the filament was not checked, nor allowance made for any deviation from black body behaviour). The single wire cores did not give a large enough image area to use this technique, but from the diamond morphology and previous experiments, the substrate temperature was estimated to be approximately 900°C.

Two substrate materials were selected. The metallic substrate used was 125 μm and 25 μm diameter W wire. The ceramic substrate used was 100 μm diameter SiC fibres, grown on a W core, supplied by DRA Sigma. All substrate materials were abraded before growth by dragging the fibres through 1-3 μm diamond grit. Coating thickness varied between 10 μm and 46 μm, giving 26 - 73% volume fraction of diamond.

Modulus testing

Elastic moduli in bending were measured for one batch of fibres using resonance equipment based on a system developed by Cranfield Research Institute [1]. The fibres were rigidly clamped at one end and the fibre excited by a frequency ramp, 10 Hz to 210 Hz. An image of the fibre was focussed on a split-diode detector. When the fibre vibrated, the difference signal contained a component at the frequency of resonance. The resonance of the clamped fibre was detected using a Hewlett Packard 35660A spectrum analyser (which also supplied the frequency ramp). In order to eliminate the effects of end corrections, 5 measurements were made on each individual fibre, shortening the fibre by 5-10 mm each time. The mean fibre diameter was determined by measurement with a micrometer at 5 different points along the fibre. Usually the first harmonic, f_1, was easier to measure than the fundamental frequency, f_0.

The equation for the fundamental bend frequency of a coated core can be re-arranged to give [1]:

$$L = \frac{1}{\sqrt{f_0}} \cdot \left[\frac{A^4[E_1 r_1^4 + E_2(r_2^4 - r_1^4)]}{16\pi^2[r_1^2 d_1 + (r_2^2 - r_1^2)d_2]} \right]^{1/4} - L_0 \qquad (1)$$

where f_0 is the fundamental frequency, A is a constant, which is 1.8752 for the system used, E_1 is the modulus of the fibre core, E_2 is the axial modulus of the diamond coating, r_2 is the radius of the coated fibre, L is the length of fibre, L_0 is the end correction to the length, d_1 is the density of the fibre core and d_2 is the density of the diamond.

The first harmonic, f_1, was assumed to have a frequency exactly 6.25 times higher than f_0, and linear plots of L versus $1/\sqrt{f_1}$ were constructed.

Tensile testing

Tensile testing of the diamond fibres was carried out using a Hounsfield H5000M universal testing machine. Single fibres were fixed to a card with a central slot 72 mm long, and once positioned in the machine, the sides of the mount were cut so that only the fibre was load bearing. Tensile loading was carried out at a constant displacement rate of 0.5 mm/min, and the load and time recorded 5 times a second. The result was discarded if fracture occurred outside the fibre gauge length. Diamond coating thickness was calculated from measuring the fibre diameter under an optical microscope.

Assuming that the core and coating remain united and intact, the strain in the fibre and coating is the same and the ratio of stress in core and coating is the same as the ratio of Young's moduli. The relation of the stress in each component and the overall fibre stress (overall load per unit area) can be estimated using a rule of mixtures. Thus, assuming that the diamond coating/core interface remains intact, the stress in the diamond coating can be calculated:-

$$\sigma_{diamond} = \frac{\sigma_{fibre}}{V_{diamond} + (1 - V_{diamond})\frac{E_{core}}{E_{diamond}}} \tag{2}$$

where σ is the stress, E is the Young's modulus and V is the volume fraction of the particular phase. $V_{core} = 1 - V_{diamond}$. Assuming that the coating fractures before the core, the diamond fracture stress can be calculated from the maximum fibre stress.

Bend radius testing

The SiC fibre core material and one other batch of fibres were tested for minimum bend radii using equipment based on that produced for the optical fibre industry [2]. A short length of fibre was bent into a wide arc between two parallel plates. One of the plates was moved towards the other at a speed of 1 mm/sec, and the separation of the plates at fracture of the fibre was recorded. The process was repeated with at least 5 fibres of each type.

However, for the length of fibre available, the bend radius of the diamond coated fibres proved to be too large for the equipment described above. Therefore the bend radius of these fibres was tested by the more simple technique of carefully bending single fibres round solid objects of progressively smaller radius of curvature until fracture occurred. An approximate bend radius was calculated by using the mean of the last successful radius and the radius around which failure occurred. Again at least 5 fibres were tested.
If a fibre is bent in the shape of a uniform circular arc, then :-

$$\frac{\sigma}{y} = \frac{E}{R} \tag{3}$$

where σ is the stress at a distance y from then neutral axis, E is the Young's modulus and R is the radius of curvature. Assuming that failure occurs at the surface of the fibre in the diamond coating, the elastic strain to failure of the diamond is given by the ratio r/R where r is the fibre radius and R is the minimum radius of curvature.

However if the curvature of the fibre is not a circular arc, then a correction must be applied. In the case of a fibre bent between two jaws, it has been shown that

$$\sigma = 1.198E\frac{d}{D-d} \tag{4}$$

where d is the fibre diameter and D is the jaw separation [2].

RESULTS

Elastic Modulus

The elastic modulus was tested using the resonance method on 3 diamond fibres from the same growth run, with 30 μm of diamond on a 95μm diameter SiC core. The results were analysed assuming a modulus of 370 GPa for the DRA Sigma fibres. This gave 877 ± 84 GPa (mean ± SD) for the modulus of the diamond coating.

Tensile strength

The fracture stress of the diamond coating was calculated assuming that the modulus of the diamond was 877 GPa (see above). The core moduli were taken as 370 GPa and 411 GPa for the SiC and W cores respectively. Mean diamond fracture strengths for various fibre batches are shown in Fig. 1. (95% confidence limits = ± 2 SE)

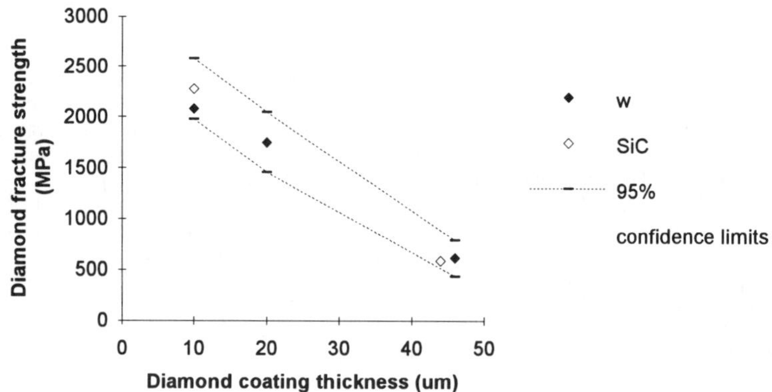

Fig. 1 Decrease in diamond fracture strength with increase in coating thickness

SiC core fibre straight from the spool gave a breaking stress of 3000 ± 232 MPa (mean ± SE). SiC fibre which had been given a abrading pre-treatment like those on which diamond was deposited gave a mean strength value of 2388 ± 132 MPa.

Bend radius

25 μm thick diamond coating failed at about 0.15 % strain whether it was deposited on a 25 μm W core (mean fibre bend radius 45 ± 3.3 μm) or a 125 μm diameter W core (mean fibre bend radius 141 ± 11.9 μm). The abraded SiC failed at about 1.4% strain.

DISCUSSION

Equivalence of tests

The tensile test procedure gave a slightly lower breaking stress for SiC core fibre straight from the spool than the quoted value of 3750 MPa. It is therefore possible that the test is under-estimating the fracture strength of diamond, due to extra bending stresses introduced when the sample is gripped. A one-tailed t-test comparing the strength of abraded SiC fibre with fibre straight from the spool suggests that the decrease in fracture strength caused by the pre-treatment of the core is significant ($p < 0.05$). However, it will not affect the elastic modulus of the core.

The bend tests carried out at DRA might have been expected to show smaller minimum bend radii than the method of bending fibres round a selection of solid objects, because of the increased possibility of surface damage during the latter tests. However, tests using both methods on abraded SiC fibre, and also on W cores (groups of 5 fibres) showed no statistically significant increase in the minimum bend radii of the fibres when using the simple test.

Mechanical properties of diamond

The resonance tests produced a value for the bend modulus of diamond that correspond well with both the quoted modulus for single crystal diamond and the range of values found for polycrystalline diamond using vibrating reed tests [3, 4].

The tensile test results suggest a fracture stress for diamond in the range of 600-2000 MPa. These values thus overlap well with the values found by bulge testing technique [5, 6]. There is a substantial decrease in fracture strength with increase in coating thickness ($r= 0.96$, $p< 0.01$), see Fig. 1, which is independent of the fibre core type. It is probable that the decrease in fracture strength is caused by an increase in grain size with increase in coating thickness, see Fig. 2. Such a relation has also been suggested by work on flat substrates [5]. However, the apparent fracture stress of the thin coatings will also tend to be higher because the sampling volume is less.

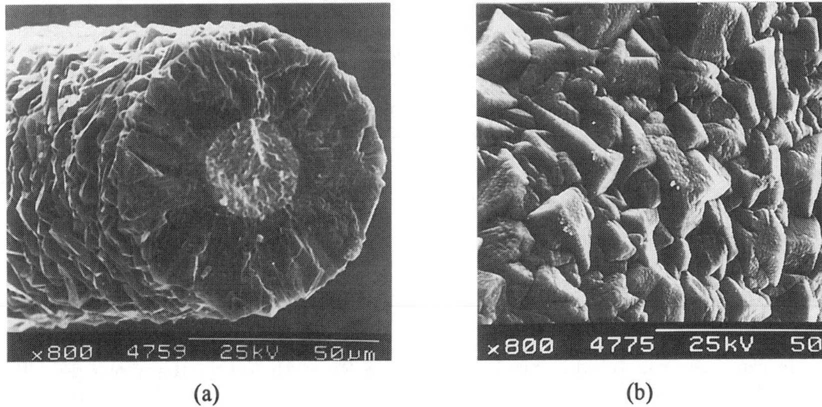

Fig. 2 Diamond coated W cores; (a) 25 μm thick film and (b) 46 μm thick film

The bend tests showed failure of the fibre at 0.15 % strain at the outer surface of 25 μm thick diamond coatings. The stress in the outer surface of the diamond at this strain is approximately 1300 MPa. This fracture strength agrees with the tensile test results within experimental error, although the sampling volume is smaller in the bend test and might therefore be expected to give a higher value. Since the bend strength is very dependent on the surface flaw deposition, the agreement between the two techniques would also suggest that failure is occurring at the diamond outer surface in the tensile test, rather than at the W or SiC / diamond interface.

An average fracture toughness (K_{1c}) for single crystal diamond is 4.2 MPa $m^{-1/2}$ [7]. If this value is combined with the fracture stress of 1300 MPa found in the bend test it gives a critical flaw length of 3 μm. The reported values of K_{1c} for CVD polycrystalline films are actually slightly higher than that for natural diamond [6, 8]), and this value for critical flaw length is therefore a conservative one. Thus an approximate critical flaw length for 25 μm thick diamond would be 3-4 μm, Using the same strain to failure the critical flaw strength is 1-2 μm for 10 μm thick film, and 16-17 μm for 45 μm thick films. (An approximate stress concentration factor of 1.12 was used for these calculations. The true stress concentrations will be determined by the actual shape of the flaws). These values for critical flaw size are of the order of the grain size of the material, see Fig. 2, and the fracture behaviour is consistent with that of other polycrystalline CVD material.

CONCLUSIONS

The diamond fibres exhibited a very high modulus of 880 GPa, and a strain to failure in bending of about 0.15%. Critical flaw size is consistent with polycrystalline ceramics produced by CVD methods. The tensile strength, 600 MPa - 2000 MPa was dependent on coating thickness, probably related to grain size and/or morphology changes.

REFERENCES

1. D. Bruce and P. Hancock, J. Inst. Metals **97**, 140 (1969).
2. M.J. Matthewson, C.R. Kurkjian and S. Gulati, J. Am. Ceram. Soc. **69**, 815 (1986).
3. S. Seino, N. Hida and S. Nagai, J. Mat. Sci. Lett. **11**, 515 (1992).
4. L. Chandra and T.W. Clyne, J. Mat. Sci. Lett. **12**, 191 (1993).
5. G.F. Cardinale and C.J. Robinson, J. Mat. Res. **7**, 1432 (1992).
6. R.S. Sussmann, J.R. Brandon, G.A. Scarsbrook, C.G. Sweeny, T.J. Valentine, A.J. Whitehead and C.J.H. Wort, Diamond and Related Materials **3**, 303 (1994).
7. J.E. Field, *The Properties of Natural and Synthetic Diamond*, (Academic Press, London, 1992).
8. M.D. Drory, C.F. Gardinier and J.S. Speck, J. Am. Ceram. Soc. **74**, 3148 (1991).

ACKNOWLEDGEMENTS

This research was supported in part by the Defence Research Agency, Farnborough, UK.

EFFECT OF MICROSTRUCTURE IN METAL REINFORCED COMPOSITE CVD DIAMOND FILMS ON MECHANICAL PROPERTIES

J.W. HOEHN*, D.F. BAHR*, J. HEBERLEIN**, E. PFENDER** AND W.W. GERBERICH*
* Chem. Engineering and Materials Sci., University of Minnesota , Minneapolis, MN 55455
** Mechanical Ennginecring, University of Minnesota, Minneapolis, MN 55455

ABSTRACT

Comparison of behavior observed during deadhesion of CVD diamond films and compositional analysis of their microstructure allows a correlation between microstructure and mechanical behavior to be made. Diamond films were grown in a DC Triple Torch Reactor using a mixture of methane and hydrogen on molybdenum substrates. To address the issue of voids at the interface, a metal binder was electroplated after the nucleation of individual diamond crystals on the substrates. The films were completed by growing a layer of diamond upon the composite layer. Interfacial composition and structure of the films are characterized by a x-ray, Auger, and Raman spectroscopy. A carbide may act to enhance the adhesion via chemical bonding between the substrate and film. Diffusion of the binder into the substrate and film is also important for mechanical properties and is confirmed by x-ray mapping. It is suggested that a diamond like carbon layer acts to enhance the adhesion of the film to the substrate.

INTRODUCTION

Diamond film growth on non-diamond substrates is being considered for possible applications of cutting tool insert coatings. An attractive cutting tool material which has been considered is tungsten carbide with a cobalt binder. Diamond film adhesion on this substrate has been shown[1] to be rather poor due to either void content at the interface or the presence of cobalt at the film substrate interface. One of the methods which has been attempted to promote adhesion is the deposition of various interlayers of material, including molybdenum and other refractory metals, which have been shown[2,3] qualitatively to be less likely to spall than from a tool material.

Adhesion on materials that form stable carbides tend to exhibit better adhesion than other substrates. Other researches[4] have found that rough interfaces act to improve adhesion, particularly by the carburization of a material. A complete characterization of the interface between a molybdenum substrate and the diamond film using various microscopy and spectroscopy techniques should provide insight towards the processing variables which may be optimized to enhance adhesion. In particular, the effect of the formation of carbides and non-diamond forms of carbon at the interfacewere examined. In addition, fractography of various films can be used to identify the surfaces which fail during adhesion testing.

EXPERIMENTAL PROCEDURE

Molybdenum plates were cut 1 mm thick by 25 mm diameter discs for use as substrates. They were then polished with 240 grit SiC paper, and cleaned in methanol in an ultrasonic bath for 10 minutes. No other treatment was done to the substrates (i.e. no rubbing with diamond powder.) The samples were then clamped to a water cooled copper holder with a stainless steel ring and placed in a DC triple torch reactor[5]. During film growth, the pressure of the chamber was kept at approximately 250 Torr. Monitoring of the temperature was accomplished using a two color optical pyrometer. The temperature was controlled by simply raising or lowering the height of the sample with respect to the plasma plume Growth times were between 30 minutes

Mat. Res. Soc. Symp. Proc. Vol. 383 © 1995 Materials Research Society

and 1 hour. For a number of films, the issue of voids at the interface was addressed by using a three step process[6,7]. In the three step process, a metal binder is electroplated after the nucleation of individual diamond crystals on the substrates. These films are then completed by growing a layer of diamond upon the composite layer.

Two different mechanical tests were performed to measure and compare the adhesive properties of the films. First, a four point bend fixture was constructed in which strips were strained to put the film in compression until spallation of the film occurred. The strips were made by cutting grooves in the substrates before deposition leaving behind 1.5 mm wide bars which were then placed in the reactor as the previous samples. Delamination was observed from the edges of the film towards the center. The second type of mechanical test used a conventional Brale indenter in a digitally controlled mechanical testing machine. The indenter tip was driven into the samples in displacement control at 20 μm per second until the desired load was reached, whereupon the indenter was withdrawn from the sample at the same rate. Load and depth were monitored during the indentation. In the one step films, a portion of the diamond film would spall off after some critical load. Accurate measurement of the spalled area was obtained by digitally capturing the images of the indents in a scanning electron microscope and tracing the edges of the delamination in an image analysis program. The spalled area was then used to calculate an effective interfacial crack length by using the square root of area divided by π.

To understand the film delamination caused by the mechanical testing, several different experimental methods were implemented for characterizing the structure including: x-ray diffraction, Auger electron spectroscopy, Raman spectroscopy, and scanning electron microscopy.

RESULTS & DISCUSSION

The response of the diamond films to the mechanical testing varied between the one step and the three step films. The three step films displayed significantly better adhesive properties than the one step films. They were found to exhibit small radial cracks, but did not spall off of the substrate. The one step films, however, consistently displayed large areas which spalled off

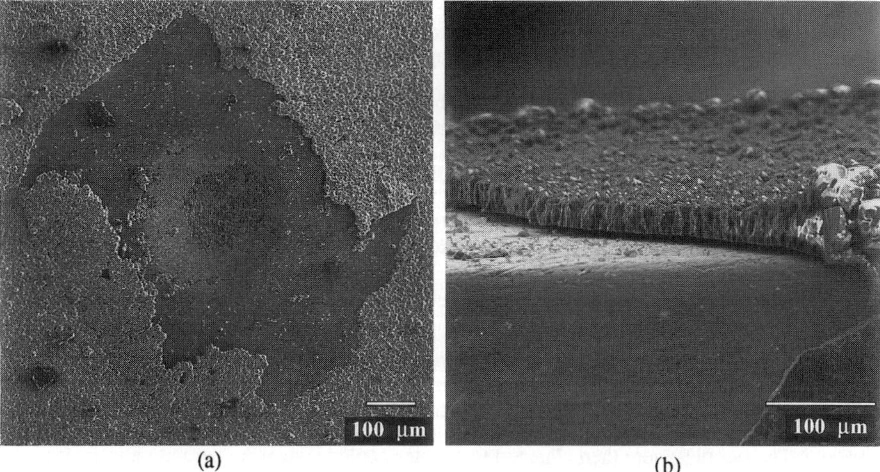

(a) (b)

Fig. 1. SEM micrographs of one step samples after two types of mechanical testing: (a) shows a depression and large spalled area resulting from Brale indentor with a 400 N load. (b) is of the edge of a remaining film on a four point bend sample after partial spallation. Note that the visible area of the film is delaminated, but remains unspalled.

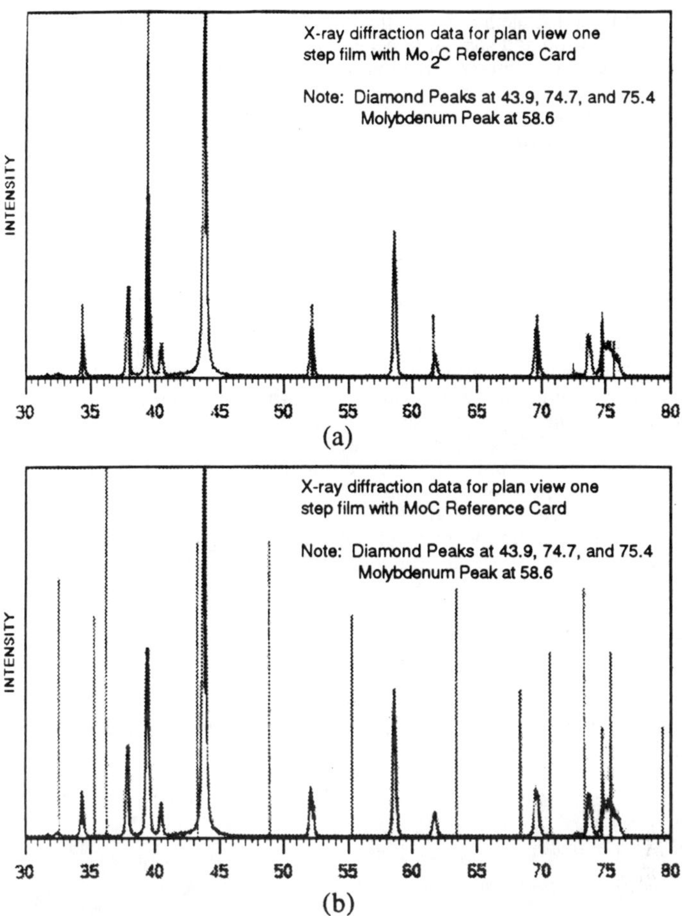

Fig. 2. X-ray diffraction data from an intact one step film. Comparison to standard files (a) and (b) displays good correlation with Mo2C but no indication of MoC.

during testing simplifying the characterization of their adhesive properties. Therefore, even though it was known and had been shown previously[6] that a three step process greatly increases the adhesive properties of the films, a significant amount of effort went into investigating the one step films. Examination of the microstructure present in each type of film gives us insight for explanation for their behavior in the mechanical tests. The discussion that follows will therefore be divided into two sections, a section dealing with the one step films: the other with the three step films.

One step films

Samples of the one step films displayed the same spallation behavior in both the bend tests and the indentation tests. Figure 1a shows a typical indent/spallation area in the one step films. Figure 1b displays the spalled area in a bend test and a portion part of the film which is delaminated from the substrate but not spalled off. It was of great interest to determine where in the films the delamination and spallation occurred in the mechanical testing. It was also of use to perform some analytical characterization on different parts of the films after testing. We were able to recover some films which had spalled off and characterize the different surfaces present using an assortment of techniques.

The initial characterization of the films was done using x-ray diffraction. Analysis was performed on the films as still adhered to the substrate. The penetrating strength of the x-rays allows information from both the film and also the substrate to be obtained. The results of the x-ray analysis, as shown in Figure 2, show a composition containing diamond, Mo_2C, and Molybdenum. Clearly missing is the presence of any MoC. The carburized layer of Mo_2C develops during deposition and is suspected to enhance the adhesion of the diamond.

Utilizing the spalling nature of the one step films, a surface analysis of the layers where the deadhesion occurs was possible. Auger electron spectroscopy was done on the three different surfaces present after spallation: the substrate left behind, back side of the spalled film which had been adhered to the substrate, and the top side of the film. The results of the Auger on the top side of the film show, unsurprisingly, that the film is carbon. The back side of the delaminated film was found to be predominantly carbon with some trace amounts (less than 5%) of molybdenum. The Auger analysis of the remaining substrate after spallation showed the layer to be 50% carbon / 50% molybdenum through a 1.5 μm depth profile. This suggests that the delamination occurs predominantly between the diamond film and a molybdenum carbide layer. However, the composition of the carburized layer is not clearly consistent with the x-ray data which would predict 67% Mo and 33% C.

Cross sections of the films were made for the scanning electron microscope by cutting through the back side of the substrate until it was sufficiently thin that it could be broken in two. In Figure 3, a cross section SEM micrograph demonstrates clearly that there is an intermediate layer between the diamond and the molybdenum which differs from both the film and the substrate. The thickness of this interlayer increased with diamond film thickness, since thicker films remained in the reactor for longer periods of time, allowing the carburized layer to increase in thickness. The carbide layer is important, because it is suspected that it increases bonding of the diamond film. At the same time, however, the thickness of the film could be a concern because an increasingly thick carbide layer might degrade the mechanical properties for which the molybdenum is used. Fortunately, the results of the mechanical testing consistently

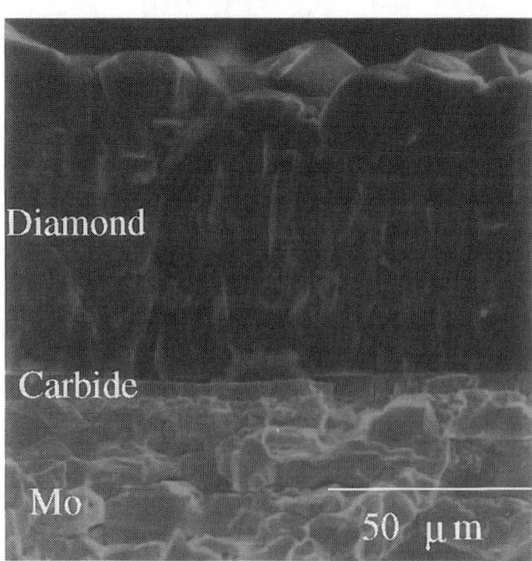

Fig. 3. Cross section SEM micrograph of a one step film. Note the different fracture surface of the carbide interlayer.

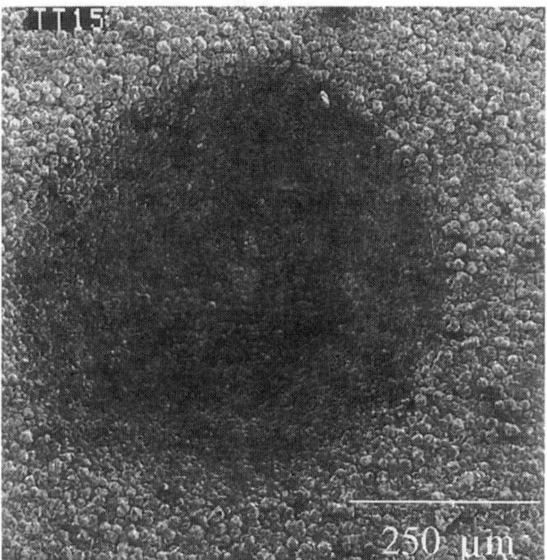

Fig. 4. SEM micrograph of depression left in a three step film from a Brale indentor with a 400 N load. Note that there is virtually no spallation.

demonstrated that delamination would occur between the carbide layer and the diamond for a range of carbide thickness.

The differences in the results of chemical analysis from the different techniques is confusing. The results from the x-ray are ambiguous because there is no way to differentiate between the diamond film and any diamond which could be present in the interlayer.

Raman spectroscopy was also performed on the film and the delaminated substrate. The results on the film surface showed a distinct diamond character. The analysis on the substrate, from which the film was delaminated, showed peaks at 1332 and 1580-1600 cm^{-1} which is characteristic of diamond like carbon. Raman spectroscopy is much less sensitive to the types of carbide structures such as suggested by the x-ray and the Auger data, but it could be argued that the carbide layer contains additional carbon in a diamond-like form. This would explain the structure exhibited by x-ray, the concentrations found using Auger, and the presence of a diamond like carbon phase in the carburized molybdenum layer as the Raman suggests. Work is currently in progress to obtain a cross section TEM analysis which should give a clear answer. Currently, it is felt that the interlayer is composed of two phases: Mo_2C and poor quality diamond. This is consistent with all of the experimental results.

Three step films

As mentioned earlier, the adhesive properties of the three step films are markedly better than those of the one step films. A thorough compositional analysis of the three step films has not been finished; however, preliminary observations correspond well with previous work. Figure 4 shows a typical depression and surrounding area resulting from an indentation. A reasonable comparison of the adhesive behavior can be made to Figure 1a since both images were captured after the same indentation load of 400 N. The spalled area of the one step film is approximately 0.7 mm^2, while the three step film under the same load has virtually zero spalled area. When the load is increased to 800 N very small spalled areas on the order of 0.025 mm^2 can be found upon close examination. Thus far, attempts to correlate the spalled regions to actual works of adhesion are progressing, but no definitive relations have been developed. Nevertheless, it is empirically clear that introducing the nickel binder into the film greatly enhances film adhesive properties. Figure 5 shows a cross section SEM micrograph of a three step film with corresponding x-ray mappings of Mo and Ni. It can be seen from the images that nickel has filled into voids in the diamond film and also diffused into the carburized interlayer. Furthermore, molybdenum has penetrated a distance into the nickel binder. The combination of this diffusion bond with the fact that nickel will absorb energy during the process of plastic deformation leads to the increase in adhesive strength of the film. Interestingly, even in the presence of the diffusion bounding, the few areas which did spall did so in the plane between the interlayer and the composite film.

CONCLUSION

It was demonstrated that using a three step process to grow a composite CVD diamond film greatly increases the adhesive properties of the film over a single growth step process. Characterizing the behavior of one step films was found to be useful because of their less complex structure. Analysis of the microstructure of the one step films shows the presence of a two to five micron carbide interlayer. Both the one step and the three step films failed at the interface between the carbide interlayer and the diamond or composite diamond film. It is felt that understanding of this system is important for further development when efforts are made to grow diamond coatings onto tungsten carbide and other tool insert materials where the addition of molybdenum as an interalyer is used to increase adhesion.

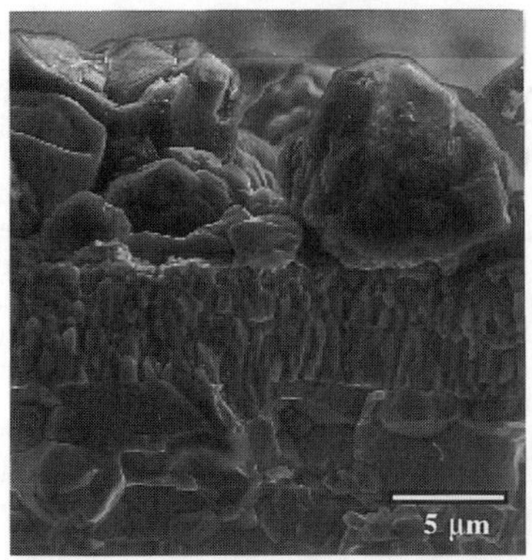

(a)

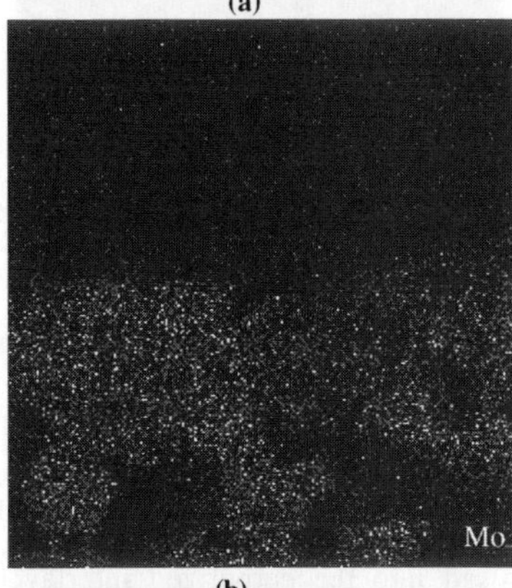

(b)

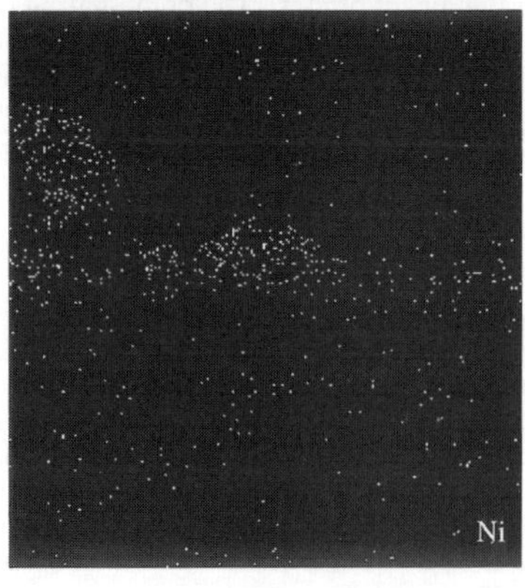

(c)

Fig. 5. Cross section SEM of a three step film showing diffusion of Ni into the substrate. (a) is a backscattered electron image of the cross section. (b) and (c) are x-ray maps of Mo and Ni respectively.

ACKNOWLEDGMENTS

The authors would like to thank V. Indajang, Dr. D. Zhuang, for there assistance and helpful discussion. Also we would like to acknowledge the support from the Basic Energy Sciences, Materials Science Division, of the Department of Energy under Contract No. DOE/DE FG02-84ER 45141 and ORTTA for funding this research.

REFERENCES

1. S. Soderberg, A. Gerendas, M. Sjostrand, Vacuum, **41**, 1317, (1990)

2. M. Nesladek, J. Spinnewyn, C. Asinari, R. Lebout, R. Lorent, Dia. Rel. Mater., **3**, 98 (1993)

3. K. Kurihara, K. Sasaki, M. Kawarada, Y. Goto, Thin Solid Films, **212**, 164 (1992)

4. S. Takatsu, K. Saijo, M. Yagi, K. Shibuki, J. Echigoya, Mat. Sci. Eng., **A140**, 747 (1991)

5. Z. Lu, L. Stachowicz, P. Kong, J. Heberlein, E. Pfender, Plasma Chem. Plasma Proc.,**11**,387 (1991)

6. C. Tsai, J.C. Nelson, W.W. Gerberich,J. Heberlein, E. Pfender, J. Mater. Res., **7**, 1967 (1992)

7. C. Tsai, J.C. Nelson, W.W. Gerberich, D.Z. Lui, J. Heberlein, E. Pfender, Thin Solid Films, **237**, 181, (1994)

DIAMOND-LIKE CARBON FILMS FOR SILICON PASSIVATION IN MICROELECTROMECHANICAL DEVICES

M. R. HOUSTON[1], R. T. HOWE[2], K. KOMVOPOULOS[3], and R. MABOUDIAN[1]
[1]Department of Chemical Engineering
[2]Department of Electrical Engineering and Computer Sciences
[3]Department of Mechanical Engineering
Berkeley Sensor and Actuator Center
University of California
Berkeley, California 94720

ABSTRACT

The surface properties of diamond-like carbon (DLC) films deposited by a vacuum arc technique on smooth silicon wafers are presented with specific emphasis given to stiction reduction in microelectromechanical systems (MEMS). The low deposition temperatures afforded by the vacuum arc technique should allow for easy integration of the DLC films into the current fabrication process of typical surface micromachines by means of a standard lift-off processing technique. Using X-ray photoelectron spectroscopy (XPS), contact angle analysis, and atomic force microscopy (AFM), the surface chemistry, microroughness, hydrophobicity, and adhesion forces of DLC-coated Si(100) surfaces were measured and correlated to the measured water contact angles. DLC films were found to be extremely smooth and possess a water contact angle of 87°, which roughly corresponds to a surface energy of 22 mJ/m^2. It is shown that the pull-off forces measured by AFM correlate well with the predicted capillary forces. Pull-off forces are reduced on DLC surfaces by about a factor of five compared to 10 nN pull-off forces measured on the RCA-cleaned silicon surfaces. In the absence of meniscus forces, the overall adhesion force is expected to decrease by over an order of magnitude to the van der Waals attractive force present between two DLC-coated surfaces. To further improve the surface properties of DLC, films were exposed to a fluorine plasma which increased the contact angle to 99° and lowered the pull-off force by approximately 20% over that obtained with as-deposited DLC. The significance of these results is discussed with respect to stiction reduction in micromachines.

INTRODUCTION

The modification of the surface structure and physical properties using carbon films has been under investigation for many years. Protective coatings for magnetic recording media and wear-resistant coatings for cutting tools are two examples where carbon films have found significant application.[1,2] Another exciting area in which passivation by a carbon film appears viable is in silicon micromachining.[3] Micromachining is the term given to a process by which micrometer-sized mechanical devices are built on a silicon substrate using standard fabrication processes from the IC industry.[4] When a micromachine is built on the silicon substrate from deposited thin films, this process is referred to as surface micromachining, as opposed to bulk micromachining in which a crystalline silicon microstructure is built by selective removal of the substrate. Surface passivation is especially critical for surface microstructures due to the very compliant structures built by this technique. Typical dimensions for these structures are on the order of 2 μm in thickness, with gaps of 1-2 μm to the substrate and other structures. To consider the forces present at these dimensions, assume a spring constant for a typical structure of 10^{-3} μN/μm and a

Mat. Res. Soc. Symp. Proc. Vol. 383 © 1995 Materials Research Society

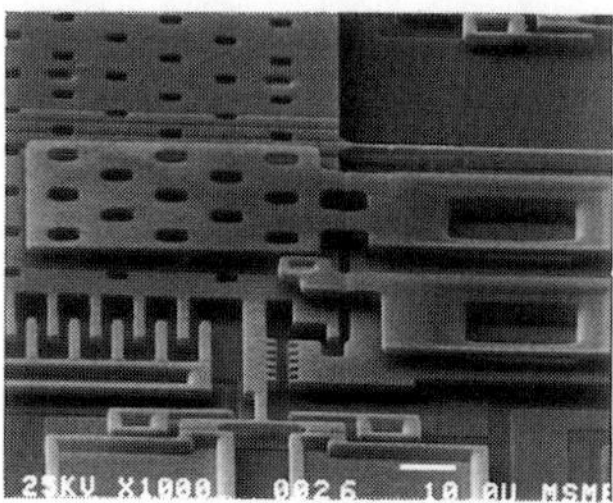

Figure 1: SEM micrograph of a micromachined accelerometer showing the complexity of the multi-layered structure and the mechanisms by which stiction can occur (see text). For reference, the bar shown in the bottom right of the figure corresponds to roughly 10 microns.

gap distance of 1 μm. In this case, the maximum available pull-off force is 1 nN! Therefore, it is necessary to produce a clean, hydrophobic, chemically passivated silicon surface in order to prevent contacting surfaces from adhering to each other due to various attractive forces.[5,6] The sticking of adjacent surfaces to one another is called *stiction*, a term originally derived out of the magnetic recording media industry which referred to any system where the friction was not proportional to the applied load due to the exceptionally strong adhesion forces.[7]

Stiction in Micromachining

Stiction is a prevalent problem among many types of micromachines, such as accelerometers, micromotors, pressure sensors, gyroscopes, and microphotonic devices.[3] In order to illustrate the concept, an SEM photograph of a simple accelerometer device is shown in Fig. 1. To measure acceleration in a direction vertical with respect to the wafer surface, the large flat plate, labeled in the picture, moves up or down in response to an acceleration (or deceleration), and its position is sensed capacitively.[8] If the device experiences an extreme acceleration or sudden shock, however, the large plate will stretch the springs holding it until it comes into contact with an adjacent silicon surface, either above or below the plate. At this juncture, the stretched springs do not generate enough restoring force to overcome the adhesion force experienced by the contacting surfaces (both covered with a native oxide layer), and the surfaces remain in contact. Hence, the accelerometer operation is disrupted. There are many other examples of stiction among the various types of micromachines, and they are all related to the strong attractive forces experienced by surfaces at these dimensions. Furthermore, in many MEMS devices the wear caused by stiction is also a concern.[9] Since the large attractive forces often provide the greatest contribution to the normal forces between two surfaces in these small devices, preventing stiction should help alleviate wear problems in many types of micromachines.

The causes of stiction have been analyzed, and the major forces existing at the dimensions of micromachines are found to be caused by capillary, van der Waals, and electrostatic attractions.[5,6] The formation of covalent bonds between two surfaces, similar to the process of wafer bonding, may also play a role in certain instances. This last phenomenon has been termed solid bridging, and will not be considered here because it appears to have been largely eliminated through the use of supercritical drying techniques.[10] It should be mentioned that forces due to gravity are lower by many orders of magnitude than the forces mentioned above, and do not play a significant role in microdevices. For an idealized system comprising two flat silicon surfaces approaching each other, the attractive forces mentioned above can be plotted as a function of their separation distance. Silicon surfaces are used in this calculation since this is the predominant fabrication material for surface micromachines. The equations used to produce these curves can be found elsewhere,[5,6] and a plot of the resultant forces is shown in Fig. 2, along with a typical spring

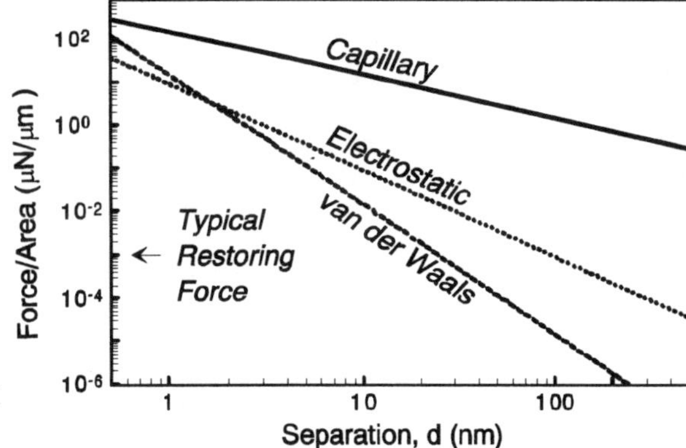

Figure 2: Comparison of attractive forces existing at the dimensions of micromachines as a function of the separation between two surfaces. Forces are given per unit contact area assuming perfectly smooth surfaces. Notice that capillary forces dominate over the entire separation range, although van der Waals forces become significant near contact. Also shown is a typical restoring force available in a micromachine device.

restoring force in MEMS devices. The cause of stiction is clearly the attractive forces present at these dimensions, which are significantly greater than the available restoring force. The desire for more compliant springs in devices such as accelerometers for increased sensitivity precludes the use of stiffer springs to produce a greater restoring force, and, in fact, increases the probability of micromachine seizure due to stiction. Hence, various surface treatments aimed at modifying surface-surface interactions must be considered.

Requirements for an Effective Surface Treatment

It can be seen from Fig. 2 that capillary forces dominate over the entire separation range, and are thus the major contributor to the attractive forces present between two hydrophilic surfaces such as silicon with their native oxide layers. In addition, it should be noted that the magnitudes of both van der Waals and electrostatic forces become significant when the two surfaces are near contact (usually taken to be a separation distance of a few Ångstroms[11]). Therefore, all of these forces must be diminished or eliminated if the adhesion force between two silicon surfaces is to be reduced below the spring restoring force.

An effective surface treatment for these micromachines must provide a hydrophobic surface in order to avoid the formation of water layers on the surface, thereby eliminating capillary forces altogether. Electrostatic charging, which directly leads to attractive electrostatic forces, is poorly understood in these devices. However, if the two surfaces coming into contact have the same composition and, hence, the same work function, then contact charging can be avoided. Further reductions in electrostatic forces will occur if the two surfaces are conductive, allowing charge dissipation to occur. Because van der Waals forces will be present in all materials, these forces can only be reduced, but not eliminated. A reduction can be achieved by using surfaces exhibiting very low surface energies, such as fluoro-ethylene polymers, for instance. Based on calculated or measured values of the Hamaker constant (the parameter used to characterize van der Waals interactions), incorporating a low surface energy material into the micromachine may reduce these forces by as much as an order of magnitude. A reduction of the actual contact area, possibly accomplished with some type of roughening technique, will be necessary to further reduce the adhesion forces in this system.

Why Diamond-Like Carbon Films?

Diamond-like carbon appears to be a promising anti-stiction coating material for micromechanical devices because it provides many of the desirable properties mentioned above. The present work is focused on DLC films deposited by a vacuum arc technique.[12] In general, these films are hard, hydrophobic, exhibit relatively low surface energies,[13] and can also be doped to enhance their conductivity characteristics.[14] Perhaps equally important, DLC films can be deposited controllably as a thin, uniform layer.[15] Other properties such as thermal stability, chemical inertness, and the ability to be deposited at or near room temperature make DLC an ideal candidate for integration into the micromachine fabrication process.

This work focuses on characterizing DLC films with respect to their promise for alleviating stiction between two contacting surfaces. Contact angle measurements have been obtained in order to quantify hydrophobicity. In some cases, X-ray photoelectron spectroscopy was used to investigate the surface composition of the films. Atomic force microscopy is capable of measuring surface-surface interactions,[16] and was performed in the present work to qualitatively and quantitatively measure the pull-off forces on the DLC surfaces. The information obtained from these techniques provides a convincing argument for the integration of DLC films into micromechanical devices for the purpose of stiction reduction.

EXPERIMENTAL

As previously mentioned, the DLC coating investigated in this work was deposited by a vacuum arc technique comprising plasma immersion ion implantation and deposition. The apparatus used for the deposition has been described in detail elsewhere.[17] This equipment generates a carbon plasma by a vacuum arc discharge to a graphite cathode, and then magnetically filters the plasma in order to remove the macroparticles. During deposition, the substrate temperature was below 50° C, the bias voltage was -100 V, and the duty cycle was fixed at 30%. It is believed that the carbon films grow by a combination of plasma condensation on the surface when the substrate is unbiased, followed by both direct and recoil implantation of carbon ions during the bias periods. The implantation of carbon ions is thought to be the necessary condition leading to films possessing diamond-like properties.[18] The physical properties of DLC films deposited by this technique have been studied as a function of deposition parameters,[13] and the conditions used to produce the films studied here were those producing the hardest and most dense films. The carbon films were approximately 100 nm thick and were deposited onto monocrystalline Si(100) substrates.

Several surfaces have been investigated thus far using the techniques already mentioned. Because silicon forms a thin oxide layer which is hydrophilic in room air, and also because silicon micromachines are often cleaned after fabrication with an oxidizing solution such as piranha or RCA, an RCA-cleaned Si(100) surface was studied. This treatment yields a surface which is hydroxyl-terminated and was completely wetted by water droplets during contact angle measurements. The produced surface was considered to be representative of the surfaces present in micromachines fabricated with conventional processes. Si(100) samples treated with an ammonium fluoride (NH_4F) solution were also used in this investigation. The ammonium fluoride treatment consists of cleaning the samples first in an RCA solution, followed by etching in ammonium fluoride for six minutes. This produces hydrophobic, hydrogen-terminated silicon surfaces which are essentially contaminant-free and have contact angles comparable to those seen on DLC. The effects of this treatment have been described in more detail elsewhere.[19] This hydrogen-terminated surface is highly reproducible, and was studied in order to provide a reference surface for our DLC films. Because hydrogen fluoride (HF) is commonly used in the microelectronics industry for cleaning and etching processes, the HF-treated silicon surface was studied also as a convenient reference. The other two surfaces investigated were the as-deposited DLC film, and the DLC film which was exposed to a fluorine plasma to raise the contact angle and further passivate the surface.

In order to quantify the degree of surface hydrophobicity, water contact angle measurements were made on various surfaces. Contact angles are related to the free energy of a surface, where a higher contact angle corresponds to a lower surface energy for the same surface topography. Thus, this information can be useful for comparing the expected van der Waals forces between two surfaces. Contact angles were measured in room air by the standard Sessile drop method,[20] using a Ramé-Hart NRL contact angle goniometer, Model A-100. The stage was backlit with a simple lamp covered with green acetate paper to minimize heating (and therefore evaporation) of the drops. A microsyringe was used to deliver 5 μl water drops. Measurements consisted of placing one drop on the surface and measuring the contact angle on each side of the drop (15-30 seconds total time for both measurements); then a second drop was placed on a different part of the surface and measured likewise. The total time elapsed from start to finish was about 2 minutes. All measurements were made at room temperature using de-ionized water (resistivity ~18 MΩ cm).

For surface chemical analysis, X-ray photoelectron spectroscopy was employed. Samples were transferred into the XPS chamber (Kratos 800-XSAM) via a differentially pumped fast insertion probe. It took approximately 15-20 minutes to introduce the samples to the ultrahigh vacuum (UHV) environment (base pressure of 5×10^{-10} Torr). A non-monochromatic Mg-K_{α} X-ray was used as the excitation source, producing an instrument resolution of 1.1 eV at a fixed analyzer transmission energy of 38 eV. The sample surface was at 75° with respect to the analyzer normal. A Kratos DS800 data acquisition and analysis system was used to obtain and process the data.

Surface interactions were measured using a Topometrix TMX 2000 Explorer scanning probe microscope. Cantilever deflection is sensed in this instrument by the reflection of a laser beam off of the cantilever and detection on a four-quadrant photodetector. For topography or pull-off force measurements, the cantilever is moved in the x, y, and z directions, and the sample remains fixed. All experiments were performed with standard silicon nitride AFM probes, which use a micromachined silicon nitride cantilever and tip for sensing. The manufacturer-quoted radius of curvature of these tips is less than 50 nm, and the nominal spring constant of the cantilever is approximately 0.032 N/m. Pull-off force data were obtained using a tip approach and retraction speed of 1 μm/sec, and the curves shown below are typical of the average force curves seen for a given surface. Samples were stored in anti-static polystyrene storage boxes in order to minimize any effects from electrostatic charging.

RESULTS AND DISCUSSION

Properties of Diamond-Like Carbon Films

Typical properties of the DLC films studied here are given in Table I. As can be seen, these films are quite dense, hard, and possess a low coefficient of friction in some environments. These properties are important in applications where wear or repetitive contacts with the film may occur. Endurance of the carbon coatings is also important, and it has been shown that magnetic recording heads modified with this DLC film maintain a low coefficient of friction for thousands of cycles, exhibiting excellent durability.[17] The diamond-like properties of these films are believed to be derived from the high fraction of sp^3 bonding between the carbon atoms, coupled with the low hydrogen content. Also of importance is the roughness of the DLC films. Since the macroparticles formed in the vacuum arc plasma are magnetically filtered in this deposition method, the DLC films are quite smooth. AFM topography images, not shown here for brevity, indicate that these films have approximately the roughness of a smooth silicon wafer, and are thus thought to conform to the substrate on which they are deposited. For stiction reduction, a rougher film is preferred in order to decrease the actual contact area between two surfaces, and, in turn, reduce the magnitude of the adhesion force. Films in which the macroparticles have not been completely filtered out may be investigated in the future as a means of surface roughening.

Table I : Properties of diamond-like carbon films synthesized by vacuum arc deposition.

Film Thickness (nm)	Density[a] (g/cm^3)	Hardness[b] (GPa)	Content of sp^3 bonds[a] (%)	Hydrogen Content[a] (%)	RMS Roughness[c] (nm)	Coefficient of Friction[b]
100	2.9	35	> 85	< 1	0.5	0.1

[a] from Ref. [21], [b] from Ref. [16], [c] this work

Quantification of Hydrophobicity

Water contact angles were measured on the diamond-like carbon films, as well as various chemically-treated silicon samples. These results are given in Table II. It can be seen that both the as-deposited DLC films and the ammonium fluoride-treated silicon surfaces exhibit similar contact angles. Using these measurements, the solid surface energy can be estimated using the equation

$$\gamma_s = \gamma_l(1+\cos\theta)^2/4\Phi^2 \qquad (1)$$

where γ_s is the free energy of the solid surface per unit area, γ_l is the surface tension of water (72 mJ/m^2), θ is the measured water contact angle, and Φ is a function of the molecular interaction properties of the two substances.[22] An average Φ value of 0.95 is used here.[23] There is a more elaborate method for calculating the surface free energies of solids which involves measuring the contact angles for a variety of different liquids.[22] Since only water was used in the present study, the values obtained for the surface energies should be considered rough estimates.

In this case the surface energy of the DLC films estimated by Eq. (1) is 22 mJ/m^2, a value significantly lower than that of water. XPS analyses of these films revealed the presence of significant amounts of oxygen confined to the DLC surface; however, the exact chemical nature of the observed oxygen could not be determined. Because the measured contact angle on these films is nearly 90°, it seems unlikely that the source of oxygen is water. Some researchers have suggested that the surface of some carbon films could actually be chemically terminated with oxygen.[24] Further experiments are under way to determine the nature of this surface oxygen. It

Table II : Water contact angles (± 1°) on DLC and chemically-treated Si(100).

Silicon Dioxide[a]	HF-Treated Si(100)	NH_4F-Treated Si(100)	As-Deposited DLC	Fluorinated DLC
0 - 30°	62°	84°	87°	99°

[a] values vary depending on growth conditions and method of cleaning.

should be mentioned that exposure to various organic, acidic, and basic cleaning solutions either had no effect or lowered the contact angle on the DLC films, with the exception of a fluorine plasma which enhanced the contact angle. This effect will be discussed next.

Fluorinated Diamond-Like Carbon Films

In an effort to further improve the surface properties of the DLC coatings, a DLC film was subjected to a SF_6 fluorine plasma. This work was based on earlier studies in which a graphite surface was fluorinated by exposure to a fluorine plasma, which significantly raised the contact angle on these surfaces.[25] The plasma conditions used here were: power of 50 Watts, total pressure of 0.222 Torr, SF_6 flow rate of 3.6 sccm, He flow rate of 30.7 sccm, and exposure time of 5 minutes. The film was not visibly damaged after the plasma treatment, although the etch rate was not measured. Interestingly, the contact angle on these surfaces increased from 87° before treatment to 99° after the fluorine plasma exposure. Thus, the fluorination process appears useful for improving the surface properties of DLC films, and at the very least, these fluorinated films serve to further illuminate the role that surface properties play in determining adhesion forces, as will be discussed below. Optimization of the fluorination process may yield even higher contact angles and degree of fluorination, an effort currently being pursued in our laboratory.

AFM Pull-Off Force Measurements

A typical AFM force curve is shown in Fig. 3, along with a schematic of the tip/cantilever position for each region of the curve. In the force curve plot, the x axis is the movement of the piezo actuator on which the cantilever is mounted, which is not the same as the actual tip-to-sample distance. The y axis is a measure of the cantilever deflection. Using both of these measurements, the relative position of the cantilever and tip can be determined. In the beginning of the force curve, the cantilever is far away from the sample surface and is not influenced by the surface forces (point (1)). As the tip approaches the surface, it jumps into contact when the gradient of the attractive force overcomes the cantilever spring constant (point (2)). The cantilever then continues to move toward the sample surface (point (3)), causing the tip to actually exert a compressive force on the sample. As the cantilever begins its motion back away from the surface, hysteresis in the piezoelectric actuator used to drive the cantilever motion causes the retraction curve to be slightly offset. When point (4) is reached, the cantilever restoring force is acting opposite to the attractive force holding the tip to the surface. Once the cantilever is stretched far enough it will exert a force sufficient to overcome this attractive force, and the tip will jump out of contact with the surface, back to its starting position (1).

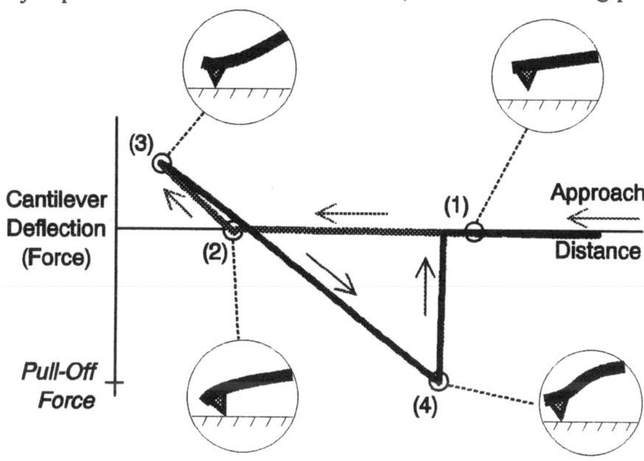

Figure 3: Example of a typical AFM force curve, where the bubbles show the relative positions of the tip, surface, and cantilever for each part of the curve. The pull-off force is measured from the difference between points (1) and (4), as shown on the y axis.

The pull-off force is the difference in the forces measured at points (1) and (4), and is believed to be a direct measure of the depth of the attractive well of the surface potential. In order to calculate the pull-off force from Fig. 3, the difference in cantilever deflection between points (1) and (4), reported in units of nA (detector current), is divided by the slope of the line leading to point (4), given in units of nA/nm. This will give the distance in nanometers that the spring was stretched just before pull-off occurred. The actual pull-off force can then be calculated by multiplying this distance by the spring constant of the cantilever beam holding the tip. These calculations have already been performed for the plots in Figs. 3 and 4.

Representative AFM force curves are shown in Fig. 4 for the surfaces discussed above, along with the value for the actual pull-off force on each surface. The same silicon nitride tip was used for all of the curves shown, and the plots are shown on the same scale for comparison purposes. Total variations in the pull-off force from day-to-day and sample-to-sample were no greater than 20%. It should be mentioned that the magnitude of the compressive force, as determined by the height of point (3) in Fig. 3, had no apparent effect on the pull-off forces on these samples. In general, much higher forces (on the order of μN) are needed in order to actually indent such hard surfaces[26] and, thus, affect the measured pull-off forces.

Theoretical values for the pull-off force can be calculated for the case of a tip in contact with a surface, assuming the tip to have a spherical shape with a known radius of curvature.[27] Because the nitride tip possesses a hydrophilic surface oxide, capillary forces are assumed to be dominant, and the pull-off force for this geometry is given by

$$F_{pull-off} = 2\pi \gamma_l R[\cos\theta_1 + \cos\theta_2] \tag{2}$$

where γ_l is the surface tension of water, R is the radius of curvature of the AFM tip at its apex, and θ_1 and θ_2 are the contact angles of the tip and sample surface, respectively. Reported contact angles for silicon nitride (with a native silicon dioxide layer) are in the range 0-30°, and are most likely dependent on the storage environment and/or any subsequent cleaning. Since these tips were not cleaned in any way, a value of 30° was used for θ_1. This is a minor point since there is little difference between the cosine of 30° and 0°. The RCA-cleaned Si(100), however, was found in the present study to be completely wetting, and a value of 0° was used for θ_2 in this case. The tip manufacturer reports a radius of curvature for these silicon nitride tips of 20 ± 15 nm. Since it was not possible to resolve the tip at its apex using SEM, a value of 15 nm was used in Eq. (2) for the estimation of pull-off forces.

In Fig. 5, theoretical values for the pull-off force calculated based on Eq. (2) are shown as a function of θ_2, the sample surface contact angle. Also shown are the data points from the pull-off force curves depicted in Fig. 4. It can be seen that the AFM data follow the predicted values fairly closely. The theoretical curve is most sensitive to the radius of curvature of the tip, which is difficult to determine accurately due to the complications in imaging such a small, nonconductive feature. Taking into account the size of the tip used in this work, the pull-off force for the RCA-treated silicon sample is in good agreement with other values found in the literature for hydrophilic samples. Specifically, the pull-off forces were estimated to be on the order of tens of nN using a tip with a radius of curvature of 50 nm,[27] and on the order of 7 nN using a 5 nm silicon tip.[16] Although the exact tip radius has not been measured in the present experiments, it is the same for all surfaces measured here since the same tip was used throughout. Furthermore, before measuring each new sample an ammonium fluoride-treated silicon sample was measured in order to test the tip to ensure there was no tip damage or contamination. The slight offset between the theory and experimental data for the more hydrophobic samples can possibly be attributed to the roughness of these surfaces. In view of Fig. 5, it appears that capillary forces

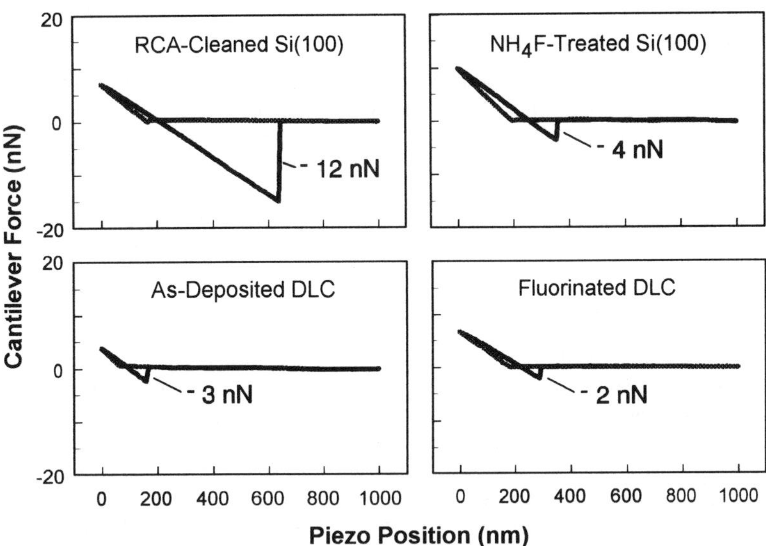

Figure 4: AFM force curves for different surfaces obtained with a silicon nitride tip. Pull-off forces are highest on the RCA-cleaned silicon surface and lowest for the fluorinated DLC films.

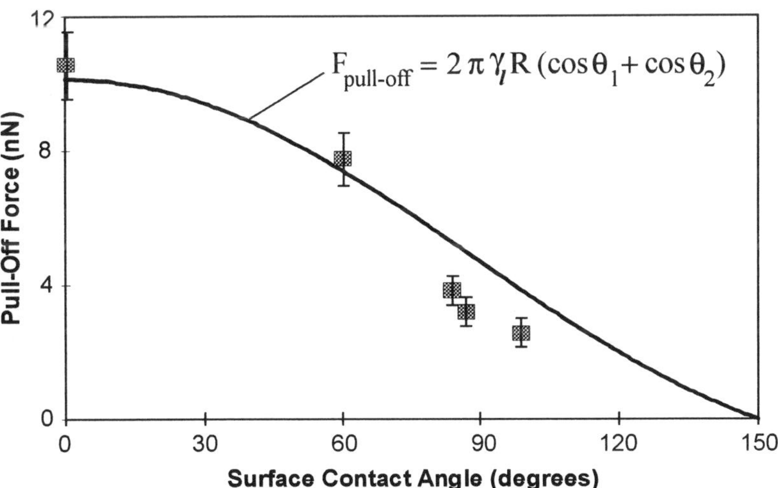

Figure 5: Comparison of theoretical and AFM pull-off force results for capillary forces. In order of increasing contact angle, the data points are for RCA-, HF-, and NH_4F-treated Si(100) surfaces, and for as-deposited and fluorinated DLC films. The vertical bars indicate the experimental scatter among 50-100 measurements.

indeed play a major role in the adhesion forces between two surfaces when at least one surface is hydrophilic.

Implications for Diamond-Like Carbon Films

Based on the results shown in Fig. 4, DLC films appear very promising for stiction reduction. Although the pull-off forces on DLC are only about a factor of 5 lower than those seen on the standard hydrophilic silicon surface, performance in an actual micromachine is expected to be enhanced considerably. This is because in that situation, a DLC-coated surface will be contacting another DLC-coated surface, unlike the case encountered in the AFM measurements where one of the surfaces was always hydrophilic (i.e., the silicon nitride tip). When two DLC surfaces come into contact, capillary forces should be eliminated altogether since neither surface will have an adsorbed water layer. Additionally, because of the relatively low surface energy of DLC films, the contributions of the van der Waals interactions will decrease as well, thereby reducing further the adhesion force.

Evidence for the reduced adhesion force to be realized in a DLC-coated micromachine can be found in our previous work using an ammonium fluoride treatment very similar to the ammonium fluoride treatment studied here in conjunction with DLC. In that study, special microstructures designed to measure the work of adhesion between two contacting surfaces were fabricated and treated with ammonium fluoride in order to produce hydrophobic, hydrogen-terminated silicon surfaces.[28] The results showed a decrease in the adhesion force by over an order of magnitude for the ammonium fluoride-treated surface compared to a standard hydrophilic silicon surface. Unfortunately, the ammonium fluoride treatment may not be suitable for many micromachining applications due to its metastable nature.[19,28] In light of the results of the present study showing that the DLC surface is even more hydrophobic and exhibits a lower pull-off force than the ammonium fluoride-treated silicon surface, a better performance for DLC in micromachines is expected.

One last topic for discussion is the actual integration of DLC into a microstructure. There are at least two possible approaches for accomplishing this objective. First, since vacuum arc deposition comprises plasma immersion ion implantation, which is a non-line-of-sight process, it may be possible to perform a blanket deposition of DLC over an entire released or partially released microstructure. In this case, a very thin DLC film would be deposited on all exposed surfaces. A second approach would be to actually pattern the DLC films, and integrate them during the actual fabrication of the device. For example, it may be possible to build small DLC bumpers on the device in certain areas which would serve as contact points where the two surfaces come together. Because the DLC deposition is performed near room temperature, a photoresist lift-off technique could be used to pattern the DLC. This integration issue will involve a substantial amount of work, both in obtaining a uniform deposition over the required wafer area, and in achieving a sufficient DLC deposition within the microstructure itself. Both of these issues are under investigation.

CONCLUSIONS

Diamond-like carbon deposited by a vacuum arc technique has proven a viable candidate as an anti-stiction coating in microelectromechanical devices. With a contact angle of nearly 90° and AFM pull-off forces significantly less than 10 nN, the surface properties of DLC are nearly ideal for reducing the adhesion forces between two contacting surfaces. It was shown that these properties can be even further improved by exposure of the DLC to a fluorine plasma. Although electrostatic forces were not investigated in this paper, doping the DLC films to enhance their conductivity should alleviate electrostatic forces, should they become significant. It should be noted that the desire for more compliant devices in future designs will continue the demand for even lower adhesion forces between contacting surfaces. One technique to achieve a further reduction in the attractive forces would be to reduce the actual contact area by roughening the surfaces, thus lowering the overall adhesion forces. Surface roughening and film integration are two important points which must be addressed in the near future. In view of these demanding issues and the results of the present work, it may be concluded that DLC is a very promising material for significant stiction reduction in both current and future generations of microelectromechanical devices.

ACKNOWLEDGMENTS

This research was partially supported by the National Science Foundation through the Industry/University Cooperative Research Center Program under Grant No. ECD-9215239 and a National Young Investigator Award (RM), as well as the Berkeley Sensor and Actuator Center at the University of California, Berkeley. MRH wishes to thank NSF for support in the form of a graduate fellowship. The DLC films were deposited using a vacuum arc system at the Lawrence Berkeley Laboratory under the directorship of I. G. Brown.

References

[1] A.H. Lettington, in *Thin Film Diamond*, edited by A. H. Lettington and J.W. Steeds, Chapman & Hall, London, (1994), p. 127.
[2] A.M. Homola, C.M. Mate, and G.B Street, *MRS Bulletin* **15**, 45 (1990).
[3] R.T. Howe, *13th Sensor Symposium, IEEE of Japan*, June 8-9, 1995.
[4] R.S. Muller, *Sensors and Actuators A (Physical)* **A21**, 1 (1990).
[5] R.L. Alley, G.J. Cuan, R.T. Howe, and K. Komvopoulos, *Proc. IEEE Solid-State Sensor and Actuator Workshop*, Hilton Head Island, SC, June 21-25, 1992, p. 202.
[6] R. Legtenberg, J. Elders, and M. Elwenspoek, *Proc. 7th Int. Conf. Solid-State Sensors and Actuators, Transducers 93*, Yokohama, Japan, June 2-5, 1993, p. 198.
[7] B. Bhushan, *Tribology and Mechanics of Magnetic Storage Devices*, Springer-Verlag, New York, (1990).
[8] G.K. Fedder, Ph.D. Thesis, Dept. of EECS, Univ. of Calif. at Berkeley, Sept. 1994.
[9] K. Deng, R.J. Collins, M. Mehregany, and C.N. Sukkenik, *Proc. IEEE Micro Electro Mechanical Systems Workshop*, Amsterdam, the Netherlands, Jan. 29 - Feb. 2, 1995, p. 368.
[10] G.T. Mulhern, D.S. Soane, and R.T. Howe, *Proc. 7th Int. Conf. Solid-State Sensors and Actuators, Transducers 93*, Yokohama, Japan, June 2-5, 1993, p. 296.
[11] J.N. Israelachvili, *Intermolecular and Surface Forces*, Academic Press, London, (1985).
[12] I.G. Brown, *Rev. Sci. Instrum.* **65**, 3061 (1994).
[13] S. Anders, A. Anders, I.G. Brown, B. Wei, K. Komvopoulos, J.W. Ager III, and K.M. Yu, *Surf. Coat. Technol.* **68/69**, 388 (1994).
[14] V.S. Veerasamy, G.A.J. Amaratunga, C.A. Davis, A.E. Timbs, W.I. Milne, and D.R. McKenzie, *J. Phys.: Cond. Matter* **5**, L169 (1993).
[15] S. Anders, A. Anders, and I. Brown, *J. Appl. Phys.* **74**, 4239 (1993).
[16] R.L. Alley, K. Komvopoulos, and R.T. Howe, *J. Appl. Phys.* **76**, 5731 (1994).
[17] K. Komvopoulos, B. Wei, S. Anders, A. Anders, and I.G. Brown, *J. Appl. Phys.* **76**, 1656 (1994).
[18] Y. Lifshitz, G.D. Lempert, and E. Grossman, *Phys. Rev. Lett.* **72**, 2753 (1994).
[19] M.R. Houston and R. Maboudian, submitted to *J. Appl. Phys.*, (1995).

[20] A.W. Neumann and R.J. Good, in *Surface and Colloid Science Vol. II: Experimental Methods*, edited by R.J. Good and R.R. Stromberg, Plenum Press, New York, (1979).

[21] Z. Feng, M.A. Brewer, K. Komvopoulos, I.G. Brown, and D.B. Bogy, *J. Mater. Res.* **10**, 165 (1995).

[22] R. Good, in *Contact Angle, Wettability, and Adhesion: Advances in Chemistry Series*, edited by F.M. Fowkes, American Chemical Society, Washington, D.C., (1964), Vol. 43.

[23] N. Watanabe, T. Nakajima, and H. Touhara, *Graphite Fluorides - Studies in Inorganic Chemistry*, Elsevier, New York, (1988), Vol. 8.

[24] G. Scott, *Polym. Eng. Sci.* **24**, 1007 (1984).

[25] G.D. Merfeld and M.A. Petrich, *J. Vac. Sci. Technol. A* **12**, 365 (1994).

[26] B. Wei and K. Komvopoulos, *J. Tribol.* **117**, (1995), in press.

[27] A. Torii, M. Sasaki, K. Hane, and S. Okuma, *Proc. IEEE Micro Electro Mechanical Systems. An Invest. of Micro Structures, Sensors, Actuators, Machines, and Systems*, Fort Lauderdale, FL, Feb. 7-10, 1993, p. 111.

[28] M.R. Houston, R. Maboudian, and R.T. Howe, *Proc. 8th Int. Conf. Solid-State Sensors and Actuators, Transducers 95*, Stockholm, Sweden, June 25-29, 1995.

ADHESION OF POLYMERS TO SYNTHETIC DIAMOND FILMS

EARL W. STROMBERG* AND MICHAEL D. DRORY**
*Lockheed Fort Worth Company, PO Box 748, Fort Worth, Texas 76101
**Crystallume, 3506 Bassett Street, Santa Clara, CA 95054

Abstract

A preliminary study of the adhesion of polymers to diamond substrates was conducted by examining the wetting characteristics of coupling agents to diamond. The contact angle between several commercial liquid coupling agents and a CVD diamond surface was measured. A silane, Z-6026, was found to wet the diamond surface. Several other coupling agents, including a titanate and an organometallic blend, wet the diamond surface significantly better than most of the other materials in this survey. It was found that water did not wet the diamond.

Introduction

The adhesion of polymers to diamond is a critical issue for mounting diamond windows, and for incorporating diamond–coated fibers in polymer matrix composites. Diamond is nearly chemically inert at room temperature making it relatively difficult to join to other materials. Although many studies have been conducted to understand the interface between diamond, ceramic and metal surfaces[1], very little work has been directed toward evaluating the interface between diamond and polymeric materials.

The measurement of the contact angle between a liquid drop and a substrate surface is typically used to evaluate the wetting and surface energy of materials[2]. The term wetting generally is defined as meaning that the contact angle between a liquid and a solid is zero or so close to zero that the liquid spreads over the solid easily. On the other hand, nonwetting means that the angle is greater than 90°. A non wetting liquid will ball up and can run off the surface easily. The objective of this study was to use a measurement of the contact angle to evaluate intermediate surface agents that might be used to create a useful interface between diamond and polymers. These surface agents may be used to improve surface wetting of the polymer on the diamond.

Experimental Procedure

The measurement procedure was as follows. The substrate was placed on the measurement stage, back-illuminated, and checked to insure that it was level and centered. The liquid to be evaluated was mixed by hand shaking in its original container. A fixed quantity of the coupling agent was then transferred directly to the specimen substrate on the measurement stage. Two techniques were used to

Mat. Res. Soc. Symp. Proc. Vol. 383 © 1995 Materials Research Society

quantitatively transfer the coupling agents. A quantitative [Eppendorf] pipettor was used for low and medium viscosity fluids. A 1-100 ml capable pipette tip was used to transfer standard 20 ml quantities of the low viscosity materials (KR-55, KR-OPPR, NZ-37, SC-3, Z-6020 and Z-6026). A larger pipette tip (101-1000 ml) was used to transfer 100 ml of the medium viscosity materials (LICA-09, LICA-38). A metal needle (16G11/2) and a 1 cc disposable tuberculin syringe was used to transfer 0.1 cc of the highest viscosity material, NZ-09. A new, clean transfer device was used for each individual test.

A single drop of liquid under evaluation was placed using the transfer technique described above onto the substrate. A stopwatch was started within 5 seconds of placing this drop on the substrate. The contact angle was measured as a function of time at 30 seconds, 1,2,3,5, 7 and 10 minutes. A comparator microscope fitted with a goniometer scale was used to measure this angle directly. Three measurements were taken for each liquid/surface combination. The reported value is the average of these three measurements.

Three substrate surfaces were evaluated: diamond, silicon, and glass. The diamond surface consisted of a thin diamond coating deposited by low–pressure (microwave) chemical vapor deposition on the surface of a thick (approximately 0.125 inch) silicon substrate. The diamond purity and continuous surface distribution was verified by Raman spectroscopy. The silicon surface consisted of a thin wafer of high purity, polished silicon. Since a typical application of coupling agents is to wet glass surfaces, a glass surface was included in this evaluation. Conventional clean microscope slides were used for this purpose.

Nine commercial coupling agents were selected for evaluation in this study: three titanates, three zirconates, two silanes, and an organometallic material. A description of these materials is presented in Tables 1 and 2. All of these materials were liquids, and most were concentrated (high percent solids) solutions. All of the coupling agents selected are compatible with generic epoxies. In addition, distilled water was also included in this study for comparison.

Results

The contact angle measured at 2 minutes after the application of the drop to each of the three substrates is presented in Figures 1,2, and 3 for titanates, zirconates, and silanes, respectively. In general, but by no means in all cases, the contact angle of the liquids on the diamond surface was lower than that of the silicon and glass surfaces. The significant exception to this observation was distilled water, which moderately wet the glass and silicon wafer surfaces (contact angles of 25 and 27 degrees, respectively), but was virtually nonwetting on the diamond surface (contact angle approximately 83 degrees). One of the silanes, Z-6026, appeared to dry out quickly on the glass and silicon substrates. This low viscosity liquid spread too fast on the diamond surface to make an accurate measurement; the contact angle was observed to be less than 2 degrees. LICA-09, a moderately high viscosity titanate, was observed to spread rapidly on the silica wafer, and at a somewhat lesser rate on the diamond.

Table 1 - List of Coupling Agents

Vendor Code	Chemical Description	Viscosity cps (@25 C)	Specific gravity (@16 C)	Solids in Solvent (%)	Flash Pt 'C
Z-6020	N-(B-aminoethyl)-G-aminoprophyltrimethoxysilane	6	1.02		49
Z-6026	amino-alkyltrimethoxysilane	2.6	0.89		16
KR 55	tetra (2,2 diallyoxymethyl)butyl, di(ditridecyl)phosphito titanate	50	0.97	99	82
LICA 09	neopentyl(dially)oxy, tri(dodecy)benzene-sulfonyl titanate	2000	1.04	80	82
LICA 38	neopentyl(dially)oxy, tri(dioctyl)pyro-phosphato titanate	5000	1.13	99	54-104
KZ OPPR	cyclo(dioctyl)pyrophosphato dioctyl zironate	110	1.10	90	71
NZ 09	neopentyl(dially)oxy, tri(dodecyl)benzene-sulfonyl zirconate	15000	1.18	95	66
NZ 37	dineopentyl(diallyl)oxy,diparamino benzoyl zirconate	100	1.11	46	96
SC 3	blend of organometallic coupling agents in n-vinyl pyrrolidone	5	1.05		82

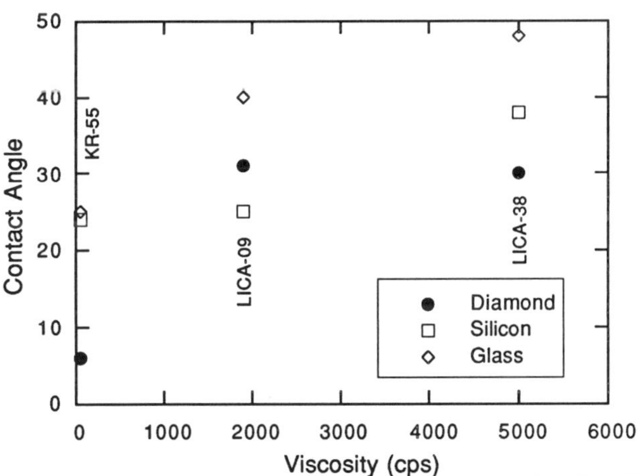

Figure 1. Measured contact angle of titanate liquids.

Table 2 - Description of Coupling Agents

KZ OPPR

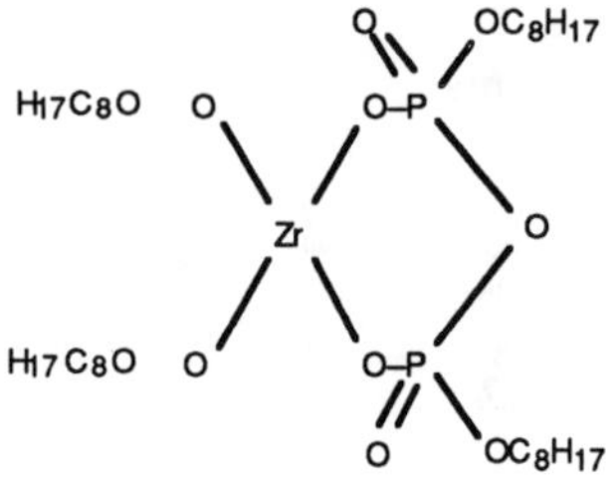

$H_{17}C_8O$ O, O–P(=O)–OC_8H_{17}

Zr, O

$H_{17}C_8O$ O, O–P(=O)–OC_8H_{17}

NZ 09

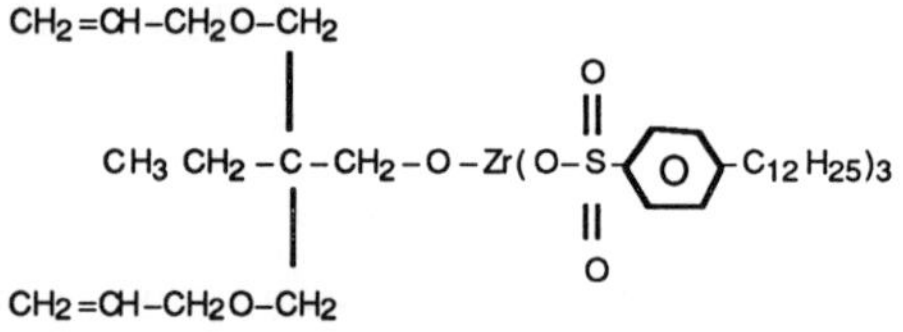

$CH_2{=}CH{-}CH_2O{-}CH_2$

$CH_3\ CH_2 - C - CH_2 - O - Zr(\,O - S(=O)_2 - C_6H_4 - C_{12}H_{25})_3$

$CH_2{=}CH{-}CH_2O{-}CH_2$

NZ 37

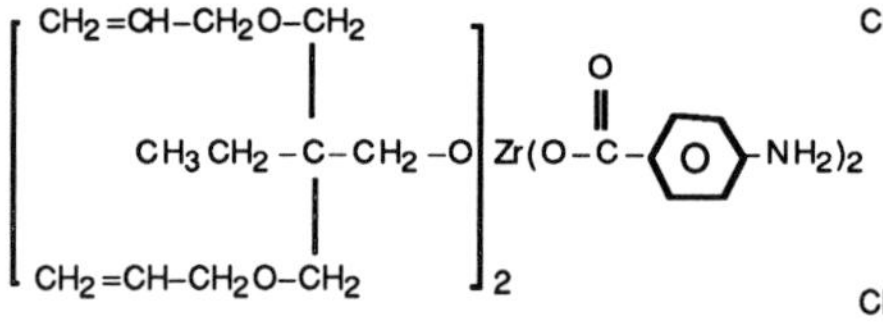

$\left[\begin{array}{c} CH_2{=}CH{-}CH_2O{-}CH_2 \\ CH_3CH_2 - C - CH_2 - O \\ CH_2{=}CH{-}CH_2O{-}CH_2 \end{array} \right]_2 Zr(O - C(=O) - C_6H_4 - NH_2)_2$

Lica 09

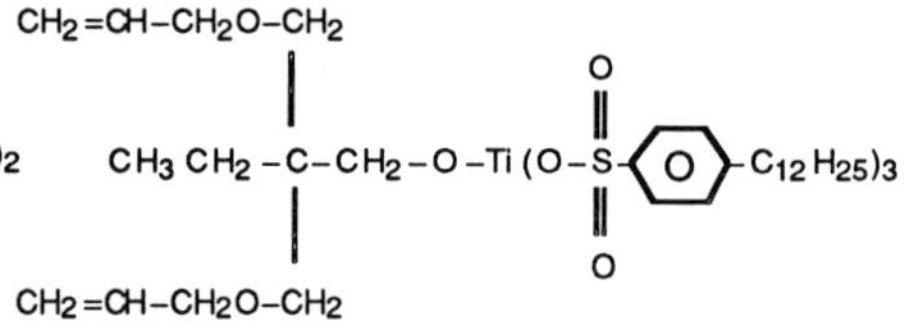

$CH_2{=}CH{-}CH_2O{-}CH_2$

$CH_3\ CH_2 - C - CH_2 - O - Ti\ (O - S(=O)_2 - C_6H_4 - C_{12}H_{25})_3$

$CH_2{=}CH{-}CH_2O{-}CH_2$

Lica 38

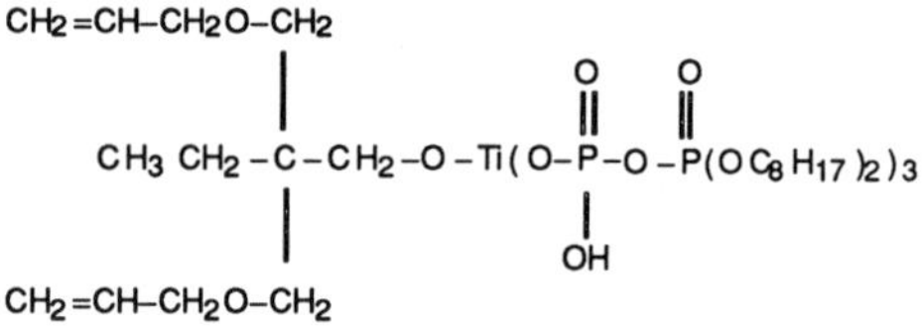

$CH_2{=}CH{-}CH_2O{-}CH_2$

$CH_3\ CH_2 - C - CH_2 - O - Ti(\,O - P(=O)(OH) - O - P(=O)(OC_8H_{17})_2)_3$

$CH_2{=}CH{-}CH_2O{-}CH_2$

KR 55

$\left[C_2H_5 - C(CH_2 - O - CH_2 - CH{=}CH_2)_2 - CH_2 - O \right]_4 Ti \cdot [HP(=O) - (O - C_{13}H_{27})_2]_2$

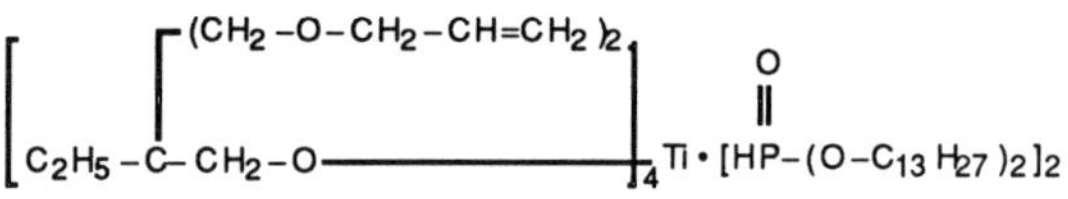

Z-6026

amino-alkyltrimethoxysilane

Z-6020

n-(β-aminoethyl)-γ-aminopropyltrimethoxysilane

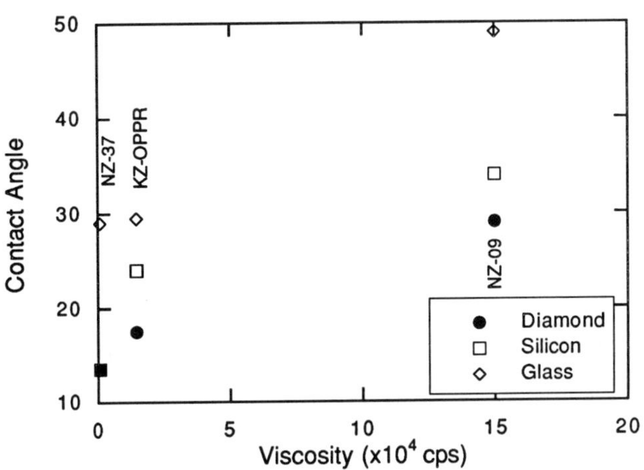

Figure 2. Measured Contact Angles for Zirconate Liquids

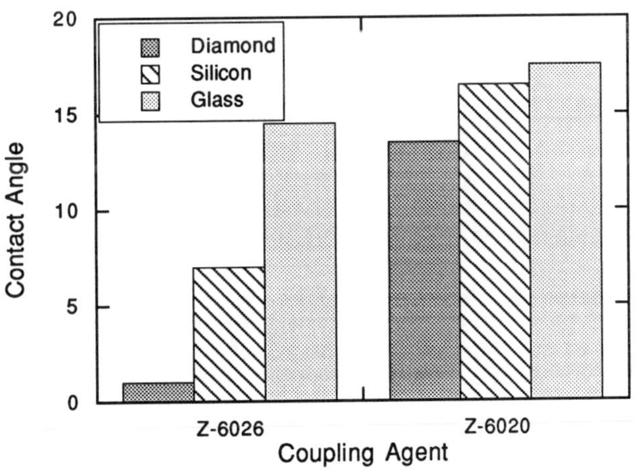

Figure 3. Measured contact angles for silane liquids.

Discussion

After placement of the drop on the test surface, two phenomena generally occurred: spreading and change in contact angle. The degree of spreading, per se, was not measured in this study. In general, a liquid placed on a solid will not wet it, but remains as a drop having a definite angle of contact between the liquid and solid phases. The liquids evaluated in this study generally performed in this manner.

The two minute comparative contact angle measurement value was used for several reasons. First, a high rate of contract angle change was observed for some materials, but the rate of change had generally slowed significantly by this time. Secondly, several materials appeared to either begin drying at the edges or reacting with the air and/or substrate by 3 minutes. Hence, a comparison past three minutes would not have been accurate for a material comparison. An example of the rate of contact angle change as a function of time is shown in Figure 4. Here, three titanates show that, for both high and low contact angle liquids, the greatest change occurred over the first 2 minutes.

A comparison of the materials by class is instructive. The silanes wet the diamond surface (see Figure 3). Both of these very low viscosity coupling agents were trimethoxysilanes containing an amino group. The contact angle of Z-6026, which had a somewhat simpler organic R-group structure than Z-6020, (see Table 2), on the diamond surface was too small to measure accurately. This effect would indicate a high degree of wetting by this coupling agent. Figure 1 is a comparison of the wetting of the three titanates on the three substrates. LICA-09 and LICA-38, which share a neopentyl(dially)oxy R group in their molecular structure, were both less able to wet the diamond, (and to a lesser extent the silicon and glass) than the KR-55 material. Since KR-55 also as a much lower viscosity than these materials, this was also compared. LICA 38, with twice the viscosity as LICA-09, actually wet the diamond better. The zirconates, on the other hand, showed an almost linear relationship between diamond wetting and viscosity. (see Figure 2) The same was not true when comparing two materials with the same functional R groups- LICA-09 and NZ-09. As seen in Figure 5, the wetting of diamond was actually lower in the material with the significantly higher viscosity. Hence, viscosity could not be considered the sole criteria in determining the wettability of a material.

The coupling agents used in this study were concentrated, typically over 90% solids. However, the typical use of these materials is in the 0.1 to 1% range. Since there were a variety of solvents and solvent percentages uses, no attempt was made during this study to calculate the actual surface energy.

The objective of this study was to compare the ability of several candidate coupling agents to wet a diamond surface. Wettability "goodness" was defined to be a low contact angle measured between the liquid coupling agent drop and the substrate surface. One material, Z-6026, wet the diamond surface extremely well. Two other materials, the titanate KR-55 and organometallic blend SC-3, exhibited significantly

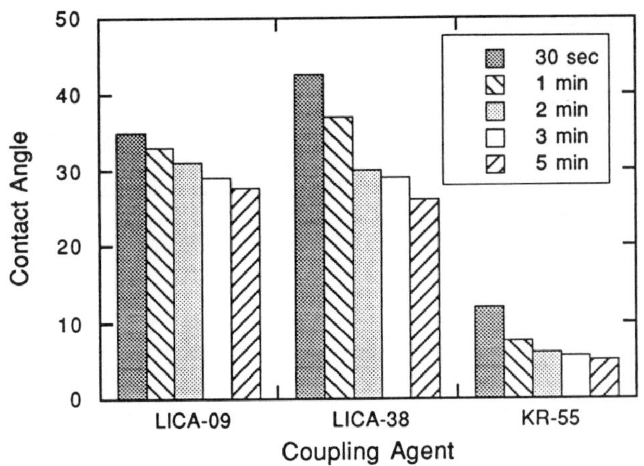

Figure 4. Contact angle of titanates on diamond as a function of time.

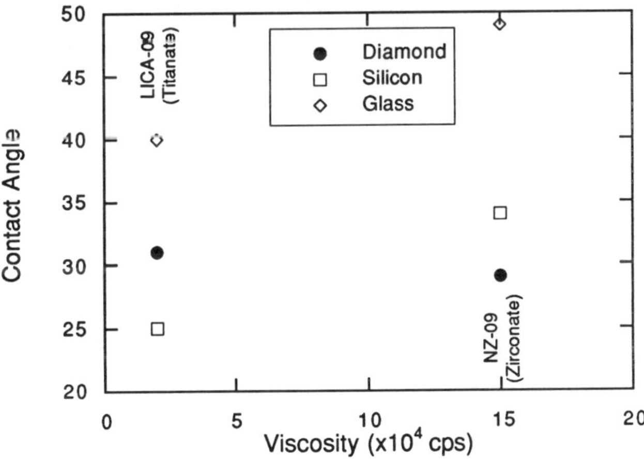

Figure 5. Comparison of titanate and zirconate with same functional groups.

better wetting than the other materials (see figure 6). This information was sufficient to downselect to these three coupling agents for further study in diamond/polymeric matrix composite systems. Although no high viscosity liquid exhibited extremely good wetting, no consistent relationship was observed between the degree of wetting and the viscosity of the liquid. It is anticipated that future studies will be conducted to examine the chemical relationship between the functional groups on the coupling agents and the diamond surface in order to better understand how wetting by these materials is effected by surface energy and short term chemical reaction.

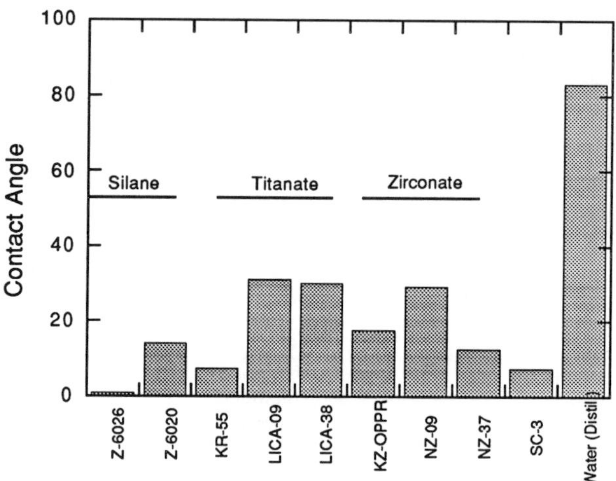

Figure 6. Summary of diamond wetting for all materials tested.

Conclusion

One material, the silane Z-6026, wet the diamond surface. Two other materials, the titanate KR-55 and the organometallic blend SC-3, exhibited significantly better wetting than the other material (see figure 6). Although no high viscosity liquid exhibited extremely good wetting, no consistent relationship was observed between the degree of wetting and the viscosity of the liquid.

References

1. C. Tsai, J. C. Nelson, W. W. Gerberich, D. Z. Lui, J. Heberlein and E. Pfender, Thin Solid Films, 237 (1994), 181-186.

2. A. W. Adamson, Physical Chemistry of Surfaces (John Wiley & Sons, Inc., New York, 1990), pp. 379-420.

DIAMOND-LIKE CARBON DEPOSITION FOR TRIBOLOGICAL APPLICATIONS AT LOS ALAMOS NATIONAL LABORATORY

K.C. Walter, M. Nastasi, H. Kung, P. Kodali, C. Munson, I. Henins, and B.P. Wood
Los Alamos National Laboratory, MS-K762, Los Alamos, NM, 87545

ABSTRACT

Diamond-like carbon (DLC) films have been deposited on silicon using two deposition techniques. Both deposition techniques used acetylene (C_2H_2) plasmas as the carbon/hydrogen source. One technique, that is relatively well known, employs an rf-plasma with an associated self-bias on the substrate. DLC films have also been deposited using a pulsed-bias method. Coatings of various thickness have been deposited using both deposition methods, and various combinations of gas pressure and bias. Coating characteristics, such as composition, density, and sp^3 content, of selected films will be presented. For each deposition technique, the correlation between coating characteristics, mechanical properties and tribological behavior is presented.

INTRODUCTION

DLC coatings are desirable for tribological applications because of their ease of fabrication, high hardness [1], low coefficients of friction and low wear rates when worn against a number of materials [2-8]. The wide variety of DLC deposition techniques have been reviewed extensively [1-2, 9-13] and will not be repeated here. This work compares the tribological performance of DLC films deposited using two deposition techniques. The rf self-bias method depends on a combination of the supplied rf power and the ratio of the cathode and anode surface areas to provide a bias on the smaller electrode. By using a cathode with a smaller surface area than the anode, a negative self bias can be imposed on the cathode. The magnitude of the bias can be tailored by adjusting the rf power and/or the cathode area.

Another DLC deposition method, that is not as well studied, uses a pulsed bias [14-15] on a target to accelerate ions from an acetylene (C_2H_2) plasma at energies low enough (<1 keV) to result in deposition instead of ion implantation. For the pulsed bias method, plasma generation and the bias on the target can be independent of the geometry and thus allows independent determination of ion deposition energy. A limited comparison of DLC deposited by cathodic arc, rf self bias, and pulsed bias methods has been previously presented [16].

This work elaborates further on the tribological properties of DLC deposited at Los Alamos National Laboratory for tribological applications through use of both rf self bias and pulsed bias deposition methods. The results are sufficient to compile relationships between important tribological parameters, such as wear rate, hardness, and elastic modulus, and deposition parameters, such as bias and pressure.

Mat. Res. Soc. Symp. Proc. Vol. 383 © 1995 Materials Research Society

EXPERIMENTAL DETAILS

The self bias coatings were deposited on silicon (0.5 mm thick) using biases ranging from -200 V to -780 V while in an acetylene plasma (C_2H_2) with neutral gas pressures ranging from 0.5 to 4.5 mTorr. The cathode electrode in the rf (13.56 MHz) circuit also served as the holder for the silicon samples. The pulsed-bias coatings were also deposited on silicon using biases ranging from -200 V to -600 V (20 μs pulses at ~12 kHz, or 25% duty cycle) while surrounded with an acetylene plasma with neutral gas pressures ranging from 0.5 to 2.5 mTorr. When using a 0.5 mTorr fill pressure, an rf plasma was generated using an independent antenna so that the plasma generation was independent of the pulse bias. At the higher fill pressure of 2.5 mTorr, a plasma was generated as the result of pulse biasing the target and plasma generation was thus coupled to the pulsed bias. Coating thickness varied from ~0.2 μm to ~11 μm depending on the deposition conditions with most coatings being about 0.5 μm thick. Following deposition, coating thickness was determined using surface profilometry, coating composition by ion beam analysis, carbon bonding states using electron energy loss spectroscopy, hardness and elastic modulus using nanoindentation, and wear and friction properties using a pin-on-disk technique.

The carbon and hydrogen concentrations of the coatings were characterized using a 3.55 MeV $^4He^+$ ion beam. Spectra of the backscattered He^+ from the carbon atoms were collected at both normal incidence and 75° to the sample surface. These conditions take advantage of the 6.5x increase in the He-C scattering cross section over what is observed using a 2 MeV He^+ beam [17]. When the sample is tilted 75° to the incident beam, hydrogen atoms are forward scattered out of the coating and were also collected. The spectra were interpreted according to the principles of elastic recoil spectroscopy (ERS). The combination of ion backscattering and ERS allowed the quantification of the atomic fractions of carbon and hydrogen in all the coatings. Since most coatings were less than 1 μm thick, combinations of He^+ backscattering and ERS were sufficient to also quantify the total number of carbon and hydrogen atoms (in units of atoms/cm^2) in the coatings. Some coatings were too thick for such analysis and in those cases, a 1.49 MeV $^1H^+$ ion beam was backscattered from the coating to quantify the total number of atoms (both carbon and hydrogen) in the coating. For all ion backscattering measurements the scattering angle was 166°. The scattering angle for the ERS measurements was 30°. The backscattered and ERS spectra were quantitatively analyzed using graphite and polystyrene as materials standards and conventional computer-aided techniques [18].

The sp^2 content of the DLC coatings was characterized using parallel electron energy loss spectroscopy (PEELS) through a Gatan PEELS system mounted on a 300 keV transmission electron microscope and cross-sectional TEM samples. The energy resolution of the PEELS spectra was about 1.7 eV, as estimated from the full-width at half-maximum of the zero-loss peak. The ratio of the sample thickness and the mean free path for plasmon scattering (t/λ_p) was determined and background subtraction for the carbon K-edge spectra was accomplished using a software package supplied by Gatan.

Nanoindentation tests were conducted using a Nano Indenter® II [19]. The hardness and elastic modulus of the films were calculated (with elastic corrections) from load/displacement curves at various depths. Average values and standard deviations were calculated from ten measurements at each depth. Hardness and elastic modulus values quoted in this work are from depths not exceeding 50% of the coating thickness.

Pin-on-disk (POD) tribological tests were conducted using 6 mm 400C stainless steel pins and a load of 0.8 N. The tests were conducted in a controlled environment where the relative humidity ranging from 27 to 32%. The maximum Herztian contact stress on the coatings depended on the elastic modulus of the coatings and varied from 0.65 to 1.1 GPa, assuming a Poisson's ratio of 0.3 for DLC [10]. The track diameter was ~3 mm and the sliding speed ~35 mm/sec. After testing, the wear track cross-sectional area was measured by surface profilometry at four positions around the track, each about 90° apart. The largest observed cross-sectional area was used to calculate the wear track volume. The wear rate, K=track volume [mm^3]/(load [N]•sliding distance [m]), was calculated using the methodology of Holmberg and Matthews [20]. During the wear test, the coefficient of friction was continuously monitored using a calibrated load cell.

RESULTS AND DISCUSSION

The DLC deposition rates (μm/hour) as a function of pressure and rf self bias are shown in Figs. 1-3. For the rf self bias method, the deposition rate clearly depends more on the acetylene (C_2H_2) pressure than the imposed rf self bias. It is speculated that this trend occurs because of two complimentary effects; (1) an increase in plasma density with the square root of the pressure, and (2) that some portion of the deposition is due to neutral radicals [21], such as CH_3^0, whose density also increases with gas pressure. Robertson [1] has reviewed the previous work that has reported the linear dependence of the deposition rate on the product of the bias and gas pressure, and the same trend, with large scatter, is seen in Fig. 3.

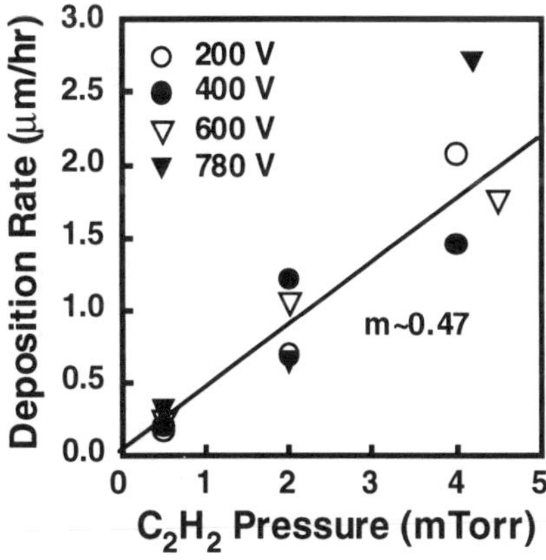

Fig. 1. DLC deposition rate is linearly dependent on the gas pressure for the rf self bias deposition method.

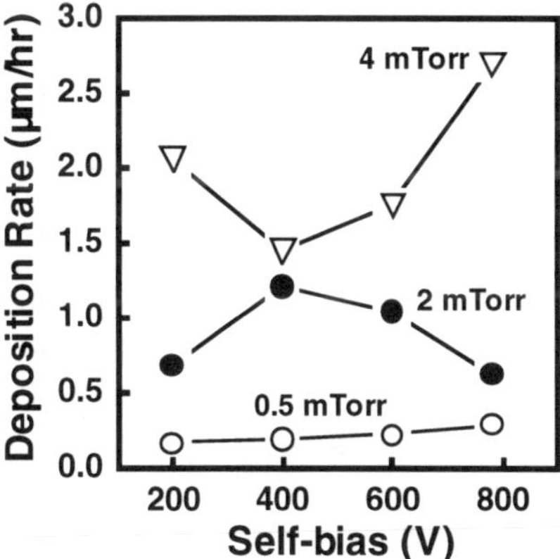

Fig. 2. The DLC deposition rate has no clear dependence on rf self-bias.

The dependence of DLC hardness and elastic modulus on the self bias and gas pressure are shown in Figs. 4-7. The DLC hardness ranges from about 10 GPa to 28 GPa (Fig. 4), and the elastic modulus ranges from 70 GPa to 230 GPa (Fig. 5), both of which are within the accepted range for DLC materials [10]. Fig. 6-7 indicate a linear dependence of hardness and elastic modulus on the gas pressure when using the rf self bias deposition method. Since the DLC hardness, elastic modulus and deposition rate are all linearly dependent on gas pressure, there is also a linear dependence of the DLC hardness and elastic modulus on deposition rate (Figs. 8-9). Finally, it is worth noting that the relationship (Fig. 10) between DLC hardness and modulus (H/E~0.12) is very close to that predicted by Robertson [10] (H/E~0.1) for diamond-like carbon materials.

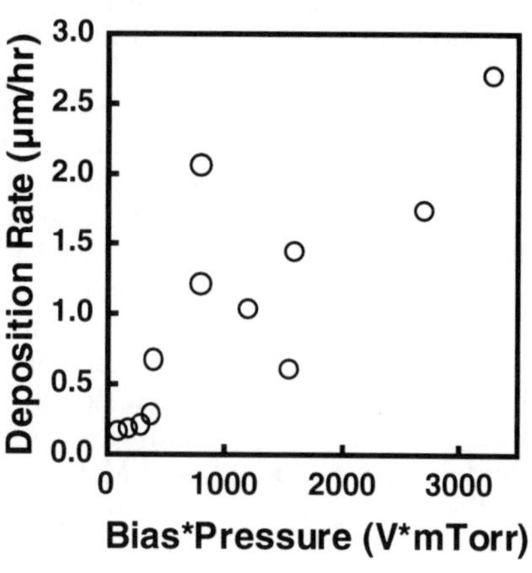

Fig. 3. The DLC deposition rate for the rf self bias method exhibits a linear trend with the product of bias and gas pressure.

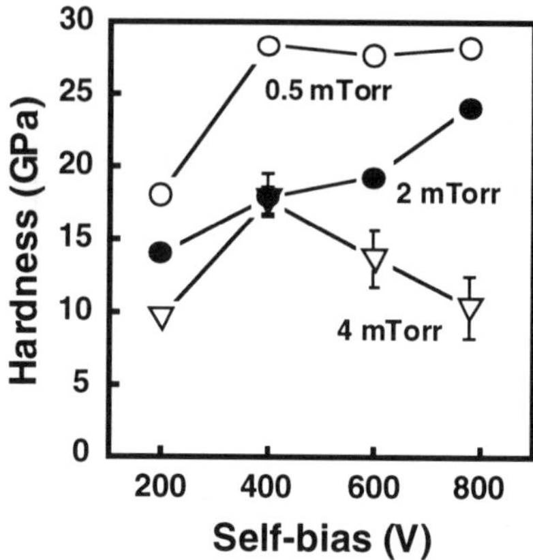

Fig. 4. Plot of the DLC hardness as a function of self bias.

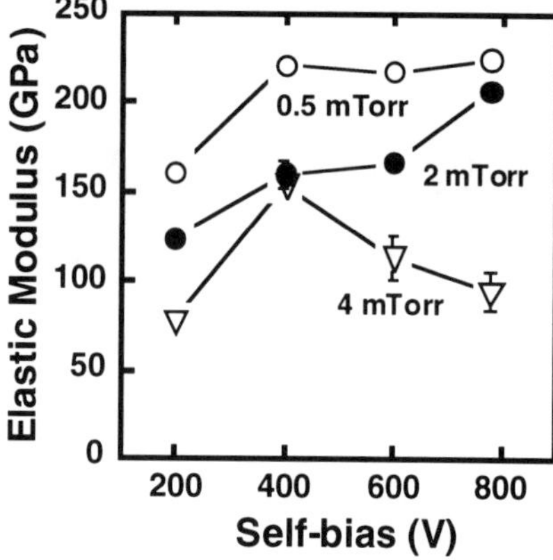

Fig. 5. Plot of the DLC elastic modulus as a function of self bias.

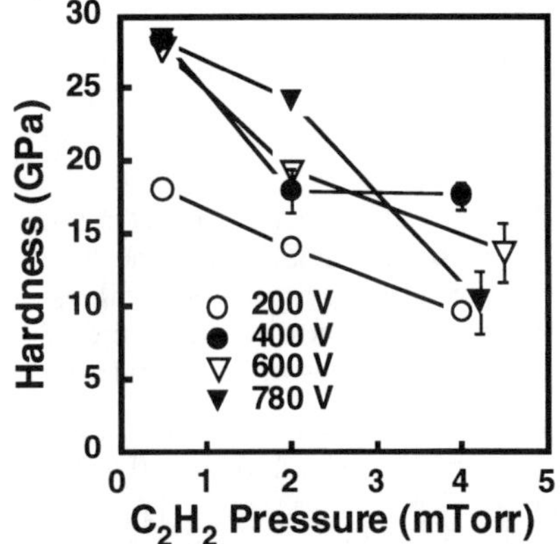

Fig. 6. Plot of the self bias DLC hardness as a function of pressure.

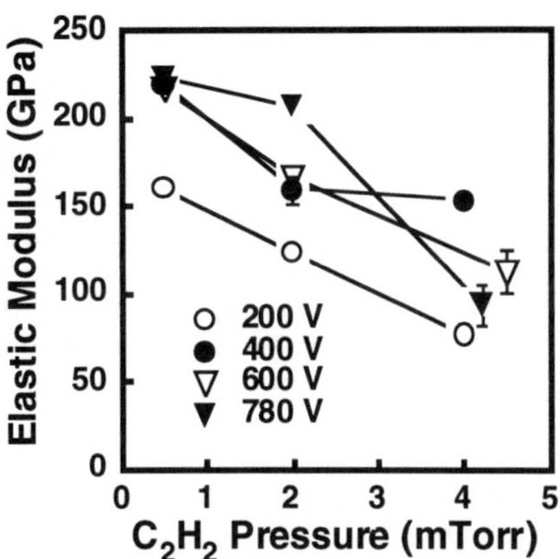

Fig. 7. Plot of the self bias DLC elastic modulus as a function of pressure.

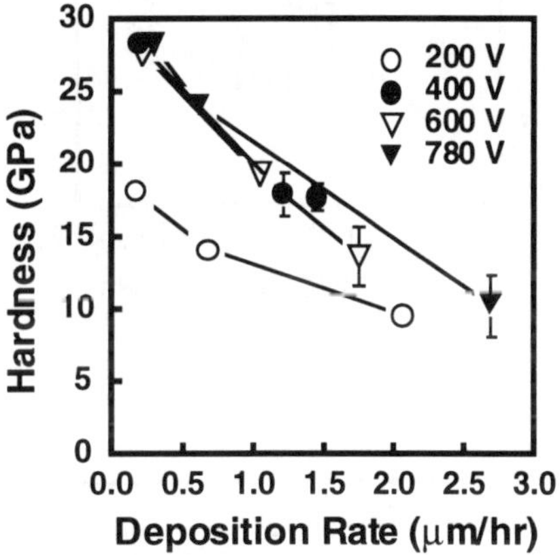

Fig. 8. Plot of the self bias DLC hardness as a function of deposition rate.

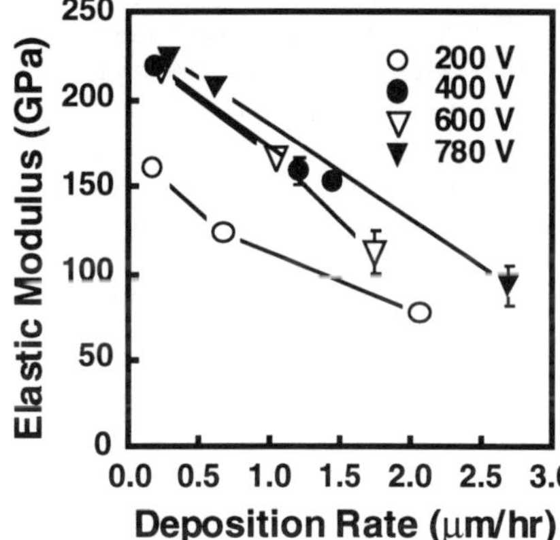

Fig. 9. Plot of the self bias DLC elastic modulus as a function of deposition rate.

The data in Figs. 1-10 can be used to predict the hardness and modulus of the self bias DLC coatings as a function of deposition rate or gas pressure. For the range of self-bias and pressure explored here, no clear dependence of DLC hardness on self-bias was observed, even though such a dependence is expected. The full set of equations are as follows:

$$dT/dt \approx 0.47\ P \tag{1a}$$

$$H \approx 27 - 3.4\ P \tag{1b}$$

$$H \approx 26 - 6.7\ (dT/dt) \tag{1c}$$

$$E \approx 217 - 26\,P \tag{1d}$$

$$E \approx 212 - 51\,(dT/dt) \tag{1e}$$

$$H \approx 0.12\,E \tag{1f}$$

where dT/dt is the deposition rate in µm/hour, P is the C_2H_2 pressure in mTorr, H is the DLC hardness in GPa, and E is the DLC elastic modulus in GPa. For the experimental conditions used in this work, the relationships in Eq. 1 gives values accurate to within ~50% for the deposition rate, within ±2 GPa for hardness and ±50 GPa for elastic modulus. Eqs. 1b &1d are valid for C_2H_2 pressures up to ~8 mTorr.

The dependence of the DLC deposition rate on the magnitude of the pulsed bias and the gas pressure is shown in Figs. 11-12. Note the deposition rate at 2.5 mTorr is approximately 50% of the deposition rate using the self bias deposition method. However, at 0.5 mTorr both deposition methods give equivalent deposition rates. It is speculated that the difference in deposition rate at 2.5 mTorr is due to the 25% duty cycle and the dynamic plasma conditions present when using the pulsed bias method. Just as in the self-bias case, the deposition rate has a clear dependence on the gas pressure (plasma density), but not the magnitude of the pulsed bias.

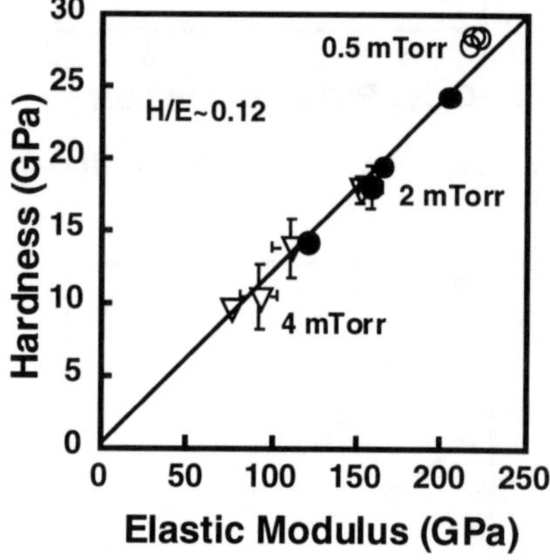

Fig. 10. The hardness and modulus of the self bias DLC is related by H/E~0.12.

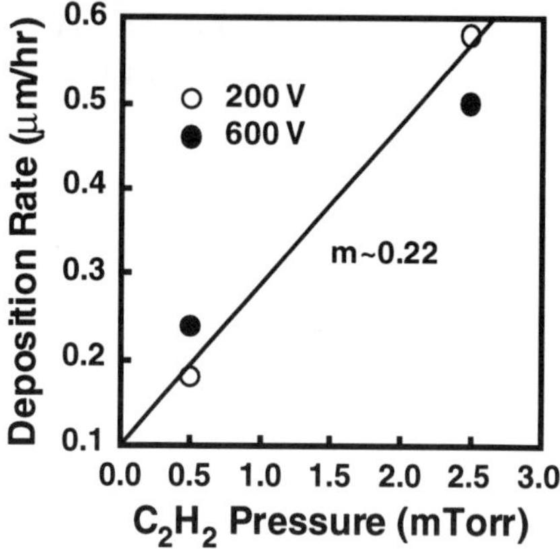

Fig. 11. Plot of the pulsed bias deposition rate as a function of acetylene pressure.

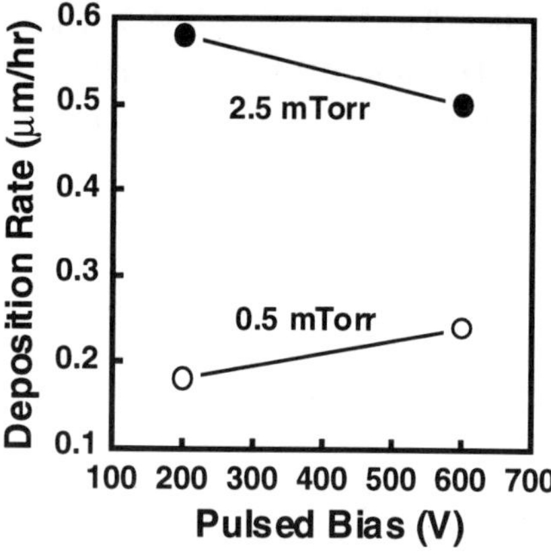

Fig. 12. Plot of the pulsed bias deposition rate as a function of pulsed bias.

The dependence of the deposition rate for the pulsed bias method on the product of bias magnitude and pressure suggests a linear trend (Fig. 13). However, the dependence on gas pressure is more obvious (Fig. 11).

The effect of pulse bias and gas pressure on DLC hardness and elastic modulus are shown in Figs. 14-17. Note that the magnitudes of the hardness (6 to 24 GPa) and moduli (50 to 200 GPa) of the DLC coatings deposited using the pulsed bias method span the same range as the coatings deposited using the self bias method, but over a narrower range of bias (200 to 600 V) and pressure (0.5 to 2.5 mTorr). It has been shown [16] that the DLC hardness will decrease if the pulsed bias is increased to 20 kV. Figs. 16-17 show the dependence of DLC coating hardness and elastic modulus on gas pressure during pulsed bias deposition. The hardness and elastic modulus are linearly dependent on the deposition rate, as shown in Figs. 18-19. The hardness and elastic modulus of the pulsed bias DLC coatings are related (Fig. 20) by H/E~0.12, just as for the self bias deposition process.

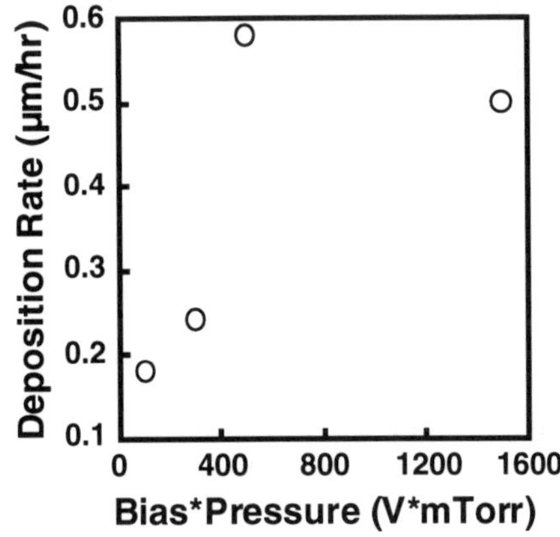

Fig. 13. The deposition rate for the pulsed bias method exhibits a linear trend with the product of bias*pressure.

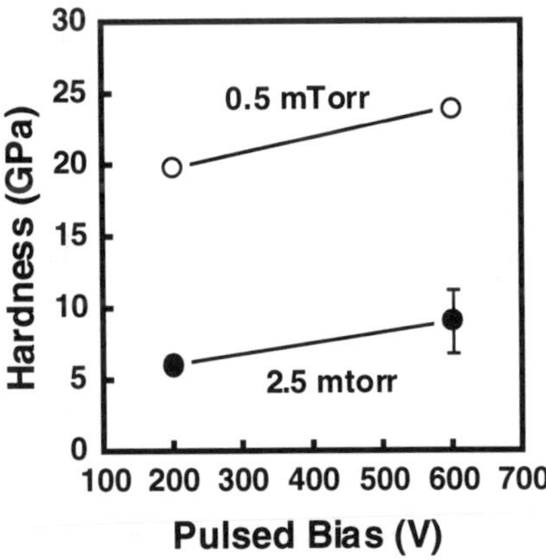

Fig. 14. Plot of the DLC hardness as a function of pulsed bias.

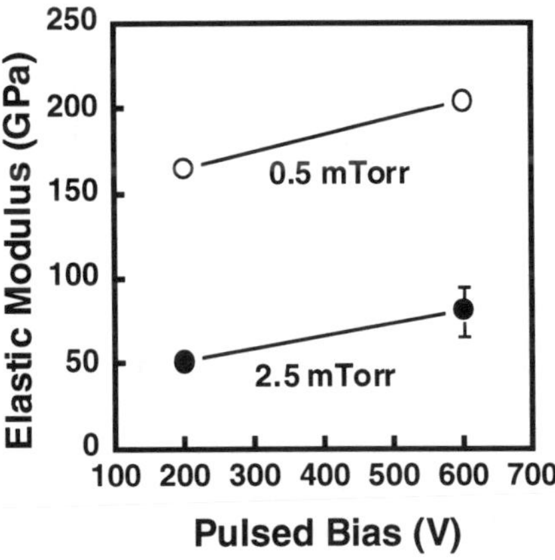

Fig. 15. Plot of the DLC elastic modulus as a function of pulsed bias.

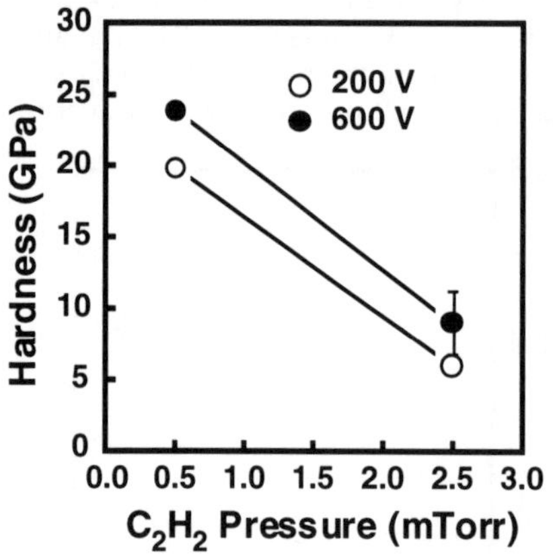

Fig. 16. Plot of the pulsed bias DLC hardness as a function of gas pressure.

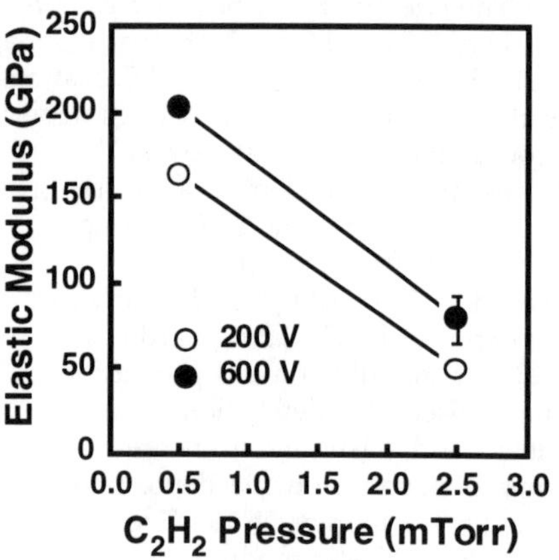

Fig. 17. Plot of the pulsed bias DLC elastic modulus as a function of gas pressure.

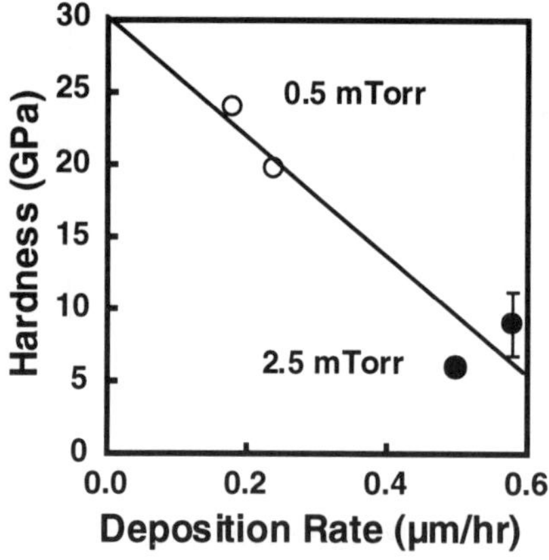

Fig. 18. Plot of the pulsed bias DLC hardness as a function of deposition rate.

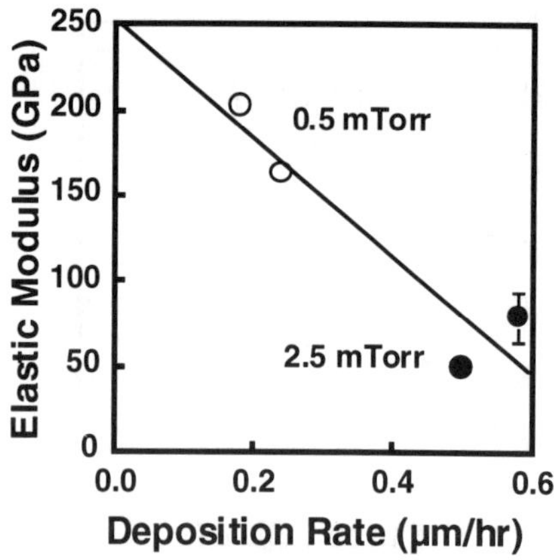

Fig. 19. Plot of the pulsed bias DLC elastic modulus as a function of deposition rate.

The data in Figs. 11-20 can be used to predict the hardness and modulus of the pulsed bias DLC coatings as a function of deposition rate or gas pressure. The full set of relationships are as follows:

$$dT/dt \approx 0.22\ P \tag{2a}$$

$$H \approx 25 - 7.2\ P \tag{2b}$$

$$H \approx 30 - 41.5\ (dT/dt) \tag{2c}$$

$$E \approx 213 - 59\ P \tag{2d}$$

$$E \approx 253 - 343\ (dT/dt) \tag{2e}$$

$$H \approx 0.12\ E \tag{2f}$$

where the variables are as defined for Eq. 1. For the experimental conditions for the pulsed bias DLC depositions, the relationships in Eq. 2 gives values accurate to within ~30% for the deposition rate, within ±3 GPa for hardness and ±40 GPa for elastic modulus. Equations 2b & 2d are valid for C_2H_2 pressures less than ~3.6 mTorr.

The hardness as a function of sp^3 content is shown in Fig. 21 for some of the coatings. As expected, the hardness increases with increasing sp^3 content and correlate well with the published dependence [10] of hardness (10 to 20 GPa) on sp^3 content (40-70%).

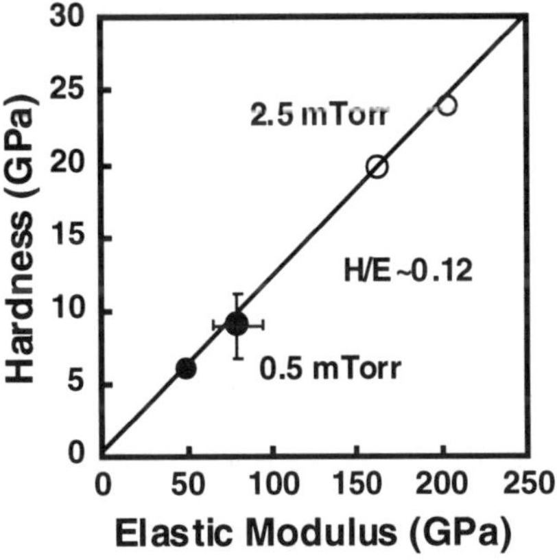

Fig. 20. The hardness and modulus of the pulsed bias DLC is related by H/E~0.12.

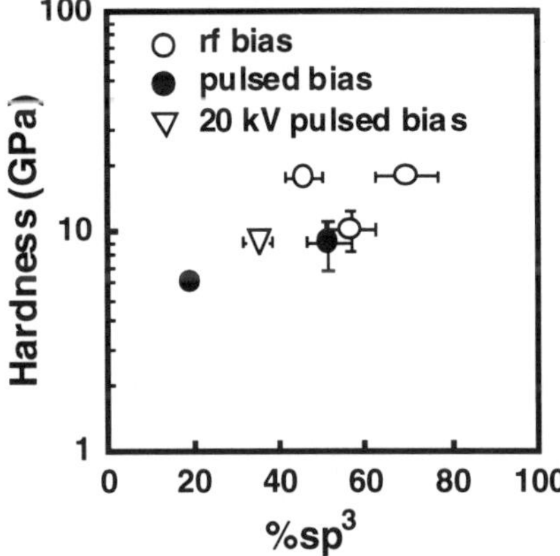

Fig. 21. Plot showing the dependence of DLC hardness on sp^3 content. The 20 kV pulsed bias point comes from Ref. 16.

Some typical ERS spectra for the DLC coatings with the minimum and maximum hydrogen concentrations are shown in Fig. 22. Similar measurements were made for all DLC coatings and the hardness dependence on hydrogen content is shown in Fig. 23. The hardness of a DLC coating decreases with increasing hydrogen content in a linear relationship that is similar for both rf self bias and pulsed bias deposition methods.

The combination of hydrogen and carbon contents and the coating thickness allowed the calculation of the DLC coating density. Although the observed density range of 2 to 3 g/cm^3 is similar to the reported range [10] of hardness (10 to 20 GPa), the error in the density measurement is about ±30%. Because of the large error, the density dependence on the deposition parameters is unclear and graphs showing the trends are not included here.

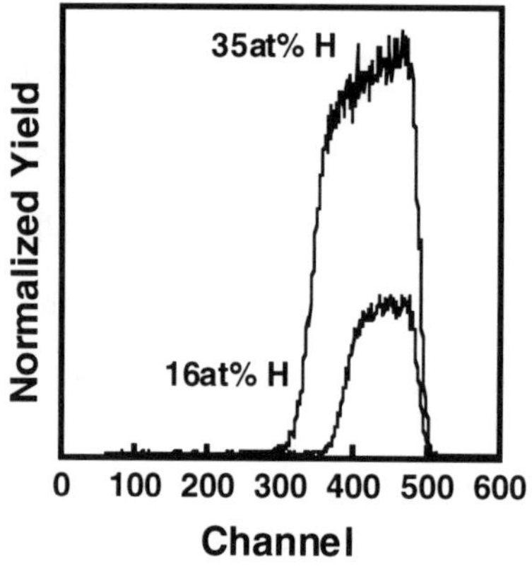

Fig. 22. The hydrogen content of DLC coatings with different hydrogen contents as measured by ERS.

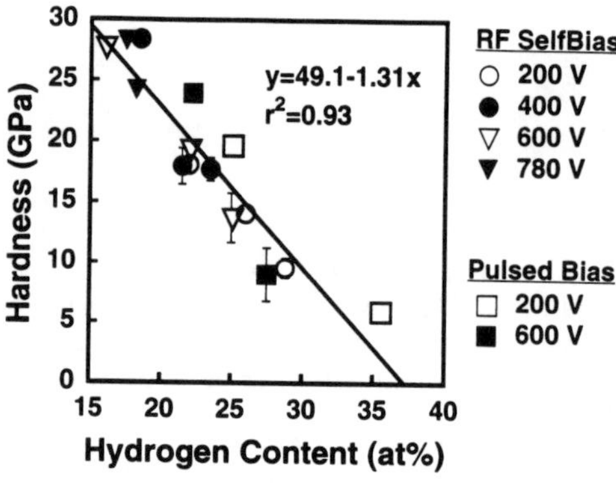

Fig. 23. Plot showing the dependence of DLC coating hardness on hydrogen content for both self bias and pulsed bias methods.

The results of the tribological tests are shown in Figs. 24-25. The wear rate, K, of DLC decreases with increasing hardness, H, for both self bias and pulsed bias deposition methods and is described by

$$K = 0.07 - 0.002\,H \tag{3}$$

The wear rates shown in Fig. 24 are a factor of 10 lower than previously reported for DLC of equivalent hardness values [16], and a factor 3 higher than other reported DLC wear rates [8] when worn against 440C stainless steel under similar tribological conditions. The discrepancy is thought to be due to variations in the 440C stainless steel pin surface and varying relative humidity amounts during the other studies. Another report of DLC deposited using a pulsed vacuum arc process [22] described larger wear rates varying from 0.05 to 0.15×10^{-6}m m^3/Nm. It should finally be noted that the wear rates reported in this work are similar to that when DLC is worn against WC [23] or when (Ti,Al)N is worn against steel [24].

The low coefficient of friction reported in Fig. 25 is similar to that reported in other works (8,16,22). It is worth noting that DLC also exhibits similar coefficients of friction when worn against alumina [25] or 52100 steel [26].

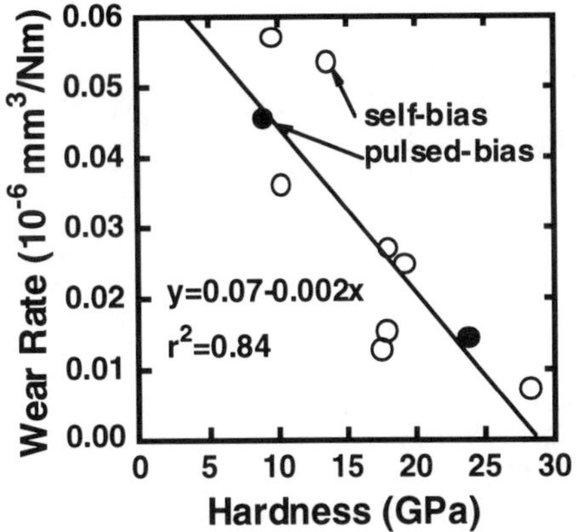

Fig. 24. Plot of the pulsed bias DLC hardness as a function of deposition rate.

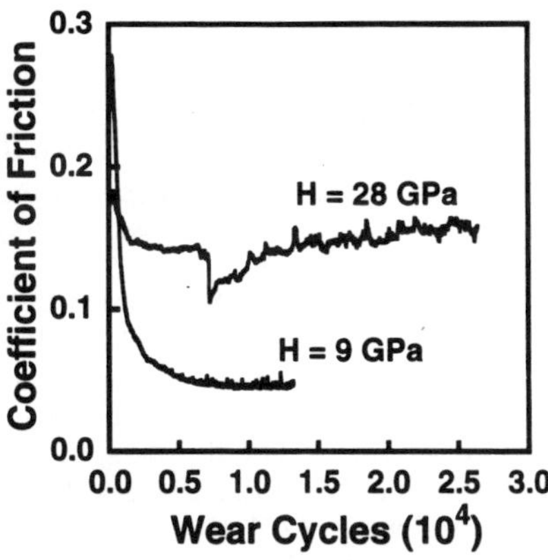

Fig. 25. Plot of the coefficient of friction for both harder and softer DLC coatings.

CONCLUSIONS

Self bias and pulsed bias methods of DLC deposition can result in coatings with comparable properties of hardness, elastic modulus, wear rate and coefficient of friction. However, the pulsed bias method has deposition rates may be lower depending on how the plasma is generated. However, DLC coatings deposited by either of these methods exhibit low wear rates and coefficients of friction and thus make excellent tribological materials.

In addition, if a desired wear rate is specified, the required hardness can be specified according to Eq. 3. Given the required hardness value, the deposition parameters for either self bias or pulsed bias deposition techniques can be specified using Eqs. 1 & 2.

ACKNOWLEDGMENTS

This work was partially supported by the United States Department of Energy Office of Basic Energy Science, Division of Materials Science. The authors wish to thank J.R. Tesmer, C.J. Maggiore, C. Evans, and M. Hollander of the Ion Beam Materials Laboratory at the Los Alamos National Laboratory for assistance with the ion beam analysis.

BIBLIOGRAPHY

1. J. Robertson, Prog. Solid St. Chem. **21**, 199 (1991).
2. J.-P. Hirvonen, R. Lappalainen, J. Koskinen, A. Antilla, T.R. Jervis, M. Trkula, J. Mater. Res. **5**, 2524 (1990).
3. K. Enke, H. Dimigen, H. Hübsch, Appl. Phys. Lett. **36**, 291 (1980).

4. R. Wei, P.J. Wilbur, A. Erdemir, F.M. Kustas, Surf. Coat. Technol. **51**, 139 (1992).
5. K. Oguri, T. Arai, J. Mater. Res. **7**, 1313 (1992), R.L.C. Wu, Surf. Coat. Technol. **57**, 258 (1992).
6. Y. Itoh, S. Hibi, T. Hioki, J. Kawamoto, J. Mater. Res. **6**, 871 (1991).
7. B. André, J.-Ph. Nabot, L. Lombard, P. Martin, NATO ASI Series E, **266**, 313 (1991).
8. A. Grill, V. Patel, B.S. Meyerson, NATO ASI Series E, **266**, 417 (1991).
9. A. Antilla, in Structure-Property Relationships in Surface-Modified Ceramics, edited by C.J. McHargue, R. Kossowsky, W.O. Hofer, (Kluwer Academic Publishers, 1989), p. 455.
10. J. Robertson, NATO ASI Series B, **266**, 331 (1991).
11. J.C. Angus, P. Koidl, S. Domitz, in Plasma Deposited Thin Films, edited by J. Mort, F. Jansen, (CRC Press, Boca Raton, Florida, 1986), p. 89.
12. M.W. Gies and M.A. Tamor, in Encyclopedia of Applied Physics, Vol. 5, edited by G.L. Trigg, (VCH Publishers, Inc., New York, 1991), p. 1.
13. Y. Catherine, NATO ASI Series B, **266**, 193 (1991).
14. G.J. Vandentop, M. Kawasaki, R.M. Nix, I.G. Brown, M. Salmeron, G.A. Somorjai, Phys. Rev. **B41**, 3200 (1990).
15. J. Chen, J.R. Conrad, R.A. Dodd, J. Mat. Engin. Perform. **2**, 839 (1993).
16. K.C. Walter, H. Kung, T. Levine, J.R. Tesmer, P. Kodali, B.P. Wood, D.J. Rej, M. Nastasi, J. Koskinen, J.-P. Hirvonen, Mat. Res. Soc. Symp. Proc., Fall 1994 MRS Meeting Symposium A, in press, 1994.
17. Handbook of Modern Ion Beam Materials Analysis, edited by J.R. Tesmer and M. Nastasi, (Materials Research Society, 1995.
18. L.R. Doolittle, Nucl. Instr. Meth. **B9**, 334 (1985).
19. Nano Instruments, Inc., Knoxville, TN.
20. K. Holmberg, A. Matthews, Coatings Tribology: Properties, Techniques and Applications in Surface Engineering, edited by D. Dowson, (Elsevier, Amsterdam, 1994), p. 53.
21. W. Müller, NATO ASI Series B: Physics 266(1991)229-241.
22. H. Ronkainen, J. Likonen, J. Koskinen, Surf. Coating Technol. 54/55(1992)570-577.
23. D.P. Monaghan, D.G. Teer, P.A. Logan, I. Efeoglu, R.D. Arnell, Surf. Coatings Technol. 60(1993)525-530.
24. B.F. Coll, P. Sathrum, R. Aharonov, J.P. Peyre, M. Benmalek, presented at 19th Internataional Conference on Metallurgical Coatings and Thin Films (ICMCTF 92), San Diego, CA, USA, October 6-10, 1992.
25. R. Wei, P.J. Wilbur, A. Erdemir, F.M. Kustas, Surf. Coatings Technol. 51(1992)139-145.
26. X. He, W. Li, H. Li, Vacuum 45(1994)977-980.

SYNTHESIS, PROPERTIES AND APPLICATIONS OF SUPERHARD AMORPHOUS COATINGS

MICHAEL A. TAMOR, Research Laboratory, SRL/MD-3439, Ford Motor Company, Dearborn, MI 48121-2053.

ABSTRACT

Low-friction/ultralow-wear coatings allow "surface engineering" for improved performance and durability, and enable use of new light weight or low cost materials. The accepted correlation of wear resistance with hardness suggests use of ceramic carbides and nitrides, with diamond being the ultimate anti-wear coating. While any of these may be deposited by chemical vapor deposition, the high cost (due to low deposition rates and high capital costs) and (usually) high deposition temperatures makes CVD coating impractical for cost-sensitive automotive applications. While rarely as hard as their crystalline counterparts, hard amorphous films exhibit similar (and occasionally superior) tribological properties and may be deposited on virtually any material at low cost. The highly nonequilibrium deposition process - conformal plasma reactive ion plating (CP-RIP) - allows tailoring of film properties and exploration of completely new compositions with no crystalline counterparts. Factors controlling the mechanical and optical properties of amorphous hard coatings, and recent progress in their application will be reviewed.

INTRODUCTION

Needs for surface engineering of rolling and sliding mechanical components fall into several broad categories: (1) reduction of wear in heavily loaded, lubricated contacts, (2) reduction of both friction and wear in situations where lubrication is limited or even undesirable, (3) reduction of friction and wear of materials that while strong, are tribologically poor, and (4) modification of light weight materials (e.g. Al, Mg) for improved durability in mechanical applications. For this discussion it is assumed that both contacting members are well finished and sufficiently hard that plastic deformation and plowing are not relevant. This leaves two dominant wear mechanisms: abrasive wear and adhesive wear. Based in part on traditional descriptions of these, as taught by Rabinowicz[1], the "ideal" tribocoating for automotive applications can be described. First, it must be at least approximately twice as hard as any prospective wear particle. Above this "2X-hardness" threshold, abrasive wear decreases extremely rapidly. In our case, this dictates twice the hardness of the hardest components in steel, or roughly 18 GPa. Second, low adhesive friction and wear dictate that the two surfaces, once brought into contact over some microscopic region, always separate with low energy (for low friction) at the original interface (for low wear); i.e. that both are highly cohesive, but not reactive. Solid film lubricants achieve the former, but not the latter. Thus a hard material with a highly stable surface chemistry is desired. Third, the coating must be conformal and very smooth. Fourth, it must not be prone to cleavage. These two virtues are intrinsic to amorphous materials. Fifth, it is helpful, but not essential that the coating be under moderate compressive stress; this tends to close fractures and can

Mat. Res. Soc. Symp. Proc. Vol. 383 © 1995 Materials Research Society

actually improve adhesion. Sixth, and most important for large-scale application in price sensitive consumer applications, it must be cheap.

"DIAMONDLIKE" CARBON

Hard amorphous carbon, also known as diamondlike carbon (DLC) and by a variety of tradenames, would appear to satisfy all the above criteria.[2] Indeed, DLC has been touted for well over a decade as a solution to any number of friction and wear problems. However, DLC suffers from three significant shortcomings which have impeded its application. (1) It is always under very high compressive stress. As shown in Figure 1, this stress increases in proportion to hardness. Even for amorphous hydrogenated carbon (AHC) films near the benchmark hardness of "only" 20 GPa, 2 GPa compressive stress places an unacceptably low limit on film thickness (roughly 1 μm) and makes adhesion unreliable. (2) It is thermally unstable; it begins to revert to graphite and lose its desirable properties during prolonged exposure to temperatures above roughly 250 C. (3) The very low friction (coefficient less than 0.1) and wear rate of AHC are obtained only in very dry conditions which cannot be assured in most applications. At high humidity, friction and wear of AHC increase so much as to leave it only comparable to a variety of other coatings.

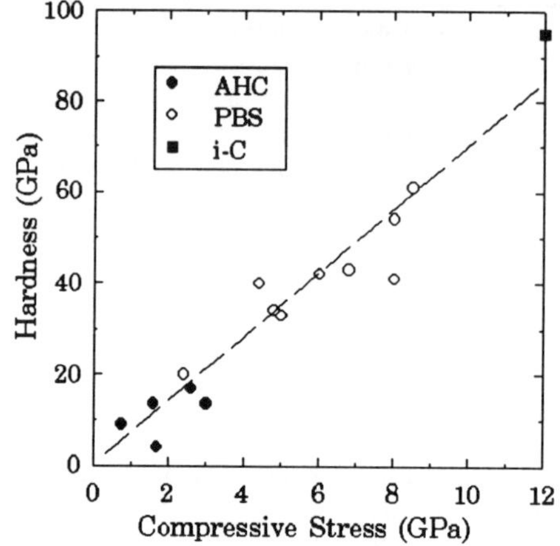

Figure 1 Hardness of various amorphous carbon films as a function of compressive stress.

Self-Stressing of Hard Amorphous Carbon

The origin, or alternatively, the necessity of compressive stress in hard amorphous carbon films has been a topic of interest for some time. Several points of view have emerged. First is the quasi-thermodynamic picture described by McKenzie, Muller and Pailthorpe.[3] They refer to the carbon phase diagram and note that at room temperature only 2 GPa of compressive stress is sufficient to lower the energy of diamond relative to graphite sufficiently that diamond is the stable phase. They propose that "amorphous diamond" is likewise stress stabilized, with the stress maximized at a particular energy for carbon atom or ion impact on the growth surface. Second, are purely kinetic models. Lifshitz, Kasi and Rabalais[4] proposed a "subplantation" regime of deposition energy in which graphitically bonded atoms are more readily displaced than those with fourfold diamondlike bonding, so preferentially accumulating the latter. They did not comment on the resulting stress and some of their assumptions were later called into question. More recently, Robertson[5] adjusted a description of the competition between densification by implantation and relaxation by diffusion of defects proposed by Davis for more conventional single-phase materials[6] to reproduce the observed relationship between growth kinetics and stress (Figure 2). While reality comprises elements from both these extremes, the essential question is: what are the prospects for low-stress "diamondlike" coatings. A purely thermodynamic picture holds that compressive stress is essential to the very existence of hard amorphous carbon, while the kinetic picture suggests that stress is merely an accident of the chosen process, and might be avoided. We approach this "chicken-or-egg" question with a *gedanken* experiment. If we imagine that one could somehow generate fully coordinated, all-sp^3 carbon film with no macroscopic stress, we can consider what might happen later. The energy associated with the requisite bond angle and length distortions is quite large (on the order of several eV per atom[7]) and lower coordination arrangements, π-bonded sp^2 pairs, are possible. Having fewer constraints, these can relax more nearly to their undistorted configurations and so lie at significantly lower energy. The barriers to relaxation of the most distorted sites are likely to be small and some will occur even at low temperature. At low density, the π-bonded pairs occupy roughly the same atomic volume as their sp^3 progenitors. However, the π-orbitals protrude significantly and as pairs accumulate the orbitals will begin to overlap. As they are highly repulsive, some of the energy gain from pairing will go into volume expansion. Thus reconfiguration of two

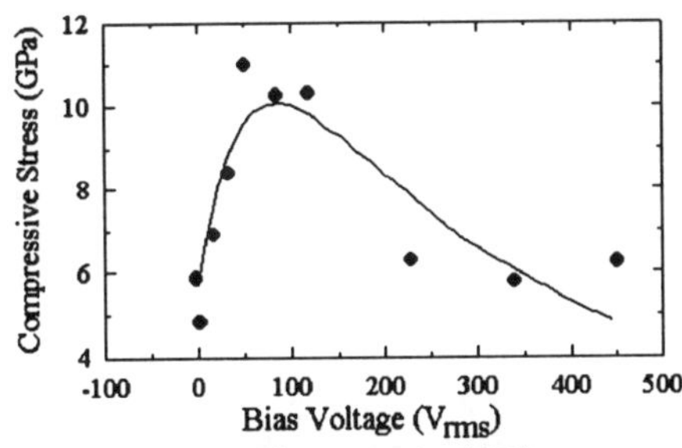

Figure 2 Compressive stress in AHC as a function of ion energy during deposition. Solid symbols are experimental, while the curve is from theory.

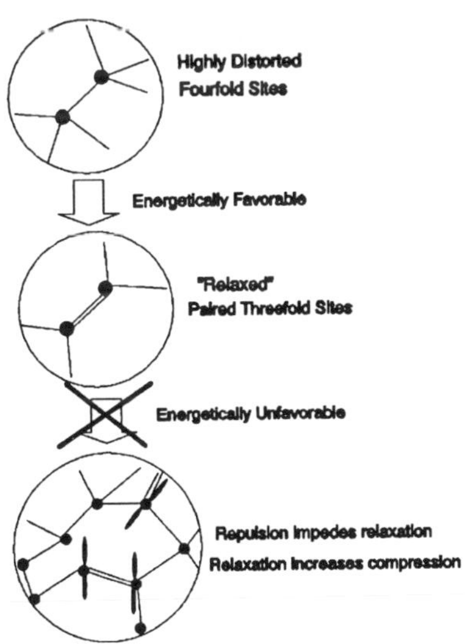

Figure 3 Cartoon of conversion of distorted sp^3-sites to π-bonded pairs.

distorted sp^3 sites to a π-bonded pair becomes analogous to inflating a small bubble in the film. This energetically favored process will generate compressive stress even if none was present when the structure was formed. In effect, hard carbon films are "self stressing." Figure 3 is a cartoon of this process. It is not clear what limits this process. Possibilities are: (1) progressively decreasing rate of conversion as the most distorted sites relax first, (2) decrease in the energetic advantage of conversion as the surrounding density of π-orbitals increases, or (3) increase in the energy required for volume expansion against the increasing compressive stress. We can make an order-of-magnitude estimate by assuming (1) a 50% volume increase (as in conversion from diamond to graphite) against a fixed bulk modulus, (2) that hardness roughly 25% of the bulk modulus (as in diamond and other hard materials) and (3) finally (and arbitrarily) that 5% of the carbon is paired. With these assumptions, hardness is roughly ten times the compressive stress, very nearly the ratio of seven shown in Figure 1. While this does not prove that self-induced stress is the dominant mechanism for strain development in amorphous carbon, it does show that it can account for the film stresses. This in turn supports the underlying proposition that compressive stress in hard amorphous carbon films is unavoidable.

Self-stressing in hard amorphous carbon has several important consequences. Obviously, film thickness is strictly limited by the strength of the interface or substrate material. It also means that truly transparent "pure amorphous diamond" cannot exist. The paired sites have an estimated optical gap of 2.5 eV, in good agreement with that measured for the hardest amorphous carbon, and far smaller than that expected for an amorphous analog to diamond. These pairs are likely to be defects within the larger band gap of "amorphous diamond" and so undermine its utility as an electronic material.

Silicon Stabilization of AHC

One approach to suppressing formation π-bonded pairs is to alloy with a carbiding element. As carbon bonded to the impurity is much more resistant to pairing (carbon in carbides is generally sp^3-bonded) the stress will be reduced accordingly. Assuming that heat has the effect of accelerating the pairing process, the alloyed carbon film should be stable at higher temperature as well. A variety of alloying elements have been tried. All tend to reduce the stress and improve thermal stability with little compromise in hardness. While they tend to make friction insensitive to humidity, the stabilized friction is usually quite high. Silicon was found to be the most effective additive. Inclusion of 5 to 20 atomic percent Si (including H) in AHC reduced compressive stress by roughly 80 %. While technologically important, this observation also supports the concept of self straining; the kinetics of the deposition process and the hydrogen content were unchanged while the sp^3-fraction (as measured by nuclear magnetic resonance) actually increased. In the kinetic picture, similar kinetics should lead to similar stress, while in the thermodynamic picture, increased sp^3 bonding , should correlate with increased stress. As predicted by the self-stressing proposal, Si simultaneously <u>increases</u> sp^3- bonding while <u>reducing</u> compressive stress.

Silicon addition stabilized the friction coefficient of AHC at roughly 0.08, the same as for AHC dry conditions. Figure 4 compares the friction of AHC without and with Si as a function of humidity. [Note that a COF of 0.1 is typical for lubricated sliding of steel against steel.] Figure 5 shows the weight loss of AHC and Si-AHC

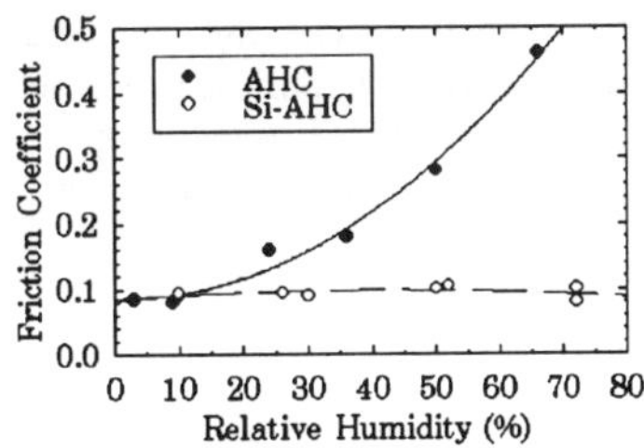

Figure 4 Coefficient of friction for unlubricated sliding of a steel pin against AHC and Si-AHC as a function of ambient humidity.

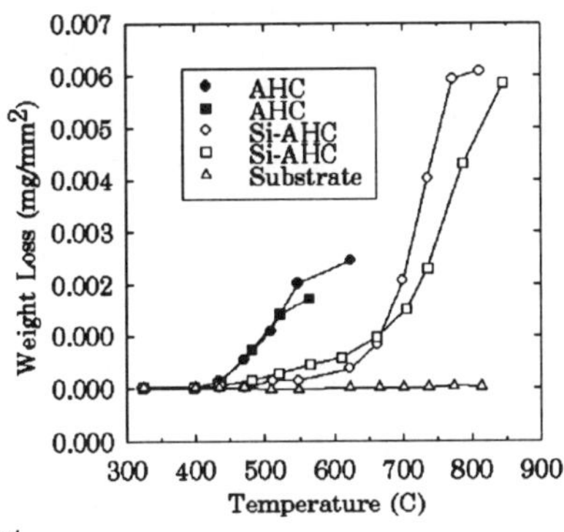

Figure 5 Weight loss of AHC and Si-AHC films after baking in air at successively higher temperatures. The Si-AHC film was roughly three times the thickness of the AHC film.

following baking in air for 2.5 hours at successively higher temperatures. The onset of weight loss was shifted upward roughly 100 C by the addition of silicon. Also, it was observed that partially decomposed AHC films were ineffective tribocoatings while the wear resistance of Si-containing films was only slightly compromised by baking in air at 500 C. While stress reduction and thermal stabilization are readily explained in terms of the increased resistance to π-bonding and oxidation resistance, the role of Si in stabilizing friction against the effects of moisture is not understood at present. In recognition of these two stabilizing effects - against thermal destruction and humidity effects - we refer to this material as "silicon-stabilized" AHC, or simply Si-AHC.

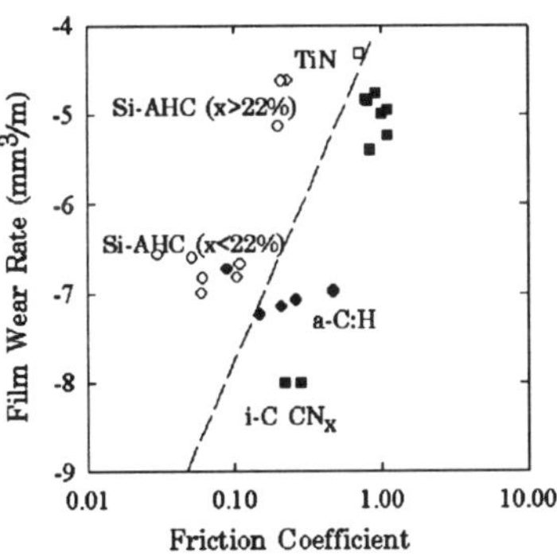

Figure 6 Log-log plot of film wear rate as a function of sliding friction against a 6 mm steel ball under 4.4 N load.

A typical wear rate of Si-AHC in unlubricated sliding against steel is 3 x 10^{-8} mm^3/N-m.

This is nearly two orders of magnitude lower than for competing tribocoatings (TiN, Ti(C,N), B_4C) measured under identical conditions and is only slightly higher than the best values reported for diamond. Figure 6 shows the wear rates for a variety of films of roughly comparable hardness in dry sliding against steel. The figure suggest a very rapid increase of film wear with friction, and little, if any correlation with hardness. This is consistent with the introductory description of the "ideal" tribocoating where it was suggested that hardness more than a few times that of prospective wear particles would be of little value. Above this threshold, it appears that the surface chemical processes that determine adhesive friction and consequent wear are dominant. It would appear that wear particles from these amorphous films are extremely small and so do not lead to abrasive self-wear.

CONCLUSIONS

Figure 7 is a somewhat lighthearted illustration of some of the findings of this work. It shows the friction (against steel) of a wide range of materials as a function of their hardness. The square root scale is used to collapse the wide range of hardness and emphasize the trends. The softest materials (tin, lead etc.) exhibit low friction, but because they shear within themselves, they are also high-wear materials. i.e. "bronze age" solid-film lubricants. For sliding of ferrous "iron age" alloys on themselves, friction is high and liquid-film lubrication is usually used. However, for "stone age" ceramics and hard AHC, friction is also low, but unlike for the soft materials, wear is also low. In effect, the easy shear of the solid film lubricant is achieved, with the shear always occurring at the original contact surface.

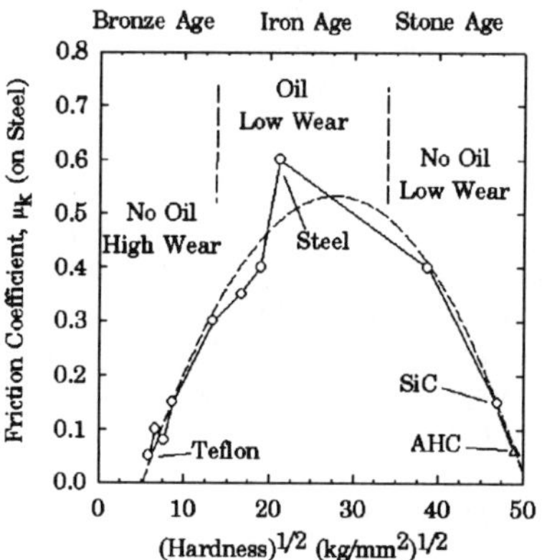

Figure 7 Friction of various materials sliding against steel as a function of hardness. The dashed curve is a parabola fit to the data to emphasize the trends.

Figure 7 re-emphasizes the contention that superior hardness does not necessarily translate to superior tribological performance; diamond is well off this hardness scale, yet there would appear to be little room for improvement over AHC. For well engineered tribosystems where substrate materials do not plastically deform under the designed loads and surfaces are smooth enough to prevent abrasive wear, effort should be devoted to understanding and controlling the surface chemistry of the coating, and its interaction with the counterface material, rather than on seeking to increase the coating hardness in the blind faith that hardness alone improves performance. Furthermore, unlike stable crystalline phases, the composition of metastable, kinetically deposited amorphous materials may be modified almost arbitrarily in this effort to master surface chemistry. In conclusion, of the expanding plethora of hard thin-film coatings, silicon-stabilized amorphous hydrogenated carbon best approximates "ideal" tribocoating: it is slippery and wear resistant under a wide range of conditions - with the possibility of further improvement, reasonably stable, smooth, and may be deposited at low temperature at acceptably low cost.

ACKNOWLEDGEMENTS

The author wishes to acknowledge the contributions of William Vassell and Timothy Potter for their efforts in developing and proving new coating technologies, and Arup Gangopadhayay for conducting the numerous tribological tests central to this work.

REFERENCES

1) E. Rabinowicz, Friction and Wear of Materials, (Wiley, NY, 1965); Tribology I: Friction, Wear and Lubrication, (MIT Center for Advanced Engineering Study, Cambridge, 1974).

2) M. W. Geis and M. A. Tamor, Encyclopedia of Applied Physics, (VCH, New York, 1993), vol. 5, pp. 1-24.

3) D. R. McKenzie, D. Muller and B. A. Pailthorpe, Phys. Rev. Lett. **67**, 773 (1991).

4) Y. Lifshitz, S. R. Kasi and J. W. Rabalais, Phys. Rev. Lett. **62**, 1290 (1989).

5) J. Robertson, Diamond Relat. Mater. **3**, 361 (1994).

6) C. A. Davis, This Solid Films **226**, 30 (1993).

7) P. C. Kelires, Phys. Rev. Lett. **68**, 1854 (1992).

MOLECULAR DYNAMICS SIMULATION OF MECHANICAL DEFORMATION OF ULTRA-THIN AMORPHOUS CARBON FILMS

J.N. GLOSLI*, M.R. PHILPOTT**, and J. BELAK*
*University of California, Lawrence Livermore National Laboratory, Livermore, CA 94550
**IBM Research Division, Almaden Research Center, 650 Harry Road, San Jose, CA 95120-6099

ABSTRACT

Amorphous carbon films approximately 20nm thick are used throughout the computer industry as protective coatings on magnetic storage disks. The structure and function of this family of materials at the atomic level is poorly understood. Recently, we simulated the growth of a:C and a:CH films 1 to 5 nm thick using Brenner's bond-order potential model with added torsional energy terms. The microstructure shows a propensity towards graphitic structures at low deposition energy (<1eV) and towards higher density and diamond-like structures at higher deposition energy (>20eV). In this paper we present simulations of the evolution of this microstructure for the dense 20eV films during a simulated indentation by a hard diamond tip. We also simulate sliding the tip across the surface to study dynamical processes like friction, energy transport and microstructure evolution during sliding.

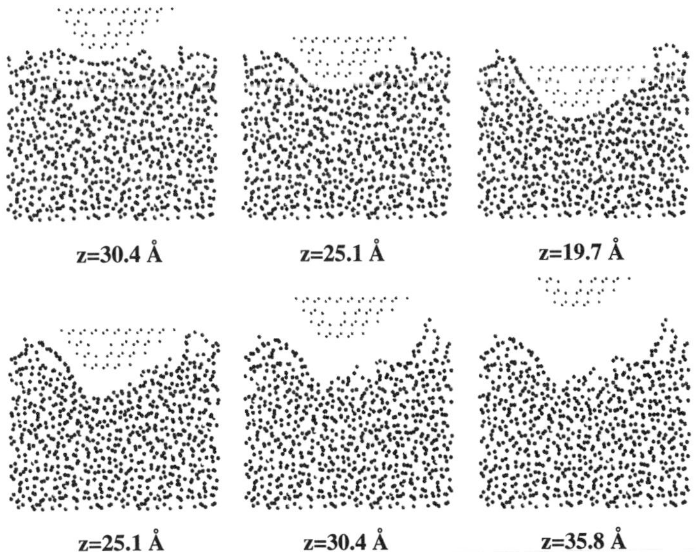

Figure 1. Six snapshots during a MD simulation of indentation and removal of a nondeformable diamond tip into an amorphous carbon substrate. The simulation cell is 4nm wide and the snapshots are 1nm thick cross-sections through the center of the tool. The indentation rate is 35 m/s.

Mat. Res. Soc. Symp. Proc. Vol. 383

INTRODUCTION

Amorphous carbon films approximately 20nm thick are used throughout the computer industry as protective coatings on magnetic storage disks [1,2]. These films contain significant sp^3 fractions and, for this reason, are mechanically hard, have low friction, and are chemically inert. Despite intense experimental [3-9] and theoretical [11-25] interest, the structure and function of these films at the atomic level is poorly understood.

Recently [26] we presented molecular dynamics (MD) simulations of the formation (both quench and deposition) of amorphous carbon films using Brenner's [27,28] empirical bond-order potential. This empirical model, based on ideas of Tersoff [29], mimics the quantum mechanics allowing carbon to form strong chemical bonds with a variety of hybridizations. We find unphysical bonding between three-fold carbon atoms without inclusion of a torsional energy barrier [28]. This model has been extensively used by Harrison [30-34] to study the tribological properties of diamond surfaces.

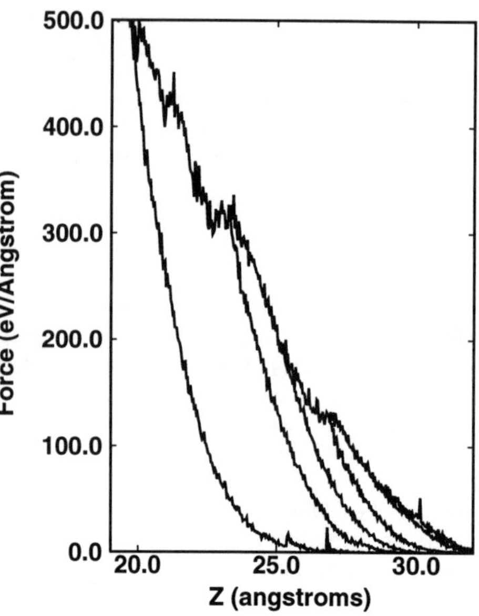

Figure 2. The loading curve for the simulation shown in Figure 1. The steps occur during plastic yielding of the surface. Reversing the tip after these yielding events leads to the hysteresis as shown.

Our results for the bonding character of the amorphous films is in qualitative agreement with the quantum mechanics based MD studies [15-22]. Three-fold carbon atoms tend to form pairs. We find a peak density as a function of incident carbon energy in the range 20-40eV, slightly less than the previous study using Tersoff's bond-order model [14]. The difference is probably due to our use of the torsional energy.

Here we present results from MD simulations of nano-indentation and sliding on these films.

METHODS

To create a larger surface simulation cell, we doubled the cell size in X and Y from our previous simulation [26] to about 4nm and continued to grow a dense film at 20eV. We stopped the deposition when the film was 4nm thick. The tribological boundary conditions are the same as in our previous simulations of metal and ceramics [35]. The bottom-most layer of atoms are fixed, the next few layers are maintained at room temperature, and periodic boundaries are applied in the surface plane. The tip is cleaved from a diamond lattice and blunted to create a radius of about 1nm. The tip atoms are held rigid during indentation and sliding and interact with the surface atoms through a truncated Lennard-Jones potential. Indentation was performed at 35 m/s and sliding at 35 and 350 m/s. These rates are comparable to the sliding speeds at the head-disk interface in magnetic recording disks.

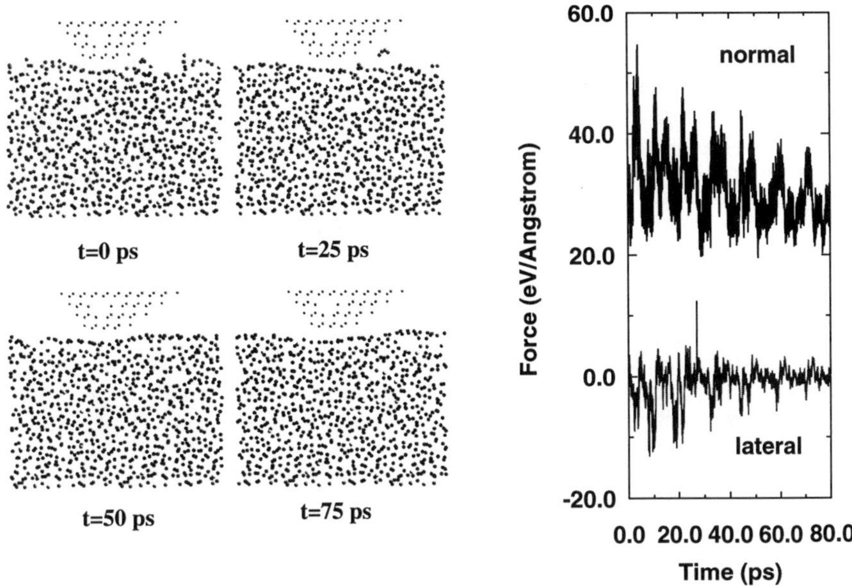

Figure 3. Four snapshots during sliding in contact under a light load. The material is moving from left to right in the figures.

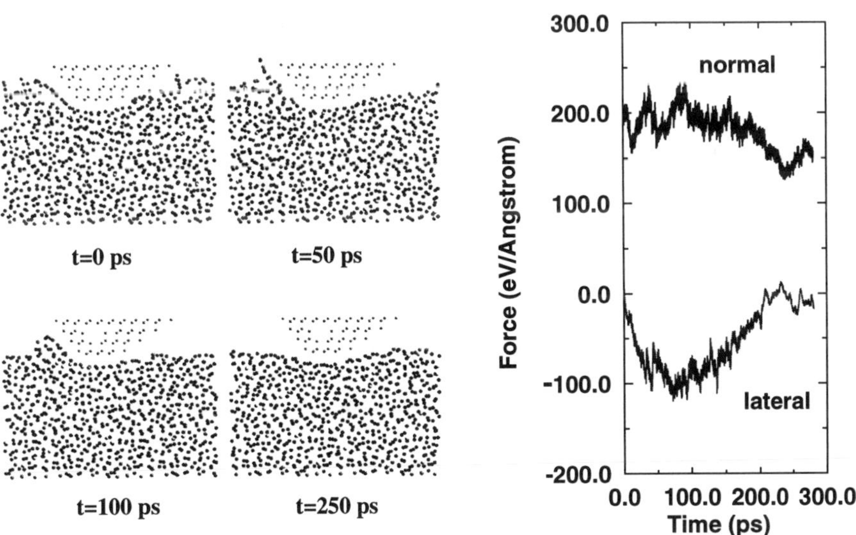

Figure 4. Four snapshots during sliding in contact under a heavy load. The material under the tip densifies leading to a final structure very similar to the starting undeformed structure.

RESULTS and DISCUSSION

Six snapshots during loading and unloading are shown in Figure 1. These snapshots are 1nm thick cross-sections through the center of the tool. Appreciable plastic deformation occurs and the first layer of surface atoms appear to align with the tool. The loading curve (force on the tool as a function of tool height) is shown in Figure 2. The steps in the curve indicate the onset of plasticity. The curve is reversible before the first step while a hysteresis loop is seen beyond the first step. This step-like structure is probably characteristic of covalently bonded materials and represents plastic flow through a rapid rearrangement of the bonding network. The lengthscale of our simulations is too small to observe the formation of cracks. The hardness calculated from this simulation is 75 +/- 25 GPa, with the error arising from our estimate of the contact area. This value agrees well with measured hardness on dense amorphous carbon films [1,2].

Sliding results are shown in Figures 3 and 4. In both figures the tool is stationary, material moves from left to right and we use periodic boundaries. In Figure 3 we apply a small normal load, less than the first plastic yield seen in Figure 2. Sliding occurs at 350 m/s. The tip conditions the surface. That is, dangling atoms are forced into the surface and the top-most layer of carbon atoms are mostly converted to sp^2 hybridization. In Figure 4 we apply a normal load greater than the first plastic yield and slide at 35 m/s. The system initially forms a chip. This chip is compressed into the surface and the density of the film increases. The resulting structure is qualitatively very similar to the initial undeformed structure.

ACKNOWLEDGEMENTS

Work performed in part under the auspices of the U.S. Department of Energy by the Lawrence Livermore National Laboratory under contract No. W-7405-ENG-48.

REFERENCES

1. H-C Tsai and D.B. Bogy, J. Vac. Sci. Technol. A **5**, 3287 (1987).
2. A. Gill, Wear **168**, 143 (1993).
3. F. Li and J.S. Lannin, Phys. Rev. Lett. **65,** 1905 (1990).
4. D.R. McKenzie, D. Muller, and B.A. Pailthorpe, Phys. Rev. Lett. **67,** 773 (1991).
5. D.R. McKenzie, Y. Yin, N.A. Marks, C.A. Davis, B.A. Pailthorpe, G.A.J. Amaratunga, V.S. Veerasamy, Diamond and Related Materials **3**, 353 (1994).
6. K.W.R. Gilkes, P.H. Gaskell, and J. Yaun, Diamond and Related Materials **3**, 369 (1994).
7. S.M. Holl, R.D. Johnson, V.J. Novotny, J.L. Williams, C.E. Caley, M. Hoinkis, and R.E. Jones, Proceedings MRS Spring 1994 Symposium, Novel Forms of Carbon II, p54-62.
8. J. Wagner, M. Ramsteiner, Ch. Wild, and P. Koidl, Phys. Rev. B **40**, 1817 (1989).
9. M. Weiler, R. Kleber, S. Sattel, K. Jung, H. Ehrhardt, G. Jungnickel, S. Deutschmann, U. Stephan, P. Blaudeck, and Th. Frauenheim, Diamond and Related Materials **3**, 245 (1994).
10. D. Beeman, J. Silverman, R. Lynds, and M.R. Anderson, Phys. Rev. B **30**, 870 (1984).
11. J. Tersoff, Phys. Rev. Lett. **61**, 2879 (1988).
12. B.A. Pailthorpe and P. Knight, Proceedings MRS Spring 1991 Symposium E.
13. B.A. Pailthorpe. J. Appl. Phys. **70**, 543 (1991).
14. H-P. Kaukonen and R.M. Nieminen, Phys. Rev. Lett **68**, 620 (1992).
15. C.Z. Wang, K.M. Ho, and C.T. Chan, Phys. Rev. Lett, **70**, 611 (1993).
16. C.Z. Wang and K.M. Ho, Phys. Rev. Lett. **71**, 1184 (1993).
17. D.A. Drabold, P.A. Fedders, and P. Strum, Phys. Rev. B **49**, 16415 (1994).

18. P. Blaudeck, Th. Frauenheim, D. Porezag, G. Seifert, and E. Fromm, J. Phys.: Condens. Matter **4**, 6389 (1992).
19. Th. Frauenheim, P. Blaudeck, U. Stephan, and G. Jungnickel, Phys. Rev. B **48**, 4823 (1993).
20. Th. Frauenheim, U. Stephan, P. Blaudeck, and G. Jungnickel, Diamond and Related Materials **3**, 462 (1994).
21. G. Jungnickel, M. Kuhn, S. Deutschmann, F. Richter, U. Stephan, P. Blaudeck, and Th. Frauenheim, Diamond and Related Materials **3**, 1056 (1994).
22. Th. Frauenheim, G. Jungnickel, Th. Kohler, and U. Stephan, J. Non-Cryst. Solids **182**, 186 (1995).
23. J. Robertson and E.P. O'Reilly, Phys. Rev. B 35, 2946 (1987).
24. J. Robertson, Diamond and Related Materials **2**, 984 (1993).
25. J. Robertson, Diamond and Related Materials **3**, 361 (1994).
26. J.N. Glosli, J. Belak, and M.R. Philpott, Proceedings MRS Fall 1994 Symposium, Thin Films: Stresses and Mechanical Properties IV.
27.D.W. Brenner, Phys. Rev. B **42,** 9458 (1990).
28. D.W. Brenner, J.A. Harrison, C.T. White, and R.J. Colton, Thin Solid Films **206,** 220 (1991).
29. J. Tersoff, Phys. Rev. B **37**, 6991 (1988).
30. J.A. Harrison, D.W. Brenner, C.T. White, and R.J. Colton, Thin Solid Films **206**, 213 (1991).
31 J.A. Harrison, C.T. White, R.J. Colton, and D.W. Brenner, Phys. Rev. B **46**, 9700 (1992).
32 J.A. Harrison, C.T. White, R.J. Colton, and D.W. Brenner, Surface Science **271**, 57 (1992).
33. J.A. Harrison, C.T. White, R.J. Colton, and D.W. Brenner, J. Phys. Chem **97**, 6573 (1993).
34. J.A. Harrison, R.J. Colton, C.T. White, and D.W. Brenner, Wear **168**, 127 (1993).
35. J. Belak, D.B. Boercker, and I.F. Stowers, MRS Bulletin **18**, 55 (1993).

STABILITY, CHEMICAL BONDING AND VIBRATIONAL PROPERTIES OF AMORPHOUS CARBON AT DIFFERENT MASS DENSITY

Th. Köhler, Th. Frauenheim and G. Jungnickel *
* Theoretische Physik III, Institut für Physik, Technische Universität, D - 09107 Chemnitz, Germany

ABSTRACT

We describe correlations between the atomic-scale structure and the global electronic and vibrational properties in amorphous carbon versus mass density. The model structures have been generated by applying different annealing regimes using a density-functional based molecular-dynamics (DF-MD) method. The stability of the amorphous modifications and the calculated vibrational density of states (VDOS) are strongly affected by the density and the annealing sequences, altering the chemical composition, the sp/sp^2/sp^3-clustering, the structure and related physical properties. A mass density of 3.0 g/cm^3 is confirmed as a magic density favoring the formation of most stable *a*-C modifications having lowest defect densities and maximum band gap. In analyzing the vibrational spectra and the localization of modes in comparison with crystalline diamond and graphite, we identify the spectral signatures for chemically different bonded species and defects, that may be used for comparison with related *Neutron-, Raman-* and *IR-* work.

I. INTRODUCTION

Molecular-dynamics (MD) modeling of atomic-scale structures on the basis of a coupling between atomic and electronic degrees of freedom in comparison with related experimental (structural, diffraction, vibrational, electronic and spectroscopic) work has become a powerful tool in elucidating structure property relations in amorphous carbon [1, 2, 3, 4].

However, the theoretical results compared with experiments in most cases represent structural averaged data of the related material properties. On the basis of these results, differential information about the local chemical bonding environment and defect distribution is difficult to discern. For practical purposes in thin film characterization, a single more local probe which may be unequivocally related to corresponding data of atomic-scale models and simultaneously is predictive of other film properties, would be highly desirable.

Vibrational spectroscopy such as *Raman* and *IR* are well developed room ambient tools available in many laboratories, which places no constraints on substrate size and shape. Furthermore, more than a hundred mostly hydrogenated carbon films have been characterized by these techniques. For a review we refer the reader to Tamor and Vassell [5] and to Silva, Amaratunga and Constantinou [6]. Due to the complete lack of systematical theoretical work in this field the obtained experimental features are analyzed in terms of effective medium theory comparing with the well-defined features in graphite and diamond.

To put the interpretation of the experimental data on a more profound theoretical basis we present the calculated vibrational spectra of amorphous models generated at densities 2.0, 3.0 and 3.52 g/cm^3 by two different annealing regimes using a DF-MD method, which is briefly described in section II. In section III we review the structural and physical properties of representative amorphous modifications and characterize their energetic stability.

Mat. Res. Soc. Symp. Proc. Vol. 383

The global band gap properties that are determined by the detailed clustering of chemically different bonded carbon atoms [7, 8] will only be touched upon in completing the discussion of the model properties. In section IV we describe the basic ideas for determining the vibrational properties including total, partial and local vibrational density of states (VDOS) completed by related localization measure. We present results obtained for all amorphous models and discuss the detailed changes in the vibrational spectra versus mass density, that are affected by the changing $sp/sp^2/sp^3$-ratio and the clustering of chemically different bonded carbon atoms. Finally, in section V we conclude and compare the calculated spectra with related Raman and IR work.

II. DENSITY-FUNCTIONAL BASED MD, SIMULATION REGIME

To model the structure formation in amorphous carbon modifications we have carried out DF based MD simulations[9] for 128 carbon atoms using fixed-volume cubic supercell arrangements of varying size in relation to the microscopic mass densities to be studied. The relaxation of the structures has been realized by applying two different cooling regimes of "dynamical quenching" character. We have performed a rapid quenching at a cooling rate of 10^{15} K/s over 2 ps as well as an extended stochastic cooling regime over 8 ps. In both cases the relaxation started from a partly equilibrated liquid state of the model structures and followed a path of exponentially decreasing temperature. As characteristics of the first regime we have rescaled the atomic velocities after each timestep τ to an "instantanous" temperature T required by the relaxation path, so that $\langle \mathbf{p}_i^2/2M \rangle = 3/2\ k_B T$ holds for the mean kinetic energy per atom. Within the second regime the rescaling of velocities is performed stochastically in time. A random generator produces equally distributed integer numbers n belonging to an intervall $[1, n_{max}]$. During a time period $n\tau$, the atoms can accelarate freely under influence of the interatomic forces. The rescaling procedure described above then is reapplied to the system. We have found that the stochastical quenching regime in any case produces more stable structures that are lower in energy than comparable systems obtained by the simple quenching dynamics.

As reliable method for the calculation of interatomic forces we used a density-functional (DF) scheme for the construction of a non-orthogonal tight-binding (TB) potentials within the LCAO formalism using the local density approximation (LDA)[10]. For details we refer to Porezag, Frauenheim, Köhler and Seifert [9]. Within this scheme the interatomic forces for the MD easily can be derived from an exact calculation of the gradients of the total energy at the considered atom sites.

III. STABILITY AND PROPERTIES OF *a*-C VERSUS MASS DENSITY

As the result of the simulation we have obtained final metastable amorphous carbon modifications at different mass densities. In all structures there is a clear tendency for the different hybrids to separate from each other and to form small interconnected subclusters. Owing to the fixed composition and constant atom number in the supercells, the cohesive energies at different densities have been compared to determine 3.0 g/cm^3 as a magic density at which most stable minimal energy amorphous carbon modifications are formed independant on the cooling regime applied. For an illustration we have plotted in Figure 1 the calculated cohesive energies of all *a*-C model structures versus density including data of

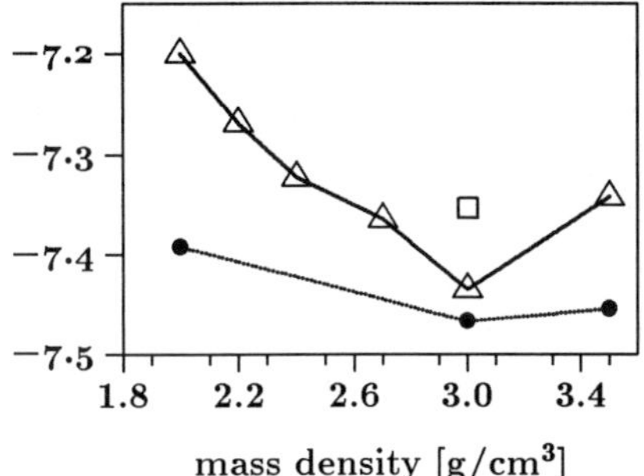

Fig. 1: Cohesive energies of amorphous carbon models versus density; rapid cooling (solid line), extended stochastic cooling (dotted line), and 64-atom supercell *ab initio* MD [4] (□).

the rapid and extended stochastically cooled structures and a 64-atom supercell obtained by the *Sankey/Drabold ab initio* MD code[4]. For reference the diamond cohesive energy per atom within the used DF-TB scheme is -8.02 eV. This confirms the stability of high density ta-C[11], which by various techniques has been recently deposited in different laboratories.

In the following discussion we will focus mainly on three densities, 2.0, 3.0 and 3.52 g/cm^3, there are the most interesting from the point of deposition and application. Further we will outline tendencies for changes in the vibrational behaviour for intermediate densities. In a previous paper we have already described how the structures and the related chemical bonding properties change with the simulation regime, and accordingly have been influenced in their energetic stability and global band gap properties [8]. Some of the most characteristic structural and electronic data of the models are listed in Table I including the fraction of 2-, 3- and 4-fold coordinated atoms, C_2, C_3, C_4, the mean bond length, R_1, bond angle, Θ, the CC coordination number, k_{CC}, and the number of rings according to a shortest path analysis, $N_{rings}^{(5,6,7)}$. We also give the information on some global electronic data, i.e. the HOMO-LUMO $\pi - \pi^*$ band gap splitting, $E_{\pi-\pi^*}$, the total number of electronic defects, $N_{defects}$, the ratios of non-bonded p- to bonding and antibonding π-states, n_p/n_π, as well as all $p + \pi$- to bonding and antibonding σ-states, $n_{p+\pi}/n_\sigma$. To complete the characterization, we have calculated the reduced neutron scattering structure factors $F(Q) = Q(S(Q) - 1)$ and the reduced radial distribution functions obtained from the atomic-scale models[8]. In comparing with neutron diffraction experiments by Li and Lannin[12] on a low-density, and Gaskell et al.[13] on a high-density a-C sample the simulated data for the extended cooled 2.0 g/cm^3 and the rapid cooled 3.0 g/cm^3 model in both, momentum and real space, agree very well with the neutron scattering data.

low-density a-C

As a general tendency, the stability and the connectivity of the model structures increases with the simulation time and the use of optimized stochastic cooling techniques. At 2.0 g/cm^3 density the number of two-fold coordinated atoms forming chain-like segments at the rapid cooling rate will be suppressed at extended stochastic cooling followed by an enhanced ring formation of favourably six-fold rings relative to odd-membered rings. The $\pi - \pi^*$ HOMO-LUMO splitting is found to be determined by the detailed π-electron clustering of undercoordinated atoms. In the rapid cooled model we obtain a high concentration of defects and well relaxed π-bonds within a mixed low-connectivity sp/sp^2-bonded matrix.

As a result the band gap opens to about 2.0 eV. In changing the relaxation to the extended stochastic cooling, the $\pi - \pi^*$ splitting decreases and the total number of defects is reduced considerably. This is due to the formation of an energetic more favourable π-cluster distribution. Simultaneously the connectivity of the network is enhanced enforcing the strain on the π-bonds from the rigid bonding environment.

high-density ta-C

The high-density, highly-tetrahedral amorphous carbon, *ta*-C of 3.0 g/cm^3 have been independently found to be stable and to exhibit interesting structural and electronic behaviour by two different *ab initio* MD schemes[3, 4], compare columns 3 to 5 in Table I. At this density the internal strain is maximally removed from the network by the separation of small, favourably even membered, π-clusters between undercoordinated sites yielding minimal defect concentration. The fraction of 4-fold coordinated atoms reaches about two thirds, which is close to the experimentally obtained values. The increase in the $\pi - \pi^*$ splitting up to 3.0 eV compared to low-density materials is mainly determined by the changing size distribution of π-clusters with respect to smaller ones and by the ability of the π-bonds to relax to a mean value of 0.6 of a C_2H_2 double bond under the constraint of a rigid sp^3 bonding environment. There is a balance between the gain of π-bonding energy due to sp^2-clustering and the residual stress in the amorphous network. If we extend the annealing time and apply stochastic cooling techniques, the sp^3 fraction is almost unaffected and at almost equal defect distribution the $\pi - \pi^*$ splitting drops down to 2.07 eV. The evolution of the $\pi - \pi^*$ gap in this structure is strongly influenced by shifting the size distribution of π-clusters to larger non-aromatic even membered clusters of up to 8 sp^2 sites, reducing the gap. Finally, two models made at diamond density are included in Table I for completing the characteristics of high-density structures, which will be analyzed in their vibrational properties. These modifications refer to hypothetical structures that are of interest for a search of minimal defect high-density wide band gap structures.

data	a-$C_{2.0}^{rap}$	a-$C_{2.0}^{ext}$	ta-$C_{3.0}^{rap}$	ta-$C_{3.0}^{ext}$	ta-$C_{3.0}^{SD}$	ta-$C_{3.52}^{rap}$	ta-$C_{3.52}^{ext}$
ΔE^{at} (eV)	0.89	0.70	0.66	0.63	0.73	0.75	0.64
$C2$ (%)	27	8	-	-	-	-	-
$C3$ (%)	64	73	36	31	25	12	8
$C4$ (%)	9	19	64	69	75	88	92
R_1 (Å)	1.43	1.47	1.54	1.54	1.56	1.53	1.54
Θ (deg)	120.6	116.2	111.0	110.9	110.2	109.6	109.4
k_{CC}	2.80	3.11	3.64	3.69	3.75	3.88	3.92
$N_{rings}^{(5,6,7)}$	9,5,8	17,28,11	36,54,32	43,58,39	22,36,38	41,79,67	46,98,6 7
$E_{\pi-\pi^*}$ (eV)	1.93	1.38	2.88	2.07	5.27	5.59	3.92
$N_{defects}$	11	4	4	4	3	8	2
n_p/n_π	0.077	0.036	0.087	0.111	0.2	1.0	0.2
$n_{p+\pi}/n_\sigma$	0.430	0.286	0.108	0.085	0.076	0.032	0.024

Tab. I: Structure, chemical bonding and global band gap properties of amorphous carbon models at various mass density; rapid cooling (*rap*), extended stochastic cooling (*ext*), and Sankey/Drabold-DFT [4](*SD*). ΔE is the calculated cohesive energy difference per carbon atom relative to the diamond energy.

IV. VIBRATIONAL PROPERTIES

For the vibrational analysis we have used fully relaxed amorphous model structures at different density obtained by conjugate gradient relaxation. Using standard methods, the vibrational eigenvalues and eigenstates have been calculated within the harmonic approximation by construction of the dynamical matrix and diagonalization.

To illustrate the changing contributions of the different hybridized atoms to the total vibrational densities of states (VDOS) we have used a projection technique. This allows us to show phonon densities of states splitted in hybrids to study the dynamical properties of the amorphous carbon networks at increasing mass density in more detail. For the discussion of the localization behavior of phonon modes we make use of inverse participation ratio $\mathrm{P_i^{-1}}$ which can be assigned to each mode i. This measure and the hybrid phonon density of states $g_{\mathcal{H}}(\omega)$ can be calculate as follows:

$$P_i^{-1} = \sum_{s=1}^{3N} |\langle \vec{\mathbf{p}}_s | \vec{\mathbf{Y}}_i \rangle|^4, \qquad g_{\mathcal{H}}(\omega) = \frac{1}{3N} \sum_{s \in \mathcal{H}} \sum_{i=1}^{3N} \delta(\omega - \omega_i) |\langle \vec{\mathbf{p}}_s | \vec{\mathbf{Y}}_i \rangle|^2.$$

$\mathcal{H}$ denotes the set $\mathcal{H} = \{\nu_i\}_{i=1}^{M \leq 3N}$ of all indices of coordinates belonging to atoms of the same hybridisation type and $\langle \vec{\mathbf{p}}_s | \vec{\mathbf{Y}}_i \rangle$ is the scalar product of a vector $\vec{\mathbf{p}}_s = (\underbrace{0, \ldots, 0}_{s-1}, 1, \underbrace{0, \ldots, 0}_{3N-s})$ with the i-th eigenvector $\vec{\mathbf{Y}}_i$. The usage of normalized eigenvectors garanties the validity of the last expression yielding the additivity of the various hybrid VDOS to the resulting total VDOS.

In Figure 2 we present the calculated total and *hybrid*-fractional vibrational spectra of 6 amorphous models generated by two different annealing regimes at 3 densities, 2.0, 3.0 and 3.52 g/cm^3. In the top of Figure 2 we have included the calculated spectra of the crystalline modifications, diamond and graphite, to point out the differences in the vibrational behaviour of the amorphous models. For reasons of comparison all spectra have been convoluted by a constant resolution function. By using the projection technique as described above, we have decomposed the total vibrational density of states into the different carbon hybrid contributions and show the corresponding curves for the sp^2- and sp^3-fractions in Figures 2, too. Additionally we have projected out structural defects by defining a localization measure of the modes, which is plotted as the inverse participation ratio in Figure 3.

The diamond spectrum is split into one dominating peak area at 1300 cm^{-1} and four less intense features at 1000, 900, 700 and 500 cm^{-1}. However, the graphite spectrum may be decomposed into two almost equally shaped main features, both linearly increasing, one from almost zero to 850 cm^{-1} and the other from 1000 cm^{-1} to about 1700 cm^{-1}. The high-frequency bound in graphite is shifted relative to diamond due to an increased sp^2-stretching force.

Comparing now the amorphous modifications, the low-density *a*-C's at 2.0 g/cm^3 are reminiscent of the graphitic type behaviour in the total broad frequency range up to 1800 cm^{-1}. Additionally, there is a considerable softening of the main low-frequency feature in the rapid cooled 2.0 g/cm^3 sample. This is due to chain-like $sp - sp^2$ segments and rings performing hindered spatially translational vibrations in a low-connectivity network. There is a well pronounced intensity drop at 1000 cm^{-1} in graphite separating high-frequency stretching vibrations from the lower-frequency spatial translational and bond-angle-changing modes. This is still obvious in the low-density structures and is smoothed out with increasing density followed by an overall compression of the total spectra to a range from 400 to 1600 cm^{-1} , where it develop a characteristic half sphere shape. This is in support of similar shaped spectra obtained by Drabold et al. [4]and Wang et al. [14]. The softening of the spectra at higher densities completely disappears and the spectra lose all reminiscence of the split spectral graphite and diamond behaviour.

Considering the fractional sp^2- and sp^3-VDOS, the intensities are correlated with the decreasing and increasing fractions of the related hybrids in the models, however, may still preserv the overall shape and frequency range at all densities. The first broad feature in the sp^2-VDOS around 600 cm^{-1} as signature for extended spatial modes of clustered sp^2-units in rings and cross-linked chain segments is still visible up to diamond density. By contrast, the intensity of the higher frequency sp^2-stretching vibrations is reduced considerably, now developing single localized modes above 1700 cm^{-1}. These modes refer to isolated sp^2-defects and paired sp^2-clusters within the rigid sp^3-matrix.

At three densities, 2.0, 3.0 and 3.52 g/cm^3 we compare the vibrational properties of models obtained at two different annealing cycles, as described in section II. With increasing simulation time and the use of stochastical cooling at 2.0 g/cm^3 the lowest frequency modes disappear as well. This is mainly due to an almost complete removal of two-fold coordinated sp-atoms followed by a suppression of the sp-fraction VDOS in support of a higher network connectivity and an increasing number of sp^3 sites. Above 1600 cm^{-1} all modes become localized representing signs of structural defects, mostly arising as undercoordinated *sp*-atoms in the higher coordinated rigid amorphous matrix. Two examples of such localized modes are the $sp-sp$-stretching at 2496 cm^{-1} and $sp-sp^2$-stretching at 2430 cm^{-1}. With more extended relaxation time the defect density is reduced in consistency with a reduced number of electronic defects listed in Table I.

Analyzing the vibrational spectra of the 3.0 g/cm^3 models obtained at the two different annealing regimes, there is almost no change in both the total shape and the sp^x-fraction VDOS. This is consistent with a similar chemical composition, compare Table I, and a favourably pairwise clustering of undercoordinated sp^2 atoms. Also both models are almost comparable in stability on the energy scale, compare Figure 1. While the localized defect states in both models are again caused by the embedding of undercoordinated sp^2-units in a rigid sp^3 bonding environment, the overall width of the defect band is considerably reduced in the more extended cooled model. By using extended stochastical cooling regimes the network stability is enhanced minimizing the internal strain. As the result inhomogenieties in the local strain are removed producing similar local environments for the embedding of isolated defects and π-bonded clusters.

Finally, the spectra of the two diamond-density amorphous models show a well pronounced half-sphere shape behaviour. Again the sp^3 fraction slightly increases with extended relaxation time, which is clearly reflected by the sp^x-fraction VDOS in Figure 2. Simultaneously, the number of defects is reduced by the formation of 3 π-bonded sp^2-pairs producing 3 localized $sp^2 - sp^2$ - stretching modes in the range 1700 to 2100 cm^{-1}. The frequency spread of these modes is due to varying local stress acting on the π-bonds, indicating a nonuniform distribution even in such small supercells at diamond-density. In the two high density extended stochastically cooled structures we find two additional localized low-frequency modes at (140-150) cm^{-1}. Both have identical origin and can be assigned to sp^3-tetrahedral breathing modes of clustered sp^3 units , which are confined within a rather stiff and rigid bonding environment. Comparing once more the spectra of the diamond-

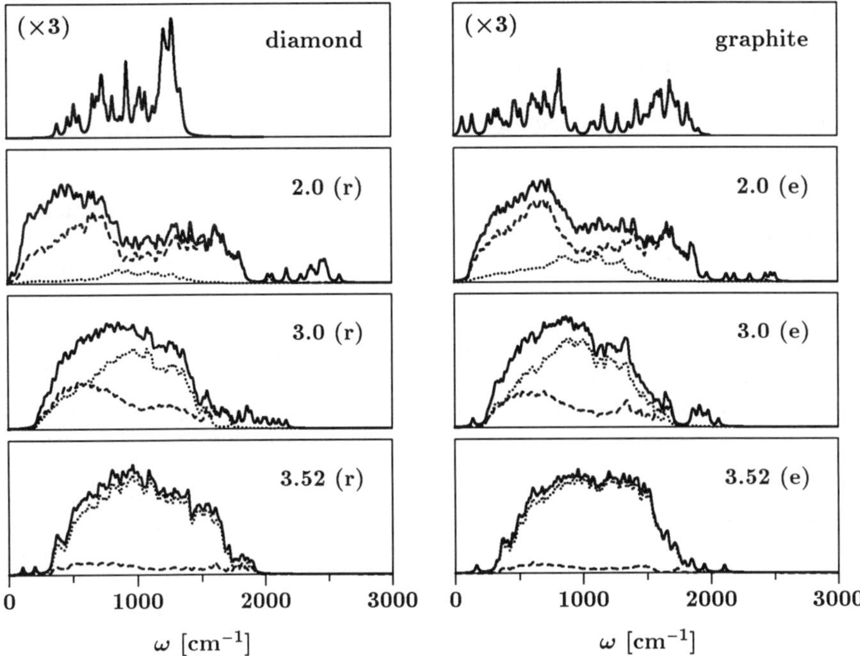

Fig. 2: Total (solid line) and hybrid-fractional (sp^2 → dashed line, sp^3 → dotted line) vibrational density of states (VDOS) of amorphous carbon modifications versus mass density compared to the spectra of diamond and graphite using identical broadenings; rapid cooling (r) , extended stochastic cooling (e).

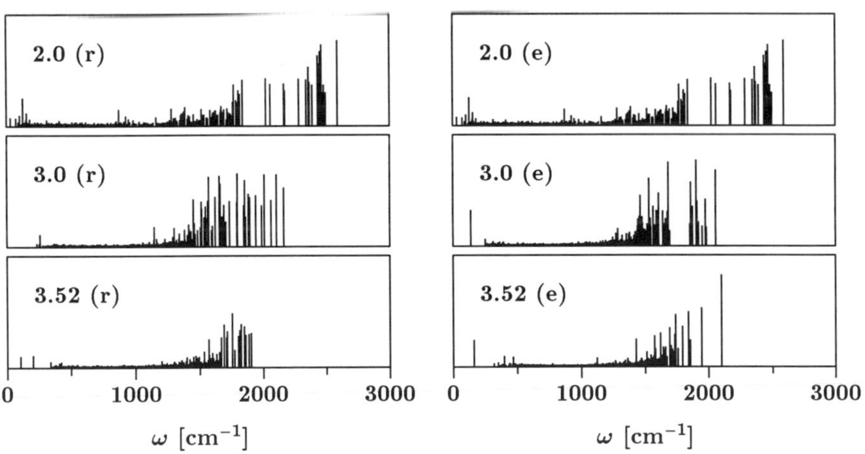

Fig. 3: Inverse participation ratio of phonon modes of amorphous carbon versus mass density; rapid cooling (r), extended stochastic cooling (e).

density amorphous carbon with that of diamond, the steep intensity drop below 1200 cm^{-1} is completely smoothed out in the amorphous modifications combined with a shift of the high-frequency edge from 1300 to 1600 cm^{-1}.

V. CONCLUSION AND COMPARISON WITH EXPERIMENTAL DATA

We have presented vibrational signatures of amorphous carbon at various densities which may be used for comparison with experimental data. While there is limited Raman data of *a*-C samples[5], almost all IR-studies have been performed on hydrogenated amorphous carbon, *a*-C:H[6]. However, a clear assignment of the modes obtained from the atomic scale models to the most characteristic experimental features can be given. As discussed by Tamor and Vassel[5], the Raman spectra of amorphous carbon films exhibit two graphite related broad features at approximately 1550 cm^{-1} (G-line) and 1350 cm^{-1} (D-line) and an additional 3-rd one centered at 600 cm^{-1}. All three have correspondingly related images in the sp^2-fraction VDOS of the models in the same frequency range, compare Figure 2. The additional 3-rd feature, that could never be observed in hydrogenated structures, recently has been discussed by Wang et al.[14] to represent characteristic out-of-plane sp^2-modes. From a localization analysis we support the findings of Wang et al. While the spatial bond-angle-changing low frequency modes of sp^2-clusters, around 600 cm^{-1} remain extended, the high frequency stretching vibrations above 1500 cm^{-1} become strongly localized due to the embedding of smaller undercoordinated atom clusters in a higher coordinated rigid bonding environment. The highest frequency localized modes at low-density structures extending to 2400-2500 cm^{-1} indicate isolated sp-clusters sparsely distributed in a higher coordinated bonding environment. By the picking of modes at well defined frequencies, we confirm Tamor's and Vassel's findings and relate the broad high frequency feature, occuring in low-density a-C around 1500 cm^{-1} to complex extended stretching modes of the sp^2-matrix. As recently discussed by Drabold et al. [4] and Wang et al. [14] these stretching modes become localized at higher density by changing from a sp^2- to an sp^3-matrix separating now smaller sp^2-clusters. While this delocalization to localization transition is obvious for the higher frequency range all modes belonging to the low-frequency feature around 600 cm^{-1} remain delocalized even at highest densities representing in all cases out-of-plane modes of cross-linked sp^2-chain segments and rings. Additionally, the upward shift of the spectral features in the extended stochastically cooled modification at 2.0 g/cm^3 in consistency with a related shift of the graphite G-line in samples deposited at higher substrate temperature[5] is a reflection of the enhanced stiffening of the host matrix combined with a reduction of the phonon mode softening.

Acknowledgements: We gratefully acknowledge support from the DFG. Sincere thanks are due to D. A. Drabold for stimulating discussions and the Physics and Astronomy Department at Ohio University, Athens for support and hospitality during the stay of two of the autors at OU.

REFERENCES

1. G. Galli, R. M. Martin, R. Car and M. Parrinello, Phys. Rev B **42**, 7470 (1990).

2. C. Z. Wang, K. M. Ho and C. T. Chan, Phys. Rev. Lett. **70**, 611 (1993).

3. Th. Frauenheim, P. Blaudeck, U. Stephan and G. Jungnickel, Phys. Rev. **B48**, 4823 (1993).

4. D. A. Drabold, P. A. Fedders, P. Stumm, Phys. Rev. B **49** (1994) 16415.

5. M. A. Tamor, W. C. Vassell, Journal of Applied Phys., in print.

6. S. R. P. Silva, G. A. J. Amaratunga, C. P. Constantinou, J. Appl. Phys. **72** (1992) 1149.

7. U. Stephan, Th. Frauenheim, P. Blaudeck and G. Jungnickel, Phys. Rev. B, **49** (1994) 1489.

8. Th. Frauenheim, G. Jungnickel, Th. Köhler, U. Stephan, Journ. Non-Cryst. Solids **182** (1995) 186.

9. D. Porezag, Th. Frauenheim, Th. Köhler, G. Seifert, R. Kaschner, Phys. Rev. B, (15-th May 1995) in print.

10. G. Seifert and R.O. Jones, Z. Phys. **D20**, 77 (1991).

11. D. R. McKenzie, D. Muller and P. A. Pailthorpe, Phys. Rev. Lett. **67** (1991) 773.

12. F. Li and J. S. Lannin, Phys. Rev. Lett. **65** (1990) 1905.

13. P. H. Gaskell, A. Saeed, P. Chieux, and D. R. McKenzie, Phys. Rev. Lett. **67** (1991) 1286.

14. C. Z. Wang, K. M. Ho, Phys. Rev. Lett. **71** (1993) 1184.

NANOINDENTATION AND NANOSCRATCHING OF HARD CARBON COATINGS FOR MAGNETIC DISKS

T.Y. TSUI*, G.M. PHARR*, W.C. OLIVER**, C.S. BHATIA***, R.L. WHITE***, S. ANDERS†, A. ANDERS†, and I. G. BROWN†
* Department of Materials Science, Rice University, Box 1892, Houston, TX 77251
** Nano Instruments, Inc., Box 14211, Knoxville, TN 37914
*** IBM Storage Systems Division, 5600 Cottle Road, San Jose, CA 95193
† Lawrence Berkeley Laboratory, Berkeley, CA 94720

ABSTRACT

Nanoindentation and nanoscratching experiments have been performed to assess the mechanical properties of several carbon thin films with potential application as wear resistant coatings for magnetic disks. These include three hydrogenated-carbon films prepared by sputter deposition in a H_2/Ar gas mixture (hydrogen contents of 20, 34, and 40 atomic %) and a pure carbon film prepared by cathodic-arc plasma techniques. Each film was deposited on a silicon substrate to thickness of about 300 nm. The hardness and elastic modulus were measured using nanoindentation methods, and ultra-low load scratch tests were used to assess the scratch resistance of the films and measure friction coefficients. The results show that the hardness, elastic modulus, and scratch resistance of the 20% and 34% hydrogenated films are significantly greater than the 40% film, thereby showing that there is a limit to the amount of hydrogen producing beneficial effects. The cathodic-arc film, with a hardness of greater than 59 GPa, is considerably harder than any of the hydrogenated films and has a superior scratch resistance.

INTRODUCTION

Conventional magnetic hard disks are composed of multilayer thin films deposited on rigid substrates. During normal hard disk operation, a slider with active magnetic read-write elements flies above the disk and occasionally makes contact with it [1]. To prevent contact damage and wear in the relatively soft magnetic layer in which the data are stored, hard overcoats are usually employed in the disk structure.

A common material for overcoat protection is amorphous carbon, which is frequently prepared by sputter-deposition in a hydrogen atmosphere in a way which incorporates some hydrogen into the film. Hydrogenated-carbon films are generally much harder than their pure carbon counterparts, but exhibit mechanical properties which depend on the hydrogen content. One objective of the current report is to document the effects of hydrogen on the hardness, elastic modulus, and scratch resistance of several sputter-deposited, hydrogenated-carbon films.

A second objective of the paper is to compare the mechanical properties of hydrogenated films with those of a new carbon material prepared by cathodic-arc deposition. This relatively pure form of carbon has a hardness approaching that of diamond and is at least four times as hard as the best hydrogenated film studied here. As such, the material has great potential for application as a new hard disk overcoat material.

The mechanical properties were measured using nanoindentation and nanoscratching techniques described in detail in a previous report [2]. To facilitate comparison, all the films were deposited on silicon substrates to a thickness of approximately 300 nm.

PROCEDURE

The films examined in this study were prepared at two different laboratories. The hydrogenated-carbon films were made at the IBM Storage Systems Division, San Jose, CA, using sputter-deposition from a carbon source onto a stationary silicon substrate in a H_2/Ar

Mat. Res. Soc. Symp. Proc. Vol. 383

mixture. The hydrogen content was varied by controlling the hydrogen partial pressure in the sputtering chamber. Sputter pressures ranging from 5 to 10 mTorr were used, and the sputter power was adjusted to achieve a deposition rate of approximately 0.5 nm/sec. Three films, all 300 nm thick, were produced, containing 20, 34, and 40 atomic % hydrogen as measured using RBS techniques.

The cathodic-arc film was made at the Lawrence Berkeley Laboratory, Berkeley, CA, by cathodic-arc deposition with a 90° bent magnetic macroparticle filter and substrate pulse-biasing. Details of the deposition system and physical properties of the film are given elsewhere [3,4]. The cathodic-arc source was operated in a pulsed mode using a pulse duration of 5 ms, a frequency of 2 Hz, and a discharge current of 300 A. During immersion in the carbon plasma, the silicon substrate was repetitively pulse-biased at a negative voltage using a pulse duration time of 2 μs and bias duty cycle of 25%. To promote film adhesion, a relatively high substrate bias of -2kV was applied in the initial stages of film growth, producing an atomically mixed interface of about 10 nm followed by a continued growth under the same conditions so as to form an amorphous carbon layer of about 30 nm thickness. The majority of the film was then grown on top of this layer at the lower substrate bias of -100V. Preliminary structural analysis using transmission electron microscopy (TEM) and electron energy loss spectroscopy (EELS) showed that the film contained a mixture of amorphous and nanocrystalline material with a sp^3 bond content of about 85% [5]. Cross-sectional TEM revealed thin amorphous layers about 30 nm thick at both the substrate-film interface and the free surface [5].

Nanoindentation and nanoscratch tests were performed at Oak Ridge National Laboratory. The hardness, H, and elastic modulus, E, of each of the films were measured with a sharp Berkovich diamond indenter using the method developed by Oliver and Pharr [6]. The nanoscratch tests were conducted with a modified nanoindentation system incorporating sensors to measure the lateral forces on the indenter as the specimen was moved laterally underneath it [2]. A blunted Berkovich diamond oriented in a face-forward direction was used to make the scratches. As detailed elsewhere [2], during each nanoscratch experiment, the diamond tip was passed along the scratch track three separate times. The first pass, called the initial scan, was performed at the light load of 40 μN and was used to map the slope and local topography of the specimen without damaging the surface. The surface profile determined in this scan was fitted to a ninth-order polynomial and subtracted from all scratch displacement data so that subtle features in the data could be observed. After the initial scan, a 1000 μm long scratch was made by linearly ramping the load on the diamond from 0 to 100 mN as the specimen was laterally displaced at a velocity of 10 μm/sec. A third scan, again at the constant light load of 40 μN and referred to as the post-scratch scan, was then made to assess the changes in surface profile resulting from the scratch. From these measurements and optical examination of the scratch track, the critical load at failure, defined as the normal load at which permanent damage first occurred, was assessed. The coefficient of sliding friction as a function of position along the scratch track was also determined from the normal and lateral forces on the diamond.

RESULTS AND DISCUSSION

Nanoindentation

Nanoindentation test results for the films are shown in Figure 1, where the hardness and elastic modulus for each film are plotted as a function of indentation contact depth. Since the films were deposited on silicon, the hardnesses and moduli tend to converge at large contact depths towards the values for bulk silicon, 12 GPa and 163 GPa, respectively [7,8]. Values more representative of the films were obtained at small depths. Estimates of the film properties are summarized in Table I.

Perhaps the most notable feature in the H and E data is the extreme hardness of the cathodic-arc film; the 59 GPa value listed in the table is approximately 4 times that of the hardest of the hydrogenated films and is nearly 2/3 of the 90 GPa value for bulk crystalline diamond [9]. Moreover, the fact that there is no clear small-depth plateau in the hardness of the cathodic-arc material suggests that the nanoindentation measurements are not substrate independent. Thus, the

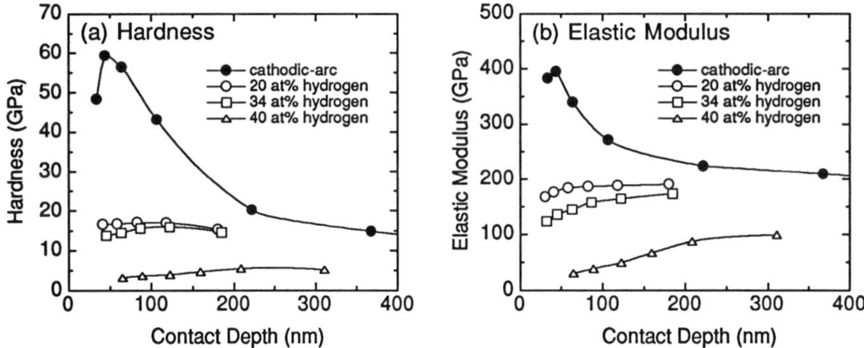

Figure 1. Nanoindentation measurements of (a) hardness and (b) elastic modulus.

true hardness of the film is probably even higher than the 59 GPa value and may approach that of diamond, consistent with the observation of at least some nanocrystalline material in the film [5]. The peak in H at small contact depths is believed to result from the amorphous surface layer observed in cross-sectional TEM [5]. Like the hardness, the elastic modulus shows a peak whose value of approximately 400 GPa also represents a lower limit. For comparison purposes, the modulus of bulk diamond is 1140 GPa [8].

Another notable feature in Figure 1 concerns the influence of hydrogen on the mechanical properties of the sputter-deposited films. The films containing 20% and 34% hydrogen have relatively high hardness and modulus, in the range H=14-17 GPa and E=135-175 GPa, but the properties of the 40% film, H=3.3 GPa and E=31 GPa, are considerably lower. The H and E values for the 40% film are in fact so low that one must question whether this is a hard-carbon material or a polymer. The properties of a sputter-deposited, hydrogen free film were not measured in this study, but from previous work we know that such films typically have hardnesses of about 12 GPa. Thus, hydrogenation produces measurable increases in H, but there are also limits to the amount of hydrogen which is beneficial.

A question arises at this point as to what the optimum hardness and elastic modulus are for a protective overcoat film in hard disk applications. While it is generally held that high hardness is important, the role played by the modulus is not quite as clear. As a first step toward answering this question, it useful to begin by assuming that overcoat films must be highly resistant to plastic deformation during contact events. This follows from the observation that many of the mechanisms of disk failure begin with or directly involve plastic deformation. In this regard, an analysis by Johnson which estimates the load, P_y, needed to initiate plastic deformation when a rigid sphere of radius, r, is pressed into contact with an elastic/plastic half-space, is useful [10]. Using Tabor's observation that the hardness of a material can be estimated as 3 times its yield strength [11], Johnson's analysis yields:

$$P_y = 0.78\, r^2 \frac{H^3}{E^2} \quad . \qquad (1)$$

This equation shows that the contact loads needed to induce plasticity are higher in materials with larger values of H^3/E^2, i.e., the likelihood of plastic deformation is reduced in materials with high hardness and low modulus, with H^3/E^2 being the controlling material parameter.

Values of the plastic resistance parameter H^3/E^2 for the films examined in this study are included in Table 1. Clearly, the highest value is that for the cathodic-arc film, and in this regard, the cathodic-arc material is very attractive material for hard disk applications. It should be noted, however, that since the H and E values for the cathodic-arc film are lower limits and therefore only approximate, there is some uncertainty as to what the value of H^3/E^2 really is for the

Table I. Summary of nanoindentation and nanoscratch measurements.

Film Type	Hydrogen content (at %)	Thickness (nm)	Hardness (GPa)	Elastic Modulus (GPa)	$\frac{H^3}{E^2}$ (GPa)	Critical Load (mN)	Friction Coefficient
sputtered	20	300	17	175	0.16	39	0.15
sputtered	34	300	14	135	0.15	42	0.09
sputtered	40	300	3.3	31	0.04	17	0.06
cathodic-arc	0	320	>59	>395	~1.3	61	0.20

cathodic-arc film. For the hydrogenated-carbon films, the 20% and 34% films should be equally resistant to plasticity based on the H^3/E^2 comparison and significantly better than the 40% film.

Nanoscratch Tests

Nanoscratch tests were conducted to examine the general scratching behavior of the films and quantify the scratch resistance for purposes of materials comparison. A complete set of nanoscratch test results for the cathodic-arc film and the hardest of the hydrogenated films (20% hydrogen) are included in Figure 2. Each set of data includes an optical micrograph of the scratch with arrows marking the beginning and end of the scratch track along with a corresponding plot of the vertical displacements of the diamond during the initial scan, the load-ramped scratch, and the post-scratch scan. It should be recalled that the initial scan profiles the unscratched surface, and the post-scratch scan is used to determine the surface damage caused by the scratch. The displacements for each of the three passes have been corrected to account for the slope and topography of the surface by subtracting from them the displacements measured in the initial scan, and for this reason, the initial scan appears as a flat line. Note that negative displacements correspond to the scratch tip being pushed into the material, and positive displacements, which appear only in the post-scratch scan, indicate that the surface has blistered outward or that debris has accumulated in the scratch track. Values for the apparent friction coefficients are also included in the plots. The friction coefficients listed in Table I were obtained from such plots by choosing the value just prior to film failure.

Fig. 2a shows the nanoscratching behavior of the 20% hydrogenated-carbon film. The scratch can be divided into three regions based on differences in the appearance of the scratch track. Starting from the left and moving to the right, the first region is defined by the first 380 µm of the scratch. In this region, the scratch is extremely smooth and shallow, in fact, so shallow that it can be optically resolved in places only with differential interference contrast. Exactly how shallow the scratch is may also be seen by comparing the post-scratch scan to the initial scan; on the relatively gross normal displacement scale plotted in the figure, the two scans are virtually indistinguishable for the first 380 µm. A closer examination revealed that there is no remnant trace of the scratch in the left-hand portion of this region, corresponding to fully recovered elastic contact. Thus, scratching in the first region may be characterized as fully elastic followed by smooth elastic/plastic ploughing in which most of the normal displacement is recovered as the diamond passes by.

The second region of the 20% hydrogenated-carbon film scratch extends from 380 to 550 µm. In this region, the film blisters by delamination at the film/substrate interface. The blistering may be observed in the optical micrograph and is also apparent in the post-scratch scan, which shows the surface to be uplifted at numerous places in this region. Given that blistering is not desirable in hard disk applications, the load at the beginning of this region, 39 mN, is defined as the critical load for film failure. It is also notable that there is a subtle change in the rate of increase in the coefficient of friction at the beginning of the second region (see Fig. 2a) which can be used to define the onset of delamination.

The third and final region of the scratch in the 20% hydrogenated-carbon film begins at 550 µm and extends to the end of the scratch. It is marked by an abrupt change in the scratch

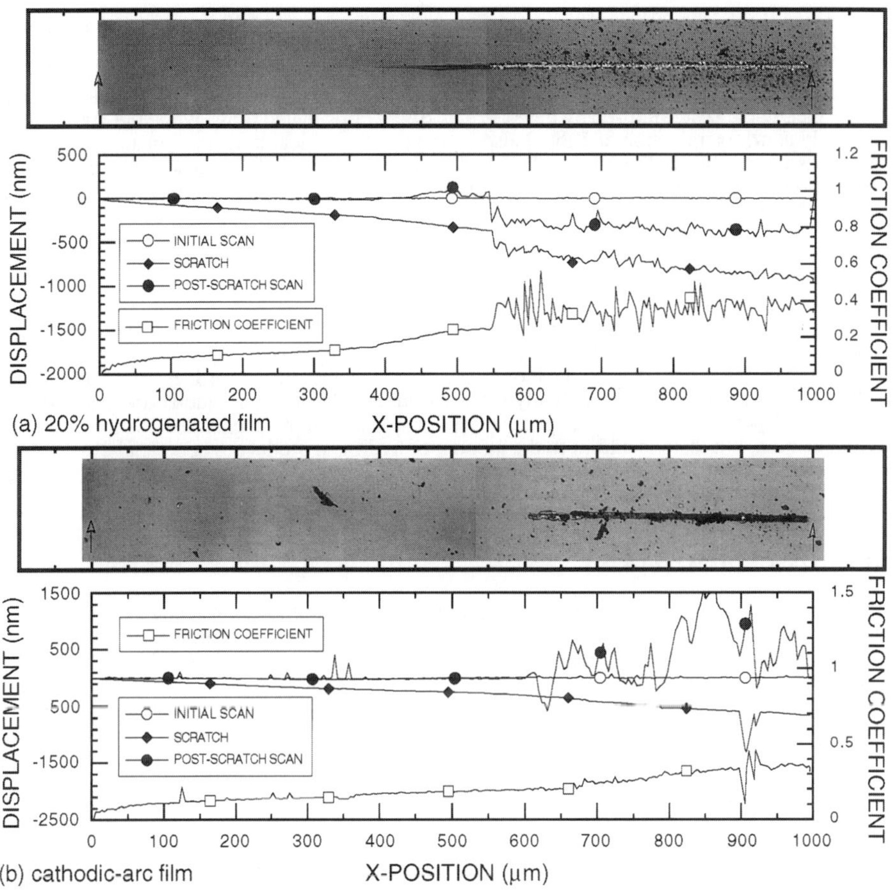

Figure 2. Optical micrographs of scratch tracks and scratch test results for: (a) the 20% hydrogenated-carbon film, and (b) the cathodic-arc carbon film.

displacement, as well as a substantial increase in the coefficient of friction. Examination of the optical micrograph reveals a large amount of small particle debris surrounding the scratch in this region, suggesting massive brittle fragmentation of the film. The post-scratch trace shows that the depth of fragmentation at the beginning of the third region is very close to the 300 nm film thickness, thus indicating complete failure and removal of the film. The overall picture which emerges is that the 20% hydrogenated film remains intact and is resistant to scratch damage at loads up to 39 mN, but increasing the load further causes the film to fail, first by delamination and blistering and then by massive brittle fragmentation.

The behavior of the cathodic-arc film is somewhat different, since as shown in Fig. 2b, only two distinct regions are observed. The first extends from 0-600 μm and is characterized by elastic contact followed by smooth elastic-plastic ploughing with nearly fully-recoverable normal displacements. At the 600 μm mark, however, corresponding to a load of 61 mN, failure begins abruptly by brittle fragmentation in the film and substrate. The brittle fragmentation is generally similar to that of the 20% hydrogenated film, but differs in that the debris particles are fewer in number and larger in size. More significantly, the critical failure load, 61 mN, is more than 50%

higher than that of the 20% hydrogenated film, thus indicating a significant increase in scratch resistance.

The critical failure loads for all four films are summarized in Table I. Using the critical load as a parameter for quantitatively assessing the scratch resistance of the materials, the cathodic-arc film is clearly the best, followed by the 34% and 20% hydrogenated films, while the 40% film is relatively poor. Examination of Table I also shows that the critical loads rank in an order which correlates reasonably well with the plastic resistance parameter, H^3/E^2, possibly suggesting a correlation between film failure in the ramped-load scratch test and the onset of plastic deformation in the film.

CONCLUSION

In conclusion, the limited results obtained here suggest that cathodic-arc carbon may offer significant advantages over conventional sputter-deposited, hydrogenated carbon as a protective overcoat material in hard disk applications. The relative values of the hardness and modulus of cathodic-arc carbon make it significantly more resistant to plastic deformation during contact, as shown in ramped-load scratching experiments. For sputtered films, the hardness is increased by hydrogenation, but there is a limit to the amount of hydrogen producing beneficial effects. For the processing conditions used here, the 20% and 34% films exhibit higher hardness and better scratch resistance than the 40% film.

ACKNOWLEDGMENTS

This research was sponsored by the Advanced Research Projects Agency as a part of the National Storage Industry Consortium program in Ultra High Density Recording; by the Division of Materials Sciences, U.S. Department of Energy, under contract DE-AC05-840R21400 with Martin Marietta Energy Systems, Inc. and through the SHaRE Program under contract DE-AC05-76OR00033 with the Oak Ridge Institute for Science and Technology; and by the U.S. Department of Energy, Division of Advanced Energy Projects, under contract No. DE-AC03-76SF00098.

REFERENCES

1. S. Chandrasekar and Bharat Bhushan, J. Tribology **112**, 1 (1990).
2. T.Y. Tsui, G.M. Pharr, W.C. Oliver, Y.W. Chung, E.C. Cutiongco, C.S. Bhatia, R.L. White, R.L. Rhodes and S.M. Gorbatkin, in Thin Films: Stresses and Mechanical Properties V, edited by S.H. Baker et al. (Mater. Res. Soc. Proc. **356**, Pittsburgh, PA), in press.
3. A. Anders, S. Anders, and I.G. Brown, Plasma Sources Sci. Technol. **4**, 1 (1995).
4. S. Anders, A. Anders, I.G. Brown, B. Wei, K. Komvopoulos, J.W. Ager, III, and K.M. Yu, Surf. Coat. Technol. **68/69**, 388 (1994).
5. G.M. Pharr, D.L. Callahan, S.D. McAdams, T.Y. Tsui, S. Anders, A. Anders, J.W. Ager III, I.G. Brown, C.S. Bhatia, S.R.P. Silva, and J. Robinson, submitted, Applied Phys. Lett.
6. W.C. Oliver and G.M. Pharr, J. Mater. Res. **7**, 1564 (1992).
7. G.M. Pharr, W.C. Oliver, and D.R. Clarke, J. Elec. Mater. **19**, 881 (1990).
8. G. Simmons and H. Wang, Single Crystal Elastic Constants and Calculated Aggregate Properties: A Handbook, 2nd ed. (MIT Press, Cambridge, MA, 1971).
9. C.A. Brookes, in The Properties of Diamond, edited by J. E. Field (Academic Press, New York, NY, 1979), pp. 383-402.
10. K.L. Johnson, Contact Mechanics, 1st ed. (Cambridge University Press, Cambridge, UK, 1985), p. 155.
11. D. Tabor, The Hardness of Metals, 1st ed. (Oxford University Press, Oxford, UK, 1951).

MECHANICAL PROPERTIES OF AMORPHOUS HARD CARBON FILMS PREPARED BY CATHODIC ARC DEPOSITION

SIMONE ANDERS*, ANDRÉ ANDERS*, JOEL W. AGER* III, ZHI WANG*, GEORGE M. PHARR**, TING Y. TSUI**, IAN G. BROWN*, AND C. SINGH BHATIA***
*Lawrence Berkeley Laboratory, 1 Cyclotron Road, Berkeley, CA 94720
**Department of Materials Science, Rice University, Houston, TX 77251-1892
***SSD/IBM, 5600 Cottle Road, San Jose, CA 95193

ABSTRACT

Cathodic arc deposition combined with macroparticle filtering of the plasma is an efficient and versatile method for the deposition of amorphous hard carbon films of high quality. The film properties can be tailored over a broad range by varying the energy of the carbon ions incident upon the substrate and upon the growing film by applying a pulsed bias technique. By varying the bias voltage during the deposition process specific properties of the interface, bulk film and top surface layer can be obtained. We report on nanoindentation and transmission electron microscopy studies as well as stress measurements of cathodic-arc amorphous hard carbon films deposited with varied bias voltage. The investigations were performed on multilayers consisting of alternating hard and soft amorphous carbon.

INTRODUCTION

Cathodic arc deposition is an emerging technology for the deposition of amorphous hard carbon films [1-7]. It has the advantage of a high deposition rate combined with the feasibility of large area deposition. Magnetic filtering of the arc plasma removes macroparticles which are produced at the cathode spots along with the plasma, and thus guarantees a high film quality [8, 9]. Amorphous hard carbon films formed by cathodic arc deposition are hydrogen-free and exhibit excellent mechanical properties such as high hardness, high mass density, and low coefficient of friction [1-7]. Applying a pulsed bias to the substrate is an easy and flexible means of modifying the energy of the incident carbon ions. The film properties such as mass density, hardness, coefficient of friction, intrinsic film stress, and elastic modulus depend strongly on the ion energy [1, 5, 6, 10, 11]. The quality of the film-substrate interface and the adhesion of the film can also be strongly influenced by the ion energy. The ion energy can be varied during the film deposition process in order to combine optimized interface qualities and desired film properties [12].

We have found in earlier experiments [6, 11, 12] that a pulsed bias voltage of - 100 V leads to the hardest films with the highest mass density and the highest stress. At a high bias voltage of -2 kV the films were softer and exhibited a much lower mass density and intrinsic compressive stress. High bias causes a deep intermixing between substrate and film and leads to superior adhesion of the films.

In the present paper we report on mechanical properties of cathodic-arc deposited hard carbon films which were formed at various bias voltages. In particular, we have formed multilayers of alternating hard and soft amorphous carbon films by varying the bias voltage during deposition. "Soft" in this case means films with hardness in the range of 15-25 GPa [6] which is much smaller than the maximum hardness of 60 GPa [12] that can be obtained for films deposited by this method but is still very hard in comparison to other thin films. The multilayers have been investigated by transmission electron microscopy and nanoindentation. Stress measurements have been performed also.

Mat. Res. Soc. Symp. Proc. Vol. 383

DEPOSITION OF HARD CARBON FILMS

A cathodic arc plasma source consisting of a 6 mm diameter graphite cathode and a cylindrical anode was used for the formation of the carbon plasma. The discharge current was 300A, the source was operated in a pulsed mode with a pulse duration of 5 ms and a repetition rate of 2 Hz. The source was connected to a 90° bent magnetic macroparticle filter. The plasma source and filter are described in detail in [8, 9]. The samples were mounted on a water-cooled sample holder keeping the sample at room temperature during the deposition. A negative pulsed bias voltage was applied to the sample with a pulse duration of 2 μs and a pulse off-time of 6 μs. The bias voltage was in the range between 0 and - 2 kV. The film thickness was measured by an oscillating quartz crystal thickness monitor. The base pressure was 10^{-4} Pa.

Three different multilayer structures have been deposited. All three structures consisted of 8 layers; the first layer at the substrate interface was a soft layer deposited at - 2 kV pulsed bias, and the top layer was a hard layer deposited at - 100 V pulsed bias. The ratio between the amount of carbon deposited at high and low bias was varied for the three structures. For the first structure the ratio was 50% soft phase/50% hard phase, for the second structure it was 10% soft phase/90% hard phase, and for the third structure it was 90% soft phase/10% hard phase. The total thickness of the multilayer structures was 250 nm. For comparison, films were deposited at high bias and low bias voltage only with the same total film thickness of 250 nm. Fig. 1 shows a simulation of the deposition for the 50% soft phase/50% hard phase structure using the code T-DYN 4.0 [13].

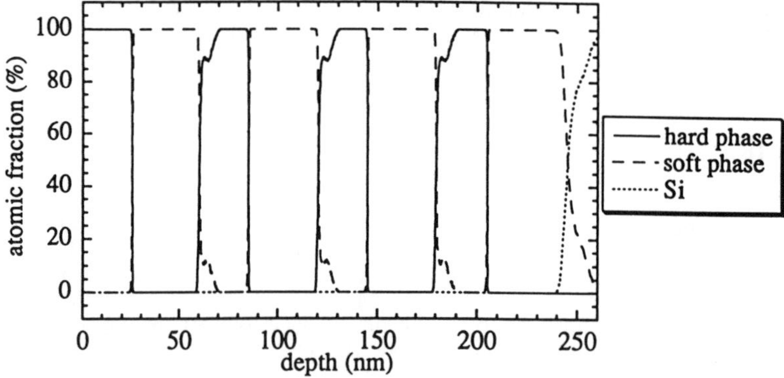

Fig. 1: Simulation of multilayer deposition for 50% soft phase/50% hard phase structure using the code T-DYN 4.0

Since the mass density of the soft (2.15 g/cm^3) and hard (3.0 g/cm^3) phases are different, the layer thickness is 25 nm for the hard phase and 35 nm for the soft phase. The soft phase shows a considerable intermixing with the underlying layers (Si or hard phase carbon) due to the high energy of the ions during the deposition whereas the hard phase shows only very small intermixing because the ion energy is low.

NANOINDENTATION

The multilayer structures were investigated using nanoindentation techniques [14] to determine the hardness and elastic modulus. The measurements were performed using a sharp Berkovich diamond indenter. Figs. 2 and 3 show the hardness and elastic modulus of the multilayer structures as a function of the indentation contact depth.

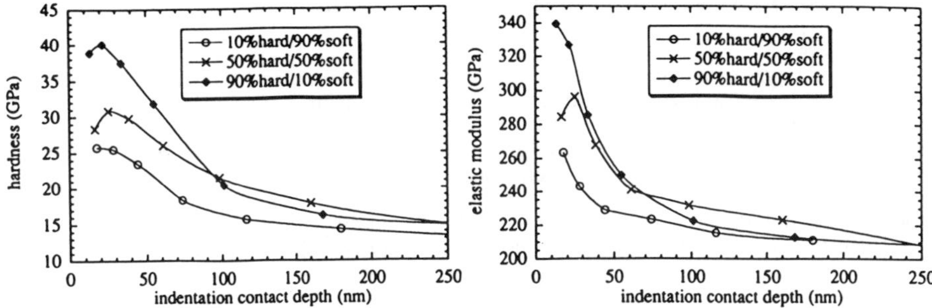

Fig. 2 Nanoindentation measurement of hardness as a function of indentation contact depth for multilayer structures of hard and soft phases of amorphous carbon on silicon. Total structure thickness 250 nm.

Fig. 3 Nanoindentation measurement of elastic modulus as a function of indentation contact depth for multilayer structures of hard and soft phases of amorphous carbon on silicon. Total structure thickness 250 nm.

For all structures the hardness and elastic modulus tend to the values for the silicon substrate at large contact depths. The values at small depth are more representative for the carbon multilayer structure. The hardness and elastic modulus are greater for structures with a higher ratio of thickness of the hard phase to thickness of the soft phase.

It is of interest to compare the hardness and elastic modulus of the multilayer structures with the properties of single layers deposited only at low or only at high bias. Figs. 4 and 5 show the peak hardness and peak elastic modulus of single layers of the same thickness as the total thickness of the multilayer structures. They demonstrate that the values for hardness and elastic modulus are almost a linear interpolation of the ratio of the hard and soft phases.

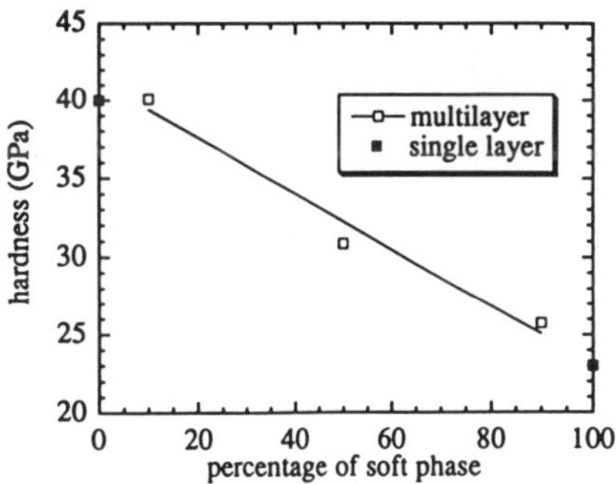

Fig. 4 Peak hardness of single layers of soft and hard carbon in comparison to multilayer structures containing soft and hard carbon layers in different thickness ratios.

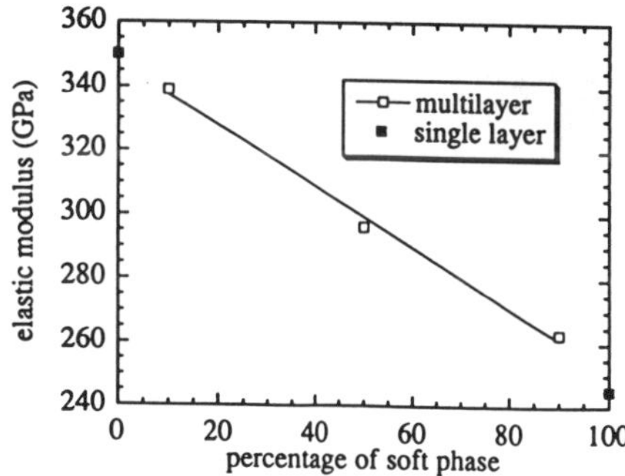

Fig. 5 Peak elastic modulus of single layers of soft and hard carbon in comparison to multilayer structures containing soft and hard carbon layers in different thickness ratios.

TRANSMISSION ELECTRON MICROSCOPY STUDIES

The multilayer structure which contained soft and hard layers of equal amounts of carbon was examined by transmission electron microscopy (TEM) using a JEOL 200 CX microscope. Fig. 6 is a cross-section image of the structure. In the upper left corner the silicon substrate is visible, on the lower right corner the glue for the sample preparation. The first dark layer at the silicon-carbon interface is probably an amorphous, atomically mixed layer containing both silicon and carbon. The light layers are phases of soft carbon whereas the dark layers are phases of hard carbon. The multilayer structure of four pairs of layers is clearly visible.

The difference in the thickness of the soft and hard layers is due to the different densities of the layers. Measurements of the mass density of soft and hard single layers determined by electron energy loss spectroscopy (EELS) [12] and by Rutherford backscattering spectroscopy and profilometry [6] result in densities of 2.1-2.2 g/cm^3 and 2.8-3.0 g/cm^3, respectively. This large difference is probably the reason for the difference in the contrast of soft and hard layers in the TEM image. The simulation (Fig. 1) predicts a ratio between the thickness of the soft to the hard layer of 1.4 based on the ratio of the densities . The TEM picture shows a ratio of almost 2; this might be due to the additional effect of intermixing and possible softening of the hard layers by ion bombardment during deposition of the following soft layer. This is supported by the fact that the top hard layer is thicker than all other hard layers inside the structure.

Fig. 7 is a cross-section image with higher magnification showing on the right hand side the silicon substrate with the corresponding electron diffraction pattern for a (112) orientation. The insert on the left side is the electron diffraction pattern for the carbon film indicating an amorphous structure.

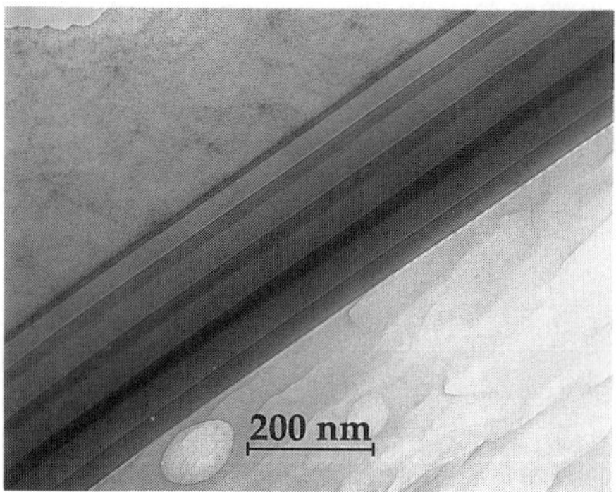

Fig. 6 TEM cross-section image of hard phase/soft phase amorphous carbon multilayer on silicon. Top left - silicon substrate, bottom right - glue for sample preparation.

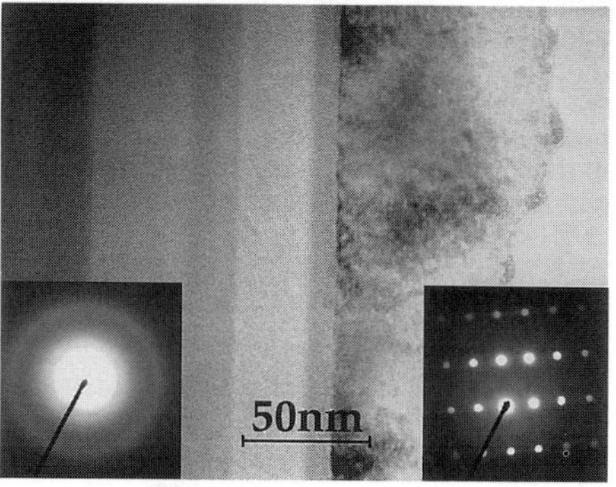

Fig. 7 TEM cross-section image of multilayer. Insert left - electron diffraction pattern of carbon material. Insert right-electron diffraction pattern of Si substrate.

STRESS MEASUREMENTS

Multilayer structures as described above were deposited on 200 μm thin Si wafers with a diameter of 25 mm. The substrate curvature was measured over the central 10 mm of the wafers before and after deposition using a profilometer of the stylus type. The film stress was calculated with the Stoney equation [15]. For comparison the stress of single layers deposited with high bias or low bias only was also determined. Fig. 8 shows that, in contrast to hardness and elastic modulus, the stress is not an interpolation between the data for high and low bias only, but is considerably reduced for the multilayer in comparison to single layers.

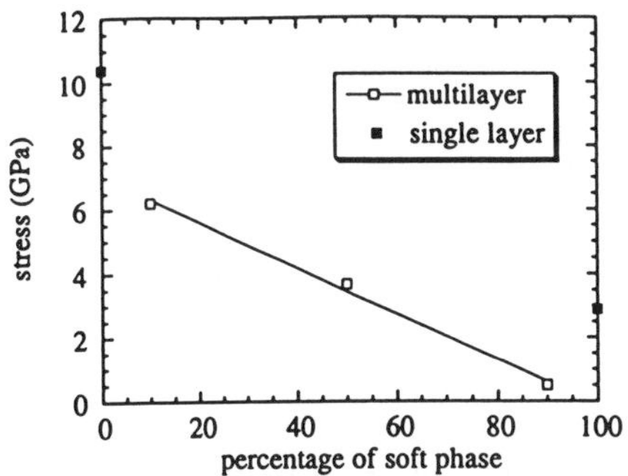

Fig. 8 Stress of single layers of soft and hard carbon in comparison to multilayer structures containing soft and hard carbon layers in different thickness ratios.

DISCUSSION AND CONCLUSIONS

It is possible to form multilayer structures containing alternating layers of soft and hard amorphous carbon by varying the pulsed bias voltage applied to the substrate during cathodic arc deposition. A computer simulation of the deposition process shows that during the deposition of a soft layer on a hard layer intermixing occurs due to the high energy of the carbon ions necessary for forming a soft layer by cathodic arc deposition. For obtaining multilayers with well-distinguished layers this limits the possible single layer thickness to a minimum of about 10 nm.

The mechanical properties of the multilayers such as hardness and elastic modulus were found to be a linear interpolation between the properties of single layers of the same thickness as the multilayer structure. This might not be the case for all multilayer structures; it is possible that the hardness can even be increased by forming multilayers [16-18] with a larger number of layers than has been investigated in this paper. It is very interesting to note that in contrast to hardness and elastic modulus the stress is not a linear interpolation between single layer properties but is considerably lower. It can be expected that the stress can be further reduced by reducing the layer thickness and increasing the number of layers in the structure. This gives the opportunity of changing to a certain degree independently the hardness and stress in the structure. It has been reported for hydrogen-free and hydrogenated amorphous carbon films [1, 5, 19, 20] that hardness (fraction of sp^3 bonds) and stress are directly correlated, and models have been developed describing the formation of sp^3 bonds as stress-induced [21]. The stress in very hard amorphous carbon films can reach high values, larger than 10 GPa [11]. Reducing the stress and containing the hardness at the same time was possible only by introducing additional chemical elements in the films as has been described for nitrogen incorporation [22]. The formation of multilayers offers an interesting alternative. It is also possible to deposit amorphous hard carbon films with a gradually varying bias voltage and to tailor in this way the film properties throughout the film the during the growth.

ACKNOWLEDGMENTS

This work was supported by the Electric Power Research Institute under Award number 8042-03, the U.S. Department of Energy, Division of Advanced Energy Projects, under contract No. DE-AC03-76SF00098; by the Center for Excellence in Synthesis and Processing - Processing for Surface Hardness, Basic Energy Sciences, U.S. Department of Energy; by the Advanced Research Projects Agency as a part of the National Storage Industry Consortium program in Ultra High Density Recording; and by the Division of Materials Sciences, U.S. Department of Energy, under contract DE-AC05-84OR21400 with Martin Marietta Energy Systems, Inc., and through the SHaRE program under contract DE-AC05-76OR00033 with the Oak Ridge Institute for Science and Education.

REFERENCES

[1] D. R. McKenzie, D. Muller, B. A. Pailthorpe, Z. H. Wang, E. Kravtchinskaia, D. Segal, P. B. Lukins, P. J. Martin, G. Amaratunga, P. H. Gaskell, and A. Saeed, Diamond Relat. Mater. **1**, 51 (1991).
[2] I. I. Aksenov and V. E. Strel'nitskii, Surf. Coat. Technol. **47**, 98 (1991).
[3] S. Falabella, D. B. Boercker, and D. M. Sanders, Thin Solid Films **236**, 82 (1993).
[4] R. Lossy, D. L. Pappas, R. A. Roy, J. J. Cuomo, and V. M. Sura, Appl. Phys. Lett. **61**, 171 (1992).
[5] P. J. Fallon, V. S. Veerasamy, C. A. Davis, J. Robertson, G. A. J. Amaratunga, W. I. Milne, and J. Koskinen, Phys. Rev. B **48**, 4777 (1993).
[6] S. Anders, A. Anders, I. G. Brown, B. Wei, K. Komvopoulos, J. W. Ager III, and K. M. Yu, Surf. Coat. Technol. **68/69**, 388 (1994).
[7] B. F. Coll, P. Sathrum, R. Aharonov, and M. A. Tamor, Thin Solid Films **209**, 165 (1992).
[8] S. Anders, A. Anders, and I. G. Brown, J. Appl. Phys. **74**, 4239 (1993).
[9] A. Anders, S. Anders, and I. G. Brown, Plasma Sources Sci. Technol. **4**, 1 (1995).
[10] J. J. Cuomo, D. L. Pappas, J. Bruley, J. P. Doyle, and K. L. Saenger, J. Appl. Phys. **70**, 1706 (1991).
[11] J. W. Ager III, S. Anders, A. Anders, and I. G. Brown, "Effect of Intrinsic Growth Stress on the Raman Spectra of Vacuum-arc-deposited Amorphous Carbon Films," Appl. Phys. Lett., to be published.
[12] G. M. Pharr, D. L. Callahan, S. D. McAdams, T. Y. Tsui, S. Anders, A. Anders, J. W. Ager III, and I. G. Brown, "Mechanical Properties and Structure of Very Hard Carbon Films Produced by Cathodic Arc Deposition," submitted to Appl. Phys. Lett.
[13] J. P. Biersack, S. Berg, and C. Nender, Nucl. Instrum. Methods Phys. Res. B **59/60**, 21 (1991).
[14] W. C. Oliver and G. M. Pharr, J. Mater Res. 7, 1564 (1992).
[15] H. Windischman, G. F. Epps, Y. Cong, and R. W. Collins, J. Appl. Phys. **69**, 2231 (1991).
[16] T. C. Chou, T. G. Nieh, T. Y. Tsui, G. M. Pharr, and W. C. Oliver, J. Mater. Res. **7**, 2765 (1992).
[17] R. C. Cammarata, T. E. Schlesinger, C. Kim, S. B. Qadri, and A. S. Edelstein, Appl. Phys. Lett. **56**, 1862 (1990).
[18] M. R. Scanlon, R. C. Cammarata, D. J. Keavney, J. W. Freeland, J. C. Walker, and C. Hayzelden, Appl. Phys. Lett. **66**, 46 (1995).
[19] M. A. Tamor, W. C. Vassel, and K. R. Carduner, Appl. Phys. Lett. **58**, 592 (1991).
[20] M. A. Tamor and W. C. Vassel, "Raman "Fingerprinting" of Amorphous Carbon Films," J. Appl. Phys., to be published.
[21] D. R. McKenzie, D. Muller, and B. A. Pailthorpe, Phys. Rev. Lett. B **67**, 773 (1991).
[22] D. F. Franceschini, C. A. Achete, and F. L. Freire, Jr., Appl. Phys. Lett. **60**, 3229 (1992).

A COMPARATIVE STUDY OF RESIDUAL STRESSES AND MICROSTRUCTURE IN a-tC FILMS

L. J. MARTÍNEZ-MIRANDA*, J. P. SULLIVAN**, T. A. FRIEDMANN**, M. P. SIEGAL**, T. W. MERCER*** AND N. J. DINARDO****.

* University of Maryland, Dept. of Materials and Nuclear Engineering, College Park, MD 20742.
**Sandia National Laboratories, Albuquerque, NM 87185
***Drexel University, Dept. of Physics, Philadelphia, PA 19104
****Drexel University, Dept. of Physics, Philadelphia, PA 19104; University of Pennsylvania, Dept. of Materials Science and Engineering, Philadelphia, PA 19104

ABSTRACT

We compare the microstructure of highly tetrahedrally-coordinated-amorphous carbon (a-tC) films prepared by pulsed laser deposition (PLD), measured using both small angle x-ray scattering (SAXS) and x-ray reflectivity, with other physical properties such as film stress and electrical resistivity. These properties are controlled by the film growth conditions and film thicknesses. Films prepared under vacuum conditions exhibit a shift in the measured mass density, as a function of laser energy density. The density for films approximately 600Å thick approach that of crystalline diamond. The measured densities for thicker, approximately 1000Å films, exhibit a smaller shift, and a lower density value. This shift correlates to observed changes in film stress and electrical resistivity. The small angle signal of the reflectivity spectra suggests the presence of layering, or in-plane density variations or a combination of both within the films.

INTRODUCTION

Amorphous highly tetrahedrally coordinated carbon (a-tC) films have potential in a variety of applications, from hard coatings to microelectronics. Therefore it is important to understand their structural, mechanical and electrical properties and how these relate to the method of film processing. These films consist of a mixture of carbon - carbon sp^2 and sp^3 bonds, grown in the absence of hydrogen. Previously reported results on a-tC films indicate the percentage of sp^3 bonds varies in the range of 77 - 90% sp^3 content [1,2]. In this study, pulsed laser deposition (PLD) is used to ablate carbon from a solid graphite target onto a silicon substrate at room temperature. The relative sp^2: sp^3 bonding ratio can be controlled depending on the growth conditions used, such as the laser wavelength, the laser energy density impinging upon the graphite target, or the background gas ambient during deposition [3,4].

Film properties obtained using Raman spectroscopy, as well as film stress [3] and electrical resistivity measurements [5], correlate with growth conditions in a manner consistent with the expected trend of sp^2: sp^3 bonding ratio. In general, film stress and electrical resistivity increase as the sp^3 content in the films increases. The films grown as described above are under high compressive stress, which varies from 1 to 6GPa as a function of increasing laser energy density. The stress is released when the films are grown in the presence of a background gas ambient [4]. Correlations made using Raman spectroscopy suggest that the effect of the gas ambient is to lower the kinetic energy of the ablated carbon species enroute to the substrate through collisional cooling.

In this paper, we present preliminary results of a reflectivity x-ray and small angle x-ray diffraction experiment on a-tC carbon films prepared using PLD in a high vacuum environment. We correlate results obtained on the density of the films with the measured stress in the films as

Mat. Res. Soc. Symp. Proc. Vol. 383 © 1995 Materials Research Society

well as to electrical resistivity measurements. The measured densities provide information that correlates with the sp^2:sp^3 ratio in the films. However, a precise measurement of this ratio is difficult due to lack of information on the densities of amorphous graphite and diamond.

EXPERIMENTAL

The a-tC films used in our study are grown by PLD, using 248nm pulsed laser radiation. A pyrolitic graphite target serves as the carbon source. The films are deposited on a rotating p-type Si (100) wafer. Details of the deposition process have been presented elsewhere [3]. The laser energy density varies between 10 and 45 J/cm^2, and the deposition rate is approximately 1Å/sec. Approximately 1/3 - 1/2 of the silicon wafer in each sample used in this experiment is left exposed. This enables the measurement and subtraction of a background scattering signal.

The stress in the films is obtained by measuring the curvature of the wafers before and after deposition, using a stylus profilometer [3]. The residual stresses are calculated using:

$$\sigma_f = \frac{E_s t_s^2}{6(1-\nu)t_f R}, \qquad (1)$$

where E_s is the Young's modulus for the substrate, ν is Poisson's ratio for the substrate, t_f and t_s are, respectively, the film and substrate thickness and R is the radius of curvature [4].

X-ray reflectivity has been used to characterize amorphous carbon films with thicknesses ranging between 200Å and 1100Å [6-8]. This technique is useful in the determination of film thickness, roughness, and density. The thickness of the films can be determined by the oscillations in the reflectivity signal. The film roughness can be determined by measuring the exponential decay of the signal. The density of the films was obtained from the reflectivity measurements using:

$$\phi_c^2 = 2 * N_o (e^2/2\pi mc^2)(Z\rho/A)\lambda^2, \qquad (2)$$

where ϕ_c is the critical angle, N_o is Avogadro's number, Z is the average atomic number, A is the average atomic mass, ρ is the mass density of the sample, and λ is the x-ray wavelength [6,7,9].The analysis of the small angle diffraction signal gives information on the presence of layering or clustering due to density variations within the films[8]. Layering gives rise to modulations in the oscillations due to film thickness in thin films. The presence of clustering or density variations can give rise to excess scattering at small angles or peaks at small angles.

In this experiment, we have combined both x-ray reflectivity and small angle x-ray scattering to determine the structure of the a-tC films. These measurements have been performed at the National Synchrotron Light Source, beamline X22A, using 1.20373Å (10.3keV) x-rays. The experimental resolution, achieved with a Si(111) monochromator and Si(111) analyzer, is $\Delta E/E \approx 1\times10^{-4}$ FWHM.

RESULTS AND DISCUSSION

Several a-tC films prepared at different laser energy densities under vacuum have been studied using the x-ray techniques described above. We have measured the x-ray reflectivity spectrum on a series of samples 250Å, 600Å and 1000Å thick prepared at laser energy densities of 11J/cm^2, 27J/cm^2 and 45J/cm^2 respectively. Figure 1 shows the x-ray reflectivity spectrum of a 600Å thick film prepared under vacuum with a laser energy density of 11J/cm^2. Note that in addition to the periodic oscillations which are proportional to the film thickness, the spectrum

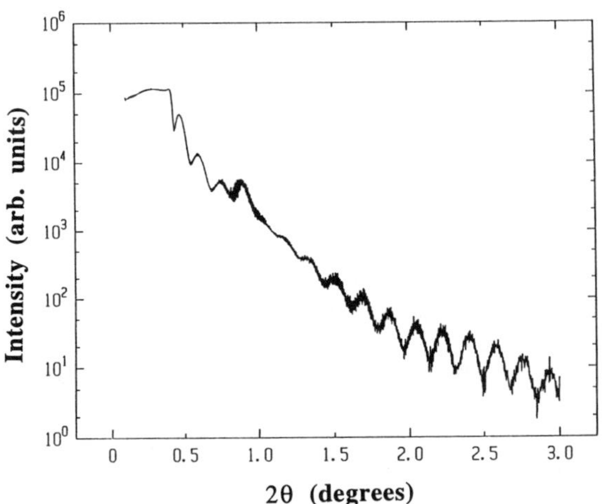

Figure 1. X-ray reflectivity curve for an at-C sample prepared at a laser energy density of 11J/cm^2 under vacuum. Note the extra scattering around $2\theta = 0.8°$.

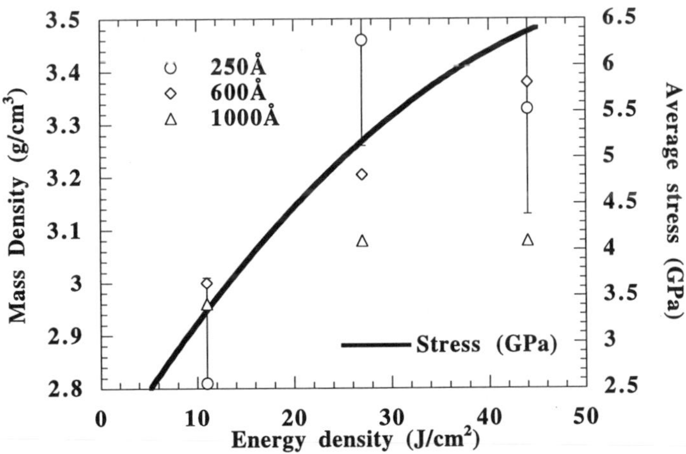

Figure 2. Mass density as a function of laser energy density for at-C films prepared in vacuum. The error bars in the measurements for the 600Å and 1000Å films are contained within the symbol. The solid line is the experimentally measured compressive stress curve (ref. 3). No fit has been performed for the mass density results as a function of stress.

shows excess scattering around $2\theta = 0.8°$, which may be related to density contrast between the film and substrate, at the film surface or to lateral density fluctuations within the film itself.

Figure 2 presents the results of a preliminary density measurement obtained from similar spectra of samples 250Å, 600Å and 1000Å thick. As a comparison, we plot the average compressive stress as a function of laser energy density [3,4]. As a function of film thickness, the mass density measurement shows a tendency toward higher values at medium and high laser energy densities. In previous studies [6], the measured densities were used to calculate roughly the percentage of sp^3 to sp^2 bonding in the film, using:

$$\rho = \left[\frac{(1-x)}{\rho_g} + \frac{x}{\rho_d}\right]^{-1}, \qquad (3)$$

where x is the sp^3 bond fraction, and ρ_g and ρ_d are the mass densities of crystalline graphite and crystalline diamond, which are 2.25g/cm^3 and 3.51g/cm^3 respectively. Using equation 3, the sp^3 bond percentage varies from approximately 77% to over 90% in the 600Å and 1000Å thick films studied. However, we note that amorphous graphite samples are typically less dense than crystalline graphite, with reported densities as low as 1.8 g/cm^3. It is likely that pure amorphous diamond would also be less dense than crystalline diamond; recent molecular dynamics calculations suggest the most stable amorphous sp^3 bonded carbon material has as density of approximately 3 g/cm^3 [11]. For this reason, it is difficult to calculate the precise sp^2:sp^3 ratio based on a linear interpolation of density using the values for both graphite and diamond, such as equation 3. However, since the density associated with sp^3 bonding is higher than that associated with sp^2 bonding, any increase in film density may be correlated with an increase in sp^3 bond content.

The functional shape as well as the position of any excess small angle scattering peaks, if present, give information on the density fluctuations within the samples. The angular position of any peaks present is proportional to a characteristic length or size of the density fluctuation. Figure 3 shows the preliminary results for the positions of the excess scattering peaks as a function of laser energy density for the 600Å and 1000Å samples respectively. These are expressed in terms of angstroms.

We now discuss the results presented above. A higher concentration of tetrahedrally coordinated carbon atoms should result in a higher compressive stress in the films and an increase in electrical resistivity. The trends observed in the average mass density of a-tC films prepared under vacuum as a function of laser energy density correlates with changes in the sp^2:sp^3 ratio inferred from Raman measurements as well as with the stress and electrical resistivity measurements [3-5].

The distribution of sp^2 and sp^3 bonded regions within the films will affect the residual stress of the films as well as the electro-optical properties of the film. As mentioned above, the intensity of the reflectivity signal as well as the presence and location of extra scattering at small angles are related to the density contrast between the film and substrate as well as to density fluctuations in different regions within the film. Figure 3 indicates that the basic length scale of such fluctuations is in the order of 100Å. These density variations may arise as a result of internal layering within the films which give rise to graphitic or low density layers at the substrate-film interface [6,7] and at the air-film interface [6,10,12]. Similarly, these variations can be a result of a distribution of alternating sp^2 and sp^3 regions in the plane of the film, which we discuss below.

Previous measurements on amorphous carbon (aC) films prepared using other methods [6,7] indicate evidence of both a contamination layer at the air-film interface [7] as well as a thin low density interfacial layer at the film-substrate boundary [6]. The results of electrical measurements on a-tC samples presented elsewhere [5] seem to indicate the existence of conductive paths or shorts through the thickness of the samples, which are consistent with regions with a higher sp^2 content than the sample average. The density difference between these regions and the bulk of the films results in a scattering signal that depends on the average of the shape, size, and the number of these regions within the sample. Regions with a characteristic size in the order of tens to hundreds of angstroms scatter X-rays at low angles, as seen in Figure 1. In addition, the presence and number of such low density regions should result in a relief of the residual stress within the film. No such measurements are available at present. A more detailed analysis of our results which will give a better picture of the internal structure of the films is in progress. This analysis will benefit from localized bond-sensitive measurements, such as high-resolution EELS, as well as from glancing x-ray scattering measurements [13].

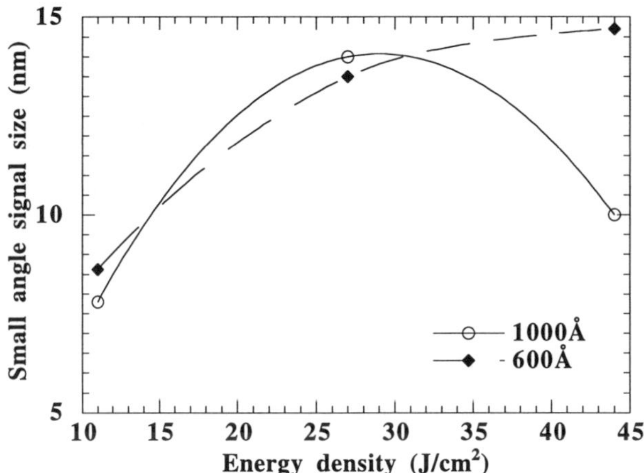

Figure 3. Small angle signal as a function of laser energy density for 1000Å and 600Å thick films prepared under vacuum.

Table I summarizes the measurements done on the 600Å and 1000Å thick films.

CONCLUSIONS

We have presented the preliminary results of an x-ray reflectivity measurement on at-C films prepared using PLD methods in vacuum. These results have been correlated to the residual stresses in the films as well as the measured electrical resistivities. For these films, the concentration of sp^3 bonds in the film, inferred from the preliminary mass density measurements correlate well with the results of the residual stress in the films as a function of laser energy density. Preliminary comparison of the reflectivity results with the resistivity measurements suggest that the films may contain in-plane density variations, as well as layering.

This work was supported through the Laboratory Directed Research and Development Program, Sandia National Laboratory, and supported by the US Dept. of Energy under contract No. DE-ACO4-94AL85000. NJD acknowledges support from the NSF under grant DMR93-

13047. We would like to thank Dr. Thomas Thurston and Dr. B. M. Ocko for the use of beamline X22A at the NSLS.

Table I. Physical properties of a-tC films grown using PLD in vacuum, measured with SAXS and x-ray reflectivity.

Sample Thickness (Å)	Laser Energy Density (J/cm^2)	Average Film Density (g/cm^3)	Small angle signal (Å)
600	11	3.00	86 ± 6
	27	3.20	135 ± 6
	44	3.38	147 ± 6
1000	11	2.96	78 ± 10
	27	3.08	140 ± 10
	44	3.08	100 ± 10

REFERENCES

1. J. J. Cuomo et al., J. Vac. Sci. ., **A 10**, 3414 (1992).
2. F. Li and J. S. Cannin, Phys. Rev. Lett., **65**, 1905 (1990).
3. M. P. Siegal , T. A. Friedmann, S. R. Kurtz, D. R. Tallant, R. L. Simpson, F. Dominguez and K. F. McCarty, Mat. Res. Soc. Symp. Proc., **349**, 507 (1994).
4. T. A. Friedmann, M. P. Siegal, D. R. Tallant, R. L. Simpson and F. Dominguez, Mat. Res. Soc. Symp. Proc., **349**, 501 (1994).
5. J. P. Sullivan , T. A. Friedmann, C. A. Apblett and M. P. Siegal, Mat. Res. Soc. Symp. Proc., **381**, in press (1995).
6. Y. Huai et al., Appl. Phys. Lett., **65**, 830 (1994).
7. M. F. Toney and S. Brennan, J. Appl. Phys., **66**, 1861 (1989).
8. L. J. Martínez-Miranda et al., submitted to Phys. Rev. **B15**, 1994.
9. M. F. Toney et al., J. Mater. Res., **3**, 351 (1988).
10. C. A. Lucas et al., Appl. Phys. Lett., **59**, 2100 (1991).
11. Th. Köhler, Th. Frauenheim, D. Porezag and D. A. Drabold, Mat. Res. Soc. Symp. Proc., **383**, in press (1995).
12. T. W. Mercer et al., Mat. Res. Symp. Proc., **358**, 863 (1995); L. J. Martínez-Miranda et al., Mat. Res. Soc. Symp. Proc., **358**, 857 (1995).
13. X. Yan and T. Egami, Phys. Rev. B, **47**, 2362 (1993); H. Chen and S. M. Heald, J. Appl. Phys., **66**, 1793 (1989).

CORRELATIONS BETWEEN THE SUBSTRATE TEMPERATURE, PROPERTIES AND TRIBOLOGICAL PERFORMANCE OF SPUTTER-DEPOSITED AMORPHOUS CARBON FILMS

E. MOUNIER*, P. JULIET*, E. QUESNEL*, Y. PAULEAU*[1], and M. DUBUS**
CEA, Nuclear Research Center (CENG), *CEREM-DEM-SGSA-LTS, **DRFMC-SPMM-PI, 17 Rue des Martyrs, 38054 Grenoble Cedex 9, France.

ABSTRACT

Amorphous carbon films have been deposited on various grounded substrates by direct current (DC) magnetron sputtering from a graphite target in an argon discharge. The argon pressure and sputtering power density were fixed at 0.25 Pa and 2.65 W cm^{-2}, respectively. The substrate temperature was varied from 50 to 350°C. The hydrogen content determined by ERDA and the electrical resistivity of films were found to be dependent on the base pressure in the deposition chamber and substrate temperature. The mass density of films evaluated from RBS data and compressive residual stresses in the films decreased with increasing substrate temperature from 2.2 to 1.4 g cm^{-3} and - 0.6 to - 0.2 GPa, respectively. The friction coefficient and wear rate of about 2-μm-thick carbon films deposited on polished stainless steel disks were determined by alumina ball-on-disk tribological measurements conducted in dry air under a load of 4.9 N with a sliding velocity of 10 m min^{-1} for 10^5 cycles. The tribological performance of carbon-coated disks are correlated with the deposition temperature and physical characteristics of carbon films.

INTRODUCTION

Amorphous carbon (a-C) films have been prepared by a variety of physical and chemical vapor deposition techniques from diverse carbon-bearing solid or gaseous source materials under various deposition conditions [1]. These films can exhibit physical properties desirable for tribological applications such as high hardness and chemical resistance, reduced residual stresses, suitable adherence to various substrates, low friction coefficient and high wear resistance; a-C and fluorinated silicon-containing carbon films are attractive candidates to serve as solid lubricant coatings for micromachines and mechanical assemblies operating in hostile environments [2]. Depending upon the deposition technique and deposition conditions, these a-C films are composed of tetrahedrally and trigonally coordinated carbon atoms with variable proportions of sp^3 and sp^2 type bonding; in addition, the hydrogen concentration in the films can vary from zero to 40 at%. The hybridization type of carbon atoms depends on the amount of energy deposited on the film surface while the film grows; high sp^3/sp^2 ratio values can be obtained from high energy carbon atoms condensed on the film surface or by exposure of the growing film to an energetic particle bombardment [3,4]. Usually, non hydrogenated a-C films are deposited by sputtering from a graphite target in argon discharges or by condensing carbon atoms from the plasma flux generated by vacuum arc discharges [5].

The tribological properties of a-C films prepared by physical and chemical vapor deposition techniques under various experimental conditions have been reviewed recently [5]. The friction and wear properties of a-C films can vary in a very wide range and are affected not only by the

[1]Also at : National Polytechnic Institute of Grenoble
ENSEEG, B.P. 75, 38402 Saint Martin d'Hères Cedex, France.

Mat. Res. Soc. Symp. Proc. Vol. 383 © 1995 Materials Research Society

tribological test conditions but also depend on the deposition method and deposition parameters. In fact, the intrinsic characteristics of the films, e.g., microstructure, morphology, mass density, residual stresses which are strongly dependent on the deposition parameters govern the mechanical properties and friction behavior of the deposited material. It is well known that the friction properties of tribological films can be controlled by the chemical reactions occurring on the materials surface in sliding contact, i.e., by the tribochemistry involved in the contact [6]. The surface chemistry which can be affected by the environment during tribological tests depends significantly on the structure and characteristics of films. No much works have been dedicated to the determination of tribological properties of sputter-deposited a-C films. The friction measurements were conducted with a-C films sputter-deposited on magnetic recording films (cobalt-based magnetic alloys) sliding against various riders (sapphire, Al_2O_3-TiC 70:30). In dry oxygen, the friction coefficient was 0.2 and increased continuously as the number of sliding cycles increased [7,8]. In dry nitrogen, the value of the friction coefficient remained constant at 0.2 up to 50 h of testing [7]. A discrepancy between literature data on the humidity effect on the friction coefficient can be noticed. According to Strom et al. [8], at low humidity (RH < 5 %), the increase in friction coefficient with increasing cycle number was smaller than that at higher humidities. Hilden et al. [9] confirmed these results for a-C films deposited on disks made of particle media (γ-Fe_2O_3 or Co-doped γ-Fe_2O_3) in organic binders while for a-C films deposited on polished NiP substrates, the friction coefficient was found to decrease from 1.3 at RH = 3 % to 0.3 at RH = 60 %. The tribological data reported in the literature do not permit to establish a clear correlation between the deposition parameters, physical properties and friction properties of sputter-deposited a-C films.

In a recent work, the pressure of the sputtering gas (argon) was found to affect the mass density, morphology, residual stresses and finally the friction and wear properties of non hydrogenated a-C films [10]. Films with improved tribological properties were produced at low argon pressures with energetic carbon atoms condensed on grounded substrates. An increase in energy deposited on the growing film seems to affect favorably the tribological properties of sputter-deposited a-C films. Besides, an increase in surface temperature would be favorable for improving surface mobility of adatoms and depositing a-C films with improved friction and wear properties. In the present work, a series of experiments were designed to investigate the effect of the substrate temperature on physical properties and friction properties of a-C films sputter-deposited on various substrates from a graphite target in an argon discharge. The hydrogen content, electrical resistivity, mass density and residual stresses of the deposited material were determined as functions of the deposition temperature. Amorphous carbon films deposited on grounded substrates at low argon pressures were submitted to alumina ball-on-disk tribological tests in dry air. The tribological properties of a-C coated disks are discussed and correlated with the deposition parameters and physical characteristics of the films.

EXPERIMENTAL PROCEDURE

Carbon films have been deposited on glass, silicon and stainless steel substrates by direct current (DC) magnetron sputtering from a graphite target in an argon discharge. The substrates were mounted on a grounded substrate holder which could be heated up to 400°C by infrared lamps. The substrate-target distance was maintained at 7 cm. The sputtering chamber was evacuated by a turbomolecular pump backed up with a mechanical pump. The base pressure value reached after 3 h of pumping was dependent on the temperature of the substrate holder which was varied from 50 to 350°C. The minimum value was about 3×10^{-4} Pa with the substrate holder at room temperature while with a substrate temperature higher than 200°C the base pressure was in the range $(1\text{-}4) \times 10^{-3}$ Pa. The surface of the graphite target (21×9) cm^2 was cleaned by ion etching for a pre-sputtering step of 5 min. During this surface cleaning treatment, the substrates were separated from the target by a movable shutter. The argon pressure and sputtering power were fixed at 0.25 Pa and 500 W, respectively.

The thickness of a-C films was determined by profilometer measurements. The films were analyzed by Rutherford backscattering spectroscopy (RBS) with a beam of 2 MeV alpha particles and a detection angle of 165°. The mass density of films was evaluated from RBS data and film thickness values. The hydrogen content in the films was obtained from elastic recoil detection analyses (ERDA) performed with a beam of 2.4 MeV alpha particles. The resistivity of a-C films deposited on glass substrates between two chromium electrodes biased to a given voltage was determined by current intensity measurements. The residual stresses in the films deposited on Si substrates were obtained from the change in the radius of curvature of substrates measured before and after deposition of films [11]. The a-C films deposited on stainless steel disks with a surface finish, R_a, of 0.05 µm were submitted to alumina ball-on-disk tribological tests conducted in dry air under a load of 4.9 N with a sliding velocity of 10 m min^{-1} for various numbers of cycles. The average distance per cycle was 7.7 cm. The wear scars in alumina balls of 8 mm in diameter with a surface roughness, R_a, of about 0.03 µm and wear tracks in a-C films were analyzed by profilometer measurements to determine the wear rate of materials in sliding contact.

RESULTS

The deposition rate of carbon films deposited at 50°C with a sputtering power of 500 W was in the range 14-22 nm min^{-1} and increased with increasing substrate temperature since the density of films decreased. In fact, the deposition rate in terms of number of carbon atoms deposited on the substrates per cm^2 and s was constant and independent of the substrate temperature. The structure of films was determined to be amorphous by x-ray and electron diffraction techniques. Generally, 2-µm-thick a-C films were tightly adherent to glass, silicon and stainless steel substrates. The RBS analyses revealed that the films deposited below 250°C were not contaminated by oxygen or other elements. The argon content in the films was less than the RBS detection limit (0.1 %). Hydrogen atoms were the major impurities non intentionally incorporated in the films. For a-C films deposited at 50°C, the hydrogen concentration increased linearly with increasing base pressure in the sputtering chamber (Fig.1). When the deposition chamber was evacuated at a base pressure of 2 x 10^{-4} Pa, the hydrogen content in the deposited material was less than the ERDA detection limit (0.5 at%), i.e., non hydrogenated a-C films could be produced under variable deposition conditions (argon pressure, sputtering power, substrate bias voltage, ...). The electrical resistivity of these a-C films deposited at 50°C was in the range 0.2-1.4 Ω cm depending on the base pressure value (Fig.1). The cross-sectional view of these a-C films examined by scanning electron microscopy revealed a densely packed structure [10].

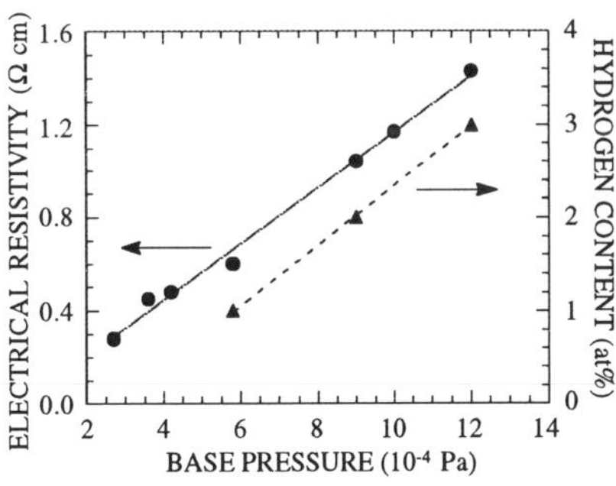

Fig.1 - Effect of the base pressure in the deposition chamber for a-C films deposited at 50°C.

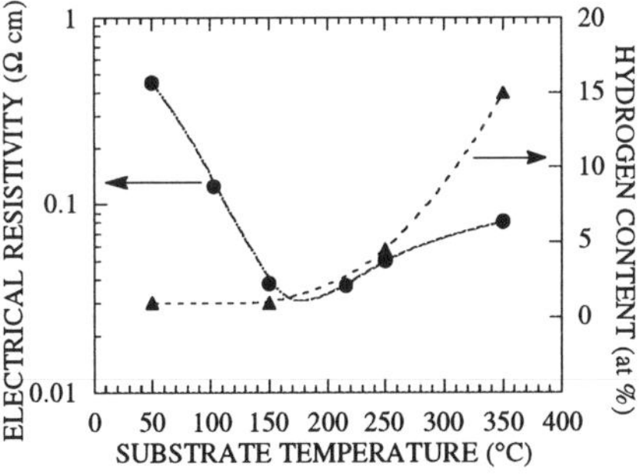

Fig.2 - Effect of the substrate temperature on the composition and electrical resistivity of a-C films.

The physical characteristics of a-C films namely hydrogen content, electrical resistivity, mass density and residual stresses were found to be dependent on the deposition temperature. The hydrogen concentration was less than 2 at% in a-C films deposited below 200°C (Fig.2). At these substrate temperatures, the base pressure in the deposition chamber prior sputter-deposition of films was less than 1 x 10^{-3} Pa. Above 200°C, the amount of hydrogen incorporated in the films increased with increasing substrate temperature; the hydrogen content was about 15 at% for a-C films deposited at 350°C (Fig.2). Under these experimental conditions, after 3 h of pumping, the base pressure in the sputtering chamber was in the range (3-4) x 10^{-3} Pa. The electrical resistivity of a-C films containing less than 1 at% of hydrogen, i.e., non hydrogenated a-C films decreased by a factor of 10 as the substrate temperature increased from 50 to 200°C. For hydrogenated amorphous carbon (a-C:H) films deposited above 200°C, the resistivity value increased progressively from 0.03 to 0.08 Ω cm (Fig.2). The mass density of non hydrogenated a-C films deposited on substrates at 50°C was close to that of bulk graphite (2.25 g cm^{-3}); for films deposited at higher temperatures, the density value decreased with increasing substrate temperature (Fig.3). The residual stresses including intrinsic stress and thermal stress in a-C films were found to be compressive (Fig.3). The stress level was found to decrease with increasing deposition temperature, i.e., with decreasing mass density of films.

The friction coefficient of a-C coated disks sliding against alumina balls was in the range 0.05-0.10 during the initial stage of tribological tests (Fig.4). A progressive rise of the friction coefficient value of a-C films deposited below 200°C was observed for 20000 sliding cycles before stabilization. The stabilized value of the friction coefficient decreased as the deposition temperature of a-C films increased from 50 to 250°C. For a-C films deposited above 300°C, the friction coefficient increased significantly during the first stage of tribological tests before stabilization at 0.2 for films deposited at 350°C; beyond 60000 sliding cycles, the friction coefficient value of the films rose sharply up to 0.8-0.9 since the alumina ball was directly in contact with the stainless steel disk.

The average value of the friction coefficient determined for the stabilization stage was relatively constant at about 0.09 for a-C films deposited below 150°C, decreased down to 0.03 for films produced at 250°C and increased rapidly up to 0.2 with increasing substrate temperature (Fig.5). The wear tracks in non hydrogenated a-C films deposited at 50°C observed after about 10^5 cycles were not very deep and exhibited a flat bottom with cracks aligned in the direction of the sliding motion. For these tribological tests, the wear rate of alumina balls was in the range (3-4) x 10^{-8} $mm^3 N^{-1} m^{-1}$ (Fig.6). The wear rates of a-C films and balls were independent of the substrate temperature in the range 50-150°C. The wear rate of a-C films was minimum for films prepared at a temperature in the range 200-250°C; above 250°C, the wear rate of films increased with

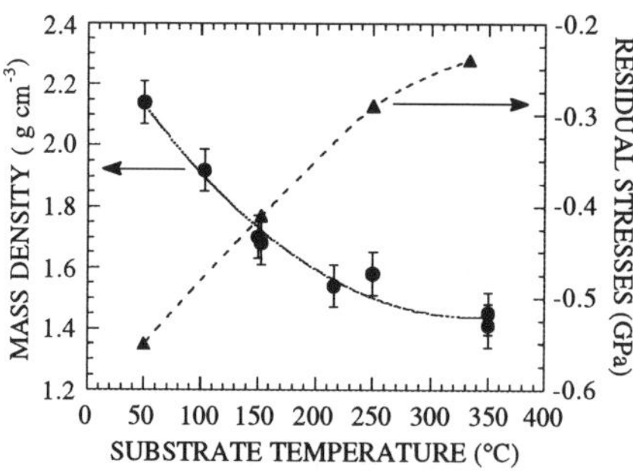

Fig.3 - Substrate temperature effect on mass density and residual stresses of a-C films

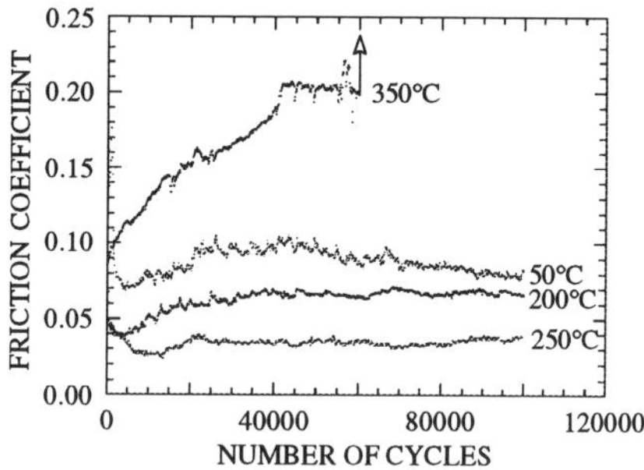

Fig.4 - Friction coefficient versus number of cycles for a-C films deposited at various substrate temperatures

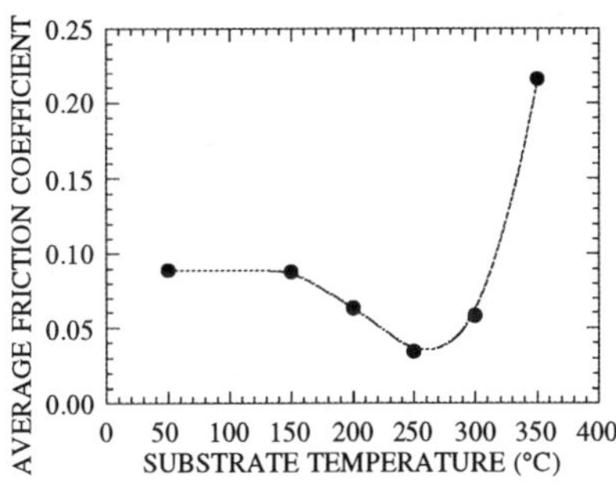

Fig.5 - Average friction coefficient versus substrate temperature

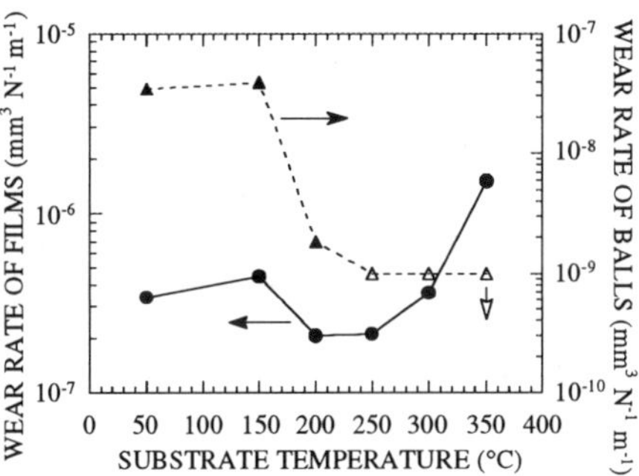

Fig.6 - Substrate temperature effect on the wear rates of a-C films and alumina balls

increasing deposition temperature.The wear rate of alumina balls was also minimum with a-C films deposited above 200°C. In fact, the wear rate of balls was less than 9 x 10^{-10} mm^3 N^{-1} m^{-1}; this value was calculated from the detection limit of the wear volume of alumina balls.

DISCUSSION

The composition of the residual atmosphere in the sputtering chamber in particular the water vapor content was directly dependent on the base pressure and substrate temperature. As a result, H atoms included in a-C films may originate from water vapor molecules outgassed from the chamber wall and dissociated in the discharge to form H and O atoms.

The electrical resistivity of a-C films was relatively low compared with that of bulk diamond carbon (> 10^{15} Ω m) and was essentially governed by the substrate temperature (Fig.2). In fact, the resistivity value lay between that of crystalline graphite in the graphitic plane (10^{-4} Ω cm) and that measured in a direction normal to the graphitic plane (1 Ω cm). These low resistivity values may be associated with both a high concentration of dangling atomic bonds and a low content of sp^3 type bonding in these films. Furthermore, the mass density of a-C films was comparable to the bulk graphite density (2.25 g cm^{-3}) and considerably lower than that of diamond (3.5 g cm^{-3}). These low values of resistivity and density tend to indicate a greater sp^2 than sp^3 bond concentration and a more probable graphite-like than diamond-like structure of films.

The decrease in resistivity with increasing substrate temperature up to 150°C may arise from a graphitization of the structure of a-C films (Fig.2). Accordingly, resistivity values lower than 0.03 Ω cm would be expected for a-C films deposited above 150°C; however, the decrease in resistivity arising from a large extent of the graphitization of films produced at high temperatures was probably compensated by the increase in resistivity resulting from saturation of dangling bonds by H atoms incorporated in a-C films deposited at a substrate temperature ranging from 150 to 350°C. In addition, since the mass density of films decreased with increasing substrate temperature, a transition from graphite-like to polymer-like structure for these a-C:H films deposited at high temperatures must not be excluded. The decrease in residual stresses observed as the substrate temperature increased resulted probably from the decrease in mass density of films. This trend is currently noticed for films produced by physical vapor deposition techniques [12]. The density of microdomains constituted of stacked atomic planes of carbon atoms can be expected to increase as the deposition temperature of a-C films increased. The progressive graphitization of the film structure with increasing substrate temperature may be responsible for the decrease in friction coefficient of films prepared at a temperature in the range 50-150°C (Fig.5). This lamellar microstructure of films interposed in sliding contact is known to be favorable to obtain reduced friction coefficient and wear rate values [13]. The graphite-like a-C films sputter-deposited at temperatures higher than 50°C exhibited a mass density lower than that of bulk graphite. As a result, a decrease in bond strength between carbon planes can be expected in these a-C films; the reduction in friction coefficient for graphite-like a-C films deposited between 150 and 250°C may originate from these reduced bond strengths in the lamellar microstructure. The increase in friction coefficient and wear rate of a-C films prepared above 250°C may be attributed to the plowing action in the sliding contact caused by a larger penetration of the alumina ball in the lubricant film. This larger penetration of the rider reflected that the structure of these films deposited on substrates at high temperatures was more polymer-like than that of films prepared below 250°C.

CONCLUSION

The substrate temperature dependence of the hydrogen content, electrical resistivity and mass density of sputter-deposited amorphous carbon films tends to indicate that graphite-like a-C films free of hydrogen can be produced below 150°C while graphite-like or polymer-like a-C:H films can be grown at higher substrate temperatures. Additional investigations are in progress to get a deeper insight into the structure of these a-C films deposited on heated substrates. The tribological properties of a-C and a-C:H films in dry air are strongly dependent on the deposition temperature. The correlation established between the substrate temperature, physical characteristics and friction properties of films demonstrates that excellent solid lubricant thin films of a-C can be prepared by sputter-deposition on substrates at a temperature in the range 200-250°C.

REFERENCES

1. A. Grill and B.S. Meyerson, in Synthetic Diamond : Emerging CVD Science and Technology, edited by K.E. Spear and J.P. Dismukes (Wiley, New York, 1993).
2. S. Miyake and R. Kaneko, Thin Solid Films **212**, 256 (1992).
3. J.J. Cuomo, J.P. Doyle, J. Bruley and J.C. Liu, Appl. Phys. Lett. **58**, 466 (1991).
4. X. He, W. Li and H. Li, J. Vac. Sci. Technol. A **11**, 2964 (1993).
5. A. Grill, Wear **168**, 143 (1993).
6. I.L. Singer, in Fundamentals of Friction : Macroscopic and Microscopic Processes, edited by I.L. Singer and H.M. Pollock, NATO-ASI Series, Serie E : Applied Sciences, Vol.220 (Kluwer Academic Publishers, Dordrecht, The Netherlands, 1992), p.237.
7. B. Marchon, N. Heiman and M.R. Khan, IEEE Trans. Magn. **26**, 168 (1990).
8. B.D. Strom, D.B. Bogy, C.S. Bhatia and B. Bhushan, ASME J. Tribol. **113**, 689 (1991).
9. M. Hilden, J. Lee, G. Ouano, V. Nayak and A. Wu, IEEE Trans. Magn. **26**, 174 (1990).
10. E. Mounier, P. Juliet, E. Quesnel and Y. Pauleau, Surf. Coat. Technol. (to be published in 1995).
11. G. Stoney, Proc. R. Soc. London Ser.A **82**, 172 (1909).
12. H. Leplan, B. Geenen, J.Y. Robic and Y. Pauleau, J. Appl. Phys. **78**, (to be published in 1995).
13. Y. Pauleau, in Materials and Processes for Surface and Interface Engineering, edited by Y. Pauleau, NATO-ASI Series, Serie E : Applied Sciences, Vol.290 (Kluwer Academic Publishers, Dordrecht, The Netherlands, 1995), p.475.

AUTHOR INDEX

SUBJECT INDEX